PROGRESS IN BRAIN RESEARCH

VOLUME 79

NICOTINIC RECEPTORS IN THE CNS
Their Role in Synaptic Transmission

Recent volumes in PROGRESS IN BRAIN RESEARCH

PROGRESS IN BRAIN RESEARCH

VOLUME 79

NICOTINIC RECEPTORS IN THE CNS
Their Role in Synaptic Transmission

EDITED BY

A. NORDBERG

Department of Pharmacology, University of Uppsala, Uppsala, Sweden

K. FUXE

Department of Histology and Neurobiology, Karolinska Institute, Stockholm, Sweden

B. HOLMSTEDT

Department of Toxicology, Karolinska Institute, Stockholm, Sweden

and

A. SUNDWALL

Pharmacia LEO Therapeutics, Helsingborg, Sweden

ELSEVIER
AMSTERDAM – NEW YORK – OXFORD
1989

LP

© 1989, Elsevier Science Publishers B.V. (Biomedical Division)

ISBN 0-444-81088-9 (volume)
ISBN 0-444-80104-9 (series)

This book is printed on acid-free paper

Published by:
Elsevier Science Publishers B.V. (Biomedical Division)
P.O. Box 211
1000 AE Amsterdam
The Netherlands

Sole distributors for the USA and Canada:
Elsevier Science Publishing Company, Inc.
655 Avenue of the Americas
New York, NY 10010
USA

Library of Congress Cataloging-in-Publication Data

Nicotinic receptors in the CNS : their role in synaptic transmission /
    edited by A. Nordberg . . . [et al.].
        p.   cm. -- (Progress in brain research ; v. 79)
        Proceedings of the International Symposium on Nicotinic Receptors
    in the CNS--Their Role in Synaptic Transmission, held in Uppsala,
    Sweden, June 19 – 21, 1988.
        Includes bibliographies and index.
        ISBN 0-444-81088-9 (U.S.)
        1. Nicotinic receptors--Congresses.   2. Nicotine--Physiological
    effect--Congresses.   I. Nordberg, A. (Agneta)   II. International
    Symposium on Nicotinic Receptors in the CNS: Their Role in Synaptic
    Transmission (1988 : Uppsala, Sweden)   III. Series.
        [DNLM: 1. Neural Transmission--congresses.   2. Nicotine--adverse
    effects--congresses.   3. Receptors, Nicotinic--congresses.
    4. Smoking--physiopathology--congresses.   W1 PR667J v. 79 / WL
    102.8 N662 1988]
    QP376.P7 vol. 79
    [QP364.7]
    612′.82 s--dc20
    [599′.0188]
    DNLM/DLC   89-12034
    for Library of Congress   CIP

Printed in The Netherlands

List of Contributors

A. Adem, Department of Pharmacology, University of Uppsala, S-751 24 Uppsala, Sweden

L.F. Agnati, Department of Human Physiology, University of Modena, Modena, Italy

A. Åhlin, Department of Psychiatry and Psychology, Karolinska Hospital, S-104 01 Stockholm, Sweden

I. Alafuzoff, Department of Pathology, Karolinska Institutet, Huddinge Hospital, S-141 86 Huddinge, Sweden

K. Andersson, Department of Histology and Neurobiology, Karolinska Institutet, S-104 01 Stockholm, Sweden

S.H. Appel, Department of Neurology, Baylor College of Medicine, Houston, TX 77030, U.S.A.

S.-M. Aquilonius, Department of Neurology, University Hospital, S-751 85 Uppsala, Sweden

D.M. Araujo, Department of Psychiatry, McGill University, Montreal, Quebec, Canada H4HIR3

D.J.K. Balfour, Departments of Pharmacology and Clinical Pharmacology, University Medical School, Ninewells Hospital, Dundee, U.K.

T. Bartfai, Department of Biochemistry, Arrhenius Laboratory, University of Stockholm, S-106 91 Stockholm, Sweden

L. Beani, Department of Pharmacology, University of Ferrara, 44100 Ferrara, Italy

N.L. Benowitz, Clinical Pharmacology Unit of Medical Service, San Francisco General Hospital Medical Center, University of California, San Francisco, CA 94143, U.S.A.

C. Bianchi, Department of Pharmacology, University of Ferrara, 44100 Ferrara, Italy

B. Bjelke, Department of Histology and Neurobiology, Karolinska Institutet, S-104 01 Stockholm, Sweden

F.A. Bobbitt, Departments of Medicine and Pharmacology, University of Kentucky, Lexington, KY 40511, U.S.A.

J. Boulter, Molecular Neurobiology Laboratory, The Salk Institute for Biological Studies, San Diego, CA 92138, U.S.A.

L. Brown, Departments of Medicine and Pharmacology, University of Kentucky, Lexington, KY 40511, U.S.A.

N. Brynne, Department of Histology and Neurobiology, Karolinska Institutet, S-104 01 Stockholm, Sweden

J. Cartaud, Institut Jacques Monod du Centre National de la Recherche Scientifique, Université-Paris VII, Paris Cedex 05, France

J.P. Changeux, Departément des Biotechnologies, Institut Pasteur, Paris Cedex 15, France

B. Clark, Department of Pathology, SIV School of Medicine, Springfield, IL-62794-9230, U.S.A.

P.B.S. Clarke, Departments of Pharmacology and Therapeutics, McGill University, Montreal, Canada H3G1Y6

B. Collier, Department of Pharmacology, McGill University, Montreal, Quebec, Canada H4HIR3

A.C. Collins, School of Pharmacy and Institute for Behavioral Genetics, University of Colorado, Boulder, CO 80309, U.S.A.

J. Connolly, Molecular Neurobiology Laboratory, The Salk Institute for Biological Studies, San Diego, CA 92138, U.S.A.

S. Consolo, Mari Negri Institute of Pharmacological Research, 20157 Milan, Italy

E. Deneris, Molecular Neurobiology Laboratory, The Salk Institute for Biological Studies, San Diego, CA 92138, U.S.A.

P. DeSarno, Department of Pharmacology, SIV School of Medicine, Springfield, IL-62794-9230, U.S.A.

L. Dong, Departments of Medicine and Pharmacology, University of Kentucky, Lexington, KY 40511, U.S.A.

T.M. Egan, Department of Biochemistry, Brandeis University, Waltham, MA 02254-9110, U.S.A.

P. Eneroth, Department of Applied Biochemistry, Huddinge University Hospital, S-141 86 Huddinge, Sweden

K.O. Fagerström, Pharmacia Leo Therapeutics AB, S-251 09 Helsingborg, Sweden

L. Ferraro, Department of Pharmacology, University of Ferrara, I-44100 Ferrara, Italy

H.C. Fibiger, Department of Psychiatry, University of British Columbia, Vancouver B.C. V6T IW5, Canada

G. Fisone, Mari Negri Institute of Pharmacological Research, 20157 Milan, Italy

B. Fontaine, Departément des Biotechnologies, Institut Pasteur, Paris Cedex 15, France

K. Fuxe, Department of Histology and Neurobiology, Karolinska Institutet, S-104 01 Stockholm, Sweden

E. Giacobini, Department of Pharmacology, SIV School of Medicine, Springfield, IL-62794-9230, U.S.A.

B.A. Giblin, Department of Pharmacology, Georgetown University School of Medicine, Washington D.C. 20007, U.S.A.

M. Goldstein, Department of Psychiatry, New York University Medical Center, New York, U.S.A.

U. Hacksell, Department of Organic Pharmaceutical Chemistry, University of Uppsala, S-751 23 Uppsala, Sweden

C. Halldin, Karolinska Pharmacy, Karolinska Hospital, S-104 01 Stockholm, Sweden

J. Hardy, Department of Biochemistry, St Mary's Hospital Medical School, London, U.K.

A. Härfstrand, Department of Histology and Neurobiology, Karolinska Institutet, S-104 01 Stockholm, Sweden

P. Hartvig, Hospital Pharmacy, University Hospital, S-751 85 Uppsala, Sweden

L. Haverkamp, Department of Neurology, Baylor College of Medicine, Houston, TX 77030, U.S.A.

S. Heinemann, Molecular Neurobiology Laboratory, The Salk Institute for Biological Studies, San Diego, CA 92138, U.S.A.

J.S. Hendry, Departments of Pharmacology and Toxicology, Virginia Commonwealth University, Richmond, VA 23298, U.S.A.

J.E. Henningfield, Addiction Research Center, National Institute on Drug Abuse, Baltimore, MD 21224, U.S.A.

M. Herrera-Marschitz, Department of Pharmacology, Karolinska Institutet, S-104 01 Stockholm, Sweden

T. Hökfelt, Department of Histology and Neurobiology, Karolinska Institutet, S-104 01 Stockholm, Sweden

V. Höllt, Physiologische Institut der Universität München, D-8000 München, F.R.G.

G. Horn, Physiologische Institut der Universität München, D-8000 München, F.R.G.

J. Irons, Department of Biochemistry, University of Bath, Bath BA27A4, U.K.

P. Jacob, III, Clinical Pharmacology Unit of Medical Service, San Francisco, General Hospital Medical Center, University of California, San Francisco, CA 94143, U.S.A.

A. Jansson, Department of Histology and Neurobiology, Karolinska Institutet, S-104 01 Stockholm, Sweden

A.M. Janson, Department of Histology and Neurobiology, Karolinska Institutet, S-104 01 Stockholm, Sweden

J. Kåhrström, Department of Medical Cell Research, University of Lund, S-223 62 Lund, Sweden

K.J. Kellar, Department of Pharmacology, Georgetown University School of Medicine, Washington D.C. 20007, U.S.A.

J. Kiritsy-Roy, Departments of Medicine and Pharmacology, University of Kentucky, Lexington, KY 40511, U.S.A.

A. Klarsfeld, Département des Biotechnologies, Institut Pasteur, Paris Cedex 15, France

C. Köhler, Department of Neurochemistry, Astra Pharmaceuticals, S-151 85 Södertälje, Sweden

Z. Lai, Department of Pharmacology, University of Uppsala, S-751 24 Uppsala, Sweden

B. Långström, Department of Organic Chemistry, University of Uppsala, S-751 21 Uppsala, Sweden

P.A. Lapchak, Department of Pharmacology, McGill University, Montreal, Quebec, Canada H4HIR3

R. Laufer, Département des Biotechnologies, Institut Pasteur, Paris Cedex 15, France

P.M. Lippiello, R.J. Reynolds Tobacco Company, Winston-Salem, NC 27102, U.S.A.

R.H. Loring, Department of Biological Chemistry and Molecular Pharmacology, Harvard Medical School, Boston, MA 02115, U.S.A.

R.J. Lukas, Division of Neurobiology, Barrow Neurological Institute, Phoenix, AZ 85013, U.S.A.

M.D. Lumpkin, Department of Biophysics, Georgetown University School of Medicine, Washington D.C. 20007, U.S.A.

G.G. Lunt, Department of Biochemistry, University of Bath, Bath BA2 7AY, U.K.

M.J. Marks, School of Pharmacy and Institute for Behavioral Genetics, University of Colorado, Boulder, CO 80309, U.S.A.

L. Marson, Departments of Medicine and Pharmacology, University of Kentucky, Lexington, KY 40511, U.S.A.

M. McIlhany, Department of Surgery, SIV School of Medicine, Springfield, IL 62794-9230, U.S.A.

J.L. McManaman, Department of Neurology, Baylor College of Medicine, Houston, TX 77030, U.S.A.

T. Melander, Department of Histology and Neurobiology, Karolinska Institutet, S-104 01 Stockholm, Sweden

C. Mellin, Department of Organic Pharmaceutical Chemistry, University of Uppsala, S-751 23 Uppsala, Sweden

L.T. Meltzer, Departments of Pharmacology and Toxicology, Virginia Commonwealth University, Richmond, VA 23298, U.S.A.

G. Mereu, Department of Pharmacology, St Louis University School of Medicine, St Louis, MO 63104, U.S.A.

B.J. Morley, Boys Town National Institute, Omaha, NE 68131, U.S.A.

L. Naes, Department of Pharmacology, St Louis University School of Medicine, St Louis, MO 63104, U.S.A.

L. Nilsson-Håkansson, Department of Pharmacology, University of Uppsala, S-751 24 Uppsala, Sweden

A. Nordberg, Department of Pharmacology, University of Uppsala, S-751 24 Uppsala, Sweden

Ö. Nordström, Department of Biochemistry, Arrhenius Laboratory, University of Stockholm, S-106 91 Stockholm, Sweden

E. Norheim-Theodorsson, Department of Clinical Chemistry, Karolinska Hospital, S-104 01 Stockholm, Sweden

H. Nybäck, Department of Psychiatry and Psychology, Karolinska Hospital, S-104 01 Stockholm, Sweden

R. Oppenheim, Department of Neurology, Baylor College of Medicine, Houston, TX 77030, U.S.A.

C. Owman, Department of Medical Cell Research, University of Lund, S-223 62 Lund, Sweden

J. Patrick, Molecular Neurobiology Laboratory, The Salk Institute for Biological Studies, San Diego, CA 92138, U.S.A.

H. Perry, Department of Pharmacology, St Louis University School of Medicine, St Louis, MO 63104, U.S.A.

A. Persson, Department of Psychiatry and Psychology, Karolinska Hospital, S-104 01 Stockholm, Sweden

K. Pierzchala, Department of Medicine and Pharmacology, University of Kentucky, Lexington, KY 40511, U.S.A.

H. Porchet, Clinical Pharmacology Unit of Medical Service, San Francisco General Hospital Medical Center, University of California, San Francisco, CA 94143, U.S.A.

R. Quirion, Department of Psychiatry, McGill University, Montreal, Quebec, Canada H4H IR3

M.J. Rand, Department of Pharmacology, University of Melbourne, Victoria, 3052, Australia

C. Rapier, Department of Biochemistry, University of Bath, Bath BA2 7AY, U.K.

C. Reavill, Department of Psychiatry, Institute of Psychiatry, London SE5 8AF, U.K.

L. Romanelli, Department of Pharmacology, University of Ferrara, I-44100 Ferrara, Italy

J.A. Rosecrans, Departments of Pharmacology and Toxicology, Virginia Commonwealth University, Richmond, VA 23298, U.S.A.

M.A.H. Russell, Institute of Psychiatry, University of London, London, U.K.

D.W. Schulz, Department of Biological Chemistry and Molecular Pharmacology, Harvard Medical School, Boston, MA 02115, U.S.A.

C.G. Swahn, Department of Psychiatry and Psychology, Karolinska Hospital, S-104 01 Stockholm, Sweden

G. Sedvall, Department of Psychiatry and Psychology, Karolinska Hospital, S-104 01 Stockholm, Sweden

K. Semba, Department of Psychiatry, University of British Columbia, Vancouver B.C. V6T IW5, Canada

P. Spalluto, Department of Pharmacology, University of Ferrara, 44100 Ferrara, Italy

W.A. Staines, Department of Histology and Neurobiology, Karolinska Institutet, S-104 01 Stockholm, Sweden

C.A. Stimler, Departments of Pharmacology and Toxicology, Virginia Commonwealth University, Richmond, VA 23298, U.S.A.

I.P. Stolerman, Department of Psychiatry, Institute of Psychiatry, London SE5 8AF, U.K.

E. Sundström, Department of Histology and Neurobiology, Karolinska Institutet, S-104 01 Stockholm, Sweden

A. Sundwall, Pharmacia Leo Therapeutics AB, S-251 09 Helsingborg, Sweden

L. Swanson, Neural Systems Laboratory, The Salk Institute for Biological Studies, San Diego, CA 92138, U.S.A.

S. Tanganelli, Department of Pharmacology, University of Ferrara, I-44100 Ferrara, Italy

B. Tinner, Department of Histology and Neurobiology, Karolinska Institutet, S-104 01 Stockholm, Sweden

B. Thorne, Department of Biochemistry, University of Bath, Bath BA2 7AY, U.K.

K. Vaca, Department of Neurology, Baylor College of Medicine, Houston, TX 77030, U.S.A.

G.R. van Loon, Departments of Medicine and Pharmacology, University of Kentucky, Lexington, KY 40511, U.S.A.

L. Vickery, Department of Pharmacology, St Louis University School of Medicine, St Louis, MO 63104, U.S.A.

G. von Euler, Department of Histology and Neurobiology, Karolinska Institutet, S-104 01 Stockholm, Sweden

E. Wada, Molecular Neurobiology Laboratory, The Salk Institute for Biological Studies, San Diego, CA 92138, U.S.A.

K. Wada, Molecular Neurobiology Laboratory, The Salk Institute for Biological Studies, San Diego, CA 92138, U.S.A.

T.C. Westfall, Department of Pharmacology, St Louis University School of Medicine, St Louis, MO 63104, U.S.A.

B. Winblad, Department of Geriatric Medicine, Karolinska Institutet, Huddinge Hospital, S-141 86 Huddinge, Sweden

S. Wonnacott, Department of Biochemistry, University of Bath, Bath BA2 7AY, U.K.

P.P. Woodson, Addiction Research Center, National Institute on Drug Abuse, Baltimore, MD 21224, U.S.A.

C.F. Wu, Department of Histology and Neurobiology, Karolinska Institutet, S-104 01 Stockholm, Sweden

K.-W. P. Yoon, Department of Neurosurgery, St Louis University School of Medicine, St Louis, MO 63104, U.S.A.

R.E. Zigmond, Department of Biological Chemistry and Molecular Pharmacology, Harvard Medical School, Boston, MA 02115, U.S.A.

Preface

The present volume constitutes the proceedings of *'The International Symposium on Nicotinic Receptors on the CNS − Their Role in Synaptic Transmission'* that was held in Uppsala, Sweden June 19−21, 1988. Sweden has a long historical tradition for research on nicotine and cholinergic mechanisms which motivated the arrangement of the symposium in Sweden. Several international meetings have earlier been arranged dealing with the different effects of nicotine on the human body. This symposium is unique in that it focuses on the central nicotinic receptors and their functional role in the CNS. In an integrative way it covers all aspects of the nicotinic receptors from molecular to clinical approaches.

Ligand binding studies with different nicotinic agonists and antagonists have shown the presence of several different subtypes of nicotinic receptors in the CNS. Studies on molecular biology using expression in the oocyte model indicate that the addition of a β_2-subunit to either an α_2-, α_3-, or α_4-subunit produces a receptor that initiates a strong response to either acetylcholine or nicotine. The β_2-subunit is expressed throughout the central nervous system and the α-subunit transcripts are expressed in separate but overlapping areas in the brain. Of special interest in the session on radioligand binding studies was the observation that neuronal bungarotoxin (NBT) but not α-bungarotoxin could antagonize the nicotinic receptor function in the brain. Another interesting tool used in the pharmacological analysis was the neosurugatoxin, a potent ganglionic nicotinic antagonist without activity at the neuromuscular junction. The meeting clearly illustrated the important role the nicotinic receptors play in the control of the release of monoamines and amino acids in the brain. It is evident that the nicotinic cholinergic recognition sites predominantly belong to the terminal parts of the cholinergic neuronal system and that these sites are especially enriched in the reticular thalamic cholinergic pathways. New anatomical findings reported at the meeting were the existence of neuropeptides such as galanin and GRF as well as dopamine in the cholinergic pathways of the brain. Studies on neurotransmitter release, neuroendocrine function and behaviour in relation to nicotine exposure revealed different types of plasticity in the central nicotinic receptor populations. Rapid desensitization as well as resistance to desensitization are common.

Human studies also indicate the development of adaptive changes in the central nicotinic receptors upon chronic nicotine exposure. Nicotine abstinence reveals that the addicted heavy smokers have more severe withdrawal effects and more subjective effects after first postabstinence cigarette. Although the half-life of nicotine in brain is short (less than 10 min) tolerance develops quickly. During the day there is an accumulation of nicotine in brain and a development of tolerance which partly can be overcome by transiently high brain levels of nicotine. Overnight abstinence allows resensitization mechanisms. The subjective and behavioural effects of nicotine differ both between and within individuals due to numerous factors such as acute and chronic tolerance, learning, conditioning and pharmacokinetic factors. Direct evidence for an adaptive change of nicotinic receptors in human brain was reported at the meeting using positron emission tomography (PET). The question whether Parkinson's disease is less common in smokers than nonsmokers is still under debate. However, basic animal data presented at this meeting indicated that chronic nicotine treatment, in part, protects dopamine nerve cells in the substantia nigra from undergoing degeneration in response to mechanical and neurotoxic injuries. In dementia disorders, such as Alzheimer's disease, a loss of high affinity nicotinic receptors has been found in the brain. It was suggested at this meeting that the loss may partly be due to an interconversion of high affinity to low affinity nicotinic binding sites which reduces the desensitization process and opens up new aspects on therapeutic approaches in Alzheimer's disease.

In our opinion, the symposium finally established that the effects of smoking are mediated by activation of central nicotinic receptors. As organizers we are indebted to the participants for their valuable contributions. We want to express our gratitude to the Swedish Tobacco Company and Pharmacia Leo Therapeutics AB for their financial support.

Uppsala, Stockholm and Helsingborg
December 1988

Agneta Nordberg
Kjell Fuxe
Bo Holmstedt
Anders Sundwall

Contents

Section IX – Human Pharmacology and Nicotine Dependence

Section X – Nicotine and Degenerative Brain Disorders

SECTION I

Introduction

A. Nordberg, K. Fuxe, B. Holmstedt and A. Sundwall (Eds.)
Progress in Brain Research, Vol. 79
© 1989 Elsevier Science Publishers B.V. (Biomedical Division)

CHAPTER 1

Neuropharmacological effects of nicotine in relation to cholinergic mechanisms

M.J. Rand

Department of Pharmacology, University of Melbourne, Victoria, 3052 Australia

Early studies on nicotine and other drugs and poisons

From time immemorial, plant materials containing substances that interact with cholinergic mechanisms have been used for hedonistic, ritual or magical purposes or for hunting, as well as for therapeutic purposes. Such materials include not only tobacco (nicotine), but also *Duboisia spp.* (nornicotine), *Lobelia inflata* (lobeline), betel (arecoline), solanaceous plants (atropine, hyoscine, etc.), *Muscaria amanita* (muscarine), esere (physostigmine), jaborandi (pilocarpine), curare (tubocurarine), and even opium (morphine), since opiates generally inhibit cholinergic transmission in many tissues. The founding fathers of pharmacology and toxicology investigated the actions of the crude substances and of the alkaloids isolated from them with enthusiasm and energy that was eventually rewarded with the development of therapeutic substances and, most importantly, with new insights into physiological mechanisms. In the latter regard, nicotine might well hold pride of place.

Elucidation of physiological mechanisms arising from observations on the actions of nicotine

Pathways of autonomic innervation

The effects of nicotine in stimulating and "paralysing" ganglia of the autonomic nervous system were discovered by Langley and his colleagues (Langley and Dickinson, 1889; Langley and Anderson, 1895; Langley, 1896, 1911). These actions were useful for determining the nature of the autonomic innervation and the location of ganglionic synapses in the nervous pathway to many organs. For example, they observed that application of nicotine to the coeliac ganglion produced a rise in blood pressure. This and similar observations on sympathomimetic responses elicited in target organs by application of nicotine to ganglia in the nervous pathway led to the deduction that nicotine stimulated ganglion cells. They also noted that after the administration of a large dose of nicotine, the response to the local application of nicotine to the ganglion was abolished, as was the response to stimulation of the preganglionic nerves; however, the response to stimulation of the postganglionic nerves persisted. These observations led to the deduction that nicotine blocked the excitation of ganglion cells.

Differentiation between types of acetylcholine receptors

Dale (1914) observed that a large dose of nicotine (30 mg) abolished the pressor action of acetylcholine in the atropinized spinal cat: this, together with the observation that the depressor action of acetylcholine, like that of muscarine, was blocked by atropine, led to the formulation of the concept of two different types of site of action of acetyl-

4

choline, which he termed "muscarinic and nicotinic" – a terminology that is still standard.

Identification of cholinergic and amino acid transmission in the central nervous system (CNS)

The most thoroughly understood action of nicotine in the CNS is inhibition of the knee-jerk reflex. This effect of nicotine was first demonstrated by Schweitzer and Wright (1938) in experiments on cats. They produced convincing evidence that it was in fact due to an action of nicotine on the spinal cord, and that it was produced by "smoking doses" (i.e. 5 μg i.v.) Subsequently, it was shown that inhibition of the knee-jerk reflex occurred in humans during the smoking of a cigarette (Clark and Rand, 1968).

In the course of studies designed to prove that acetylcholine is a transmitter in the CNS, the action of nicotine was investigated on a spinal neurone, the Renshaw cell. This was the first use of the technique of electrophoretic application of drugs in the CNS (Curtis and Eccles, 1958). The technique, which allowed the precise application of drugs to identifiable neurones, was extended to a range of other substances and led to the discovery of the depressant action of gamma-aminobutyric acid (GABA) and the excitant actions of glutamic and aspartic acids on Renshaw cells and other neurones (Curtis et al., 1959, 1960). Although it was not at first recognized that these amino acids could be neurotransmitters, the discoveries that sprang from studies on nicotine were of crucial importance in the chain of investigations which established that they were in fact the most common neurotransmitters in the CNS (Johnston, 1978; Watkins and Evans, 1981).

Recognition of non-adrenergic, non-cholinergic transmission

Observations on the pharmacological actions of nicotine contributed to a more recent development in neuroscience, namely the recognition of non-adrenergic, non-cholinergic (NANC) nerves. Thus Burnstock et al. (1966) noted that the smooth muscle of the taenia of the guinea-pig caecum was innervated by neurones having their cell bodies in the Auerbach's plexus which, when excited by electrical stimulation or ganglion-stimulating drugs such as nicotine or DMPP, produced relaxation which was entirely different from the relaxation produced by excitation of noradrenergic nerves. In a subsequent analysis of the actions of nicotine and other ganglia stimulants on the taenia of the guinea-pig caecum, it was confirmed that they acted on nicotinic receptors of both cholinergic and NANC neurones (Hobbiger et al., 1969). Further studies on NANC transmission pioneered by Burnstock led to the acceptance of peripheral neurotransmitter systems other than the "classical" cholinergic and (nor)adrenergic systems (for an historical account, see Rand and Mitchelson, 1986).

Molecular biology of receptors

The nicotinic cholinoceptors of electroplaxes from the electric organ of the *Torpedo* were the first receptors to be isolated and characterized. The quaternary structure and the relationship between acetylcholine recognition sites and functional responses have been determined, and the various transitional states have been identified. There have been several authoritative reviews on these subjects (Claudio et al., 1986; Karlin et al., 1986; Changeux et al., 1987; Colquhoun et al., 1987).

Location of nicotinic receptors

The main sites at which nicotinic receptors have been identified are listed in Table 1. The nicotinic cholinoceptors clearly have a functional role in transmission through autonomic ganglionic synapses and adrenal medullary chromaffin cells, and there is little new information to add to the extensive corpus of knowledge on these subjects.

Likewise, the functional role of nicotinic cholinoceptors of the electroplax and motor end-plate is clearly established: in fact, these are the most thoroughly studied of all receptors. However, a full understanding of the process of transmission at the neuromuscular junction must take into account the presence of prejunctional nicotinic receptors on the motoneurone nerve terminals as well as the more familiar postjunctional receptors. Convincing evidence for the existence of prejunctional cholinoceptors on motoneurone terminals, and their functional role in facilitating transmitter release, has been summarized recently by Bowman et al. (1988). It appears that these receptors are normally activated by acetylcholine released from the motoneurone terminals, and they function to maintain the availability of transmitter stores of acetylcholine for release during a train of repetitive nerve stimulation. Both the postjunctional and prejunctional nicotinic cholinoceptors of the neuromuscular junction are blocked by tubocurarine and other clinically used muscle relaxants, whereas agents such as α-bungarotoxin and erabutoxin b are selective for the postjunctional cholinoceptors, and hexamethonium is a selective blocker of prejunctional receptors. However, the prejunctional nicotinic cholinoceptors of the neuromuscular junction differ from ganglionic nicotinic cholinoceptors, since they are not affected by the ganglion blocking drug trimetaphan. Agonists acting on the prejunctional nicotinic cholinoceptors of the neuromuscular junction include acetylcholine, nicotine, DMPP and carbachol. Bowman et al. (1988) suggested that the prejunctional nicotinic cholinoceptors of the neuromuscular junction resemble certain nicotinic receptors in the central nervous system which may also subserve a positive feedback function in the release of transmitter acetylcholine (Clarke, 1987).

The effects of nicotine and other nicotinic agonists on noradrenergic sympathetic nerve terminals have been comprehensively reviewed (Fozard, 1977; Starke, 1977; Löffelholz, 1979; Balfour, 1982; Su, 1982; Nedergaard, 1988; Rand and Story, 1988). Their effect in releasing noradrenaline from sympathetic nerve terminals is blocked not only by hexamethonium and related drugs but also by a diverse range of other substances, including the α-adrenoceptor antagonists phenoxybenzamine and phentolamine (Su and Bevan, 1970; Starke and Montel, 1974; Collett and Story, 1984), but not prazosin or yohimbine (Collett and Story, 1984), the β-adrenoceptor antagonist propranolol (Collett and Story, 1984), the blockers of neuronal amino uptake cocaine and desipramine (Su and Bevan, 1970; Westfall and Brasted, 1972, 1974; Collett and Story, 1984), and anaesthetic agents (Göthert, 1974; Göthert et al., 1976). Furthermore, nicotine blocks its own action in stimulating noradrenergic nerve terminals: the phenomenon has been described as autoinhibition (Löffelholz, 1970) or desensitization (Steinsland and Furchgott, 1975). However, in contrast to the blockade of transmission that accompanies the loss of responsiveness to nicotine in the neuromuscular junction or in autonomic ganglia, there is no functional impairment of noradrenergic transmission. In addition to releasing noradrenaline, nicotine has been reported to facilitate and to inhibit stimulation-induced release of noradrenaline (for reviews see: Nedergaard, 1988; Rand and Story, 1988). In most cases, there is no obvious source of acetylcholine for activation of nicotinic cholinoceptors of sympathetic nerve terminals, and their functional significance remains obscure, despite the striking effects that can be produced by activating them.

The nicotinic cholinoceptors of sensory nerve terminals, and especially the chemoceptive ter-

TABLE I

Sites of action of nicotine

Autonomic neurones
   Ganglion cells
   Noradrenergic nerve terminals
Adrenal medullary and other chromaffin cells
Neuromuscular junction
   Motor end-plate (and electroplax)
   Motoneurone nerve terminals
Sensory nerve terminals
Renshaw cells and other CNS neurones
Synaptic terminals in CNS

minals associated with the heart and lungs, are activated by nicotine in lower concentrations than are those at most other locations, and the consequences of their activation on the central nervous system and in reflexly elicited responses of the cardiovascular system are considerable (see Ginzel, 1988). However, blockade of these receptors does not affect the functional integrity of sensory nerve terminals; furthermore, there is no apparent source of acetylcholine or any other endogenous ligand to activate them. Therefore, they, like the nicotinic cholinoceptors of sympathetic nerve terminals, do not appear to be involved in physiological processes.

As for the nicotinic cholinoceptors in the CNS, a functional role has been rigorously established only at a few sites, notably the Renshaw cells in the spinal cord. It remains to be determined which of the binding sites for nicotinic cholinoceptor ligands in the CNS are involved in physiological functions, and which, if any, are like the nicotinic cholinoceptors of sympathetic and sensory nerve terminals in that they do not appear to subserve a functional role, at least within the present framework of knowledge.

Actions of nicotine relevant to amounts arising from tobacco use

Studies on the relationship between the pharmacokinetic handling of drugs and their pharmacodynamic actions have generally focussed on the concentration of the drug in plasma as an index of that available at the specific sites of action. In many instances, the relationship is sufficiently good to provide guidelines for therapeutic management, for example, with anti-epileptic drugs and cardiac glycosides; on the other hand, with some drugs there is little or no relationship between plasma concentrations and pharmacodynamic actions. In the case of nicotine, measurements of venous plasma concentrations have provided useful insights, particularly in arriving at doses and regimes that are relevant to smoking and other forms of tobacco use, and to the use of nicotine

chewing gum. Russell (1988) has pointed out the importance, not only of using doses of nicotine that are relevant to those taken during smoking, but of taking into account the rate of change of nicotine concentrations, and the development of tolerance.

However, it has been suggested that arterial plasma levels of nicotine, particularly the peak levels that arise from the absorption of nicotine after inhalation of cigarette smoke, might provide a better relationship with its peripheral actions. Furthermore, it is clear that a knowledge of the concentrations of nicotine produced in the CNS by smoking would be useful as a guide in arriving at relevant concentrations to use in experimental systems. An approach to obtaining these data was expounded by Benowitz (1988). He measured the steady-state distribution of nicotine in various organs after prolonged administration to rats and rabbits, and found that the brain:blood ratio was about 3. He used his experimental data in a model of human tissue perfusion to estimate nicotine concentrations that would arise after intrapulmonary administration of 1.5 mg of nicotine over 10 min, to simulate cigarette smoking, and arrived at peak values of about 40 ng/ml for arterial blood and 75 ng/ml for brain nicotine concentrations. As it happens, we have made direct measurements of nicotine concentrations in arterial plasma and cerebrospinal fluid after administration of nicotine intravenously and during and after smoking in animal experiments with dogs and in human experiments. The experiments were carried out some years ago, in the early 1970s, but the findings were not published except in the form of theses for higher degrees. In the light of Benowitz's (1988) paper, it seems worthwhile to take this opportunity to make our findings more widely known.

Nicotine concentrations in arterial plasma

The arterial blood concentration of nicotine rose sharply to a peak of $0.9 - 1.54 \ \mu M$ within $10 - 20$ s of administration of one 35 ml puff of cigarette smoke in the first third of an inspiratory cycle in dogs. The peak levels, and the basal level, gradually increased with successive puffs of smoke from the same cigarette (Isaac and Rand, 1969).

Qualitatively similar observations were made in my laboratory by Giles (1975), when nicotine concentrations in human arterial plasma were measured during smoking, but quantitatively the levels were lower, as would be expected with the 3 – 5 times greater body mass of the human subjects compared to the dogs. After inhalation of a single puff of cigarette smoke, the level rose to a peak of about 100 ng/ml (ca. 0.6 μM) in about 20 s and fell rapidly back to the basal level, then a second small recirculation peak appeared about 80 s after the single puff (Fig. 1A). With inhalation of successive puffs at 1 min intervals, the peaks and the basal levels gradually increased (Fig. 1B).

Responses to intra-arterial nicotine

In most studies on the vascular actions of nicotine in man, nicotine has been given by the medium of tobacco smoke or by intravenous injection. However, Fewings et al. (1966) observed the effects of nicotine administered into the brachial artery on blood flow in the hand and forearm of human volunteers. They saw both vasodilator and vasoconstrictor components in the response to nicotine, and each of these components was produced in two different ways. Thus, vasoconstriction in skin blood vessels was produced reflexly by stimulation of sensory nerve endings, and occurred not only in the limb receiving the intra-arterial injection but also in the opposite limb, except when the sensory pathway was blocked by infiltration of a local anaesthetic; vasoconstriction confined to the injected regions also occurred as a result of stimulation of sympathetic nerve terminals. Vasoconstrictor responses, whether reflexly or locally produced, were absent after α-adrenoceptor blockade, and did not occur in sympathectomized limbs. Dilatation of muscle blood vessels was produced reflexly by sympathetic excitation, and occurred in both the injected and the opposite limb, except in sympathectomized subjects. A long-lasting vasodilator response was obtained in the injected limb of sympathectomized subjects, and also in one subject with a totally denervated arm (as a result of avulsion of the shoulder joint), and this finding led to the conclusion that nicotine had a direct vasodilator action on smooth muscle. Although the study by Fewings et al. (1966) illustrates the com-

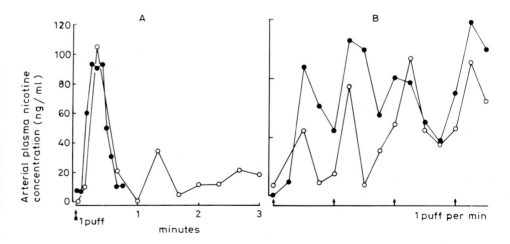

Fig. 1. Nicotine concentration in human arterial plasma. Samples of blood from a catheter in the brachial artery were taken into tubes in a fraction collector for assay of nicotine by GLC. (A) Samples collected at 5 and 20 s intervals from two volunteer subjects commencing with inhalation of one puff of cigarette smoke. (B) Samples collected at 15 s intervals while four puffs of cigarette smoke were inhaled at 1 min intervals by two further volunteer subjects. (Data from Giles, 1975.)

8

plexity of the actions of nicotine, its relevance to the vascular actions of "smoking doses" of nicotine is distant. The dose used was 0.5 mg of nicotine bitartrate and it was injected in 5 – 10 s. Taking into account the rate of blood flow in the brachial artery, the concentration of nicotine in the blood was in the range of 50 – 100 μM, or about 100 times the peak levels of nicotine appearing in arterial blood after inhalation of cigarette smoke (Giles, 1975).

Levels in CNS

The entry of nicotine into the brain has generally been studied with [14C]nicotine, which provides a sensitive means for determining the distribution of nicotine in tissues, including the brain, although it does not allow a discrimination between nicotine and its metabolites. The distribution of radioactivity in the brains of rats, mice and cats after the administration of [14C]nicotine was particularly studied by Schmiterlöw and his colleagues (Appelgren et al., 1962; Schmiterlöw and Hansson, 1962, 1965; Hansson and Schmiterlöw, 1962; Schmiterlöw et al., 1967), and by Ganz et al. (1951), Stalhandske (1970) and Turner (1971). It was established that more nicotine accumulates in gray than in white matter and that the highest levels occur in basal ganglia and in nuclei of the diencephalon and medulla oblongata (Appelgren et al., 1962; Schmiterlöw and Hansson, 1962, 1965; Schmiterlöw et al., 1967; Turner, 1971). The amount of radioactivity entering the brain depends not only on the dose but also on the regime of administration of [14C]nicotine (Schmiterlöw et al., 1967; Stalhandske, 1970; Turner, 1971).

It is obviously not a practical proposition with currently available technology to make dynamic measurements of nicotine in brain tissue during and after smoking or simulated smoking. However, a reasonable approach to this is feasible with measurement of nicotine concentrations in cerebrospinal fluid (CSF). The relationship between the concentrations of nicotine in CSF and brain tissue is not known. However, nicotine is

taken up into nervous tissue; for example, by cells in the cat's superior cervical ganglion, giving an intracellular:extracellular ratio 6:1, after intraarterial injection (Brown et al., 1969, 1971), and it is also taken up into sympathetic nerve terminals of the rabbit aorta (Bevan and Su, 1972; Nedergaard and Schrold, 1977).

In my laboratory, Jellinek (1974) studied the appearance of nicotine in the CSF of dogs after the intravenous administration of 32.5 μg/kg of nicotine given in a single dose or as 10 divided doses at 30 s intervals to simulate smoking. Samples of CSF from the cisterna magnum and of venous plasma were taken at 10 min intervals after the start of administration of nicotine. Nicotine concentrations in CSF and plasma were higher when it was given in a single dose than in divided doses, and were higher in CSF than in plasma with either regime of administration. The findings are summarized in Fig. 2. Similar findings were obtained when [3H]nicotine was administered and radioactivity in CSF and plasma was measured. It was established that the entry of nicotine into the CSF was not affected by pretreatment of the dogs with mecamylamine. In two dogs, a sample of CSF was taken by lumbar puncture at the same time as a sample was taken from the cisterna magnum; in

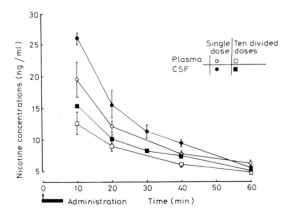

Fig. 2. Nicotine concentrations in samples of plasma and CSF, assayed by GLC, after administration of nicotine (32.5 μg/kg, i.v.), either in a single dose or in 10 divided doses at 5 s intervals. (Data from Jellinek, 1974.)

9

both, the concentration of nicotine was lower in the spinal CSF than in CSF from the fourth ventricle.

In a second study in my laboratory by Giles (1975), the presence of nicotine in the CSF of humans after smoking a cigarette was determined. The subjects were patients who were undergoing diagnostic pneumoencephalography, in whom large volumes of CSF are withdrawn and replaced with air, and patients in whom CSF levels of tricyclic antidepressant drugs were being measured to determine whether there was evidence for a correlation between clinical responses and drug levels. Patients in these groups who were habitual smokers were invited to participate in the smoking study, which did not involve any modification of the invasive procedures. Subjects abstained from smoking on the day of the trial, except for the cigarette smoked just before the beginning of procedures for inserting the puncture needle. The interval between the end of smoking and the moment

of obtaining the sample of CSF was recorded, but was not under our control. The nicotine concentrations in the samples were assayed by GLC. In pneumoencephalography patients, repeated 10 ml samples of CSF were taken, up to a total of 100–150 ml. However, only the first sample was used, as subsequent samples gave widely discordant results. In the other patients, 5 ml of CSF was taken, and 1 ml was available for assay of nicotine. The results are summarized in Fig. 3. The CSF samples were taken by lumbar puncture, and the nicotine concentrations are probably lower than they would have been in CSF from the cerebral ventricles, to judge by Jellinek's (1974) findings with dogs. In two subjects samples of venous blood taken at the same time as the CSF samples were available; in both, the nicotine concentration was higher in the CSF than in the plasma, which is in accord with Jellinek's (1974) findings.

Responses to injection of nicotine into the cerebral ventricles

The mean peak nicotine concentration in the CSF of dogs after giving 32.5 μg/kg in 10 divided doses at 30 s intervals (to mimic smoking to some extent) was 15 ng/ml (= ca. 100 nM). This is about two orders of magnitude less than the amount required to elicit behavioural and cardiovascular effects when injected into the cerebral ventricles of the dog in the studies of Lang and Rush (1973) in my department. They observed that 20–100 μg of nicotine bitartrate (8–24 μg nicotine base; ca. 3–15 μM) injected into the lateral ventricle produced licking, swallowing, and signs of alertness, together with an immediate transient rise in blood pressure and a slight increase in heart rate, then there was a secondary prolonged increase in blood pressure and heart rate, restlessness and vomiting. Acute tolerance developed after the first dose, and responsiveness to i.c.v. nicotine did not reappear until one or two days later. The effects of i.c.v. nicotine were completely abolished by mecamylamine (250 μg, i.c.v.). It is worth noting that similar responses to acetylcholine chloride (10–20

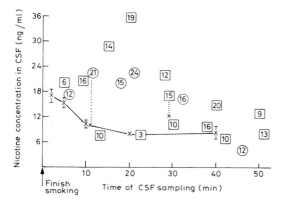

Fig. 3. Nicotine concentrations in human CSF at various times after smoking a cigarette. Each value represents the value from a single subject. The circles are from patients in whom CSF was withdrawn for assay of antidepressant drugs. The squares are from patients undergoing diagnostic pneumoencephalography. The numbers within each symbol are estimates of the intake of nicotine (in μg/kg), based on observations of smoking behaviour and the nature of the cigarette smoked. The large crosses are mean plasma nicotine concentrations after ad libitum smoking of a cigarette in another group of subjects. In two cases, plasma samples were taken from patients at the same time as the CSF sample, as indicated by the small crosses. (Data from Giles, 1975.)

μg) were completely abolished by atropine and unaffected by mecamylamine. Previous studies on the effects of i.c.v. nicotine had been in cats, with divergent results: one group usually saw a fall in blood pressure (Armitage and Hall, 1967); another usually saw a rise in blood pressure (Pradhan et al., 1967).

Summary and Conclusions

Studies with nicotine have, from the outset, been of fundamental importance for the elucidation of physiological mechanisms. However, many nicotinic cholinoreceptors do not appear to subserve a physiological role, at least within the present framework of knowledge. The main challenge in the immediate future is to determine the functional role of nicotinic cholinoceptors in the CNS. This knowledge will not only add to the physiological mechanisms that have been unravelled as a result of studies with nicotine, but will also provide a major contribution to the understanding of tobacco use.

References

Appelgren, L.E., Hansson, E. and Schmiterlöw, C.G. (1962) The accumulation and metabolism of ^{14}C-labelled nicotine in the brain of mice and rats. *Acta Physiol. Scand.,* 56: 249 – 257.

Armitage A.K. and Hall, G.H. (1967) Effects of nicotine on the systemic blood pressure when injected into the cerebral ventricles of cats. *Int. J. Pharmacol.,* 6: 143 – 149.

Balfour, D.J.K. (1982) The effects of nicotine on brain neurotransmitter systems. *Pharmacol. Ther.,* 16: 269 – 282.

Benowitz, N.L. (1988) Pharmacokinetics and pharmacodynamics of nicotine. In M.J. Rand and K. Thurau (Eds.), *Pharmacology of Nicotine,* ICSU Press, New York, pp. 2 – 18.

Bevan, J.A. and Su, C. (1972) Uptake of nicotine by the sympathetic nerve terminals in the blood vessel. *J. Pharmacol. Exp. Ther.,* 182: 419 – 426.

Bowman, W.C., Marshall, I.G., Gibb, A.J. and Harborne, A.J. (1988) Feedback control of transmitter release at the neuromuscular junction. *Trends Pharmacol. Sci.,* 9: 16 – 20.

Brown, D.A., Hoffman, P.C. and Toth, L.J. (1969) [^{3}H]Nicotine in cat superior cervical and nodose ganglia after close-arterial injection in vivo. *Br. J. Pharmacol.,* 35: 406 – 417.

Brown, D.A., Halliwell, J.V. and Scholfield, C.N. (1971) Uptake of nicotine and extracellular space markers by isolated rat ganglia in relation to receptor activation. *Br. J. Pharmacol.,* 42: 100 – 113.

Burnstock, G., Campbell, G. and Rand, M.J. (1966) The inhibitory innervation of the taenia of the guinea-pig caecum. *J. Physiol.* (Lond.), 182: 504 – 526.

Changeux, J.-P., Giraudat, J. and Dennis, M. (1987) The nicotinic acetylcholine receptor: molecular architecture of a ligand-regulated ion channel. *Trends Pharmacol. Sci.,* 8: 459 – 465.

Clark, M.S.G. and Rand, M.J. (1968) Effect of tobacco smoke on the knee-jerk reflex in man. *Eur. J. Pharmacol.,* 3: 294 – 302.

Clarke, P.B.S. (1987) Recent progress in identifying nicotinic cholinoceptors in mammalian brain. *Trends Pharmacol. Sci.,* 8: 32 – 35.

Claudio, T. (1986) Recombinant DNA technology in the study of ion channels. *Trends Pharmacol. Sci.,* 7: 308 – 312.

Collett, A.R. and Story, D.F. (1984) Effects of adrenoceptor antagonists and neuronal uptake inhibitors on dimethylphenylpiperazinium-induced release of catecholamines from the rabbit isolated adrenal gland and guinea-pig atria. *J. Pharmacol. Ther.,* 231: 379 – 386.

Colquhoun, D., Ogden, D.C. and Mathie, A. (1987) Nicotinic acetylcholine receptors of nerve and muscle: functional aspects. *Trends Pharmacol. Sci.,* 8: 465 – 472.

Curtis, D.R. and Eccles, R.M. (1958) The excitation of Renshaw cells by pharmacological agents applied electrophoretically. *J. Physiol. (Lond.),* 141: 435 – 445.

Curtis, D.R., Phillis, J.W. and Watkins, J.C. (1959) The depression of spinal neurones by γ-amino-n-butyric acid and β-alanine. *J. Physiol. (Lond.),* 146: 185 – 203.

Curtis, D.R., Phillis, J.W. and Watkins, J.C. (1960) The chemical excitation of spinal neurones by certain amino acids. *J. Physiol. (Lond.),* 150: 656 – 682.

Fewings, J.D., Rand, M.J., Scroop, G.C. and Whelan, R.F. (1966) The action of nicotine on the blood vessels of the hand and forearm in man. *Br. J. Pharmacol.,* 26: 567 – 579.

Fozard, J. (1979) Cholinergic mechanisms in adrenergic function. In S. Kalsner (Ed.), *Trends in Autonomic Pharmacology, Vol. 1,* Urban and Schwarzenberg, Baltimore and Munich, pp. 145 – 194.

Ganz, A., Kelsey, F.E. and Geiling, E.M.K. (1951) Excretion and tissue distribution studies on radioactive nicotine. *J. Pharmacol. Exp. Ther.,* 102: 209 – 214.

Giles, M.P. (1975) *Nicotine Pharmacokinetics in Man.* Ph. D. Thesis, University of Melbourne.

Ginzel, K.H. (1988) The lungs as sites of origin of nicotine-induced skeletomotor relaxation and behavioural and electrocortical arousal in the cat. In M.J. Rand and K. Thurau (Eds.), *Pharmacology of Nicotine,* ICSU Press, New York, pp. 267 – 292.

Göthert, M. (1974) Effects of halothane on the sympathetic

nerve terminals. *Naunyn-Schmiedeberg's Arch. Pharmacol.,* 286: 125 – 143.

Göthert, M., Kennerknecht, E. and Thielecke, G. (1976) Inhibition of receptor-mediated noradrenaline release from the sympathetic nerves of the isolated rabbit heart by anaesthetics and alcohols in proportion to their hydrophobic property. *Naunyn-Schmiedebergs's Arch. Pharmacol.,* 292: 145 – 152.

Hansson, E. and Schmiterlöw, C.G. (1962) Physiological distribution and fate of C^{14} labelled nicotine in mice and rats. *J. Pharmacol. Ther.,* 137: 91 – 101.

Hobbiger, F., Mitchelson, F. and Rand, M.J. (1969) The actions of some cholinomimetic drugs on the isolated taenia of the guinea-pig caecum. *Br. J. Pharmacol.,* 36: 53 – 69.

Isaac, P.F. and Rand, M.J. (1969) Blood levels of nicotine and physiological effects after inhalation of tobacco smoke. *Eur. J. Pharmacol.,* 8: 269 – 283.

Jellinek, P. (1974) *Studies on the Pharmacokinetics and Behavioural Effects of Nicotine.* Ph.D. Thesis, University of Melbourne.

Johnston, G.A.R. (1978) Neuropharmacology of amino acid inhibitory transmitters. *Ann. Rev. Pharmacol. Toxicol.,* 18: 269 – 289.

Karlin, A., Kao, P.N. and DiPaola, M. (1986) Molecular pharmacology of the nicotinic acetylcholine receptor. *Trends Pharmacol. Sci.,* 7: 304 – 308.

Langley, J.N. (1896) On the nerve cell connections of the splanchnic nerve fibres. *J. Physiol. (Lond.),* 20: 223 – 246.

Langley, J.N. (1911) The effect of various poisons on the response to nervous stimuli chiefly in relation to the bladder. *J. Physiol. (Lond.),* 43: 125 – 181.

Langley, J.N. and Anderson, H.K. (1895) The innervation of the pelvic and adjoining viscera. *J. Physiol. (Lond.),* 19: 71 – 139.

Langley, J.N. and Dickinson, W.L. (1889) On the local paralysis of peripheral ganglia, and on the connection of different classes of nerve fibres with them. *Proc. R. Soc., London,* 46: 423 – 431.

Löffelholz, K. (1970) Autoinhibition of nicotinic release of noradrenaline from postganglionic sympathetic nerves. *Naunyn-Schmiedeberg's Arch. Pharmacol.,* 267: 49 – 63.

Löffelholz, K. (1979) Release induced by nicotinic agonists. In D.M. Paton (Ed.), *The Release of Catecholamines from Adrenergic Neurones,* Pergamon Press, Oxford, pp. 275 – 301.

Nedergaard, O.A. (1988) Effect of nicotine on neuroeffector transmission in blood vessels. In M.J. Rand and K. Thurau (Eds.), *Pharmacology of Nicotine,* ICSU Press, New York, pp. 143 – 162.

Nedergaard, O.A. and Schrold, J. (1977) The mechanism of action of nicotine on vascular adrenergic neuroeffector transmission. *Blood Vessels,* 14: 325 – 347.

Pradhan, S.N., Bhattacharya, I.C. and Atkinson, K.S. (1967) The effects of intraventricular administration of nicotine on blood pressure and some somatic reflexes. *Ann. N.Y. Acad.*

Sci., 142: 50 – 56.

Rand, M.J. and Mitchelson, F. (1986) The guts of the matter: contribution of studies on smooth muscle to discoveries in pharmacology. In M.J. Parnham and J. Bruinvels (Eds.), *Discoveries in Pharmacology, Vol. 3,* Elsevier Science Publishers, Amsterdam, pp. 19 – 61.

Rand, M.J. and Story, D.F. (1988) Modulation of noradrenaline release from sympathetic nerve terminals by cholinomimetic drugs and cholinergic nerves. Freund Publishing House, Tel Aviv (in press).

Russell, M.A.H. (1988) Nicotine intake by smokers: are rates of absorption or steady-state levels more important? In M.J. Rand and K. Thurau (Eds.), *Pharmacology of Nicotine,* ICSU Press, New York, pp. 375 – 402.

Schmiterlöw, C.G. and Hansson, E. (1962) Physiological disposition and fate of nicotine labelled with carbon-14 in mice. *Nature (Lond.),* 194: 298 – 299.

Schmiterlöw, C.G. and Hansson, E. (1965) Tissue distribution of ^{14}C-nicotine. In U.S. von Euler (Ed.), *Tobacco Alkaloids and Related Substances,* Pergamon Press, London, pp. 75 – 86.

Schmiterlöw, C.G., Hansson, E., Andersson, G., Appelgren, L.E. and Hoffman, P.C. (1967) Distribution of nicotine in the central nervous system. *Ann. N.Y. Acad. Sci.,* 142: 1 – 14.

Schweitzer A. and Wright, S (1938) Action of nicotine on the spinal cord. *J. Physiol. (Lond.),* 94: 136 – 147.

Stalhandske, T. (1970) Effects of increased liver metabolism of nicotine on its uptake, elimination and toxicity in mice. *Acta Physiol. Scand.,* 80: 222 – 234.

Starke, K. (1977) Regulation of noradrenaline release by presynaptic receptor systems. *Rev. Physiol. Biochem. Pharmacol.,* 77: 1 – 124.

Starke, K. and Montel, H. (1974) Influence of drugs with affinity for α-adrenoceptors on noradrenaline release by potassium, tyramine and dimethylphenylpiperazinium. *Eur. J. Pharmacol.,* 27: 273 – 280.

Steinsland, O.S. and Furchgott, R.F. (1975) Desensitization of the adrenergic neurons of the isolated rabbit ear artery to nicotinic agonists. *J. Pharmacol. Exp. Ther.,* 193: 138 – 148.

Su, C. (1982) Actions of nicotine and smoking on circulation. *Pharmacol. Ther.,* 17: 129 – 141.

Turner, D.M. (1971) Metabolism of small multiple doses of C^{14} nicotine in the cat. *Br. J. Pharmacol.,* 41: 521 – 529.

Watkins, J.C. and Evans, R.H. (1981) Excitatory amino acid transmitters. *Ann. Rev. Pharmacol. Toxicol.,* 21: 165 – 204.

Westfall, T.C. and Brasted, M. (1972) The mechanism of action of nicotine on adrenergic neurons in the perfused guinea-pig heart. *J. Pharmacol. Exp. Ther.,* 182: 409 – 418.

Westfall, T.C. and Brasted, M. (1974) Specificity of blockade of the nicotine-induced release of ^{3}H-norepinephrine from adrenergic neurons of the guinea-pig heart by various pharmacological agents. *J. Pharmacol. Exp. Ther.,* 189: 659 – 664.

Molecular Biology of the Nicotinic Cholinergic Receptor

A. Nordberg, K. Fuxe, B. Holmstedt and A. Sundwall (Eds.)
Progress in Brain Research, Vol. 79
© 1989 Elsevier Science Publishers B.V. (Biomedical Division)

CHAPTER 2

Molecular biology of acetylcholine receptor long-term evolution during motor end-plate morphogenesis

Jean-Pierre Changeux[a], Bertrand Fontaine[a], André Klarsfeld[a], Ralph Laufer[a] and Jean Cartaud[b]

[a] URA CNRS 0210, "Neurobiologie Moléculaire", Département des Biotechnologies, Institut Pasteur, 25 rue du Docteur Roux, 75724 Paris Cédex 15, and [b] Microscopie Electronique et Biologie Cellulaire des Membranes, Institut Jacques Monod du Centre National de la Recherche Scientifique, Université Paris VII, 2 Place Jussieu, 75251 Paris Cédex 05, France

Introduction

The acetylcholine nicotinic receptor (AChR) from vertebrate neuromuscular junction and fish electromotor synapse remains one of the best known receptors for neurotransmitters (for review, see Popot and Changeux, 1984; Hucho, 1986; Changeux et al., 1987; Lindstrom et al., 1987). Moreover, recent studies with nicotinic AChR from neural tissue legitimize the extension of many of the basic properties uncovered with muscle AChR (though not necessarily all of them) to brain AChR (see Boulter et al., this volume). Also, the striking structural homologies found between muscle AChR and the recently cloned and sequenced brain receptor for glycine (Grenningloh et al., 1987) and for γ-aminobutyric acid (Schofield et al., 1987) suggest that the known neurotransmitter-gated ion channels belong to a gene family of transmembrane proteins which display several important functional properties in common.

In particular, receptors undergo two distinct categories of allosteric transitions elicited by the neurotransmitter: (1) the fast opening of the ion channel and (2) slower conformational changes, which take place in the 0.1 s to minute time scale, leading to high affinity refractory states, referred to as "desensitized" (review Changeux et al., 1984). Such slower transitions might be affected by reversible allosteric effectors and/or covalent modifications such as phosphorylation (Hopfield et al., 1988), and may serve as a model for the short-term regulation of synapse efficacy at the postsynaptic level (review Changeux and Heidmann, 1987; Mulle et al., 1988).

In recent years, complementary DNA (cDNA) probes and chromosomic genes coding for several of the basic components of the motor end-plate have been cloned and sequenced (Noda et al., 1982, 1983a, b, c; Sumikawa et al., 1982; Claudio et al., 1983; Devillers-Thiéry et al., 1983; Numa et al., 1983). The analysis of motor end-plate "long-term" development with the tools of recombinant DNA technology thus became possible. In this short chapter, some of its recent advances, which primarily concern the basic component of the postsynaptic domain, the acetylcholine receptor, are presented.

Molecular architecture of the motor end-plate

The motor end-plate (review Salpeter and Loring, 1985) results from the juxtaposition and complex association of two elements (Fig. 1a bottom): the ending of the motor nerve and a subneural domain, separated by a 50 – 100 nm cleft. The nerve terminal contains 30 – 60 nm clear vesicles filled with acetylcholine, which make linear arrays on top of membrane specializations referred to as "active zones". On the opposite side of the cleft, the muscle membrane makes repeated foldings with, at the top of the folds, thickenings located in front of the active zones and composed of closely packed AChR molecules (about 10 000 molecules per μm^2). A few microns away from the end-plate,

the density of AChR drops to less than 10 molecules per μm^2. The postsynaptic membrane is covered, on the cleft side, by an electron dense layer, the basal lamina, which contains collagen, heparan sulfate proteoglycan, laminin, several end-plate-specific antigens and the tailed forms of acetylcholinesterase. On the cytoplasmic side, a peripheral protein of molecular weight 43 000 underlies the AChR molecules to which it tightly binds with a 1-to-1 stoichiometry (review Changeux et al., 1984). The whole subsynaptic domain is anchored to a complex network of filaments which consists, in the case of the electromotor synapse, of desmin-rich intermediate filaments and includes, in the case of the motor

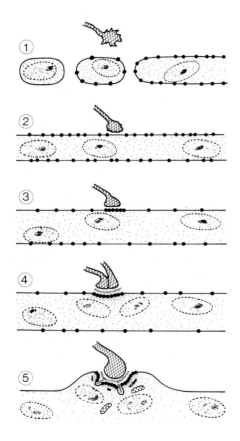

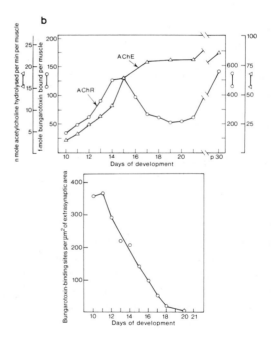

Fig. 1(a) Schematic representation of the evolution of the AChR in the course of the formation of the neuromuscular junction (●, AChR). (1) fusion of myoblasts into myotubes; AChR biosynthesis is enhanced; (2) the exploratory motor axon approaches; (3) the growth cone contacts the myotube, a subneural cluster of AChR forms; (4) several motor nerve endings converge on the subneural cluster of AChR; (5) one motor nerve ending becomes stabilized, subneural folds develop: interactions with the cytoskeleton become apparent. (From Changeux et al., 1987b.) (b) Evolution of AChR content of the *posterior latissimus dorsi* muscle in the chick. (Modified from Betz et al., 1980.)

end-plate, actin, α-actinin, vinculin and filamin (review Salpeter and Loring, 1985).

The cytoplasm of the muscle fiber makes a small eminence at the level of the terminal arborization of the motor nerve where mitochondria and 4 to 8 muscle nuclei named "fundamental nuclei" by Ranvier (1875) are accumulated.

The postsynaptic domain of the motor end-plate thus represents a highly organized supra-molecular differentiation of muscle surface. To test for its stability, denervation experiments were carried out with the electric organ of *Electrophorus electricus*. Within a week after denervation, all the nerve endings disappeared, yet up to 52 days later high density AChR patches were still detected by autoradiography after labelling with snake α-toxins, with the distribution, shape and dimension of former subneural areas (Bourgeois et al., 1973, 1978b). A similar persistence of subneural AChR aggregates was also reported with the denervated motor end-plate in a number of vertebrate species (Frank et al., 1975, ref. in Salpeter and Loring, 1985). These experiments illustrate that the organization of the adult postsynaptic domain is a rather stable structure and this raises the question of how such a sophisticated supramolecular architecture arises.

Evolution of AChR distribution and properties during end-plate formation

The development of the postsynaptic domain involves a complex sequence of molecular processes (Fig. 1a) which include regulation of gene transcription and multiple post-transcriptional phenomena (review Salpeter and Loring, 1985; Merlie and Smith, 1986; Changeux et al., 1987a, b; Schuetze and Role, 1987). In chick or rat, this evolution lasts for several weeks during embryonic development and after birth. Figure 1b illustrates the evolution of the total content in AChR of the fast muscle *posterior latissimus dorsi* followed

with snake venom α-toxins (Betz et al., 1980). The initial increase of AChR number (step 1) coincides with the proliferation of myoblasts and their fusion into myotubes. At this stage, the AChR becomes diffusely distributed all along the myotubes at a density of $100-500$ molecules per μm^2, exhibits significant rotational and lateral motion, undergoes rapid metabolic turnover (half-life $17-22$ h) and possesses a mean channel open time of $3-10$ ms. As soon as the growth cone of the exploratory motor axon contacts the surface of the myotube, the AChR molecules start to cluster under the nerve endings to a density of $1-2000$ molecules per μm^2, become immobile and electrical signals are transmitted. When the first neuromuscular contacts become functional, the "contrast" between the surface densities of junctional and extrajunctional AChR increases as a consequence of a sharp decline of extrajunctional AChR. This decrease (step 2) is significant enough to cause a drop in the total content of muscle AChR. Meanwhile, around birth in the rat (or more than 3 weeks after hatching in the chick), the metabolic half-life of the AChR molecule increases up to a value of about 11 days in adult muscle (step 3). Then, about a week after birth in mammals (but not in birds), the mean duration of channel openings becomes $3-5$ times shorter (Schuetze and Role, 1987). Throughout this development, the overall number of AChR molecules in the postsynaptic domain steadily augments and this increase can be as large as 15 to 20-fold after birth (step 4).

A differential regulation of AChR distribution thus takes place in junctional and extrajunctional areas of the developing muscle fiber. From early stages of embryonic development (3.5 day in the chick) the spinal cord motor neurons are spontaneously active and movements of the embryo take place (Hamburger, 1970). An important question is whether this nerve-evoked muscle activity plays a critical role in any of the above mentioned steps.

Biosynthesis of the AChR in the developing muscle fiber analyzed by the methods of molecular genetics

The initial increase in AChR number which takes place in the muscle primordium (step 1) can, at least in part, be reproduced in vitro with primary muscle cultures or myogenic cell lines (review Merlie and Smith, 1986). Under the conditions where this increase occurs, radioactive or heavy isotope labelled amino acids become incorporated into the AChR protein without change of AChR degradation rate. A de novo synthesis of AChR molecules thus takes place and the state of activity of the muscle fibers does not seem to control this synthesis either in vivo or in vitro.

The use of cloned cDNA and/or chromosomic genes coding for the α, β, γ, and δ-subunits of muscle AChR in several vertebrate species (including humans) (Numa et al., 1983; Merlie and Smith, 1986; Moss et al., 1987) provided important insights into the mechanisms involved in this enhanced biosynthesis. The regulation could take place at the level of the transcription of AChR genes into messenger RNA (mRNA), of the translation of the mRNA into protein or of any of the subsequent steps which yield a functional receptor exposed at the surface of the muscle cell. It was shown that the steady-state levels of the four subunit mRNAs dramatically increase during the terminal stages of muscle differentiation, both in vivo and in vitro. This results from an over 8-fold increase of their rates of transcription, at least for the genes coding for the α- and δ-subunits [as shown by nuclear run on experiments in the C2 mouse cell line (Merlie and Smith, 1986)].

Most nuclear genes in higher eukaryotes consist of alternating exons and non-coding intron sequences. The latter are present within the gene primary transcript and subsequently spliced to yield the *mature* mRNA which is then transported from the nucleus to the cytoplasm. The regulation of gene transcription has been shown to involve DNA segments located on both sides (5′ *up*stream or 3′ *down*stream) of the transcribed sequence and even within the non-coding introns. To look for such regulatory DNA sequences near the chicken AChR α-subunit gene, the upstream flanking region of this gene (850 base pairs from its 5′-end) was inserted in front of a "reporter gene" which codes for the bacterial enzyme chloramphenicol acetyltransferase (CAT). This "chimeric" con-

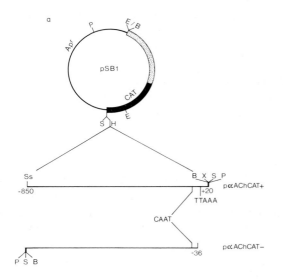

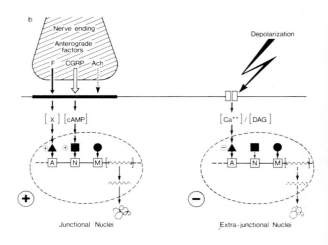

Fig. 2. (a) Diagram of the recombinant plasmid used to demonstrate the presence of transcriptional regulatory elements in the 5′ flanking region of the chicken α-subunit gene. (From Klarsfeld et al., 1987.) (b) Model of regulation of AChR α-subunit gene expression in subneural and extrajunctional areas of the developing neuromuscular junction. (Modified from Changeux et al., 1987a, b.)

struction (Fig. 2) was introduced into cultured cells and the expression of the reporter gene under the command of the AChR regulatory sequences was investigated by assaying CAT activity in cell extracts prepared a few days later. Interestingly, this construction directed high CAT gene expression in myotubes, but not in cultured myoblasts, nor in fibroblasts (Klarsfeld et al., 1987). Hence, this 5'-flanking region of the α-subunit gene contains DNA elements which selectively control the expression of AChR α-subunit in muscle cells during myotube differentiation.

Another approach toward defining the regions of the AChR chromosomic gene involved in the regulation of transcription is to look for chromatin regions which are exposed to the transcription machinery and, as a consequence, are more vulnerable to degradation by enzymes such as DNAse than the rest of the chromosome. For the mouse δ-subunit gene (Crowder and Merlie, 1986), two such hypersensitive sites were found in the chromatin prepared from muscle cells, but not in that extracted from other cell types. One lies close to the 3'-end of the transcribed domain, and another close to its 5'-end: precisely in the same domain as that identified by the chicken α-subunit chimeric gene experiments.

The activity-dependent repression of extrajunctional AChR biosynthesis

Chronic paralysis of the chick embryo in ovo by snake venom α-toxins or curare-like agents does not affect the initial onset of AChR biosynthesis but prevents the subsequent decrease of AChR content, which corresponds to the elimination of extrajunctional AChR (step 2). This effect takes place without any change of the AChR metabolic degradation rate and therefore represents an activity-dependent repression of AChR biosynthesis (Betz et al., 1977; Burden, 1977a, b; Bourgeois et al., 1978a). Interestingly, it can be reproduced to some extent with primary cultures of myotubes which exhibit spontaneous (nonneurogenic) firing (review Salpeter and Loring,

1985). Blocking this spontaneous electrical activity by tetrodotoxin, an inhibitor of the voltage-dependent Na^+ channel results after 24 and 48 h of treatment, in a 1.5 to 2-fold increase in surface AChR accompanied by a respective 4.5 and 13-fold increase in α-subunit mRNA levels (Klarsfeld and Changeux, 1985). Actinomycin D, an inhibitor of transcription, blocks the effect of tetrodotoxin which, thus, predominantly (though perhaps not exclusively) corresponds to a stimulation of transcription.

This regulation can also be studied by following the increase of AChR levels caused by denervating the adult muscle (Klarsfeld and Changeux, 1985; Merlie and Smith, 1986: Evans et al., 1987; Moss et al., 1987). With this system, the direct assay of the α-subunit primary (unspliced) gene transcript, which includes the introns, demonstrates that this regulation (which can be reversed by electrically stimulating the muscle) takes place at the level of transcription of the chromosomic gene into mRNA (Shieh et al., 1987).

Intracellular signals linking electrical activity to gene expression

The intracellular "second messengers" which link the electrical activity of the muscle cell to the transcription of AChR genes are not yet securely identified. Since firing of the action potentials is accompanied by mechanical contraction of the muscle fiber, biochemical events associated with the mechanical contraction may regulate AChR biosynthesis. This possibility is ruled out by the analysis of muscles from the mouse mutant *muscular dysgenesis* (Powell and Friedman, 1977) where cells fire action potentials but do not contract. Yet, AChR biosynthesis is repressed by electrical activity even in such dysgenic myotubes. It is therefore the electrical activity of the myotubes and *not* their mechanical contraction which regulates AChR biosynthesis.

A second messenger postulated to link electrical activity to the expression of AChR genes is Ca^{2+} (ref. in Salpeter and Loring, 1985; Changeux et al.,

1987a, b). Upon depolarization of the cell membrane, the cytoplasmic Ca^{2+} concentration transiently rises as a result of Ca^{2+} release from the sarcoplasmic reticulum. Ca^{2+} ions may also enter the cell via voltage-sensitive Ca^{2+} channels or via the ion channel associated with the AChR. Several lines of evidence, mostly pharmacological, support the notion that a rise of cytoplasmic Ca^{2+} level shuts down AChR gene transcription. Moreover, the finding that a known activator of protein kinase C, the phorbol ester TPA, selectively blocks the increase of α-subunit mRNA levels caused by tetrodotoxin (Fontaine et al., 1987), raises the possibility that the physiological activation of this enzyme by intracellular Ca^{2+} and diacylglycerol constitutes an intermediate step in the repression of AChR genes.

Subneural regulation of AChR gene expression

At the time when AChR is disappearing from extra junctional areas, the number of AChR molecules still steadily increases under the nerve ending. How does the accumulation of subneural AChR react to the repression caused by the electrical activity of the muscle cell membrane? A first approach to this question consisted in the identification of soluble factors from neural tissues which increase AChR number on cultured myotubes in vitro (see Salpeter and Loring, 1985). A trypsin-sensitive polypeptide of about 42 kDa purified from chick brain differentially enhances AChR synthesis with a time course similar to that observed at developing synapses (Usdin and Fischbach, 1986). Among lower molecular weight factors, a 10 kDa polypeptide and ascorbic acid elicit similar effects without affecting AChR metabolic degradation rate (ref. in Salpeter and Loring, 1985; Schuetze and Role, 1987).

Other potentially important compounds are the neuropeptides known to coexist (Hökfelt et al., 1986) with ACh in spinal cord motor neurons. One of them is calcitonin gene-related peptide (CGRP) (Rosenfeld et al., 1983), a peptide of 37 amino acids which exerts several biological actions such as cardiovascular effects, inhibition of gastric secretion and ingestive behavior and may act as a neurotransmitter in the sensory tract. CGRP has also been localized in spinal cord motor neurons and at the level of the motor nerve endings. Moreover, it is released by depolarization from cultures of rat trigeminal neurons, and it enhances skeletal muscle contraction (ref. in Fontaine et al., 1986, 1987; Laufer and Changeux, 1987). Addition of CGRP to chick muscle cells in primary culture increases the level of surface AChR and the content in α-subunit mRNA by about $30-50\%$ and 300%, respectively (Fontaine et al., 1986, 1987; New and Mudge, 1986). Cholera toxin, an activator of adenylate cyclase, produces the same effect, which is not additive with that of CGRP, suggesting that CGRP may act by stimulating the production of cAMP in skeletal muscle cells. In support of this hypothesis, CGRP was shown to increase cAMP levels in cultured myotubes and isolated mouse diaphragm and to activate adenylate cyclase in muscle membranes (Takami et al., 1986; Laufer and Changeux, 1987). Interestingly, the stimulatory effect of CGRP or cholera toxin on AChR number (or α-subunit mRNA level) persists when the spontaneous activity of the cultured myotubes is blocked by tetrodotoxin (Fontaine et al., 1986, 1987). Moreover, the phorbol ester TPA, an activator of protein kinase C, does not significantly modify the increase of α-subunit mRNA by CGRP or cholera toxin but abolishes the effect of tetrodotoxin on AChR biosynthesis and α-subunit mRNA levels (Fontaine et al., 1987). It thus appears that the regulation of AChR biosynthesis by CGRP involves an intracellular pathway which is different from that utilized by membrane depolarization or activation of protein kinase C. However, at this stage, it is *not* known whether CGRP is physiologically involved in end-plate morphogenesis and/or maintenance. Still, it remains a plausible candidate along with other neural "anterograde" trophic factors, which may act synergically and/or sequentially during development.

Differential regulation of AChR α-subunit genes in subneural and extrajunctional nuclei

The experiments with CGRP and tetrodotoxin raise the possibility of a differential regulation of AChR biosynthesis in subneural vs. extrajunctional areas. First evidence in favor of such a possibility came from the comparison of the steady-state levels of AChR mRNA in hand-dissected synapse-free and synapse-rich samples of mouse diaphragm. An enrichment of 2.9 ± 0.5-fold of α-subunit mRNA and of 9 to 14-fold of δ-subunit mRNA was found in synapse-rich samples (Merlie and Sanes, 1985).

Detection of α-subunit mRNA by in situ hybridization (Fontaine et al., 1988) (Fig. 3) further demonstrates that in 15-day-old chick muscles, a differential accumulation of this message occurs around the end-plate nuclei referred to as "fundamental". Moreover, in extrajunctional areas the labelling was found negligible but dramatically increased after denervation. A contribution of mRNA stabilization and/or transport cannot be excluded at this stage, although the absence of a mRNA gradient from extrajunctional vs. junctional areas revealed by in situ hybridization does not favor this hypothesis. Since an enhanced rate of AChR gene transcription is known to occur after denervation, the in situ

hybridization data most likely reveal the state of transcription of the nuclear genes as recently established, for instance, by in situ hybridization with intronic probes.

The results are consistent with a model (Fig. 2) (Fontaine and Changeux, 1989; Changeux et al., 1987a, b) according to which: (a) in the adult muscle the "fundamental" nuclei which underlie the nerve endings are in a state of AChR gene expression different from those located in extrajunctional areas and (2) the development and maintenance of this pattern is compartmentalized and involves different intracellular signalling pathways. For instance, CGRP (or another neural factor) co-released with acetylcholine by the motor nerve ending would increase the expression of the genes coding for the AChR via the cAMP cascade in subneural areas while, in extrajunctional areas, the electrical activity of the muscle membrane would repress their transcription by an intracellular pathway involving calcium and more speculatively, the activation of protein kinase C.

Accordingly, a set of distinct DNA regulatory sequences and their corresponding DNA-binding regulatory proteins, sensitive to the second messengers mentioned above, would be involved in the differential regulation of AChR genes in subneural vs. extrajunctional nuclei (Changeux et al., 1987a, b). An important outcome of these studies thus becomes the identification of these regulatory elements.

Discoordinate regulation of the genes coding for AChR subunits

Such regulatory DNA sequences might differ for the four AChR subunit genes. In the mouse, the genes coding for the four subunits of the AChR have been located on three different chromosomes (Heidmann et al., 1986) and, therefore, cannot be co-regulated via one and the same set of cis-acting DNA regulatory sequences. A discoordination of AChR gene expression following denervation has been reported for rat and chick muscle (Evans et al., 1987; Moss et al., 1987). It became particularly

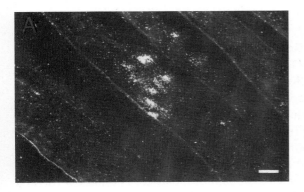

Fig. 3. Detection by in situ hybridization of AChR α-subunit mRNA in 15-day-old chick muscle. (From Fontaine et al., 1988.)

striking in the case of the postnatal evolution of the subneural AChR in mammals. The mRNA of a fifth subunit referred to as ϵ (ref. in Witzemann et al., 1987), was initially identified by its cDNA in calf muscle. Expression of combinations of subunit-specific mRNAs in *Xenopus* oocytes shows that the AChR molecules which have incorporated the ϵ-subunit display *adult* short mean channel open time and conductances, while those with γ-subunit possess *embryonic* channel conductance and gating properties. Interestingly, the level of ϵ-subunit mRNA was shown to increase in the adult while that of γ-subunit mRNA decreased, and a correlation between the changes in γ- and ϵ-subunit mRNAs abundance and the relative density of the two classes of AChR channels was suggested in the rat (Witzemann et al., 1987). A switch of gene expression may thus differentially affect at least *one* AChR subunit gene in mammals. Yet, it takes place at late stages of end-plate maturation many days after the first functional synaptic contacts have been formed and the postsynaptic AChR clustered.

Post-transcriptional regulation of AChR evolution

The transcription of AChR structural genes into mRNA represents only the first of a long sequence of events which account for AChR expression (Merlie and Smith, 1986). The overall synthesis of the subunit polypeptides takes about 1 min. Then, the subunits progressively acquire their correct conformation, such as the ability to bind snake α-toxin and cholinergic ligands (in the case of the α-subunit). They assemble into the α_2-, β-, γ-, δ-oligomer and are transported to the cell surface. The whole process lasts more than 2 h. Throughout this sequence nearly 70% of the synthesized α-subunits are degraded. A drastic process of molecular selection thus accompanies the biosynthesis of the AChR (Merlie and Smith, 1986). Regulatory processes affecting post-transcriptional events are expected to occur but regulation of mRNA stability has not been described so far. On

the other hand, in cultured mouse BC3H1 cells, calf serum decreases the number of cell surface AChR without affecting α-subunit mRNA levels. Moreover, TPA decreases the total number of α-bungarotoxin binding sites in chick muscle cultures without affecting AChR α-subunit mRNA content (Fontaine et al., 1987). Finally, electrical activity was shown to inhibit the association of AChR subunits in primary cultures of embryonic rat muscle (Carlin et al., 1986).

One of the latest steps of AChR evolution, which is also one of the earliest in the development of the motor end-plate, is the clustering of surface AChR under the immobilized growth cone (ref. in Salpeter and Loring, 1985; Schuetze and Role, 1987). In *Xenopus,* this local increase of AChR density unambiguously forms out of a pre-existing surface pool of functional AChR by lateral redistribution in the plasma membrane. On the other hand, in chick nerve-muscle co-cultures, a local insertion of newly formed AChR molecules rapidly becomes the dominant process.

The clustering of AChR can develop in the absence of nerve contacts but is elicited by electrical fields, basic polypeptide-coated latex beads and protein(s) extracted from the basal lamina referred to as agrin (ref. in Schuetze and Role, 1987). A regulation of the interaction of AChR with the cytoskeleton has been suggested to play a role in cluster formation. Yet, at an early stage of the differentiation of *Torpedo marmorata* electrocyte (45 nm embryo), a large AChR cluster develops at one pole of the cell in the absence of nerve and without a detectable association of the 43 kDa protein which appears only at a later stage (Kordeli et al., 1988). Consequently, the 43 kDa protein does not seem necessary for the aggregation of the AChR but may consolidate the aggregate once formed and for instance promote its interaction with the cytoskeleton.

In conclusion, the development of the postsynaptic domain of the motor end-plate involves a complex sequence of events at both transcriptional and post-transcriptional levels. Several, but not all, of these steps are regulated by the state of activity of the postsynaptic cell. Several of these processes are not expected to occur in mononucleated nerve cells such as the differential regulation of gene transcription in such synaptic and extrasynaptic areas. Yet, their identification is expected to provide new insights into the mechanisms of synapse formation and its long-term regulation by activity in neuronal networks.

Acknowledgements

This work was supported by grants from the Muscular Dystrophy Association of America, the Collège de France, the Ministère de la Recherche, INSERM (contract no. 872 004). and DRET (contract no. 87/211).

References

Betz, H., Bourgeois, J.P. and Changeux, J.P. (1977) *FEBS Lett.* 77: 219–224.

Betz, H., Bourgeois, J.P. and Changeux, J.P. (1980) Evolution of cholinergic proteins in developing slow and fast skeletal muscles from chick embryo. *J. Physiol.,* 302: 197–218.

Bourgeois, J.P., Popot, J.L., Ryter, A. and Changeux, J.P. (1973) Consequences of denervation on the distribution of the cholinergic (nicotinic) receptor sites from *Electrophorus Electricus* revealed by high resolution autoradiography. *Brain Res.,* 62: 557–563.

Bourgeois, J.P., Betz, H. and Changeux, J.P. (1978a) *C.R. Acad. Sci.,* 286D: 773–776.

Bourgeois, J.P., Popot, J.L., Ryter, A. and Changeux, J.P. (1978b) Quantitative studies on the localization of the cholinergic receptor protein in the normal and denervated electroplaque from *Electrophorus Electricus. J. Cell Biol.,* 79: 200–216.

Burden, S. (1977a) Development of the neuromuscular junction in the chick embryo. The number, distribution and stability of the acetylcholine receptor. *Dev. Biol.,* 57: 317–329.

Burden, S. (1977b) Acetylcholine receptors at the neuromuscular junction: developmental change in receptor turnover. *Dev. Biol.,* 61: 79–85.

Carlin, B.E., Lawrence, Jr., J.C., Lindstrom, J.M. and Merlie, J.P. (1986) Inhibition of acetylcholine receptor assembly by activity in primary cultures of embryonic rat muscle cells. *J. Biol. Chem.,* 261: 5180–5186.

Changeux, J.P. (1986) Coexistence of neuronal messengers and molecular selection. In T. Hokfelt, K. Fuxe and B. Pernow (Eds.) *Progess in Brain Research, Vol. 68,* Elsevier, New York, pp. 373–403.

Changeux, J.P. and Danchin, A. (1976) Selective stabilization of developing synapses as a mechanism for the specificication of neuronal networks. *Nature, (Lond.),* 264: 705–712.

Changeux, J.P., Devillers-Thiéry, A. and Chemouilli, P. (1984) Acetylcholine receptor: an allosteric protein. *Science,* 225: 1335–1345.

Changeux, J.P. and Heidmann, T. (1987) Allosteric receptors and molecular models of learning. In G. Edelman, W.E. Gall and W.M. Cowan (Eds.), *Synaptic Function,* John Wiley, New York, pp. 549–601.

Changeux, J.P., Klarsfeld, A. and Heidmann, T. (1987a) The acetylcholine receptor and molecular models for short and long term learning. Dahlem Konferenzen. In J.P. Changeux and M. Konishi (Eds.), *The Cellular and Molecular Bases of Learning,* Wiley, London, pp. 31–83.

Changeux, J.P., Devillers-Thiéry, A., Giraudat, J., Dennis, M., Heidmann, T., Revah, F., Mulle, C., Heidmann, O., Klarsfeld, A., Fontaine, B., Laufer, R., Nghiêm, H.O., Kordeli, E. and Cartaud, J. (1987b) The acetylcholine receptor: functional organisation and evolution during synapse formation. In O. Hayaishi (Ed.), *Strategy and Prospects in Neuroscience,* Taniguchi Symposia on Brain Sciences no. 10, Japan Scientific Societies Press Tokyo VNU Science Press BV, Utrecht, pp. 29–76.

Claudio, T., Ballivet, M., Patrick, J. and Heinemann, S. (1983) Nucleotide and deduced amino acid sequences of *Torpedo californica* acetylcholine receptor gamma-subunit. *Proc. Natl. Acad. Sci. U.S.A.,* 80: 1111–1115.

Crowder, C.M. and Merlie, J.P. (1986) DNAse I-hypersensitive sites surround the mouse acetylcholine receptor δ-subunit gene. *Proc. Natl. Acad. Sci. U.S.A.,* 83: 8405–8409.

Devillers-Thiery, A., Giraudat, J., Bentaboulet, M. and Changeux, J.P. (1983) Complete mRNA coding sequence of the acetylcholine binding alpha subunit of *Torpedo marmorata* acetylcholine receptor: a model for the transmembrane organization of the polypeptide chain. *Proc. Natl. Acad. Sci. U.S.A.,* 80: 2067–2071.

Evans, S., Goldman, D., Heinemann, S. and Patrick, J. (1987) Muscle acetylcholine receptor biosynthesis. *J. Biol. Chem.,* 262: 4911–4916.

Fontaine, B. and Changeux, J.P. (1989) *J. Cell. Biol.,* 108: 1025–1037.

Fontaine, B., Klarsfeld, A., Hökfelt, T. and Changeux, J.P. (1986) Calcitonin gene-related peptide, a peptide present in spinal cord motoneurons, increases the number of acetylcholine receptors in primary cultures of chick embryo myotubes. *Neurosci. Lett.,* 71: 59–65.

Fontaine, B., Klarsfeld, A. and Changeux, J.P. (1987) Calcitonin gene-related peptide and muscle activity regulate acetylcholine receptor α-subunit mRNA levels by distinct intracellular pathways. *J. Cell Biol.,* 105: 1337–1342.

Fontaine, B., Sassoon, D., Buckingham, M. and Changeux, J.P. (1988) Detection of the nicotinic acetylcholine receptor α-subunit mRNA by in situ hybridization at neuromuscular junctions of 15-day old chick striated muscles. *EMBO J.,* 7: 603 – 609.

Frank, E., Gautvik, K. and Sommer-Schild, H. (1975) Cholinergic receptors at denervated mammalian endplates. *Acta Physiol. Scand.,* 95: 66 – 76.

Hamburger, V. (1970) Embryonic mobility in vertebrates. In F.O. Schmitt (Ed.) *The Neuroscience: Second Study Program,* Rockefeller University Press, New York, pp. 141 – 151.

Grenningloh, G., Rienitz, A., Schmitt, B., Methfessel, C., Zensen, M., Beyreuther, K., Gundelfinger, E.D. and Betz, H. (1987) The strychnine-binding subunit of the glycine receptor shows homology with nicotinic acetylcholine receptors. *Nature (Lond.),* 328: 215 – 220.

Heidmann, O., Buonanno, A., Goeffroy, B., Robert, B., Guénet, J.L., Merlie, J.P. and Changeux, J.P. (1986) Chromosomal localization of the nicotinic acetylcholine receptor genes in the mouse. *Science.,* 234: 866 – 868.

Hökfelt, T., Holets, V.R., Staines, W., Meister, B., Melander, T., Schalling, M., Schultzberg, M., Freedman, J., Björklund, H., Olson, L., Lindk, B., Elfvin, L.G., Lundberg, J., Lindgren, J.A., Samuelsson, B., Terenius, L., Post, C., Everitt, B. and Goldstein, M. (1986) Coexistence of neuronal messengers – an overview. *Prog. Brain Res.,* 68: 33 – 70.

Hopfield, J.F., Tank, D.W., Greengard, P. and Huganir, R.L. (1988) Functional modulation of the nicotinic acetylcholine receptor by tyrosine phosphorylation. *Nature,* 336: 677 – 680.

Hucho, F. (1986) The nicotinic acetylcholine receptor and its ion channel. *Eur. J. Biochem.,* 158: 211 – 226.

Klarsfeld, A. and Changeux, J.P. (1985) Activity regulates the level of acetylcholine receptor alpha-subunit mRNA in cultured chick myotubes. *Proc. Natl. Acad. Sci. U.S.A.,* 82: 4558 – 4562.

Klarsfeld, A., Daubas, P., Bourachot, B. and Changeux, J.P. (1987) A 5′ flanking region of the chicken acetylcholine receptor alpha-subunit gene confers tissue-specificity and developmental control of expression in transfected cells. *Mol. Cell Biol.,* 7: 951 – 955.

Kordeli, E., Cartaud, J., Nghiêm, H.O., Devillers-Thiéry, A. and Changeux, J.P. (1988) Asynchronous assembly of the acetylcholine receptor and of the 43 kD-*v*-protein in the postsynaptic membrane of developing *Torpedo marmorata* electrocyte. *J. Cell. Biol.,* 108: 127 – 139.

Laufer, R. and Changeux, J.P. (1987) Calcitonin gene-related peptide elevates cyclic AMP levels in chick skeletal muscle: possible neurotrophic role for a coexisting neuronal messenger. *EMBO J.,* 6: 901 – 906.

Lindstrom, J., Schoepfer, R. and Whiting P. (1987) Molecular studies of the neuronal nicotinic acetylcholine receptor family. *Mol. Neurobiol.,* 1: 281 – 337.

Merlie, J. and Sanes, J.R. (1985) Concentration of acetylcholine receptor mRNA in synaptic regions of adult muscle fibers. *Nature (Lond.),* 317: 66 – 68.

Merlie, J.P. and Smith, M.M. (1986) Synthesis and assembly of acetylcholine receptor, a multisubunit membrane glycoprotein. *J. Memb. Biol.,* 91: 1 – 10.

Moss, S.J., Beeson, D.M., Jackson, J.F., Darlison, M.G. and Barnard, E.A. (1987) Differential expression of nicotinic acetylcholine receptor genes in innervated and denervated chicken muscle. *EMBO J.,* 6: 3917 – 3921.

Mulle, C., Benoit, P., Pinset, C., Roa, M. and Changeux, J.P. (1988) Calcitonin gene-related peptide enhances the rate of desensitization of the nicotinic acetylcholine receptor in cultured mouse muscle cells. *Proc. Natl. Acad. Sci. U.S.A.,* 85: 5728 – 5732.

New, H.V. and Mudge, A.W. (1986) Calcitonin gene-related peptide regulates muscle acetylcholine receptor synthesis. *Nature (Lond.),* 323: 809 – 811.

Noda, M., Takahashi, H., Tanabe, T., Toyosato, M., Furutani, Y., Hirose, T., Asai, M., Inayama, S., Miyata, T. and Numa, S. (1982) Primary structure of alpha-subunit precursor of *Torpedo californica* acetylcholine receptor deduced from cDNA sequence. *Nature (Lond.),* 299: 793 – 797.

Noda, M., Takahashi, H., Tanabe, T., Toyosato, M., Kikyotani, S., Hirose, T., Asai, M., Takashima, H., Inayama, S., Miyata, T. and Numa, S. (1983a) Primary structures of beta and delta-subunit precursors of *Torpedo californica* acetylcholine receptor deduced from cDNA sequences. *Nature (Lond.),* 301: 251 – 255.

Noda, M., Takahashi, H., Tanabe, T., Toyosato, M., Kikyotani, S., Furutani, Y., Hirose, T., Takashima, H., Inayama, S., Miyata, T. and Numa, S. (1983b) Structural homology of *Torpedo californica* acetylcholine receptor subunits. *Nature (Lond.),* 302: 528 – 532.

Noda, M., Furutani, Y., Takahashi, H., Toyosato, M., Tanabe, T., Shimizu, S., Kikyotani, S., Kayano, T., Hirose, T., Inayama, S. and Numa, S. (1983c) Cloning and sequence analysis of calf cDNA and human genomic DNA encoding α-subunit precursor of muscle acetylcholine receptor. *Nature (Lond.),* 305: 818 – 823.

Numa, S., Noda, M., Takahashi, H., Tanabe, T., Toyosato, M., Furutani, Y. and Kikyotani, S. (1983) Molecular structure of the nicotinic acetylcholine receptor. *Cold Spring Harb. Symp. Quant. Biol.,* 48: 57 – 69.

Popot, J.L. and Changeux, J.P. (1984) The nicotinic receptor of acetylcholine: structure of an oligomeric integral membrane protein. *Physiol. Rev.,* 64: 1162 – 1184.

Powell, J.A. and Friedman, B. (1977) Electrical membrane activity: effect on distribution, incorporation and degradation of acetylcholine receptors in membranes of cultured muscle. *J. Cell Biol.,* 75: 323a.

Ranvier, L. (1875). *Traité Technique d'Histologie,* Savy, Paris.

Rosenfeld, M.G., Mermod, J.J., Amara, S.G., Swanson, L.W., Sawchenko, P.E., Rivier, J., Vale, W.W. and Evans, R.M. (1983) Production of a novel neuropeptide encoded by the calcitonin gene via tissue-specific RNA processing. *Nature (Lond.),* 304: 129 – 135.

Salpeter, M. and Loring, R.H. (1985) Nicotinic acetylcholine receptors in vertebrate muscle: properties, distribution and neural control. *Prog. Neurobiol.,* 25: 297 – 325.

Schofield, P.R., Darlison, M.G., Fujita, N., Burt, D.R., Stephenson, F.A., Rodriguez, H., Rhee, L.M., Ramachandran, J., Reale, V., Glencorse, T.A., Seeburg, P.H. and Barnard, E.A. (1987) Sequence and functional expression of the GABA$_A$ receptor shows a ligand-gated receptor super family. *Nature (Lond.),* 328: 221 – 227.

Schuetze, E. and Role, L. (1987) Developmental regulation of nicotinic acetylcholine receptors. *Ann. Rev. Neurosci.,* 10: 403 – 457.

Shieh, B.H., Ballivet, M. and Schmidt, J. (1987) Quantitation of an alpha subunit splicing intermediate: evidence for transcriptional activation in the control of acetylcholine receptor expression in denervated chick skeletal muscle. *J.*

Cell Biol., 104: 1337 – 1341.

Sumikawa, K., Houghton, M., Smith, J.C., Bell, L., Richards, B.M. and Barnard, E.A. (1982) The molecular cloning and characterization of cDNA coding for the alpha subunit of the acetylcholine receptor. *Nucleic Acid Res.,* 10: 5809 – 5822.

Takami, K., Hashimoto, K., Uchida, S., Tohyama, M. and Yoshida, H. (1986) Effect of calcitonin gene-related peptide on the cyclic AMP level of isolated mouse diaphragm. *Jpn. J. Pharmacol.,* 42: 345 – 350.

Usdin, T.B. and Fischbach, G.D. (1986) Purification and characterization of a polypeptide from chick brain that promotes the accumulation of acetylcholine receptors in chick myotubes. *J. Cell Biol.,* 103: 493 – 507.

Witzemann, V., Barg, B., Nishikawa, Y., Sakmann, B. and Numa, S. (1987) Differential regulation of muscle acetylcholine receptor γ and ϵ-subunit mRNAs. *FEBS Lett.,* 223: 104 – 112.

A. Nordberg, K. Fuxe, B. Holmstedt and A. Sundwall (Eds.)
Progress in Brain Research, Vol. 79
© 1989 Elsevier Science Publishers B.V. (Biomedical Division)

CHAPTER 3

Structure and function of neuronal nicotinic acetylcholine receptors deduced from cDNA clones

Jim Patrick[a], Jim Boulter[a], Evan Deneris[a], Keiji Wada[a], Etsuko Wada[a], John Connolly[a], Larry Swanson[b] and Steve Heinemann[a]

[a] Molecular Neurobiology Laboratory and [b] Neural Systems Laboratory, The Salk Institute for Biological Studies, San Diego, CA 92138, U.S.A.

Cholinergic synaptic transmission in the central nervous system (CNS) is both muscarinic and nicotinic (for review see Goodman and Gillman, 1975). It was generally thought that the muscarinic system is the major contributor to cholinergic transmission in the brain. This may have been due, in part, to the availability of good high affinity ligands for the muscarinic receptor and, in part, to the focus on the nicotinic receptor in the peripheral nervous system. Electrophysiological studies of neuronal nicotinic receptors focussed primarily upon the peripheral ganglia with less emphasis on the CNS. There remains considerable disagreement about the nature and extent of nicotinic cholinergic transmission in the brain.

The popular dependence upon nicotine obtained from tobacco is sometimes used as an argument for a role for nicotinic cholinergic transmission in the brain. It is not clear, however, that nicotinic acetylcholine receptors are the only sites of action of nicotine. It may be the case that there are several different classes of molecules that bind nicotine, some of which may not be cholinergic neurotransmitter receptor molecules. There have been many elegant studies of nicotine binding to CNS tissue, both in extracts and in brain slices (Clarke et al., 1985; Wonnacott, 1987). These studies show that there are both high and low affinity binding sites (Marks et al., 1986; Wonnacott, 1987), that the

binding sites are prevalent (Clarke et al., 1984), and that they are widely distributed in the brain (Clarke et al., 1985). They do not, however, show that the binding site is a functional nicotinic acetylcholine receptor.

The toxin α-bungarotoxin played a major role in the great progress made on the nicotinic acetylcholine receptor present at the neuromuscular junction (for review see Fambrough, 1976). Because the toxin also bound to brain membrane preparations and to cells known to contain neuronal nicotinic acetylcholine receptors, it was widely believed that the toxin bound to neuronal nicotinic acetylcholine receptors. Although it was shown in several systems that the toxin did not block activation of these receptors (Patrick and Stallcup, 1977b; Duggan et al., 1976) it seemed possible that the toxin bound to them in a way that did not affect function. This idea was made unlikely in one system by pharmacological experiments (Patrick and Stallcup, 1977b) and by the demonstration that antibodies that block receptor function do not recognize the molecule that binds toxin (Patrick and Stallcup, 1977a). It was further shown that the toxin-binding molecule had a distribution in the synapse unlike that expected for a receptor (Jacob and Berg, 1983). Finally, a new toxin, that had contaminated earlier bungarotoxin preparations (Chippinelli and Zigmond, 1978), was isolated and shown to block activation of the neuronal nicotinic receptor found on peripheral neurons (Ravdin and Berg, 1979). The function of the α-bungarotoxin-binding component in neural tissue is still unknown.

Antibodies that were raised against the muscle type nicotinic acetylcholine receptor recognize proteins in the CNS and have been used to determine the distribution of the antigenic determinants in the brain (Swanson et al., 1983). They have also been used to purify the molecules that carry the antigenic determinants (Whiting and Lindstrom, 1986). One subunit of these oligomeric molecules binds the affinity labelling reagent MBTA (Whiting and Lindstrom, 1987). However, the molecules purified on the basis of their immunological cross-reactivity with the muscle nicotinic receptor have not been shown to be functional acetylcholine receptors. There is, however, good evidence in chick ciliary ganglion neurons that at least one of the antibodies recognizes a functional receptor (Smith et al., 1986). It is also the case that the amino terminal sequence of one of the proteins corresponds to the sequence of a functional neuronal nicotinic receptor deduced from an expressible cDNA clone (Whiting et al., 1987).

Our approach to the neuronal nicotinic acetylcholine receptor was based on the idea that nucleic acids encoding muscle nicotinic acetylcholine receptor subunits would hybridize, at low stringency, to nucleic acids encoding neuronal nicotinic acetylcholine receptor subunits. Our first indication that this approach might be successful came from a Northern blot in which a probe prepared from a clone encoding a mouse muscle nicotinic receptor α-subunit was hybridized at low stringency to poly (A)⁺ RNA prepared from the PC12 cell line. This probe recognized two bands, one of about 2 kb and one of about 3.5 kb (see Fig. 1). The RNA encoding the muscle α-subunit is about 2 kb. We knew that the probe was not hybridizing to an identical sequence because the hybridization signal was lost when the stringency was increased by raising the temperature. This result suggested that we could identify cDNA clones encoding

receptor subunits by hybridization with clones encoding the muscle α-subunit. We therefore first screened a cDNA library prepared from RNA from the PC12 cell line and later screened libraries prepared from rat brain. This approach resulted first in an α-subunit encoding clone from the PC12 cell line (Boutler et al., 1986) and subsequently in the isolation of clones encoding a variety of neuronal nicotinic acetylcholine receptor subunits (Deneris et al., 1987; Goldman et al., 1987; Wada et al., 1988). A similar approach was taken by Dr. Marc Ballivet and his collaborators using chicken genomic libraries as the source of receptor coding sequences (Kao et al., 1984; Nef et al., 1987, 1988). The clones isolated and sequenced by our two groups define a family of genes that encode sub-

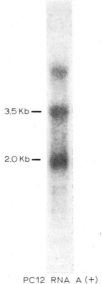

3.5 Kb —

2.0 Kb —

PC12 RNA A (+)
Probe : pMAR α 15

Fig. 1. Autoradiograph of Northern blot hybridization using radiolabeled cDNA encoding the mouse muscle α-subunit and poly(A)⁺ RNA isolated from the PC12 cell line. Eighteen micrograms of poly(A)⁺ RNA were electrophoresied on a 1% agarose gel. Hybridization was carried out in 5 X SSPE (1X SSPE is 180 mM NaCl, 9 mM Na₂HPO₄, 0.9 mM NaH₂PO₄, 1 mM EDTA pH 7.4), 50% formamimide, 1% sodium Dodecyl sulfate (SDS), 1 X Denhardt's at 42°C, and the blot was washed in 5 X SSPE and 0.05% SDS at 65°C.

units of functional neuronal nicotinic acetylcholine receptors. The following paragraphs present the properties of the proteins encoded by these clones, demonstrate that they are, in fact, subunits of functional acetylcholine receptors, and discuss their distribution in the rat brain.

Figure 2 shows four protein sequences deduced from the nucleotide sequences of cDNA clones. Inspection of these sequences shows remarkable similarity to the subunits of the muscle nicotinic acetylcholine receptor. Each protein has a leader sequence, followed by a stretch of about 210 amino acids unremarkable in their hydropathy. These sequences are followed by three hydrophobic domains that are highly conserved, both among the sequences shown in Fig. 2 and among the sequences of the muscle receptor subunits. These three domains are thought to form membrane spanning regions, the second of which, in the case of the muscle receptor, is thought to be part of the ion channel. The cytoplasmic domain that follows membrane spanning region III is noticeably not conserved and is variable in length. There is a fourth potential membrane spanning region at the carboxy terminal end of the protein that, in this model, would place the carboxy terminal sequences in the extracellular space. The amino terminal extracellular domain has the conserved cysteines found in subunits of all ligand-gated ion channels sequenced to date. The subunits identified as α_2, α_3, and α_4 also have the two contiguous cysteines that are labelled by the affinity reagent MBTA (Boutler et al., 1987). These residues are missing in the subunit identified as β_2 and are also missing in all the non-α-muscle receptor subunits.

We proved that these proteins are subunits of

NEURONAL NICOTINIC ACETYLCHOLINE RECEPTOR SUBUNITS

```
ALPHA2        MTLSHSALQFWTHLYLWCLLLVPAVLTQQGSHTHAEDRLFKHL  FGGYNRWARPVPNTSDVVIVRFGLSIAQLIDVDEKNQMMTTNVWLKQEWNDY
ALPHA3             MGVVLLPPPLSMLMLVLMLLPAASASEAEHRLFQYL  FEDYNEIIRPVANVSHPVIIQFEVSMSQLVKVDEVNQIMETNLWLKQIWNDY
ALPHA4      MEIGGPGAPPPLLLLPLLLLLGTGLLPASSHIETRAHAEERLLKRL  FSGYNKWSRPVGNISDVVLVRFGLSIAQLIDVDEKNQMMTTNVWVKQEWHDY
BETA2            MLACMAGHSNSMALFSFSLLWLCSGVLGTDTEERLVEHLLDPSRYNKLIRPATNGSELVTVQLMVSLAQLISVHEREQIMTTNVWLTQEWEDY
            ----------SIGNAL  PEPTIDE----------

ALPHA2      KLRWDPAEFGNVTSLRVPSEMIWIPDIVLYNNADGEFAVTHMTKAHLFFTGTVHWVPPAIYKSSCSIDVTFFPFDQQNCKMKFGSWTYDKAKIDLEQMER
ALPHA3      KLKWKPSDYQGVEFMRVPAEKIWKPDIVLYNNADGDFQVDDKTKALLKYTGEVTWIPPAIFKSSCKIDVTYFPFDYQNCTMKFGSWSYDKAKIDLVLIGS
ALPHA4      KLRWDPGDYENVTSIRIPSELIWRPDIVLYNNADGDFAVTHLTKAHLFYDGRVQWTPPAIYKSSCSIDVTFFPFDQQNCTMKFGSWTYDKAKIDLVSIHS
BETA2       RLTWKPEDFDNMKKVRLPSKHIWLPDVVLYNNADGMYEVSFYSNAVVSYDGSIFWLPPAIYKSACKIEVKHFPFDQQNCTMKFRSWTYDRTEIDLVLKSD

                                    **
ALPHA2      TVDLKDYWESGEWAIINATGTYNSKKYDCCAEIYPDVTYYFVIRRLPLFYTINLIIPCLLISCLTVLVFYLPSECGEKITLCISVLLSLTVFLLLITEII
ALPHA3      SMNLKDYWESGEWAIIKAPGYKHEIKYNCCEEIYQDITYSLYIRRLPLFYTINLIIPCLLISFLTVLVFYLPSDCGEKVTLCISVLLSLTVFLLLITEII
ALPHA4      RVDQLDFWESGEWVIVDAVGTYNTRKYECCAEIYPDITYAFIIRRLPLFYTINLIIPCLLISCLTVLVFYLPSECGEKVTLCISVLLSLTVFLLVITETI
BETA2       VASLDDFTPSGEWDIIALPGRRNENPDDST   YVDITYDFIIRRKPLFYTINLIIPCVLITSLAILVFYLPSDCGEKMTLCISVLLALTVFLLLISKIV
                             ACETYLCHOLINE                MEMBRANE SPANNING I         MEMBRANE SPANNING II

ALPHA2      PSTSLVIPLIGEYLLFTMIFVTLSIVITVFVLNVHHRSPSTHN MPNWVRVALLGRVPRWLMMNRPLPPMELHGSPDLKLSPSYHWLETNMDAGEREETE
ALPHA3      PSTSLVIPLIGEYLLFTMIFVTLSIVITVFVLNVHYRTPTTHT MPTWVKAVFLNLLPRVMFMTRPTSGEGDTPKTRTFYGAELSNLNCFSRADSKSCKE
ALPHA4      PSTSLVIPLIGEYLLFTMIFVTLSIVITVFVLNVHHRSPRTHT MPAWVRRVFLDIVPRLLFMKRPSVVKDNCRRLIESMHKMANAPRFWPEPVGEPGIL
BETA2       PPTSLDVPLVGKYLMFTMVLVTFSIVTSVCVLNVHHRSPTTHT MAPWVKVVFLEKLPTLLFLQQPRHRCARQRLRLRRRQEREGEAVFFREGPAADPC
                    MEMBRANE SPANNING III                                            ----CYTOPLASMIC REGION----

ALPHA2      EEEEEEDENICVCAGLPDSSMGVLLYGHGGLHLRAMEPETKTPSQA
ALPHA3      GYPCQDGTCGYCHHRRVKISNFSAANLTRSSSSESVNAVL
ALPHA4      SDICNQGLSPAPTFCNPTDTAVETTQPTCRSPPLEVPDLKTSEVEKASPCPSPGSCPPPKSSSGAPMLIKARSLSVQHVPSSQEAAEDGIRCRSRSIQYC
BETA2       TCFVNPASVQGLAGAFRAEPTAAGPGR
            -------------------------------CYTOPLASMIC REGION-----------------------------------

ALPHA2                                                                         SEILLSPQIQKALEGVHYIADRLRSEDADSSVKEDWKY
ALPHA3                                                                         SLSALSPEIKEAIQSVKYIAENMKAQNVAKEIQDDWKY
ALPHA4      VSQDGAASLADSKPTSSPTSLKARPSQLPVSDQASPCKCTCKEPSPVSPVTVLKAGGTKAPPQHLPLSPALTRAVEGVQYIADHLKAEDTDFSVKEDWKY
BETA2                                                                          SVGPCSCGLREAVDGVRFIADHMRSEDDDQSVREDWKY
            ---------------------CYTOPLASMIC REGION----------------------             AMPHIPATHIC HELIX

ALPHA2      VAMVVDRIFLWLFIIVCFLGTIGLFLPPFLAGMI*
ALPHA3      VAMVIDRIFLWVFILVCILGTAGLFLQPLMARDDT*
ALPHA4      VAMVIDRIFLWMFIIVCLLGTVGLFLPPWLAGMI*
BETA2       VAMVIDRLFLWIFVFVCVFGTVGMFLQPLFQNYTATTFLHPDHSAPSSK*
                  MEMBRANE SPANNING IV
```

Fig. 2. Amino acid alignment of the β_2-subunit with the rat neuronal α-subunit. Aligned with the β_2- (Deneris et al., 1987) subunit are the neuronal α_2- (Wada K. et al., 1988), α_3- (Boutler et al., 1986), α_{4-1}- (Goldman et al., 1987) subunits.

functional nicotinic acetylcholine receptors using the *Xenopus* oocyte expression system (Wada et al., 1988). Each cDNA clone was placed downstream of the SP6 promoter and capped RNA was prepared in vitro using the SP6 polymerase. These RNA transcripts were injected singly, and in pairwise combinations, into *Xenopus* oocytes. The oocytes were incubated for 1 – 5 days and then tested for the presence of functional receptors by measuring either voltage response to applied acetylcholine. Because the oocytes may express muscarinic receptors, the bathing solution contained atropine to block these responses. Only RNA derived from α_4 gave a response, small but reproducible, to acetylcholine when injected singly.

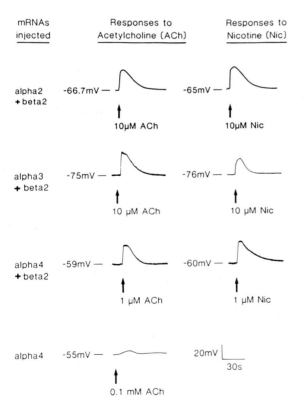

Fig. 3. This figure shows voltage traces obtained from *Xenopus* oocytes injected with RNA derived from the neuronal α- and β-genes. The RNA combinations injected are shown on the left and representative responses to applied acetylcholine and nicotine are shown on the right.

However, addition of the β_2-subunit to either α_2, α_3, or α_4 resulted in a robust response to either acetylcholine or nicotine (see Fig. 3). These responses were blocked by d-tubocurarine and were absent in uninjected or sham-injected oocytes. We interpret these experiments to mean that the clones encode subunits of nicotinic acetylcholine receptors and that the β_2-subunit acts in concert with either the α_2-, α_3-, or α_4-subunit. There are, however, other interpretations. We have not ruled out the idea that there are cryptic receptors in the oocyte membrane that are activated by some coordinate activity of the β_2- and α-subunits. We have not ruled out the idea that a protein present in the oocyte membrane combines with the β_2- and α-subunits to make a functional receptor. We have not ruled out the idea that the β_2-subunit acts catalytically to turn the α-subunits into functional receptor homo-oligomers.

The simplest interpretation, however, is that the β_2-subunit combines with either α_2, α_3, or α_4 to make a functional receptor. We know that a receptor is produced in the oocyte in response to the injection of RNA encoding these subunits. We do not know however if we are missing a subunit that is used in vivo. There is certainly precedence for creation of functional receptors without a full complement of subunits. In addition, although the β_2 works in concert with each of the α's in the oocyte, we do not know if it does so in vivo. The best evidence for such an interaction comes from the PC12 cell line where we find RNA encoding both β_2 and α_3. There is, therefore, the potential for the interaction of these two subunits in this cell line.

We know that the genes encoding these clones are expressed in the brain because we isolated the clones from cDNA libraries prepared from RNA isolated from rat hypothalamus, hippocampus, and cerebellum. We determined the distribution of transcripts encoding the different subunits using the technique of in situ hybridization (Goldman et al., 1986). Because the transcripts are homologous in their coding regions and might hybridize with heterologous probes, we used the 3' untranslated

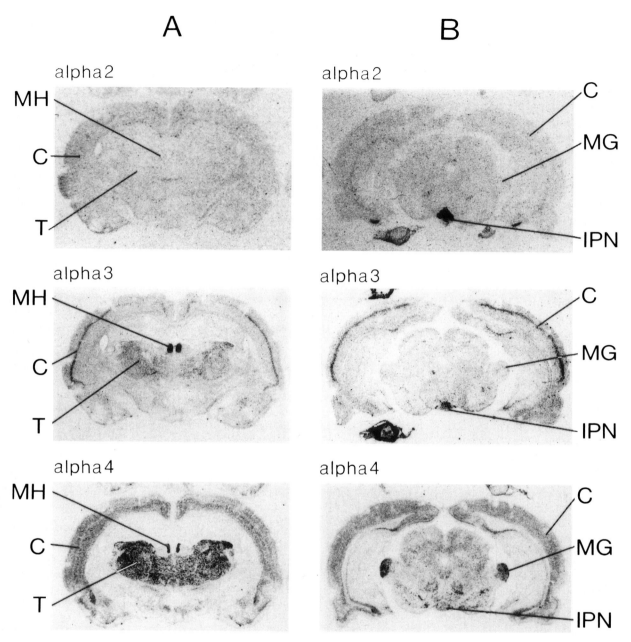

Fig. 4. Comparison of the distribution of α_2-, α_3- and α_4-transcripts by in situ hybridization histochemistry. Serial coronal sections through the medial habenula (A) and the interpeduncular nucleus (B) were hybridized with the probes for α_2, α_3 and α_4. In (B), slides contain sections of the trigeminal ganglion. Abbreviations: C = cortex; IPN = interpeduncular nucleus; MH = medial habenula; MG = medial geniculate nucleus; T = thalamus. Tissue preparation and hybridization were performed as previously described (Wada et al., 1988), with minor modifications. Briefly, paraformaldehyde-fixed rat brain sections (25 μm) were mounted on poly-L-lysine-coated slides, digested with proteinase K(10 μg/ml, 37°C, 30 min), acetylated, and dehydrated. Hybridization with ^{35}S-radiolabelled RNA probe (5 – 10 × 10^6 cpm/ml) was performed at 55°C for 12 – 18 h in a solution containing 50% formamide, 0.3M NaCl, 10 mM Tris (pH 8.0), 1 mM EDTA, 0.05% tRNA, 10 mM DTT, 1 X Denhardt's solution and 10% dextran sulfate. Because of the high sequence similarities in the protein coding regions of the cDNAs, 3'-untranslated sequences were used to make probes. The Eco RI/3' end, Bal I/3' end and Bgl I 3' end fragments derived from α_2, α_3 (Boutler et al., 1986) and α_{4-2} (Goldman et al., 1987) cDNA clones, respectively, were subcloned into the plasmid, pSP65 and used to synthesize antisense RNA probes in vitro. After hybridization, sections were treated with RNaseA (20 μg/ml, 37°C, 30 min) and washed in 0.1 X SSC at 55°C. Dehydrated slides were exposed to X-ray films for 3 – 16 days at 4°C. (From Wada K. et al., 1988.)

sequences to prepare transcript-specific probes. The results of these studies showed that the β_2-gene is expressed throughout the CNS, and the α-subunit transcripts are expressed in separate but overlapping sets of nuclei in the brain (see Fig. 4). The α_4-transcript is the most widespread of the three, followed by α_3 and α_2. It is important to note that these experiments show the location of the RNA transcripts, and not of the protein. Although is seems reasonable to assume that RNA is translated into protein and that the resulting protein is used for a physiological function, this has not been demonstrated. Neither is it known, for example, whether some subunit combinations are targeted for axons and some are targeted for dendrites. There are however good techniques available for generating protein from these clones in forms suitable for the production of antibodies.

References

Boulter, J., Evans, K., Goldman, D., Martin, G., Treco, D., Heinemann, S. and Patrick, J. (1986) Isolation of a cDNA clone coding for a possible neural nicotinic acetylcholine receptor alpha-subunit. *Nature (Lond.),* 319: 368 – 374.

Boulter, J., Connolly, J., Deneris, E., Goldman, D., Heinemann, S. and Patrick, J. (1987) Functional expression of two neuronal nicotinic acetylcholine receptors from cDNA clones identifies a gene family. *Proc. Natl. Acad. Sci. U.S.A.,* 84: 7763 – 7767.

Brown, D.A. and Fumagalli, L. (1977) Dissocation of alpha-bungarotoxin binding and receptor block in the rat superior cervical ganglion. *Brain Res.,* 129: 165 – 168.

Carbonetto, S.T., Fambrough, D.M. and Moller, K.J. (1978) Nonequivalence of alpha-bungarotoxin receptors and acetylcholine receptors in chick sympathetic neurons. *Proc. Natl. Acad. Sci. U.S.A.,* 72: 1016 – 1028.

Chiappinelli, V.A. and Zigmond, R.E. (1978) Alpha bungarotoxin blocks nicotinic transmission in the avian ciliary ganglion. *Proc. Natl. Acad. Sci. U.S.A.,* 75: 2999 – 3003.

Clarke, P.B.S., Pert, C.B. and Pert, A. (1984) Autoradiographic distribution of nicotinic receptors in rat brain. *Brain Res.,* 323: 390 – 395.

Clarke, P.B.S., Schwartz, R.D., Paul S.M., Pert, C.B. and Pert, A. (1985) Nicotinic binding in rat brain: autoradiographic comparison of [³H]acetylcholine, [³H]alicotine, and [¹²⁵I]-alpha-bungarotoxin. *J. Neurosci.,* 5: 1307 – 1315.

Deneris, E.S., Connolly, J., Boulter, J., Patrick, J. and Heinemann, S. (1987) Identification of a gene that encodes a non-alpha subunit of neuronal nicotinic acetylcholine receptors. *Neuron,* 1: 45 – 54.

Duggan, A.W., Hall, J.G. and Lee, C.Y. (1976) Alpha-bungarotoxin, cobra neurotoxin and excitation of Renshaw cells by acetylcholine. *Brain Res.,* 107: 166 – 170.

Fambrough, D.M. (1976) Development of cholinergic innervation of skeletal, cardiac, and smooth muscle. In A.M. Goldberg and I. Hannin (Eds.), *Biology of Cholinergic Function* Raven Press, New York.

Goldman, D., Simmons, D., Swanson, L.W., Patrick, J. and Heinemann S. (1986) Mapping brain areas expressing RNA homologous to two different acetylcholine receptor alpha subunit cDNAs. *Proc. Natl. Acad. Sci. U.S.A.,* 83: 4076 – 4080.

Goldman, D., Deneris, E., Luyten, W., Kohchar, A., Patrick, J. and Heinemann, S. (1987) Members of a nicotinic acetylcholine receptor gene family are expressed in different regions of the mammalian central nervous system. *Cell,* 48: 965 – 973.

Goodman, L.S. and Gillman, A.G. (1975) *The Pharmacological Basis of Therapeutics,* MacMillan Publishing Co., Inc. New York.

Jacob, M.H. and Berg, D.K. (1983) The ultrastructural localization of alpha-bungarotoxin binding sites in relation to synapses on chick ciliary ganglion neurons. *J. Neurosci.* 3: 260 – 271.

Kao, P.N., Dwork, A.J., Kaldany, R.J., Silver, M.L., Wideman, J., Stein, S. and Karlin, A. (1984) Identification of two alpha-subunit half-cystines specifically labeled by an affinity reagent for the acetylcholine binding site. *J. Biol. Chem.,* 259: 1162 – 1165.

Marks, J., Stitzel, J., Romm, E., Wehner, J. and Collins, A. (1986) Nicotinic binding sites in rat and mouse brain: comparison of acetylcholine, nicotine, and alpha bungarotoxin. *Mol. Pharmacol.,* 30: 427 – 436.

Nef, P., Oneyser, C., Barkas, T. and Ballivet, M. (1987) In A. Maelicke (Ed.), *Nicotinic Acetylcholine Receptor Structure and Function,* Springer Verlag, Berlin, pp. 417 – 429.

Nef, P., Oneyser, C., Alliod, C., Couturier, S. and Ballivet, M. (1988) Genes expressed in the brain define three distinct neuronal nicotinic acetylcholine receptors. *EMBO J.,* 7: 595 – 601.

Patrick, J. and Stallcup, W. (1977a) Immunological distinction between acetylcholine receptor and the alpha-bungarotoxin-binding component on sympathetic neurons. *Proc. Natl. Acad. Sci. U.S.A.,* 74: 4689.

Patrick, J. and Stallcup, W. (1977b) Alpha-bungarotoxin binding and cholinergic receptor function on a rat sympathetic nerve line. *J. Biol. Chem.,* 252: 8629.

Ravdin, P. and Berg, D.K. (1979) Inhibition of neuronal acetylcholine sensitivity by alpha toxins from *Bungarus multicinctus* venom. *Proc. Natl. Acad. Sci. U.S.A.,* 76: 2072 – 2076.

Smith, M.A., Margiotta, J.F., Franco, A. Jr., Lindstrom, J.M.

and Berg, D.K. (1986) Cholinergic modulation of an acetylcholine receptor-like antigen on the surface of chick ciliary ganglion neurons in cell culture. *J. Neurosci.,* 6: 946 – 953.

Swanson, L.W., Lindstrom, J., Tzartos, S., Schmued, L.C., O'Leary, D.M. and Cowan, W.M. (1983) Immunohisto-chemical localization of monoclonal antibodies to the nicotinic acetylcholine receptor in chick midbrain. *Proc. Natl. Acad. Sci. U.S.A.,* 80: 4532 – 4536.

Wada, E., Wada, K., Boulter, E., Deneris, E.S., Heinemann, Patrick, J. and Swanson, L. (1988) The distribution of alpha2, alpha3, alpha4 and beta2 neuronal nicotinic receptor subunit mRNAs in the central nervous system. A hybridiza-tion histochemical study in the rat. *J. Comp. Neurol.* (sub-mitted).

Wada, K., Ballivet, M., Boulter, J., Connolly, J., Wada, E.,

Deneris, E.S., Swanson, L.W., Heinemann, S. and Patrick, J. (1988) Functional expression of a new pharmacological subtype of brain nicotinic acetylcholine receptor. *Science,* 240: 330 – 334.

Whiting, P.J. and Lindstrom, J.M. (1986) Purification and characterization of a nicotinic acetylcholine receptor from chick brain. *Biochemistry,* 25: 2082 – 2093.

Whiting, P. and Lindstrom, J. (1987) Affinity labelling of neuronal acetylcholine receptors localizes acetylcholine-bind-ing sites to the beta-subunits. *FEBS Lett.,* 213: 55 – 60.

Whiting, P., Esch, F., Shimasaki, S. and Lindstrom, J. (1987) Neuronal nicotinic acetylcholine receptor beta subunit is cod-ed for by the cDNA clone alpha4. *FEBS Lett.,* 219: 459 – 463.

Wonnacott, S. (1987) Brain nicotine binding sites. *Human Tox-icol.,* 6: 343 – 353.

SECTION III

Histochemical and Physiological Analysis of the Cholinergic System and their Associated Nicotinic Receptors

A. Nordberg, K. Fuxe, B. Holmstedt and A. Sundwall (Eds.)
Progress in Brain Research, Vol. 79
© 1989 Elsevier Science Publishers B.V. (Biomedical Division)

CHAPTER 4

Organization of central cholinergic systems

Kazue Semba and Hans C. Fibiger*

Division of Neurological Sciences, Department of Psychiatry, University of British Columbia, Vancouver, B.C. V6T 1W5, Canada

Introduction

The past decade has witnessed substantial advances in understanding the complex anatomy of central cholinergic systems. The reasons for this can easily be appreciated. Methodological advances such as the pharmacohistochemical procedure for acetylcholinesterase (AChE; Lynch et al., 1972) and the recognition that this could be used to identify some populations of cholinergic neurons (Lehmann and Fibiger, 1979), provided early opportunities to study the organization of these cells (Fibiger, 1982). More importantly, the development of specific antibodies to choline acetyltransferase (ChAT) and their use in immunohistochemistry provided for the first time a definitive method for identifying and studying the distribution of cholinergic neurons in the central nervous system (Kimura et al., 1981; Armstrong et al., 1983; Houser et al., 1983). When used in combination with the many retrograde and anterograde tract-tracing techniques that are now part of the neuroanatomist's armamentarium, AChE pharmacohistochemistry and ChAT immunohistochemistry have provided a rich supply of information about the projections of these neurons. At the same time, the recognition that cholinergic neurons in the basal forebrain contribute to the pathophysiology of Alzheimer's disease has gener-

ated continued widespread interest in the structure and function of cholinergic systems (Coyle et al., 1983). The aim of this chapter is to provide an update and overview of the anatomy of central cholinergic systems, and to discuss some areas about which controversy and uncertainty still exist.

Cholinergic basal nuclear complex

Cholinergic neurons in the basal forebrain are distributed across several classically defined nuclei, including the medial septal nucleus, the nucleus of the diagonal band of Broca (vertical and horizontal limbs), the magnocellular preoptic area, the substantia innominata, and the globus pallidus (Fig. 1). They are large (20 – 50 μm), multipolar neurons which have multiple (3 – 8) primary dendrites, extensive dendritic fields, and frequent dendritic spines (Fig. 2; Semba et al., 1987b). Many laboratories have studied the anatomy of these magnocellular neurons and there appears to be widespread agreement about their distribution in the basal forebrain. In contrast, there is currently little consensus concerning an appropriate nomenclature for this group of neurons. On the basis of cytoarchitectonic, biochemical and connectional considerations, Saper (1984) proposed that these widely dispersed magnocellular cells in the basal forebrain are in fact a single nucleus, "the magnocellular basal nucleus". A recent study using computer-aided data acquisition and three-dimensional reconstruction provided clear evi-

* To whom correspondence should be addressed.

baboon

rat

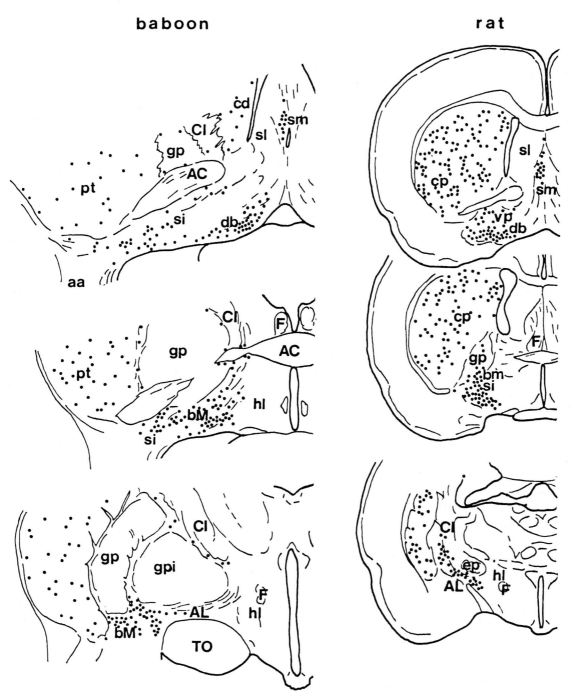

Fig. 1. The distributions of ChAT-immunoreactive neurons at three levels of the forebrain in the baboon (left) and rat (right). Each dot represents one labelled neuron. The illustrations are modified from Vincent et al. (1986b) taken originally from Satoh and Fibiger (1985) for the baboon data, and Satoh et al. (1983a) for the rat data. Abbreviations: AC = anterior commissure; AL = ansa lenticularis; CI = internal capsule; F = fornix; TO = optic tract; aa = anterior amygdaloid area; bm = nucleus basalis magnocellularis; bM = nucleus basalis of Meynert; cd = caudate; cp = caudate-putamen; db = nuclei of the diagonal band; ep = entopeduncular nucleus; gp = globus pallidus; gpi = globus pallidus, internal lamina; hl = lateral hypothalamic area; pt = putamen; si = substantia innominata; sm = medial septal nucleus; sl = lateral septal nucleus; vp = ventral pallidum.

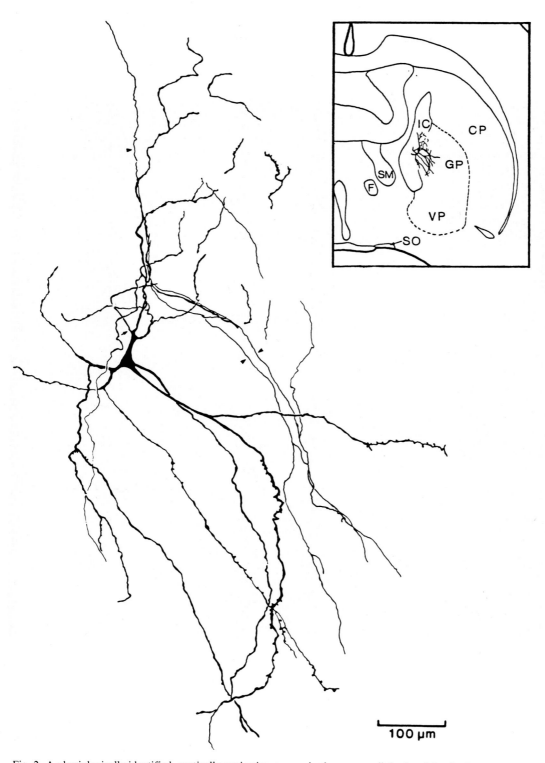

Fig. 2. A physiologically identified, cortically projecting neuron in the magnocellular basal forebrain reconstructed from seven consecutive 100 μm sections following antidromicity testing and intracellular iontophoresis of HRP. As indicated in inset, this neuron

→

dence that these cholinergic neurons do indeed form a continuum (Fig. 3; Schwaber et al., 1987), and this confirmed earlier speculations of Satoh et al. (1983a) who referred to the continuum as the "rostral cholinergic column" to distinguish it from another apparent continuum of cholinergic neurons in the mesopontine tegmentum, the "caudal cholinergic column" (see below). The findings of

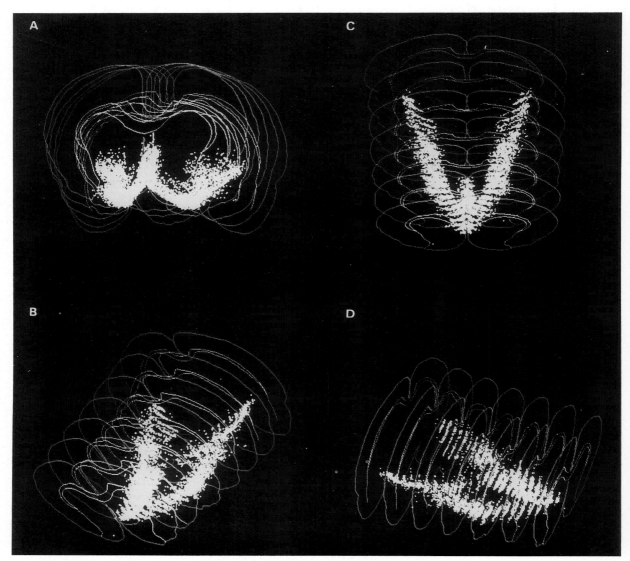

Fig. 3. Three-dimensional displays of the distribution of cholinergic neurons in the magnocellular basal forebrain from several viewpoints. (Adopted from Schwaber et al., 1987, the original in color.)

was located at the border between the internal capsule and the globus pallidus. The arrow indicates the presumptive axon of this neuron, and some branches of this presumptive axon are indicated by arrowheads. (Modified from Semba et al., 1987b.) Abbreviations: CP = caudate-putamen; F = fornix; GP = globus pallidus; IC = internal capsule; SM = stria medullaris; SO = supraoptic nucleus; VP = ventral pallidum.

Schwaber et al. (1987) thus provided strong support for Saper's (1984) proposal that magnocellular neurons in the basal forebrain, the majority of which are cholinergic, form a single nucleus. Schwaber et al. (1987) referred to this continuum as the "cholinergic basal nuclear complex" (CBC).

An alternative nomenclature has been proposed by Mesulam et al. (1983b). On the basis of connectivity patterns, these investigators proposed that cholinergic neurons in the basal forebrain can be subdivided into four groups. Ch1 and Ch2 represent cholinergic neurons within the medial septal nucleus and the vertical limb of the diagonal band and provide the major source of cholinergic afferents to the hippocampus. Ch3 is contained primarily within the lateral part of the horizontal limb of the diagonal band, including the magnocellular preoptic area (De Olmos et al., 1978), and is the major source of cholinergic activity in the olfactory bulb. Ch4 consists of cholinergic neurons that are defined by the fact that they project to the neocortical mantle. According to this definition, Ch4 neurons are dispersed throughout the rostrocaudal extent of the CBC, and this in fact represents one of the limitations of the Ch nomenclature. Another is that as Mesulam et al. (1983b) point out, the boundaries between the various Ch1 – 4 subdivisions, particularly in the rat, are difficult to establish due to the considerable overlap in the boundaries of these groups, defined as they are by their efferent connections.

Presently, a satisfactory nomenclature for the components of the CBC, if indeed they exist in a functionally or anatomically significant sense, has yet to be achieved. Whether this will be possible will probably depend largely on the results of future experiments. For example, when sufficiently detailed information becomes available on the afferent connections of the CBC, it may be possible to subdivide this continuum on this basis. Also, the extent to which various neuropeptides are colocalized in various groups of neurons in the CBC may also provide grounds for defining subgroups of this complex. As has been discussed so elegantly by Nieuwenhuys (1985), the problem of incorporating neurotransmitter-specified neuronal groups as revealed by modern immunohistochemistry into the fabric of classical neuroanatomy is clearly not unique to cholinergic systems and unifying concepts seem far from being achieved.

Cortical projections

Despite the now considerable literature that describes the projections of the basal forebrain upon the neocortex, some of this work is limited by uncertainty concerning the cholinergic identity of the retrogradely labelled neurons. It is now well established that throughout the rostrocaudal extent of the CBC, non-cholinergic neurons are often interspersed with the cholinergic cells and that the former also have telencephalic projections (Mesulam et al., 1983b; Rye et al., 1984). Therefore, studies in which neither AChE pharmacohistochemistry nor ChAT immunohistochemistry was used to identify the retrogradely labelled cells as cholinergic, pose potential interpretative difficulties. This concern is somewhat alleviated by the finding that cholinergic and non-cholinergic pathways from the CBC seem to follow similar trajectories to their targets (Woolf et al., 1986).

Although efferents from the caudal pole of the CBC, the nucleus basalis magnocellularis (nBM; Lehmann et al., 1980), have generally been considered to be the major source of cholinergic activity in the neocortex, this is an oversimplification and it should be recognized that neurons dispersed throughout the rostrocaudal extent of the CBC contribute to this projection (Bigl et al., 1982; McKinney et al., 1983; Saper, 1984; Henderson, 1987). In the species that have been investigated to date, there is evidence that the cortical projections of the CBC are organized topographically. In the rat, cholinergic neurons mostly in the vicinity of the ventromedial globus pallidus and substantia innominata project to the frontal and parietal cortex, while the occipital (medial) and cingulate cortices appear to be innervated primarily by more rostrally located cells in the horizontal limb of the diagonal band (Bigl et al., 1982; McKinney et al.,

1983). It should be emphasized, however, that an injection into a particular cortical site has never been observed to label exclusively a particular, currently recognized subdivision of the CBC (medial septum, nucleus of the diagonal band, magnocellular preoptic area, etc.). At this level of analysis, therefore, the CBC shows only a very loose topographical relationship with the neocortex.

Saper (1984) suggested that in the rat, the CBC is organized in a manner such that magnocellular neurons that project to a particular cortical area are located primarily among the corticodiencephalic descending fibers arising from that cortical region. He therefore proposed that neurons in the CBC are well placed to receive reciprocal inputs from the cortical regions that they innervate, and that this may explain the organization of the CBC more fully than the earlier attempts that were based simply on rostrocaudal and mediolateral considerations (Wenk et al., 1980; Bigl et al., 1982; McKinney et al., 1983; Price and Stern, 1983). A subsequent ultrastructural study provided some support for the hypothesis that there are reciprocal connections between the cerebral cortex and the CBC. Thus, after wheat germ agglutinin-horseradish peroxidase injections into either the insular or cingulate cortex, Lemann and Saper (1985) observed numerous cases of anterogradely labelled axon terminals making symmetric synaptic contacts with retrogradely labelled neurons in the basal forebrain. It should be noted, however, that other findings are not consistent with this hypothesis. For example, Mesulam and Mufson (1984) did not obtain evidence for projections to Ch4 from the motor or sensory cortical areas in rhesus monkey, but did observe such projections from the orbitoinsular and medial temporal lobes. In addition, experiments where tracers have been injected into different regions of the CBC in the rat have generally reported retrogradely labelled cells in only restricted parts of the neocortex (Lamour et al., 1984; Haring and Wang, 1986; Semba and Fibiger, in preparation). In both the rat and primate, therefore, it appears that only limited parts of the neocortex have the potential to have reciprocal connections with the CBC.

Saper (1984) also proposed a second organizational principle of the CBC, that being that its axons are confined to two major efferent pathways, the medial and lateral projection systems. The medial pathway arises from neurons in the medial septal and diagonal band nuclei, as well as in medial substantia innominata and globus pallidus. Axons from these neurons ascend through the septum, pass around the genu of the corpus callosum and terminate in medial cortical structures. The lateral pathway arises from more laterally located neurons in the substantia innominata, globus pallidus and magnocellular preoptic area, and runs laterally through the ventral striatum into the external capsule. From there the axons exit and terminate in lateral aspects of the cerebral hemisphere. Using brain lesions combined with AChE histochemistry the existence of the medial and lateral efferent pathways was confirmed by Kristt et al. (1985). These authors also obtained evidence for a third, "anterior" pathway that innervates the anterior frontal cortex. In the monkey there is evidence that the CBC may innervate the telencephalon through as many as five distinct fiber pathways (Kitt et al., 1987).

In the rat and primate, neurons in the CBC appear to innervate relatively small, circumscribed cortical fields. Although there is no unanimity, the majority of attempts to determine the extent to which neurons in the CBC send collaterals to diverse regions of the rat neocortex have failed to find evidence for widely spaced collateralization and indicate that individual neurons in the CBC have terminal fields that are restricted to a single cytoarchitectonic area, usually less than 1.5 mm in diameter (Bigl et al., 1982; Price and Stern, 1983; Saper, 1984; Kristt et al., 1985; see Saper, 1984 for discussion of McKinney et al., 1983). However, other evidence obtained from the cat brain suggests that single neurons in the basal forebrain can have axon collaterals that innervate widely separated cortical regions (Adams et al., 1986; Boylan et al., 1986). The extent to which these disparate findings are due to species differences or technical factors cannot be ascertained at present. It is also important to recognize that in some of the above studies, the chemical identity of the retrogradely labelled forebrain neurons was not determined.

Anterograde tracing and ChAT immunohistochemical studies both suggest that most cholinergic fibers and terminals in the rat cerebral cortex arise from neurons in the CBC and that there are significant regional differences with respect to their degree of lamination, this ranging from pronounced to undetectable (Saper, 1984; Luiten et al., 1987; Eckenstein et al., 1988). Although cholinergic terminals are typically found in all cortical layers, layer V appears to be particularly heavily innervated in motor and most sensory cortical areas (McDonald et al., 1987; Eckenstein et al., 1988). In general, those regions with the most pronounced cytoarchitectonic definition of the cortical layers show the clearest evidence of lamination of cholinergic fibers and terminals (Eckenstein et al., 1988). Individual cholinergic fibers of CBC origin show frequent local branching in at least some parts of the cortex and appear to form terminals in different cortical layers (Luiten et al., 1987; Eckenstein et al., 1988).

Other projections of the CBC

In addition to its well-known projections to the cerebral cortex, neurons in the CBC innervate other telencephalic, diencephalic and mesencephalic structures as well. These other targets of the CBC include the hippocampal formation, the olfactory bulb, the amygdala, certain thalamic nuclei and the interpeduncular nucleus.

The septo-hippocampal projection is perhaps the first central system about which widespread consensus was reached with regard to its cholinergic nature (for review see Fibiger, 1982). The origins of this projection are cholinergic neurons in the medial septal nucleus and the vertical limb of the nucleus of the diagonal band. Not all neurons in these nuclei that innervate the hippocampal formation have been identified as cholinergic. Thus, depending on the site in the hippocampal formation into which retrograde tracers are injected, anywhere from about 20 to 80% of the retrogradely labelled cells also stain for ChAT (Rye et al., 1984; Amaral and Kurz, 1985). It appears, therefore, that a sizeable portion of the septo-hippocampal projection is not cholinergic. Köhler et al. (1984a) have obtained evidence that at least part of the non-cholinergic septo-hippocampal projection is GABAergic. The cholinergic innervation of the hippocampal formation has a distinct laminar organization and is topographically ordered. Cholinergic axons reach the hippocampal formation mainly by way of the dorsal fornix and fimbria and appear to terminate throughout the hippocampal formation, and heavily in (1) the strata oriens and radiatum adjacent to the pyramidal cells in fields CA2 – 3 of the hippocampus, and (2) the infra- and supragranular zones of the dentate gyrus (Mellgren and Srebo, 1973; Lynch et al., 1977; Frotscher and Léránth, 1985; Nyakas et al., 1987; Matthews et al., 1987). Cholinergic fibers appear to terminate on pyramidal, granule and non-pyramidal cells in the hippocampal formation (Frotscher and Léránth, 1985). Small numbers of intrinsic ChAT-containing neurons are also scattered throughout the hippocampal formation (Houser et al., 1983; Frotscher and Léránth, 1985, 1986; Matthews et al., 1987; see below). They do not appear to contribute significantly to the laminar pattern of cholinergic activity in this structure. Some cholinergic neurons of the septo-hippocampal projection contain the neuropeptide galanin (Melander et al., 1986). This finding and its physiological significance are discussed in detail elsewhere in this volume (see Chapter 7 by Melander et al.). At present, other neuropeptides have not been identified in neurons of the CBC.

The amygdala, particularly the basolateral nucleus, receives a significant input from cholinergic neurons in the basal forebrain. In a combined retrograde tracer, ChAT immunohistochemical study, Carlsen et al. (1985) found many double labelled neurons in the caudal part of the CBC (magnocellular preoptic area and nucleus basalis) after injections into the basolateral amygdaloid nucleus. Control injections into the piriform-entorhinal cortex labelled more rostrally located cholinergic neurons in the medial septum and nuclei of the diagonal band. The latter observation clarifies earlier accounts of the cholinergic innervation of the amygdala which suggested that neurons throughout the rostrocaudal extent of the CBC project to the amygdala (Nagai et al., 1982; Woolf and Butcher, 1982). Biochemical measures of ChAT activity also indicate that the basolateral nucleus, particularly the rostral part, contains the heaviest cholinergic innervation of the amygdaloid nuclei (Hellendall et al., 1986). On the other hand, while all amygdaloid nuclei contain measurable levels of ChAT activity, the medial nucleus is particularly low (Hellendall et al., 1986). According to double label retrograde transport studies, different populations of neurons in the CBC innervate the amygdala and neocortex respectively (Carlsen et al., 1985).

The interpeduncular nucleus is richly innervated by cholinergic neurons. Combined lesion and biochemical studies as well as combined retrograde transport and ChAT immunohistochemical studies indicate that the CBC is one source of this input (Fibiger, 1982; Woolf and Butcher, 1985). Thus, injections of retrograde tracers into the interpeduncular nucleus label many cholinergic neurons in the vertical and horizontal limbs of the diagonal band and the magnocellular preoptic area (Woolf and Butcher, 1985). Axons from these neurons may reach the interpeduncular nucleus via the stria medullaris and the fasciculus retroflexus (see Fibiger, 1982). However, other findings suggest that these fibers descend through the medial forebrain bundle (Groenewegen et al., 1986). The medial habenular nucleus and the pedunculopontine tegmental nucleus also contribute to the cholinergic innervation of the interpeduncular nucleus (Woolf and Butcher, 1985; see below). A detailed understanding of the complex cholinergic innervation of the interpeduncular nucleus has not yet been achieved (see Contestabile et al., 1987).

Several other forebrain regions are also innervated by the CBC. These include the olfactory bulb which receives its cholinergic afferents from neurons located primarily in the nucleus of the horizontal limb of the diagonal band and the magnocellular preoptic area (Záborszky et al., 1986) and the reticular nucleus of the thalamus which is innervated by neurons in the nucleus basalis and magnocellular preoptic area (Hallanger et al., 1987). Collaterals of the CBC-thalamic projection have recently been shown to innervate the neocortex (Jourdain, Semba and Fibiger, unpublished observations). In contrast to the reticular nucleus, in the rat other thalamic nuclei are not innervated by the CBC and receive their cholinergic innervation from the mesopontine tegmentum (see below). In the cat and monkey, however, the mediodorsal thalamic nucleus is also innervated by the CBC (Steriade et al., 1987). The major projections of the CBC are summarized in Fig. 4.

Afferent connections of the CBC

A detailed understanding of the innervation of cholinergic neurons in the basal forebrain is an important prerequisite for any sophisticated appreciation of the functions of this group of cells. Unfortunately, at present, information about this important issue can only be described as rudimentary. Although there are reports dealing with the afferent connections of the various basal forebrain regions that contain cholinergic neurons, these data are of limited value because non-cholinergic

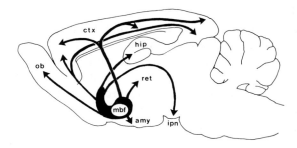

Fig. 4. A schematic summary of the projections of the cholinergic basal nuclear complex. Abbreviations: amy = amygdala; ctx = cortex; hip = hippocampus; ipn = interpeduncular nucleus; ob = olfactory bulb; ret = reticular nucleus of the thalamus.

neurons are interspersed with cholinergic neurons through the whole extent of the CBC. Only ultrastructural approaches are capable of providing definitive information about the specific innervation of the cholinergic cells and at present few investigators have attempted such studies (Ingham et al., 1985; Armstrong, 1986; see below). An appreciation of the afferent connections of the CBC is important for another reason and this pertains to the question as to whether the CBC should be viewed as a single nucleus or as a series of contiguous but separate nuclear groups (Saper, 1984; Schwaber et al., 1987). If, for example, the rostral aspects of the CBC shared substantially the same afferents as the caudal part, then this would support a unitary view of this nuclear complex. If, on the other hand, different elements of the CBC varied significantly with respect to their afferent connections then this might provide a basis for delineating different components of the CBC.

Retrograde tracing studies in the rat indicate that a region containing the caudal aspects of the CBC, the nBM, is innervated heavily by the dorsolateral frontal cortex, the caudate-putamen, the intralaminar thalamic nuclei, the hypothalamus and the subthalamic nucleus (Haring and Wang, 1986). The projection from the caudate-putamen seems to be directed at cholinergic cells, although evidence for direct synaptic connections with these cholinergic neurons has not been obtained (Grove et al., 1986). In contrast, the projection from the amygdala has been shown to terminate, at least in part, directly upon cholinergic neurons (Záborszky et al., 1984). Basal forebrain afferents arising in the neocortex, amygdala and midline thalamic nuclei may be glutamatergic or aspartatergic (Russchen et al., 1985; Fuller et al., 1987). Major brainstem afferents to the nBM include the pedunculopontine nucleus, the laterodorsal tegmental nucleus, the dorsal raphe nucleus, the ventral tegmental area and the substantia nigra, pars compacta (Haring and Wang, 1986; Semba et al., 1988a). Ultrastructural studies have revealed that substance P-containing (Bolam et al., 1986) and GABA-containing (Záborszky et al., 1986) axonal terminals make synaptic contact with cholinergic neurons in the nBM, although the origins of these terminals are not known.

The afferent connections of the primate nBM have been studied by a number of laboratories and show many similarities with the rat. Thus, there is consensus that the nBM is innervated by restricted regions of the cortex, these including orbitofrontal, rostral insular, temporal, prepiriform and entorhinal cortices (Mesulam and Mufson, 1984; Russchen et al., 1985; Irle and Markowitsch, 1986). Subcortical afferents include the septum, amygdala, midline thalamic nuclei, and hypothalamus, while brainstem nuclei that innervate the nBM include the dorsal and median raphe nuclei, the ventral tegmental area, the substantia nigra, pars compacta, the locus coeruleus, the parabrachial nuclei and the pedunculopontine nucleus (Mesulam and Mufson, 1984; Russchen et al., 1985; Irle and Markowitsch, 1986). It is evident, therefore, that cholinergic neurons in the nBM may potentially receive inputs from a wide variety of limbic, cortical, diencephalic and brainstem structures, most of which are regions of polysensory convergence and integration. The nBM does not appear to receive direct afferents from primary sensory or motor regions.

The question concerning the extent to which parts of the CBC have common or divergent efferent connections cannot be answered satisfactorily at present. On the basis of the limited anatomical information that is available, the medial septum, nuclei of the diagonal band, magnocellular preoptic area and nucleus basalis appear to share many common afferents, particularly from the brainstem (Fig. 5; Swanson and Cowan, 1979; Haring and Wang, 1986; Semba et al., 1988a; Vertes, 1988). Some of these afferents have been demonstrated to make direct synaptic contacts with cholinergic neurons in these regions (Záborszky and Cullinan, 1989). However, some differences may exist between the telencephalic and diencephalic structures that innervate cholinergic neurons in these regions. For example, while the horizontal limb of the diagonal band and magno-

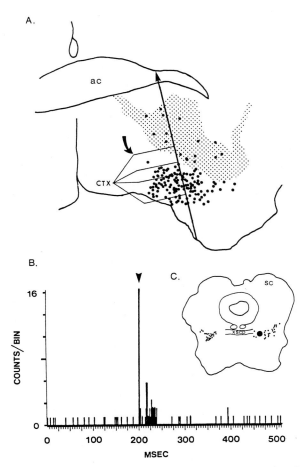

A.

B.

C.

Fig. 5. Physiological and anatomical data to illustrate the effects of the activation of one afferent to a cortically projecting neuron in the basal forebrain. (A) This neuron was recorded at the location indicated by a curved arrow on the electrode track (the straight arrow pointing to the anterior commissure, ac). The reconstruction was made from adjacent, alternate sections which were processed to visualize ChAT-immunoreactive somata (dots), or substance P neuropil (stippling), in combination with micrometer readings and a dye spot at the end of the track. ChAT-immunoreactive neurons demarcate the magnocellular basal forebrain, and substance P neuropil defines the ventral pallidum (Heimer and Wilson, 1975; Haber and Nauta, 1983). Including this neuron, a total of four neurons were antidromically activated from the cortex (ctx) in this track. (B) A peristimulus time histogram of the same neuron in response to electrical stimulation (arrowhead) in the PPT (shown in C). Note the initial excitatory followed by inhibitory response displayed by this spontaneously active neuron. (C) The site of the stimulating electrode was confirmed to be in the vicinity of PPT neurons labelled with NADPH-disphorase (small dots), a specific histochemical marker for cholinergic neurons in the mesopontine tegmentum (Vincent et al., 1983a). (Taken from Semba et al., 1988a.)

cellular preoptic area are richly innervated by the nucleus accumbens, ventral pallidum and septum, the region of the nucleus basalis is selectively innervated by intralaminar thalamic nuclei (Semba and Fibiger, in preparation). In addition, the horizontal limb of the diagonal band-magnocellular preoptic region, but not the nucleus basalis, receives a non-cholinergic projection from the contralateral basal forebrain (Semba et al., 1988b). However, until the afferents to the cholinergic neurons in these regions can be established, the question as to the possible differential innervation of the subdivisions of the CBC will remain unanswered. Given the extensive dendritic tree of these magnocellular neurons (Semba et al., 1987b), as well as the current limitations of immunohistochemistry in demonstrating immunoreactivity in distal dendrites, this represents a formidable challenge for future research.

Mesopontine tegmental cholinergic system

The presence of an ascending cholinergic pathway arising in the midbrain and pons was first suggested by Shute and Lewis (1967) in their classic work using AChE histochemistry. Because of parallels between the location of AChE-labelled neurons and the brainstem sites where electrical stimulation could induce cortical EEG activation (Moruzzi and Magoun, 1949; see also Steriade, 1981), Shute and Lewis (1967) suggested that this cholinergic projection is the anatomical substrate of the ascending reticular activating system proposed by Moruzzi and Magoun (1949). Although this hypothesis still remains to be fully substantiated, the original anatomical findings of Shute and Lewis (1967) have been confirmed and further extended by recent studies using more specific cholinergic markers and modern tract-tracing techniques.

Cells of origin

The cholinergic projection neurons in the tegmentum of the midbrain and pons are delimited rostrally just caudal to the substantia nigra, and caudally by the locus coeruleus, as illustrated in

47

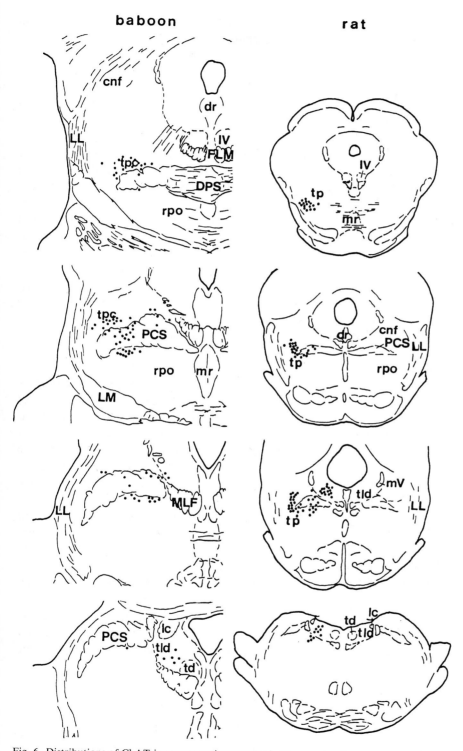

Fig. 6. Distributions of ChAT-immunoreactive neurons in the mesopontine tegmentum of the baboon (left) and rat (right). The data are modified from Vincent et al. (1986b) originally taken from Satoh and Fibiger (1985) for the baboon, and Satoh et al. (1983a)

→

Fig. 6 for the baboon and rat. The regions containing these neurons are often referred to as the pedunculopontine tegmental nucleus (PPT) and the laterodorsal tegemental nucleus (LDT). At some levels, these two nuclei are continuous with each other, and cholinergic neurons are distributed in these nuclei more or less as a continuum.

The term LDT was originally used by Castaldi (1926) in his Golgi studies of the mesencephalon and pons. The LDT is currently defined as a cluster of neurons lateral to the dorsal tegmental nucleus of Gudden, rostromedial to Barrington's nucleus and the locus coeruleus (Tohyama et al., 1978). Many neurons in the region of the LDT are cholinergic, but non-cholinergic neurons are interspered with cholinergic neurons.

The PPT was first defined cytoarchitechtonically in the human brain as a cluster of large, darkly stained neurons which distribute in close association with the superior cerebellar peduncle (Jacobsohn, 1909; Olszweski and Baxter, 1954). The PPT of the rat has been studied by Rye et al. (1987). These authors observed that all large neurons constituting the PPT are cholinergic. However, as discussed above in the case of the CBC, here again the nomenclature used by different authors has varied considerably; some authors have followed the original definition of the PPT as consisting only of large, cholinergic neurons (e.g. Hallanger et al., 1987; Rye et al., 1987), whereas the fact that these neurons are interspersed with smaller, non-cholinergic neurons has prompted a less restrictive use of the term PPT to include all the cells within the region demarcated by cholinergic neurons (e.g. Isaacson and Tanaka, 1986; Woolf and Butcher, 1986; Goldsmith and van der Kooy, 1988; Semba et al., 1988a). Mesulam et al. (1983b) designated cholinergic neurons in the PPT and LDT as sectors Ch5 and Ch6, respectively. In some early studies,

the PPT was also referred to as the cuneiform nucleus or the parabrachial nucleus (Shute and Lewis, 1967; Palkovits and Jacobowitz, 1974). Clearly this is another area where the standardization of nomenclature is urgent.

Projections

Anatomical studies during the past few years have firmly established that cholinergic neurons in the mesopontine tegmentum of the rat give rise to major ascending pathways that terminate in various forebrain and upper brainstem structures, as well as a descending pathway that innervates the medulla. Cholinergic neurons in PPT and LDT innervate some of the same brain regions. However, some differences also appear to exist between the projections of these two nuclei; there is a tendency for the LDT to project to brain regions involved in "limbic" and autonomic functions. The ascending course of the mesopontine cholinergic pathway has been studied using AChE histochemistry (Shute and Lewis, 1967; Wilson, 1985) and anterograde transport (Satoh and Fibiger, 1986). A degree of axonal branching of PPT and LDT cholinergic neurons appears to exist between selective target structures (Woolf and Butcher, 1986; Scarnati et al., 1987a; see also Steriade et al., 1988).

Thalamic projection
The presence of major cholinergic innervation of the thalamus can be easily appreciated by heavy staining of this structure for AChE (Shute and Lewis, 1967; Jacobowitz and Palkovits, 1974) and ChAT (Mesulam et al., 1983b; Sofroniew et al., 1985). In the rat thalamus, the highest densities of ChAT immunoreactive fibers are found in the anteroventral, reticular, lateral mediodorsal, and intralaminar nuclei, while the other nuclei contain

---

for the rat. Abbreviations: DPS = decussation of the superior cerebellar peduncle; FLM = medial longitudinal fasciculus; IV = trochlear nucleus; LL = lateral lemniscus; LM = medial lemniscus; PCS = superior cerebellar peduncle; cnf = nucleus cuneiformis; dr = dorsal raphe nucleus; lc = locus coeruleus; mr = median raphe nucleus; mV = mesencephalic nucleus of the trigeminal nerve; rpo = pontine reticular formation, oral part; td = dorsal tegmental nucleus of Gudden; tld = laterodorsal tegmental nucleus; tp = nucleus tegmenti pedunculopontinus; tpc = nucleus tegmenti pedunculopontinus, subnucleus compacta.

light to moderate densities (Levey et al., 1987a). AChE staining also displays a similar pattern, except that the anterodorsal nucleus, which contains little ChAT immunoreactivity, displays intense AChE staining (Levey et al., 1987a). Virtually all cholinergic activity in the thalamus must derive extrathalamically, since no cholinergic somata have been observed in the thalamus with the exception of the medial habenula (rat: Houser et al., 1983; Ichikawa and Hirata, 1986; Levey et al., 1987a; cat: Vincent and Reiner, 1987; see below).

A number of studies have used a combination of lesions and biochemical measurements of ChAT activity or AChE histochemistry to obtain information on the organization of brainstem cholinergic afferents to discrete regions of the thalamus (Hoover and Jacobowitz, 1979; Hoover and Baisden, 1980; Rotter and Jacobowitz, 1981). The projection from the PPT to the intralaminar nuclei was demonstrated by radioactive amino acid autoradiography (rat: Sugimoto and Hattori, 1984; cat: Moon Edley and Graybiel, 1983). Retrograde labelling with [³H]choline has also been used to examine the cholinergic nature of the mesopontine tegmental projection to the thalamus (Wiklund and Cuénod, 1984).

More precise information has been provided by studies that combined the more recently developed tract-tracing techniques and either AChE histochemistry (Hoover and Baisden, 1980; Mesulam et al., 1983b; Woolf and Butcher, 1986) or ChAT immunohistochemistry (rat: Sofroniew et al., 1985; Satoh and Fibiger, 1986; Woolf and Butcher, 1986; Hallanger et al., 1987; cat: Paré et al., 1988; Smith et al., 1988; Steriade et al., 1988; dog: Isaacson and Tanaka, 1986; monkey: Steriade et al., 1988). Important conclusions made by these studies are as follows: firstly, a majority of cholinergic neurons in the mesopontine tegmentum project to the thalamus. For example, following large HRP injections into the anterior thalamus, 91% of the HRP labelled neurons in the ipsilateral PPT and LDT regions were ChAT-immunoreactive, and about 60% of ChAT-immunoreactive neurons in the same region were HRP-labelled (Sofroniew et al., 1985). In a series of studies in the cat, the total of percentages of cholinergic tegmental neurons projecting to different thalamic nuclei reached 230% (Paré et al., 1988; Steriade et al., 1988). These findings indicate that although mesopontine cholinergic neurons project to other brain regions as well (see below), the thalamus is clearly a major target of these neurons.

Secondly, the cholinergic projection from the brainstem is a major afferent to the thalamus, and exceeds those of brainstem aminergic neurons, at least in terms of cell numbers. Of the brainstem neurons retrogradely labelled following HRP injections into various thalamic nuclei, those that were also ChAT-immunoreactive were greater in number than those situated in the locus coeruleus or the raphe nuclei (rat: Hallanger et al., 1987, cat: Paré et al., 1988; Steriade et al., 1988). However, in the case of the intralaminar and associational nuclei of the thalamus, there also appears an additional major projection from the mesopontine reticular formation which is neither cholinergic nor aminergic (Hallanger et al., 1987; Paré et al., 1988; Steriade et al., 1988).

Thirdly, the cholinergic innervation of the thalamus by the PPT and LDT is not uniform. While cholinergic neurons in the PPT innervate all thalamic nuclei, the cholinergic innervation by the LDT appears to be more selective, involving predominantly limbic, and some associational and intralaminar parts of the thalamus, including the anterior, laterodorsal, central medial, and mediodorsal nuclei in both rat (Hallanger et al., 1987) and cat (Paré et al., 1988; Smith et al., 1988; Steriade et al., 1988). Thus, the cholinergic innervation of the sensory and motor relay nuclei are provided mainly by the PPT. In the cat, the PPT afferent predominates over that of the LDT in all the thalamic nuclei except for the anterior nuclei, where the two inputs are about equal (Steriade et al., 1988). Some data which are available with the monkey also suggest the possibility of selective or topographical innervation of the thalamus by PPT and LDT cholinergic neurons (Steriade et al., 1988).

Finally, the reticular nucleus of the thalamus appears to be unique in that it receives dual cholinergic innervation, from both the brainstem tegmentum (PPT and LDT) and the basal forebrain in both rat (Hallanger et al., 1987) and cat (Steriade et al., 1987; Paré et al., 1988; see above). This is also the case with the mediodorsal nucleus in the cat (Steriade et al., 1987) but not the rat (Levey et al., 1987b). The significance of this dual cholinergic innervation of some thalamic nuclei is currently unknown.

Thus, these recent studies have revealed selective organization of cholinergic afferents to the thalamus from the brainstem tegmentum including the PPT and LDT. However, little information is currently available about the synaptic connectivity of these cholinergic afferents (de Lima et al., 1985). Together, these anatomical data provide a basic framework in which the physiological significance of cholinergic input to the thalamus from the mesopontine tegmentum can be studied.

Projections to other forebrain regions and the upper brainstem

Like the projection to the thalamus, ascending projections of the PPT and LDT to extrathalamic structures appear to overlap to some extent. Those brain structures innervated by both PPT and LDT cholinergic neurons include the pretectal area, lateral hypothalamus, superior colliculus, and lateral preoptic area (Satoh and Fibiger, 1986; Woolf and Butcher, 1986).

The ascending projections of the PPT and LDT, however, also differ in some respects. The regions which are preferentially innervated by the LDT include the medial prefrontal cortex, lateral septum, and magnocellular basal forebrain regions (including the medial septum, diagonal band, and magno-

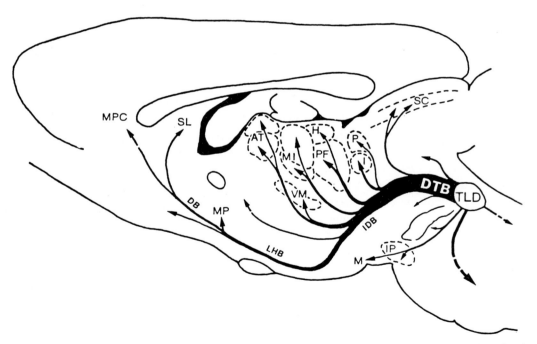

Fig. 7. A diagram of the ascending projections of the laterodorsal tegmental nucleus (TLD). Adopted from Satoh and Fibiger (1986). Abbreviations: AT = anterior thalamic nuclei; DB = diagonal band; DTB = dorsal tegmental bundle; H = lateral habenula; I = interstitial magnocellular nucleus of the posterior commissure; IP = interpeduncular nucleus; LHB = lateral hypothalamic bundle; M = lateral mammillary nucleus; MI = mediodorsal (lateral part) and intralaminar thalamic nuclei; MP = magnocellular preoptic area; MPC = medial prefrontal cortex; P = anterior pretectal area; PF = parafascicular thalamic nucleus; SC = superior colliculus; SL = lateral septum; VM = ventromedial thalamic nucleus.

cellular preoptic area), lateral mammillary nucleus, interpeduncular nucleus, and median raphe nucleus (Vincent et al., 1983b; Woolf and Butcher, 1985, 1986; Satoh and Fibiger, 1986). The LDT projections to the medial prefrontal cortex (Sakanaka et al., 1983; Vincent et al., 1983) and the lateral septum (Sakanaka et al., 1981) also contain substance P (see below). The LDT provides a cholinergic input to the interpeduncular nucleus (Woolf and Butcher, 1985; Satoh and Fibiger, 1986), which is also innervated by cholinergic neurons in the magnocellular basal forebrain (Woolf and Butcher, 1985; see above). A summary of the ascending projections of the LDT is presented in Fig. 7.

Those forebrain structures preferentially innervated by the PPT include the supraoptic nucleus, nucleus accumbens, and olfactory tubercle, although the cholinergic nature of these projections remains to be confirmed (Satoh and Fibiger, 1986).

The PPT has long been thought to be a part of extrapyramidal circuitry, with reciprocal connections with a number of extrapyramidal structures including the substantia nigra, subthalamic nucleus, entopeduncular nucleus, and striatum (e.g. Graybiel, 1977; Nomura et al., 1980; Jackson and Crossman, 1981; Saper and Loewy, 1982; Hammond et al., 1983; Moon Edley and Graybiel, 1983; Sugimoto and Hattori, 1984; Scarnati et al., 1987b). Of these connections, the PPT projection to the substantia nigra has been a subject of considerable research and discussion, because it raises the possibility of a direct cholinergic modulation of dopaminergic neurons. The cholinergic nature of the PPT projection to the substantia nigra has been supported in studies using [^{3}H]leucine (Sugimoto and Hattori, 1984) or ChAT immunohistochemistry combined with retrograde transport (Woolf and Butcher, 1986; Clarke et al., 1987). Some ChAT-immunoreactive neuropil has also been seen in the caudal pars reticulata of the rat (Gould and Butcher, 1986), and in the pars compacta of the ferret (Henderson and Greenfield, 1987). Tokuno et al. (1988) observed that degener-

ating terminals after lesions of the PPT make synaptic contacts with compacta neurons that were retrogradely labelled after HRP injections in the caudate-putamen. In support of these anatomical data, Clarke et al. (1987) have reported excitation of compacta neurons following kainate injections into the PPT, this being blocked by systemic injections of the nicotinic antagonist mecamylamine.

However, Rye et al. (1987) have investigated the distribution of anterograde and retrograde labelling following HRP injections into the substantia nigra or globus pallidus in relation to the cholinergic perikarya in the PPT, and have concluded that it is not the PPT but an area adjacent to it (termed "midbrain extrapyramidal area") that has reciprocal connections with these extrapyramidal structures. This finding raises questions about some of the earlier findings. This may explain the report by Sugimoto and Hattori (1984) that although labelled fibers were present in the compacta region of the substantia nigra following [^{3}H]leucine injections into the PPT, no labelled terminals were seen to form synaptic contacts. Consistent with these anatomical data, Scarnati et al. (1986) found that iontophoretically administered ACh excited only a few compacta neurons. Obviously future studies involving the stimulation of or tracer injections into the PPT should include confirmation of stimulation/injection sites in relation to histochemically or immunohistochemically visualized cholinergic neurons of the PPT. It would be of interest to determine if there are synaptic contacts between ChAT-immunoreactive axon terminals and tyrosine hydroxylase- or dopamine-immunoreactive dendrites and somata in the compacta region.

Projections to the lower brainstem and spinal cord
In addition to extensive ascending projections as described above, cholinergic neurons in the mesopontine tegmentum give rise to some descending projections. Studies using radioactive amino acid autoradiography and retrograde tracing have shown that the PPT contains neurons projecting to the raphe nuclei, the pontine and medullary

reticular formation (Garcia-Rill et al., 1983; Moon Edley and Graybiel, 1983), and the spinal cord (Nomura et al., 1980). Of these projections, however, only that to the medullary reticular formation appears to contain a significant cholinergic component (Butcher and Woolf, 1986; Rye et al., 1988), and this arises primarily in the PPT (Rye et al., 1988). The cholinergic innervation of the spinal cord is limited (Jones et al., 1986; Rye et al., 1988). However, many non-cholinergic neurons in the PPT region have spinal projections (Goldsmith and van der Kooy, 1988; Rye et al., 1988). As discussed below, cholinergic neurotransmission has been implicated in the pontomedullary mechanisms of REM sleep as well as sensorimotor functions, and the cholinergic PPT projection to the medullary reticular formation provides an interesting area for future research in relation to these functions.

Afferent connections

Information on the afferent connections of PPT and LDT cholinergic neurons are at present fragmentary. Afferents from the forebrain and upper brainstem to the LDT have been identified using both retrograde and anterograde tracing techniques in the rat: major sources of afferents include the midbrain reticular formation, central gray, zona incerta, lateral habenula, various hypothalamic nuclei, nucleus of the diagonal band, bed nucleus of the stria terminalis, and medial prefrontal cortex (Satoh and Fibiger, 1987). Although no systematic studies have been conducted specifically to study afferents to the PPT, there is some evidence to suggest that the PPT and LDT have some common afferents. A degree of selectivity also appears to exist, however. For example, the lateral habenula projects to the LDT, but not to the PPT (Herkenham and Nauta, 1979). The zona incerta appears to project more heavily to the LDT than the PPT (Ricardo, 1981). In the hypothalamus, both the paraventricular nucleus (Conrad and Pfaff, 1976b) and the medial preoptic area (Simerly and Swanson, 1988) appear to project equally to

the PPT and LDT, whereas the anterior hypothalamus projects more heavily to the LDT (Conrad and Pfaff, 1976b), and the lateral hypothalamus more heavily to the PPT (Hosoya and Matsushita, 1981).

There is some evidence to suggest that both PPT and LDT receive afferents from the lower brainstem. For example, both PPT and LDT appear to receive a major input from the reticular formation in the pons and medulla (Jones and Yang, 1985; Vertes et al., 1986). In addition, part of the projections from the dorsal column nuclei appears to terminate in the area corresponding to the PPT (Massopust et al., 1985).

Although the neurotransmitters contained in these afferents are yet to be identified, there is some evidence to suggest that the afferents from the lateral hypothalamus and zona incerta may contain α-melanocyte-stimulating hormone; neurons containing this hormone are present in these regions, and the axons of some of these neurons descend to innervate the inferior colliculus and spinal cord (Köhler et al., 1984b; Shiosaka et al., 1985). Although neurons in the basal forebrain innervate the mesopontine tegmentum, this projection does not appear to be cholinergic (Semba et al., 1986). Recent immunohistochemical observations by Sutin and Jacobowitz (1988) suggest that some of the afferents to LDT may contain cholecystokinin, glutamate decarboxylase, tyrosine hydroxylase, serotonin, and/or neuropeptide Y.

As discussed above for the afferents to the CBC, all the findings at the light microscopic level point to possible afferents to cholinergic neurons in the PPT and LDT, and final confirmation rests on the examination at the electron microscopic level.

Colocalization

Cholinergic neurons in the mesopontine tegmentum contain not only acetylcholine, but other neuroactive chemicals as well. Vincent et al. (1983) have demonstrated that ChAT-immunoreactive neurons in the PPT and LDT also stain histochemically for NADPH-diaphorase, and that approximately 30% of these diaphorase-positive neurons in the same regions are immunoreactive for substance P. Furthermore, most of the diaphorase/substance P/cholinergic neurons also contain corticotropin-releasing factor and bombesin/gastrin-releasing peptide (Standaert et al., 1986; Vincent et al., 1986).

Standaert et al. (1986) have shown colocalization of atrial natriuretic peptide in all cholinergic neurons in the PPT and LDT following colchicine treatment in the rat. The region of the LDT also contains neurons that are immunoreactive for galanin, calcitonin gene-related peptide, neurotensin (Sutin and Jacobowitz, 1988), and dynorphin B (Fallon and Leslie, 1986; Sutin and Jacobowitz, 1988). The extent to which these peptides are found in cholinergic neurons in the LDT as well as its physiological significance remains to be investigated.

Cholinergic interneurons

Caudate-putamen

The caudate-putamen contains a population of cholinergic neurons which constitute about 1 – 2% of the total cell population (Fibiger, 1982; Groves, 1983; Phelps et al., 1985; Semba et al., 1987a). These neurons are distributed evenly throughout the striatum. Although their number is modest, striatal cholinergic neurons have large somata and dendritic fields; their dendrites, usually 3 – 4 in number and sparsely spiny, extend sometimes over 400 μm from the soma as described in studies using AChE histochemistry (Lehmann and Fibiger, 1979; Satoh et al., 1983b), ChAT immunohistochemistry (Kimura et al., 1981; Satoh et al., 1983a; Phelps et al., 1985), and the intracellular HRP technique (Bishop et al., 1982). On the basis of these morphological features, striatal cholinergic neurons have been thought to correspond to the large aspiny neurons described in Golgi studies (e.g. DiFiglia et al., 1980; Chang and Kitai, 1982; Takagi et al., 1984). This conclusion has been confirmed in species including rat (Satoh et al., 1983a,

b), cat (Kimura et al., 1981; Vincent and Reiner, 1987), and primate (Satoh and Fibiger, 1985). Unlike cholinergic neurons in the mesopontine tegmentum, striatal cholinergic neurons do not contain NADPH-diaphorase; it is present, however, in somatostatin-containing interneurons in the striatum (Vincent et al., 1983a, b).

The absence of retrograde labelling of cholinergic, large aspiny neurons following tracer injections into various brain regions including the substantia nigra and globus pallidus suggests that these neurons are intrinsic to the striatum (Parent et al., 1980; Woolf and Butcher, 1981). The synaptic connectivity of these neurons has been investigated using ChAT immunohistochemistry, AChE histochemistry, and Golgi techniques. Cholinergic interneurons appear to receive occasional symmetric and asymmetric synaptic contacts on the soma and dendrites (Satoh et al., 1983b; Bolam et al., 1984); increasingly greater numbers of contacts are seen at the distal dendrites (Phelps et al., 1985). In addition, the initial segment of the axon of these neurons has been observed to receive symmetric synapses (Phelps et al., 1985). Some of the presynaptic terminals have been demonstrated to contain substance P, and these contacts are symmetric (Bolam et al., 1986). Tyrosine hydroxylase-immunoreactive, probably dopaminergic, terminals also make synaptic contacts with ChAT-immunoreactive somata and proximal dendrites (Kubota et al., 1987). Axons of these cholinergic neurons make synaptic contact with the somata and dendrites of striatal neurons (Wainer et al., 1984), some of which appear to be the medium spiny, projection neurons of the caudate-putamen (Bolam et al., 1984; Phelps et al., 1985).

Ventral striatum

The ventral striatum consists of those dopamine-rich basal forebrain structures including the nucleus accumbens, substriatal gray, olfactory tubercle, and striatal cell bridges, which share similarities with the caudate-putamen or the dorsal striatum (Heimer and Wilson, 1975). These regions also contain cholinergic neurons, which are thought to be interneurons (Fibiger, 1982). The ventral striatum contains among the highest levels of ChAT (Palkovits et al., 1974) and AChE (Jacobowitz and Palkovits, 1974) activities in the brain. ChAT-immunoreactive somata in the ventral striatum are similar in morphology to ChAT-positive neurons in the dorsal striatum, except perhaps for a generally smaller somal size (rat: Sofroniew et al., 1982; Armstrong et al., 1983; Satoh et al., 1983a; Ichikawa and Hirata, 1986; cat: Kimura et al., 1981; Vincent and Reiner, 1987; monkey: Mesulam et al., 1984; Satoh and Fibiger, 1985). Ultrastructural features of ChAT-immunoreactive somata and terminals are also similar to those of the dorsal striatal ChAT-positive neurons (Phelps and Vaughn, 1986).

Cerebral cortex

Although the existence of cholinergic neurons in the cerebral cortex was long a matter of controversy (Fibiger, 1982), currently available data based on immunohistochemistry using different ChAT antibodies indicate that these neurons exist at least in the rat; observations in other species including cat (Kimura et al., 1981; de Lima and Singer, 1986; Vincent and Reiner, 1987) and monkey (Mesulam et al., 1983a; Campbell et al., 1987) have been negative.

The presence of cholinergic neurons in the rat cortex was first reported in an immunohistochemical study by McGeer et al. (1974). More recent demonstrations were provided by Eckenstein and Thoenen (1983) and Houser et al. (1983). Although AChE-positive neurons are also present in the rat cortex, Levey et al. (1984) demonstrated that all ChAT-immunoreactive cortical neurons lack strong AChE staining, and vice versa, indicating that ChAT and AChE are contained in different cortical cell populations. ChAT-immunoreactive neurons in the rat isocortex are generally small, bipolar cells, and most numerous in layers II and III, suggesting that these are intrinsic neurons (Levey et al., 1984; Houser et al., 1985; Ichikawa

and Hirata, 1986; Parnavelas et al., 1986; Eckenstein et al., 1988). In the forebrain of the rat, however, ChAT-immunoreactive cells are distributed in deeper layers, including layer V where these neurons correspond to pyramidal neurons (Nishimura et al., 1988). The ChAT-positive neurons in the motor and somatosensory cortices have been observed to receive both symmetric and asymmetric synapses on their somata and dendrites (Houser et al., 1985; Parnavelas et al., 1986). Over 80% of ChAT-immunoreactive bipolar neurons of the rat cortex have been reported to contain vasoactive intestinal polypeptide as well (Eckenstein and Baughman, 1984).

In the allocortex including the entorhinal and pyriform cortices, ChAT-immunoreactive neurons are usually small and multipolar (Levey et al., 1984). In the rat hippocampal formation, only a small number of ChAT-immunoreactive neurons are present (Houser et al., 1983); these neurons have been described as small and multipolar (Levey et al., 1984) or pleomorphic belonging to several cell types (Matthews et al., 1987). These are most numerous in the stratum lacunosum-moleculare of the regio superior, and in the granule cell layer and stratum moleculare of the dentate gyrus (Frotscher and Léránth, 1985; Matthews et al., 1987; but see Wainer et al., 1984).

Thus, these data indicate that, despite some early negative reports (Sofroniew et al., 1982; Armstrong et al., 1983; Satoh et al., 1983a), cholinergic neurons do exist in the cortex of adult rat. In the monkey, some cortical neurons express ChAT immunoreactivity early during embryonic development; these neurons are located in the white matter and deep layers of the cortex (Hendry et al., 1987).

Other cholinergic pathways and cholinergic neurons with unknown projections

In addition to the cholinergic projection neurons and interneurons described above, there are cholinergic neurons which appear to give rise to less extensive cholinergic pathways, as well as cholinergic neurons with unknown projections. Although all the motoneurons in both somatic and visceral cranial nerve nuclei are cholinergic, the reader is referred to other reviews for the discussion of these neurons (Vincent et al., 1986b; Fibiger and Vincent, 1987).

Habenulo-interpeduncular pathway

Cholinergic neurons in the habenula have been a subject of research and debate (Fibiger, 1982). Early studies using ChAT immunohistochemistry indicated that there are no cholinergic neurons in the habenula (rat: Armstrong et al., 1983; Satoh et al., 1983a; cat: Kimura et al., 1981). However, more recent studies have demonstrated that there are indeed ChAT-immunoreactive neurons in the ventral portion of the medial habenula in the rat (Houser et al., 1983; Ichikawa and Hirata, 1986; Eckenrode et al., 1987), cat (Vincent and Reiner, 1987), and monkey (Mesulam et al., 1984). A few ChAT-immunoreactive neurons were also seen in the lateral habenula of the cat (Vincent and Reiner, 1987). A strongly ChAT-positive fiber bundle has been observed to arise from the ventrolateral aspect of the medial habenula to descend in the fasciculus retroflexus and enter the interpeduncular nucleus (Houser et al., 1983; Ichikawa and Hirata, 1986). Organotypic cultures containing the habenula of the rat have also been shown to contain ChAT-immunoreactive neurons (Keller et al., 1984). Retrograde HRP studies have indicated that an afferent to the interpeduncular nucleus arises in the medial habenula (Herkenham and Nauta, 1979; Marchand et al., 1980). These data suggest that there is a cholinergic projection from the ventromedial habenula to the interpeduncular nucleus.

In addition to this cholinergic projection, there is a second habenulopeduncular pathway which arises from neurons located dorsally to the cholinergic neurons, i.e. in the dorsal half of the medial habenula, and these neurons contain substance P (Contestabile et al., 1987). However, the cholinergic projection appears to be segregated from this substance P projection in terms of laterality

and terminal fields within the interpeduncular nucleus (Contestabile et al., 1987). The interpeduncular nucleus also receives a major cholinergic input, via the habenula, from the magnocellular basal forebrain (see above).

Cholinergic neurons in the parabigeminal nucleus: parabigemino-collicular pathway?

The parabigeminal nucleus in the pons has been reported to contain ChAT-immunoreactive neurons in the cat (Jones and Beaudet, 1987; Vincent and Reiner, 1987) and mouse (Mufson et al., 1986), but not in the rat (Satoh et al., 1983a) or primate (Satoh and Fibiger, 1985). In the mouse, these cholinergic neurons have been reported to project to the superior colliculus (Mufson et al., 1986). Although this projection exists in the rat as well (Beninato and Spencer, 1986; Schümann, 1987), it does not appear to be cholinergic (Beninato and Spencer, 1986); the cholinergic innervation of the superior colliculus in the rat originates primarily in the PPT and LDT (Beninato and Spencer, 1986; Satoh and Fibiger, 1986; Woolf and Butcher, 1986).

Cholinergic neurons in the hypothalamus

Although the involvement of cholinergic neurotransmission in some hypothalamic functions has been indicated in biochemical and pharmacological studies, the anatomical source or sources of hypothalamic acetylcholine has been the subject of some controversy (Fibiger, 1982). Early immunohistochemical studies failed to identify ChAT-immunoreactive neurons in the rat hypothalamus (Armstrong et al., 1983; Satoh et al., 1983a), suggesting the extrinsic origin of hypothalamic acetylcholine. More recently, however, Vincent and Reiner (1987) have reported that in the cat weakly ChAT-immunoreactive neurons are present in the lateral hypothalamus, surrounding the fornix and extending medially to the arcuate nucleus. These neurons were much smaller than the ChAT-immunoreactive neurons in the magnocellular basal forebrain. Using a polyclonal ChAT antibody, Tago et al. (1987) have reported a more widespread distribution of small ChAT-immunoreactive neurons, including the tubero-infundibular system, in the rat, monkey, and human hypothalamus. Similar observations have been reported in the rat by Rao et al. (1987), who used the same monoclonal ChAT antibody as was used by Vincent and Reiner (1987). These findings in a number of different species suggest that cholinergic neurons are present in the hypothalamus. The projections and specific functions of these cholinergic neurons remain to be investigated.

Other cholinergic neurons

Recent studies using sensitive immunohistochemical techniques have reported the presence of ChAT-immunoreactive neurons in some brain regions which were not previously thought to contain cholinergic neurons. These areas include the amygdala (rat: Carlsen and Heimer, 1986; Sofroniew et al., 1987; cat: Vincent and Reiner, 1987), substantia nigra (rat: Gould and Butcher, 1986), parabrachial nuclei (cat: Jones and Beaudet, 1987; Vincent and Reiner, 1987), and superior colliculus (cat: Vincent and Reiner, 1987). The vestibular nuclei have also been reported to contain ChAT-immunoreactive neurons in the monkey but not in the rat (Carpenter et al., 1987).

The medullary reticular formation also contains ChAT-immunoreactive neurons (cat: Kimura et al., 1981; Jones and Beaudet, 1987; Vincent and Reiner, 1987). It should be noted, however, that at least some of these ChAT-immunoreactive neurons may be motoneurons of the accessory motor nuclei of the cranial nerve or preganglionic parasympathetic neurons; these somatic and autonomic motoneurons tend to be scattered within the medullary reticular formation (Székely and Matesz, 1982; Semba and Egger, 1986; Semba, Joseph, and Egger, unpublished observations).

The spinal cord contains some ChAT-immunoreactive neurons which are neither somatic motoneurons, nor preganglionic autonomic neurons. In the rat, these include (1) small to medium-sized neurons in laminae III–V of the dorsal horn, (2) medium to large multipolar neurons in lamina VII which delineate the boundary between the dorsal and the ventral horns, and (3) small neurons in the gray matter surrounding the central canal (lamina X) (Houser et al., 1983, Barber et al., 1984; Borges and Iversen, 1986). Similar observations have been reported in the cat (Vincent and Reiner, 1987). The projections of these neurons are currently unknown.

Conclusions

The distribution and projections of central cholinergic neurons are becoming reasonably well understood. The major remaining anatomical challenge is to determine the afferent connections of these groups of neurons. This represents a formidable challenge for future research. Another important question concerns the extent to which cells that are faintly stained by some ChAT antibodies, and perhaps contain only a small number of ChAT molecules, are indeed cholinergic; that is, do all of these cells release acetylcholine? Given the great sensitivity of current immunohistochemical techniques, this is becoming an increasingly important question. The anatomical studies reviewed above, together with the recent publication of maps of the distribution of central nicotinic receptors (Clarke et al., 1985; Deutch et al., 1987; Swanson et al., 1987), also raise the question as to which central cholinergic systems transmit information via postsynaptic nicotinic receptor mechanisms. Neuroscience now appears to be in an excellent position to address this issue. For example, the high concentrations of nicotinic receptors in the thalamus, together with the fact that most thalamic nuclei are innervated exclusively by cholinergic neurons in the LDT and PPT, make this projection system a strong candidate for examining nicotinic transmission. Electrophysiological and pharmacological studies on this question will be of considerable interest.

References

Adams, C.E., Cepeda, C., Boylan, M.K., Fisher, R.S., Hull, C.D., Buchwald, N.A., Wainer, B.H. and Levine, M.S. (1986) Basal forebrain neurons have axon collaterals that project to widely divergent cortical areas in the cat. *Brain Res.*, 297: 365–371.

Amaral, D.G. and Kurz, J. (1985) An analysis of the origins of the cholinergic and noncholinergic septal projections to the hippocampal formation of the rat. *J. Comp. Neurol.*, 240: 37–59.

Armstrong, D.M. (1986) Ultrastructural characterization of choline acetyltransferase containing neurons in the basal forebrain of rat: evidence for a cholinergic innervation of intracerebral blood vessels. *J. Comp. Neurol.*, 250: 81–92.

Armstrong, D.M., Saper, C.B., Levey, A.I., Wainer, B.H. and Terry, R.D. (1983) Distribution of cholinergic neurons in rat brain: demonstrated by the immunocytochemical localization of choline acetyltransferase. *J. Comp. Neurol.*, 216: 53–68.

Barber, R.P., Phelps, P.E., Houser, C.R., Crawford, G.D., Salvaterra, P.M. and Vaughn, J.E. (1984) The morphology and distribution of neurons containing choline acetyltransferase in the adult rat spinal cord: an immunocytochemical study. *J. Comp. Neurol.*, 229: 329–346.

Beninato, M. and Spencer, R.F. (1986) A cholinergic projection to the rat superior colliculus demonstrated by retrograde transport of horseradish peroxidase and choline acetyltransferase immunohistochemistry. *J. Comp. Neurol.*, 253: 525–538.

Bigl, V., Woolf, N.J. and Butcher, L.L. (1982) Cholinergic projections from the basal forebrain to frontal, parietal, temporal, occipital, and cingulate cortices: a combined fluorescent tracer and acetylcholinesterase analysis. *Brain Res. Bull.*, 8: 727–749.

Bishop, G.A., Chang, H.T. and Kitai, S.T. (1982) Morphological and physiological properties of neostriatal neurons: an intracellular horseradish peroxidase study in the rat. *Neuroscience*, 7: 179–191.

Bolam, J.P., Wainer, B.H. and Smith, A.D. (1984) Characterization of cholinergic neurons in the rat neostriatum. A combination of choline acetyltransferase immunocytochemistry, Golgi-impregnation and electron microscopy. *Neuroscience*, 12: 711–718.

Bolam, J.P., Ingham, C.A., Izzo, P.N., Levey, A.I., Rye, D.B., Smith, A.D. and Wainer, B.H. (1986) Substance P-containing terminals in synaptic contact with cholinergic neurons in the neostriatum and basal forebrain: a double immunohistochemical study in the rat. *Brain Res.*, 397: 279–289.

Borges, L.F. and Iversen, S.D. (1986) Topography of choline acetyltransferase immunoreactive neurons and fibers in the rat spinal cord. *Brain Res.*, 362: 140–148.

Boylan, M.K., Fisher, R.S., Hull, C.D., Buchwald, N.A. and Levine, M.S. (1986) Axonal branching of basal forebrain projections to the neocortex: a double-labeling study in the cat. *Brain Res.*, 375: 176–181.

Brodie, M.S. and Proudfit, H.K. (1984) Hypoalgesia induced by the local injection of carbachol into the nucleus raphe magnus. *Brain Res.*, 291: 337–342.

Butcher, L.L. and Woolf, N.J. (1986) Central cholinergic systems: synopsis of anatomy and overview of physiology and pathology. In A.B. Scheibel and A.F. Wechsler (Eds.), *The Biological Substrates of Alzheimer's Disease,* Academic Press, New York, pp. 73–86.

Campbell, M.J., Lewis, D.A., Foote, S.L. and Morrison, J.H. (1987) Distribution of choline acetyltransferase-, serotonin-, dopamine-β-hydroxylase-, tyrosine hydroxylase-immunoreactive fibers in monkey primary auditory cortex. *J. Comp. Neurol.*, 261: 209–220.

Carlsen, J. and Heimer, L. (1986) A correlated light and electron microscopic immunocytochemical study of cholinergic terminals and neurons in the rat amygdaloid body with special emphasis on the basolateral amygdaloid nucleus. *J. Comp. Neurol.*, 244: 121–136.

Carlsen, J., Záborszky, L. and Heimer, L. (1985) Cholinergic projections from the basal forebrain to the basolateral amygdaloid complex: a combined retrograde fluorescent and immunohistochemical study. *J. Comp. Neurol.*, 234: 155–167.

Carpenter, M.B., Chang, L., Pereira, A.B. and Hersh, L.B. (1987) Comparisons of the immunocytochemical localization of choline acetyltransferase in the vestibular nuclei of the monkey and rat. *Brain Res.*, 418: 403–408.

Castaldi, L. (1926) Studi sulla struttura e sullo sviluppo del mesencefalo. Ricerche in Cavia cobaya. III. *Arch. Ital. Anat. Embriol.*, 23: 481–609.

Chang, H.T. and Kitai, S.T. (1982) Large neostriatal neurons in the rat: an electron microscopic study of gold-toned Golgi-stained cells. *Brain Res. Bull.*, 8: 631–643.

Clarke, P.B.S., Schwartz, R.D., Paul, S.M., Pert, C.B. and Pert, A. (1985) Nicotinic binding in rat brain: autoradiographic comparison of [^{3}H]acetylcholine, [^{3}H]nicotine, and [^{125}I]-α-bungarotoxin. *J. Neurosci.*, 5: 1307–1315.

Clarke, P.B.S., Hommer, D.W., Pert, A. and Skirboll, L.R. (1987) Innervation of substantia nigra neurons by cholinergic afferents from pedunculopontine nucleus in the rat: neuroanatomical and electrophysiological evidence. *Neuroscience*, 23: 1011–1019.

Conrad, L.C.A. and Pfaff, D.W. (1976a) Efferents from medial basal forebrain and hypothalamus in the rat. I. An autoradiographic study of the medial preoptic area. *J. Comp. Neurol.*, 169: 185–220.

Conrad, L.C.A. and Pfaff, D.W. (1976b) Efferents from medial basal forebrain and hypothalamus in the rat. II. An autoradiographic study of the anterior hypothalamus. *J. Comp. Neurol.*, 169: 221–262.

Contestabile, A., Villani, L., Fasolo, A., Franzoni, M.F., Gribaudo, L., Oktedalen, O. and Fonnum, F. (1987) Topography of cholinergic and substance P pathways in the habenulo-interpeduncular system of the rat. An immunocytochemical and microchemical approach. *Neuroscience*, 21: 253 – 270.

Coyle, J.T., Price, D.L. and DeLong, M.R. (1983) Alzheimer's disease: a disorder of cortical cholinergic innervation. *Science*, 219: 1184 – 1189.

De Lima, A.D. and Singer, W. (1986) Cholinergic innervation of the cat striate cortex: a choline acetyltransferase immunocytochemical analysis. *J. Comp. Neurol.*, 250: 324 – 338.

De Lima, A.D., Montero, V.M. and Singer, W. (1985) The cholinergic innervation of the visual thalamus: an EM immunocytochemical study. *Exp. Brain Res.*, 59: 206 – 212.

De Olmos, J., Hardy, H. and Heimer, L. (1978) The afferent connections of the main and the accessory olfactory bulb formations in the rat: an experimental HRP-study. *J. Comp. Neurol.*, 181: 213 – 244.

Deutch, A.Y., Holliday, J., Roth, R.H., Chun, L.L.Y. and Hawrot, E. (1987) Immunohistochemical localization of a neuronal nicotinin acetylcholine receptor in mammalian brain. *Proc. Natl. Acad. Sci. U.S.A.*, 84: 8697 – 8701.

DiFiglia, M., Pasik, T. and Pasik, P. (1980) Ultrastructure of Golgi-impregnated and gold-toned spiny and aspiny neurons in the monkey neostriatum. *J. Neurocytol.*, 9: 471 – 492.

Eckenrode, T.C., Barr, G.A., Battisti, W.P. and Murray, M. (1987) Acetylcholine in the interpeduncular nucleus of the rat: normal distribution and effects of deafferentation. *Brain Res.*, 418: 273 – 286.

Eckenstein, F.P. and Baughman, R.W. (1984) Two types of cholinergic innervation in cortex, one co-localized with vasoactive intestinal polypeptide. *Nature (Lond.)*, 309: 153 – 155.

Eckenstein, F.P. and Thoenen, H. (1983) Cholinergic neurons in the rat cerebral cortex demonstrated by immunohistochemical localization of choline acetyltransferase. *Neurosci. Lett.*, 36: 211 – 215.

Eckenstein, F.P., Baughman, R.W. and Quinn, J. (1988) An anatomical study of cholinergic innervation in rat cerebral cortex. *Neuroscience,* 25: 457 – 474.

Fallon, J.H. and Leslie, F.M. (1986) Distribution of dynorphin and enkephalin peptides in the rat brain. *J. Comp. Neurol.*, 249: 293 – 336.

Fibiger, H.C. (1982) The organization and some projections of cholinergic neurons of the mammalian forebrain. *Brain Res. Rev.*, 4: 327 – 388.

Fibiger, H.C. and Vincent, S.R. (1987) Anatomy of central cholinergic neurons. In H.Y. Meltzer (Ed.), *Psychopharmacology. The Third Generation of Progress,* Raven Press, New York, pp. 211 – 218.

Frotscher, M. and Léránth, C. (1985) Cholinergic innervation of the rat hippocampus as revealed by choline acetyltransferase immunocytochemistry: a combined light and electron microscopic study. *J. Comp. Neurol.*, 239: 237 – 246.

Frotscher, M. and Léránth, C. (1986) The cholinergic innervation of the rat fascia dentata: identification of target structures on granule cells by combining choline acetyltransferase immunocytochemistry and Golgi impregnation. *J. Comp. Neurol.*, 243: 58 – 70.

Fuller, T.A., Russchen, F.T. and Price, J.L. (1987) Sources of presumptive glutamergic/aspartergic afferents to the rat ventral striatopallidal region. *J. Comp. Neurol.*, 258: 317 – 338.

Garcia-Rill, E., Skinner, R.D., Gilmore, S.A. and Owings, R. (1983) Connections of the mesencephalic locomotor region (MLR). II. Afferents and efferents. *Brain Res. Bull.*, 10: 63 – 71.

Goldsmith, M. and van der Kooy, D. (1988) Separate non-cholinergic descending projections and cholinergic ascending projections from the nucleus tegmenti pedunculopontinus. *Brain Res.*, 445: 386 – 391.

Gould, E. and Butcher, L.L. (1986) Cholinergic neurons in the rat substantia nigra. *Neurosci. Lett.*, 63: 135 – 319.

Graybiel, A.M. (1977) Direct and indirect preoculomotor pathways of the brainstem: an autoradiographic study of the pontine reticular formation in the cat. *J. Comp. Neurol.*, 175: 37 – 78.

Groenewegen, H.J., Ahlenius, S., Haber, S.N., Kowall, N.W. and Nauta, W.J.H. (1986) Cytoarchitecture, fiber connections, and some histochemical aspects of the interpeduncular nucleus in the rat. *J. Comp. Neurol.*, 249: 65 – 102.

Grove, E.A., Domesick, V.B. and Nauta, W.J.H. (1986) Light microscopic evidence of striatal input to intrapallidal neurons of cholinergic cell group Ch4 in the rat: a study employing the anterograde tracer *Phaseolus vulgaris* leucoagglutinin (PHA-L). *Brain Res.*, 367: 379 – 384.

Groves, P.M. (1983) A theory of the functional organization of the neostriatum and the neostriatal control of voluntary movement. *Brain Res. Rev.*, 5: 109 – 132.

Haber, S.N. and Nauta, W.J.H. (1983) Ramifications of the globus pallidus in the rat as indicated by patterns of immunohistochemistry. *Neuroscience,* 9: 245 – 260.

Hallanger, A.E., Levey, A.I., Lee, H.J., Rye, D.B. and Wainer, B.H. (1987) The origins of cholinergic and other subcortical afferents to the thalamus in the rat. *J. Comp. Neurol.*, 262: 105 – 124.

Hammond, C., Rouzaire-Dubois, B., Féger, J., Jackson, A. and Crossman, A.R. (1983) Anatomical and electrophysiological studies on the reciprocal projections between the subthalamic nucleus and nucleus tegmenti pedunculopontinus in the rat. *Neuroscience,* 9: 41 – 52.

Haring, J.H. and Wang, R.Y. (1986) The identification of some sources of afferent input to the rat nucleus basalis magnocellularis by retrograde transport of horseradish peroxidase. *Brain Res.*, 366: 152 – 158.

Heimer, L. and Wilson, R.D. (1975) The subcortical projections of the allocortex: similarities in the neural associations of the hippocampus, the piriform cortex, and the neocortex. *Golgi Centennial Symposium, Proceedings*, M. Santini (Ed.), Raven Press, New York.

Hellendall, R.P., Godfrey, D.A., Ross, C.D., Armstrong, D.M. and Price, J.L. (1986) The distribution of choline acetyltransferase in the rat amygdaloid complex and adjacent cortical areas, as determined by quantitative micro-assay and immunohistochemistry. *J. Comp. Neurol.*, 249: 486 – 498.

Henderson, Z. (1987) Source of cholinergic input to ferret visual cortex. *Brain Res.*, 412: 216 – 268.

Henderson, Z. and Greenfield, S.A. (1987) Does the substantia nigra have a cholinergic innervation? *Neurosci. Lett.*, 73: 109 – 113.

Hendry, S.H.C., Jones, E.G., Killackey, H.P. and Chalupa, L.M. (1987) Choline acetyltransferase-immunoreactive neurons in fetal monkey cerebral cortex. *Dev. Brain Res.*, 37: 313 – 317.

Herkenkam, M. and Nauta, W.J.H. (1979) Efferent connections of the habenular nuclei in the rat. *J. Comp. Neurol.*, 187: 19 – 48.

Hoover, D.B. and Baisden, R.H. (1980) Localization of putative cholinergic neurons innervating the anteroventral thalamus. *Brain Res. Bull.*, 5: 519 – 524.

Hoover, D.B. and Jacobowitz, D.M. (1979) Neurochemical and histochemical studies of the effect of a lesion of the nucleus cuneiformis on the cholinergic innervation of discrete areas of the rat brain. *Brain Res.*, 170: 113 – 122.

Hosoya, Y. and Matsushita, M. (1981) Brainstem projections from the lateral hypothalamic area in the rat, as studied with autoradiography. *Neurosci. Lett.*, 24: 111 – 116.

Houser, C.R., Crawford, G.D., Barber, R.P., Salvaterra, P.M. and Vaughn, J.E. (1983) Organization and morphological characteristics of cholinergic neurons: an immunocytochemical study with a monoclonal antibody to choline acetyltransferase. *Brain Res.*, 266: 97 – 119.

Houser, C.R., Crawford, G.D., Salvaterra, P.M. and Vaughn, J.E. (1985) Immunocytochemical localization of choline acetyltransferase in rat cerebral cortex: a study of cholinergic neurons and synapses. *J. Comp. Neurol.*, 234: 17 – 34.

Ichikawa, T. and Hirata, Y. (1986) Organization of choline acetyltransferase-containing structures in the forebrain of the rat. *J. Neurosci.*, 6: 281 – 292.

Ingham, C.A., Bolam, J.P., Wainer, B.C. and Smith, A.D. (1985) A correlated light and electron microscopic study of identified cholinergic basal forebrain neurons that project to the cortex in the rat. *J. Comp. Neurol.*, 239: 176 – 192.

Irle, E. and Markowitsch, H.J. (1986) Afferent connections of the substantia innominata/basal nucleus of Meynert in carnivores and primates. *J. Hirnforsch.*, 27: 343 – 367.

Isaacson, L.G. and Tanaka, Jr., D. (1986) Cholinergic and non-cholinergic projections from the canine pontomesencephalic tegmentum (Ch5 area) to the caudal intralaminar thalamic nuclei. *Exp. Brain Res.*, 62: 179 – 188.

Jackson, A. and Crossman, A.R. (1981) Basal ganglia and other afferent projections to the peribrachial region in the rat: a study using retrograde and anterograde transport of horseradish peroxidase. *Neuroscience*, 6: 1537 – 1549.

Jacobowitz, D.M. and Palkovits, M. (1974) Topographic atlas of catecholamine and acetylcholinesterase-containing neurons in the rat brain. I. Forebrain (telencephalon, diencephalon). *J. Comp. Neurol.*, 157: 13 – 28.

Jacobsohn, L. (1909) *Uber die Kerne des Menschlichen Hirnstamms*, Verlag der Konigl Akademie der Wisenschaftern, Berlin.

Jones, B.E. and Beaudet, A. (1987) Distribution of acetylcholine and catecholamine neurons in the cat brainstem: a choline acetyltransferase and tyrosine hydroxylase immunohistochemical study. *J. Comp. Neurol.*, 261: 15 – 32.

Jones, B.E. and Yang, T.-Z. (1985) The efferent projections from the reticular formation and the locus coeruleus studied by anterograde and retrograde axonal transport in the rat. *J. Comp. Neurol.*, 242: 56 – 92.

Jones, B.E., Paré, M. and Beaudet, A. (1986) Retrograde labeling of neurons in the brain stem following injections of [^{3}H]choline into the rat spinal cord. *Neuroscience*, 18: 901 – 916.

Keller, F., Rimvall, K. and Waser, P.G. (1984) Slice cultures confirm the presence of cholinergic neurons in the rat habenula. *Neurosci. Lett.*, 52: 299 – 304.

Kimura, H., McGeer, P.L., Peng, J.H. and McGeer, E.G. (1981) The central cholinergic system studied by choline acetyltransferase immunohistochemistry in the cat. *J. Comp. Neurol.*, 200: 151 – 201.

Kitt, C.A., Mitchell, S.J., DeLong, M.R., Wainer, B.H. and Price, D.L. (1987) Fiber pathways of basal forebrain cholinergic neurons in monkeys. *Brain Res.*, 406: 192 – 206.

Köhler, C., Chan-Palay, V. and Wu, J.-Y. (1984a) Septal neurons containing glutamic acid decarboxylase immunoreactivity project to the hippocampal region in the rat brain. *Anat. Embryol.*, 169: 41 – 44.

Köhler, C., Haglund, L. and Swanson, L.W. (1984b) A diffuse αMSH-immunoreactive projection to the hippocampus and spinal cord from individual neurons in the lateral hypothalamic area and zone incerta. *J. Comp. Neurol.*, 223: 501 – 514.

Kristt, D.A., McGowan, Jr., R.A., Martin-MacKinnon, N. and Solomon, J. (1985) Basal forebrain innervation of rodent neocortex: studies using acetylcholinesterase histochemistry, Golgi and lesion strategies. *Brain Res.*, 337: 19 – 39.

Kubota, Y., Inagaki, S., Shimada, S., Kito, S., Eckenstein, F. and Tohyama, M. (1987) Neostriatal cholinergic neurons receive direct synaptic inputs from dopaminergic axons. *Brain Res.*, 413: 179 – 184.

Lamour, Y., Dutar, P. and Jobert, A. (1984) Cortical projections of the nucleus of the diagonal band of Broca and of the substantia innominata in the rat: an anatomical study using the anterograde transport of a conjugate of wheat germ agglutinin and horseradish peroxidase. *Neuroscience*, 2: 395 – 408.

Lehmann, J. and Fibiger, H.C. (1979) Acetylcholinesterase and the cholinergic neuron. *Life Sci.*, 25: 1939 – 1947.

Lehmann, J., Nagy, J.I, Atmadja, S. and Fibiger, H.C. (1980) The nucleus basalis magnocellularis: the origin of a cholinergic projection to the neocortex of the rat. *Neuroscience*, 5: 1161 – 1174.

Lemann, W. and Saper, C.B. (1985) Evidence for a cortical projection to the magnocellular basal nucleus in the rat: an electron microscopic axonal transport study. *Brain Res.*, 334: 339 – 343.

Levey, A.I., Wainer, B.H., Rye, D.B., Mufson, E.J. and Mesulam, M.-M. (1984) Choline acetyltransferase-immunoreactive neurons intrinsic to rodent cortex and distinction from acetylcholinesterase-positive neurons. *Neuroscience*, 13: 341 – 353.

Levey, A.I., Hallanger, A.E. and Wainer, B.H. (1987a) Choline acetyltransferase immunoreactivity in the rat thalamus. *J. Comp. Neurol.*, 257: 317 – 332.

Levey, A.I., Hallanger, A.E. and Wainer, B.H. (1987b) Cholinergic nucleus basalis neurons may influence the cortex via the thalamus. *Neurosci. Lett.*, 74: 7 – 13.

Luiten, P.G.M., Gaykema, R.P.A., Traber, J. and Spencer, Jr., D.G. (1987) Cortical projection patterns of magnocellular basal nucleus subdivisions as revealed by anterogradely transported *Phaseolus vulgaris* leucoagglutinin. *Brain Res.*, 413: 229 – 250.

Lynch, G.S., Lucas, P.A. and Deadwyler, S.A. (1972) The demonstration of acetylcholinesterase containing neurones within the caudate nucleus of the rat. *Brain Res.*, 45: 617 – 629.

Lynch, G.S., Rose, G. and Gall, C. (1977) Anatomical and functional aspects of the septo-hippocampal projections. In: *Functions of the Septo-Hippocampal System, Ciba Foundation Symposium*, Elsevier, Amsterdam, pp. 5 – 24.

McDonald, J.K., Speciale, S.G. and Parnavelas, J.G. (1987) The laminar distribution of glutamate decarboxylase and choline acetyltransferase in the adult and developing visual cortex of the rat. *Neuroscience*, 21: 825 – 832.

McGeer, P.L., McGeer, E.G., Singh, V.K. and Chase, W.H. (1974) Choline acetyltransferase localization in the central nervous system by immunohistochemistry. *Brain Res.*, 81: 373 – 379.

McKinney, M., Coyle, J.T. and Hedreen, J.C. (1983) Topographic analysis of the innervation of the rat neocortex and hippocampus by the basal forebrain cholinergic system. *J. Comp. Neurol.*, 217: 103 – 121.

Marchand, E.R., Riley, J.N. and Moore, R.Y. (1980) Interpeduncular nucleus afferents in the rat. *Brain Res.*, 193: 339 – 352.

Massopust, L.C., Hauge, D.H., Ferneding, J.C., Doubek, W.G. and Taylor, J.J. (1985) Projection systems and terminal localization of dorsal column afferents: an autoradiographic and horseradish peroxidase study in the rat. *J. Comp. Neurol.*, 237: 533 – 544.

Matthews, D.A., Salvaterra, P.M., Crawford, G.D., Houser, C.R. and Vaughn, J.E. (1987) An immunocytochemical study of choline acetyltransferase-containing neurons and axon terminals in normal and partially deafferented hippocampal formation. *Brain Res.*, 402: 30 – 43.

Melander, T., Staines, Wm.A. and Rökaeus, A. (1986) Galanin-like immunoreactivity in hippocampal afferents in the rat, with special reference to cholinergic and noradrenergic inputs. *Neuroscience*, 19: 223 – 240.

Mellgren, S.I. and Srebro, B. (1973) Changes in acetylcholinesterase and distribution of degenerating fibers in the hippocampal region after septal lesions in the rat. *Brain Res.*, 52: 19 – 36.

Mesulam, M.-M. and Mufson, E.J. (1984) Neural inputs into the nucleus basalis of the substantia innominata (Ch4) in the rhesus monkey. *Brain*, 107: 253 – 274.

Mesulam, M.-M., Mufson, E.J., Levey, A.I. and Wainer, B.H. (1983a) Cholinergic innervation of cortex by the basal forebrain: cytochemistry and cortical connections of the septal area, diagonal band nuclei, nucleus basalis (substantia innominata), and hypothalamus in the rhesus monkey. *J. Comp. Neurol.*, 214: 170 – 197.

Mesulam, M.-M., Mufson, E.J., Wainer, B.H. and Levey, A.I. (1983b) Central cholinergic pathways in the rat: an overview based on an alternative nomenclature (Ch1 – Ch6). *Neuroscience*, 10: 1185 – 1201.

Mesulam, M.-M., Mufson, E.J., Levey, A.I. and Wainer, B.H. (1984) Atlas of cholinergic neurons in the forebrain and upper brainstem of the macaque based on monoclonal choline acetyltransferase immunohistochemistry and acetylcholinesterae histochemistry. *Neuroscience*, 12: 669 – 686.

Moon Edley, S. and Graybiel, A.M. (1983) The afferent and efferent connections of the feline nucleus tegmenti pedunculopontinus, pars compacta. *J. Comp. Neurol.*, 217: 187 – 215.

Moruzzi, G. and Magoun, H.W. (1949) Brain stem reticular formation and activation of the EEG. *Electroencephalogr. Clin. Neurophysiol.*, 1: 455 – 473.

Mufson, E.J., Martin, T.L., Mash, D.C., Wainer, B.H. and Mesulam, M.-M. (1986) Cholinergic projections from the parabigeminal nucleus (Ch8) to the superior colliculus in the mouse: a combined analysis of horseradish peroxidase transport and choline acetyltransferase immunohistochemistry. *Brain Res.*, 370: 144 – 148.

Nagai, T., Kimura, H., Maeda, T., McGeer, P.L., Peng, F. and McGeer, E.G. (1982) Cholinergic projections from the basal forebrain of rat to the amygdala. *J. Neurosci.*, 2: 513 – 520.

Nieuwenhuys, R. (1985) *Chemoarchitecture of the Brain*, Springer-Verlag, Berlin – Heidelberg – New York – Tokyo, pp. 1 – 246.

Nishimura, Y., Natori, M. and Mato, M. (1988) Choline acetyltransferase immunopositive pyramidal neurons in the rat frontal cortex. *Brain Res.,* 440: 144 – 148.

Nomura, S., Mizuno, N. and Sugimoto, T. (1980) Direct projections from the pedunculopontine tegmental nucleus to the subthalamic nucleus in the cat. *Brain Res.,* 196: 223 – 227.

Nyakas, C., Luiten, P.G.M., Spencer, D.G. and Traber, J. (1987) Detailed projection patterns of septal and diagonal band efferents to the hippocampus in the rat with emphasis on innervation of CA1 and dentate gyrus. *Brain Res. Bull.,* 18: 533 – 545.

Olszewski, J. and Baxter, D. (1954) *Cytoarchitecture of the Human Brain Stem,* J.B. Lippincott Company, Philadelphia, Montreal.

Palkovits, M. and Jacobowitz, D.M. (1974) Topographic atlas of catecholamine and acetylcholinesterase-containing neurons in the rat brain. II. Hindbrain (mesencephalon, rhombencephalon). *J. Comp. Neurol.,* 157: 29 – 42.

Palkovits, M., Saavedra, J.M., Kobayashi, R.M. and Brownstein, M. (1974) Choline acetyltransferase content of limbic nuclei of the rat. *Brain Res.,* 79: 443 – 450.

Paré, D., Smith, Y., Parent, A. and Steriade, M. (1988) Projections of brainstem core cholinergic and non-cholinergic neurons of cat to intralaminar and reticular thalamic nuclei. *Neuroscience,* 25: 69 – 86.

Parent, A., O'Reilly-Fromentin, J. and Boucher, R. (1980) Acetylcholinesterase-containing neurons in cat neostriatum: a morphological and quantitative analysis. *Neurosci. Lett.,* 20: 271 – 276.

Parnavelas, J.G., Kelly, W., Franke, E. and Eckenstein, F. (1986) Cholinergic neurons and fibers in the rat visual cortex. *J. Neurocytol.,* 15: 329 – 336.

Phelps, P.E. and Vaughn, J.E. (1986) Immunocytochemical localization of choline acetyltransferase in rat ventral striatum: a light and electron microscopic study. *J. Neurocytol.,* 15: 595 – 617.

Phelps, P.E., Houser, C.R. and Vaughn, J.E. (1985) Immunocytochemical localization of choline acetyltransferase within the rat neostriatum: a correlated light and electron microscopic study of cholinergic neurons and synapses. *J. Comp. Neurol.,* 238: 286 – 307.

Price, J.L. and Stern, R. (1983) Individual cells in the nucleus basalis-diagonal band complex have restricted axonal projections to the cerebral cortex in the rat. *Brain Res.,* 269: 352 – 356.

Rao, Z.R., Yamano, M., Wanaka, A., Tatehata, T., Shiosaka, S. and Tohyama, M. (1987) Distribution of cholinergic neurons and fibers in the hypothalamus of the rat using choline acetyltransferase as a marker. *Neuroscience,* 20: 923 – 934.

Ricardo, J.A. (1981) Efferent connections of the subthalamic region in the rat. II. The zona incerta. *Brain Res.,* 214: 43 – 60.

Rotter, A. and Jacobowitz, D.M. (1981) Neurochemical identification of cholinergic forebrain projection sites of the nucleus tegmentalis dorsalis lateralis. *Brain Res. Bull.,* 6: 525 – 529.

Russchen, F.T., Amaral, D.G. and Price, J.L. (1985) The afferent connections of the substantia innominata in the monkey, *Macaca fascicularis. J. Comp. Neurol.,* 242: 1 – 27.

Rye, D.B., Wainer, B.H., Mesulam, M.-M., Mufson, E.J. and Saper, C.B. (1984) Cortical projections arising from the basal forebrain: a study of cholinergic and noncholinergic components employing combined retrograde tracing and immunohistochemical localization of choline acetyltransferase. *Neuroscience,* 13: 627 – 643.

Rye, D.B., Saper, C.B., Lee, H.J. and Wainer, B.H. (1987) Pedunculopontine tegmental nucleus of the rat: cytoarchitecture, cytochemistry, and some extrapyramidal connections of the mesopontine tegmentum. *J. Comp. Neurol.,* 259: 483 – 528.

Rye, D.B., Lee, H.J., Saper, C.B. and Wainer, B.H. (1988) Medullary and spinal efferents of the pedunculopontine tegmental nucleus and adjacent mesopontine tegmentum in the rat. *J. Comp. Neurol.,* 269: 315 – 341.

Sakanaka, M., Shiosaka, S., Takatsuki, K., Inagaki, S., Takagi, H., Senba, E., Kawai, Y., Hara, Y., Iida, H., Minagawa, H., Matsuzaki, T. and Tohyama, M. (1981) Evidence for the existence of a substance P-containing pathway from the nucelus laterodorsalis tegmenti (Castaldi) to the lateral septal area of the rat. *Brain Res.,* 230: 351 – 355.

Sakanaka, M., Shiosaka, S., Takatsuki, K. and Tohyama, M. (1983) Evidence for the existence of a substance P-containing pathway from the nucleus laterodorsalis tegmenti (Castaldi) to the medial frontal cortex of the rat. *Brain Res.,* 259: 123 – 126.

Saper, C.B. (1984) Organization of cerebral cortical afferent systems in the rat. II. Magnocellular basal nucleus. *J. Comp. Neurol.,* 222: 313 – 342.

Saper, C.B. and Loewy, A.D. (1982) Projections of the pedunculopontine tegmental nucleus in the rat: evidence for additional extrapyramidal circuitry. *Brain Res.,* 252: 367 – 372.

Satoh, K. and Fibiger, H.C. (1985) Distribution of central cholinergic neurons in the baboon (Papio papio). II. A topographic atlas correlated with catecholamine neurons. *J. Comp. Neurol.,* 236: 215 – 233.

Satoh, K. and Fibiger, H.C. (1986) Cholinergic neurons of the laterodorsal tegmental nucleus: efferent and afferent connections. *J. Comp. Neurol.,* 253: 277 – 302.

Satoh, K., Armstrong, D.M. and Fibiger, H.C. (1983a) A comparison of the distribution of central cholinergic neurons as demonstrated by acetylcholinesterase pharmacohistochemistry and choline acetyltransferase immunohistochemistry. *Brain Res. Bull.,* 11: 693 – 720.

Satoh, K., Staines, W.A., Atmadja, S. and Fibiger, H.C. (1983b) Ultrastructural observations of the cholinergic

62

neuron in the rat striatum as identified by acetylcholinesterase pharmacohistochemistry. *Neuroscience,* 10: 1121 – 1136.

Scarnati, E., Proia, A., Campana, E. and Pacitti, C. (1986) A microiontophoretic study on the nature of the putative synaptic neurotransmitter involved in the pedunculopontine-substantia nigra pars compacta excitatory pathway of the rat. *Exp. Brain Res.,* 62: 470 – 478.

Scarnati, E., Gasbarri, A., Campana, E. and Pacitti, C. (1987a) The organization of nucleus tegmenti pedunculopontinus neurons projecting to basal ganglia and thalamus: a retrograde fluorescent double labeling study in the rat. *Neurosci. Lett.,* 79: 11 – 16.

Scarnati, E., Proia, A., Di Loreto, S. and Pacitti, C. (1987b) The reciprocal electrophysiological influence between the nucleus tegmenti pedunculopontinus and the substantia nigra in normal and decorticated rats. *Brain Res.,* 423: 116 – 124.

Schümann, R. (1987) Course and origin of the crossed parabigeminotectal pathways in rat. A retrograde HRP-study. *J. Hirnforsch.,* 28: 585 – 590.

Schwaber, J.S., Rogers, W.T., Satoh, K. and Fibiger, H.C. (1987) Distribution and organization of cholinergic neurons in the rat forebrain demonstrated by computer-aided data acquisition and three-dimensional reconstruction. *J. Comp. Neurol.,* 263: 309 – 325.

Semba, K. and Egger, M.D. (1986) The facial "motor" nerve of the rat: control of vibrissal movement and examination of motor and sensory components. *J. Comp. Neurol.,* 247: 144 – 158.

Semba, K., Reiner, P.B., Disturnal, J.E., Atmadja, S., McGeer, E.G. and Fibiger, H.C. (1986) Evidence for descending basal forebrain projections: anatomical and physiological studies. *Soc. Neurosci. Abst.,* 12: 904.

Semba, K., Fibiger, H.C. and Vincent, S.R. (1987a) Neurotransmitters in the mammalian striatum: neuronal circuits and heterogeneity. *Can. J. Neurol. Sci.,* 14: 386 – 394.

Semba, K., Reiner, P.B., McGeer, E.G. and Fibiger, H.C. (1987b) Morphology of cortically projecting basal forebrain neurons in the rat as revealed by intracellular iontophoresis of horseradish peroxidase. *Neuroscience,* 20: 637 – 651.

Semba, K., Reiner, P.B., McGeer, E.G. and Fibiger, H.C. (1988a) Brainstem afferents to the magnocellular basal forebrain studied by axonal transport, immunohistochemistry, and electrophysiology in the rat. *J. Comp. Neurol.,* 267: 433 – 453.

Semba, K., Reiner, P.B., McGeer, E.G. and Fibiger, H.C. (1988b) Non-cholinergic basal forebrain neurons project to the contralateral basal forebrain in the rat. *Neurosci. Lett.,* 84: 23 – 28.

Shiosaka, S., Kawai, Y., Shibasaki, T. and Tohyama, M. (1985) The descending α-MSHergic (α-melanocyte-stimulating hormone-ergic) projections from the zona incerta and lateral hypothalamic area to the inferior colliculus and spinal cord in the rat. *Brain Res.,* 338: 371 – 375.

Shute, C.C.D. and Lewis, P.R. (1967) The ascending cholinergic reticular system: neocortical, olfactory and subcortical projections. *Brain,* 90: 497 – 520.

Simerly, R.B. and Swanson, L.W. (1988) Projectons of the medial preoptic nucleus: a *Phaseolus vulgaris* leucoagglutinin anterograde tract-tracing study in the rat. *J. Comp. Neurol.,* 270: 209 – 242.

Smith, Y., Paré, D., Deschenes, M., Parent, A. and Steriade, M. (1988) Cholinergic and non-cholinergic projections from the upper brainstem core to the visual thalamus in the cat. *Exp. Brain Res.,* 70: 166 – 180.

Sofroniew, M.V., Eckenstein, F., Thoenen, H. and Cuello, A.C. (1982) Topography of choline acetyltransferase-containing neurons in the forebrain of the rat. *Neurosci. Lett.,* 33: 7 – 12.

Sofroniew, M.V., Priestley, J.V., Consolazione, A., Eckenstein, F. and Cuello, A.C. (1985) Cholinergic projections from the midbrain and pons to the thalamus in the rat, identified by combined retrograde tracing and choline acetyltransferase immunohistochemistry. *Brain Res.,* 329: 213 – 223.

Sofroniew, M.V., Pearson, R.C.A. and Powell, T.P.S. (1987) The cholinergic nuclei of the basal forebrain of the rat: normal structure, development and experimentally induced degeneration. *Brain Res.,* 411: 310 – 331.

Standaert, D.G., Saper, C.B., Rye, D.B. and Wainer, B.H. (1986) Colocalization of atriopeptin-like immunoreactivity with choline acetyltransferase- and substance P-like immunoreactivity in the pedunculopontine and laterodorsal tegmental nuclei in the rat. *Brain Res.,* 382: 163 – 168.

Steriade, M. (1981) Mechanisms underlying cortical activation: neuronal organization and properties of the midbrain reticular core and intralaminar thalamic nuclei. In O. Pompeiano and C.A. Marsan (Eds.), *Brain Mechanisms and Perceptual Awareness,* Raven Press, New York, pp. 327 – 377.

Steriade, M., Parent, A., Paré, D. and Smith. Y. (1987) Cholinergic and non-cholinergic neurons of rat basal forebrain project to reticular and mediodorsal thalamic nuclei. *Brain Res.,* 408: 372 – 376.

Steriade, M., Paré, D., Parent, A. and Smith, Y. (1988) Projections of cholinergic and non-cholinergic neurons of the brainstem core to relay and associational thalamic nuclei in the cat and macaque monkey. *Neuroscience,* 25: 47 – 67.

Sugimoto, T. and Hattori, T. (1984) Organization and efferent projections of nucleus tegmenti pedunculopontinus pars compacta with special reference to its cholinergic aspects. *Neuroscience,* 11: 932 – 946.

Sutin, E.L. and Jacobowitz, D.M. (1988) Immunocytochemical localization of peptides and other neurochemicals in the rat laterodorsal tegmental nucleus and adjacent area. *J. Comp. Neurol.,* 270: 243 – 270.

Swanson, L.W. and Cowan, W.M. (1979) The connections of the septal region in the rat. *J. Comp. Neurol.,* 186: 621 – 656.

Swanson, L.W., Simmons, D.M., Whiting, P.J. and Lindstrom, J. (1987) Immunohistochemical localization of neuronal nicotinic receptors in the rodent central nervous system. *J. Neurosci.*, 7: 3334–3342.

Székely, G. and Matesz, C. (1982) The accessory motor nuclei of the trigeminal, facial, and abducens nerves in the rat. *J. Comp. Neurol.*, 210: 258–264.

Tago, H., McGeer, P.L., Bruce, G. and Hersh, L.B. (1987) Distribution of choline acetyltransferase-containing neurons of the hypothalamus. *Brain Res.*, 415: 49–62.

Takagi, H., Somogyi, P. and Smith, A.D. (1984) Aspiny neurons and their local axons in the neostriatum of the rat: a correlated light and electron microscopic study of Golgi-impregnated material. *J. Neurocytol.*, 13: 239–265.

Tohyama, M., Satoh, K., Sakumoto, T., Kimoto, Y., Takahashi, Y., Yamamoto, K. and Itakura, T. (1978) Organization and projections of the neurons in the dorsal tegmental area of the rat. *J. Hirnforsch.*, 19: 165–176.

Tokuno, H., Morizumi, T., Kudo, M. and Nakamura, Y. (1988) A morphological evidence for monosynaptic projections from the nucleus tegmenti pedunculopontinus pars compacta (TPC) to nigrostriatal projection neurons. *Neurosci. Lett.*, 85: 1–4.

Vertes, R.P. (1988) Brainstem afferents to the basal forebrain in the rat. *Neuroscience*, 24: 907–935.

Vertes, R.P., Martin, G.F. and Waltzer, R. (1986) An autoradiographic analysis of ascending projections from the medullary reticular formation in the rat. *Neuroscience*, 19: 873–898.

Vincent, S.R. and Reiner, P.B. (1987) The immunohistochemical localization of choline acetyltransferase in the cat brain. *Brain Res. Bull.*, 18: 371–415.

Vincent, S.R., Satoh, K., Armstrong, D.M. and Fibiger, H.C. (1983a) NADPH-diaphorase: a selective histochemical marker for the cholinergic neurons of the pontine reticular formation. *Neurosci. Lett.*, 43: 31–36.

Vincent, S.R., Satoh, K., Armstrong, D.M. and Fibiger, H.C. (1983b) Substance P in the ascending cholinergic reticular system. *Nature (Lond.)*, 306: 688–691.

Vincent, S.R., Staines, W.A. and Fibiger, H.C. (1983c) Histochemical demonstration of separate populations of somatostatin and cholinergic neurons in the rat striatum. *Neurosci. Lett.*, 35: 111–114.

Vincent, S.R., Satoh, K., Armstrong, D.M., Panula, P., Vale, W. and Fibiger, H.C. (1986a) Neuropeptides and NADPH-diaphorase activity in the ascending cholinergic reticular system of the rat. *Neuroscience*, 17: 167–182.

Vincent, S.R., Satoh, K. and Fibiger, H.C. (1986b) The localization of central cholinergic neurons. *Prog. Neuro-Psychopharmacol. Biol. Psychiatr.*, 10: 637–656.

Wainer, B.H., Levey, A.I., Mufson, E.J. and Mesulam, M.-M. (1984) Cholinergic systems in mammalian brain identified with antibodies against choline acetyltransferase. *Neurochem. Int.*, 6: 163–182.

Wenk, H., Bigl, V. and Meyer, U. (1980) Cholinergic projections from magnocellular nuclei of the basal forebrain to cortical areas in rats. *Brain Res. Rev.*, 2: 295–316.

Wiklund, L. and Cuénod, M. (1984) Differential labelling of afferents to thalamic centromedian-parafascicular nuclei with [^{3}H]choline and D-[^{3}H]aspartate; further evidence for transmitter specific retrograde labelling. *Neurosci. Lett.*, 46: 275–281.

Wilson, P.M. (1985) A photographic perspective on the origins, form, course and relations of the acetylcholinesterase-containing fibres of the dorsal tegmental pathway in the rat brain. *Brain Res. Rev.*, 10: 85–118.

Woolf, N.J. and Butcher, L.L. (1981) Cholinergic neurons in the caudate-putamen complex proper are intrinsically organized: a combined Evans Blue and acetylcholinesterase analysis. *Brain Res. Bull.*, 7: 487–507.

Woolf, N.J. and Butcher, L.L. (1982) Cholinergic projections to the basolateral amygdala: a combined Evans Blue and acetylcholinesterase analysis. *Brain Res. Bull.*, 8: 751–763.

Woolf, N.J. and Butcher, L.L. (1985) Cholinergic systems in the rat brain. II. Projections to the interpeduncular nucleus. *Brain Res. Bull.*, 14: 63–83.

Woolf, N.J. and Butcher, L.L. (1986) Cholinergic systems in the rat brain. III. Projections from the pontomesencephalic tegmentum to the thalamus, tectum, basal ganglia, and basal forebrain. *Brain Res. Bull.*, 16: 603–637.

Woolf, N.J., Hernit, M.C. and Butcher, L.L. (1986) Cholinergic and non-cholinergic projections from the rat basal forebrain revealed by combined choline acetyltransferase and *Phaseolus vulgaris* leucoagglutinin immunohistochemistry. *Neurosci. Lett.*, 66: 281–286.

Záborszky, L., Leranth, C. and Heimer, L. (1984) Ultrastructural evidence of amygdalofugal axons terminating on cholinergic cells of the rostral forebrain. *Neurosci. Lett.*, 52: 219–225.

Záborszky, L., Carlsen, J., Brashear, H.R. and Heimer, L. (1986a) Cholinergic and GABAergic afferents to the olfactory bulb with special emphasis on the projection neurons in the nucleus of the horizontal limb of the diagonal band. *J. Comp. Neurol.*, 243: 488–509.

Záborszky, L., Heimer, L., Eckenstein, F. and Leranth, C. (1986b) GABAergic input to cholinergic forebrain neurons. An ultrastructural study using retrograde tracing of HRP and double immunolabling. *J. Comp. Neurol.*, 250: 282–295.

Záborszky, L. and Cullinan, W.E. (1989) Hypothalamic axons terminate on forebrain cholinergic neurons: an ultrastructural double-labeling study using PHA-L tracing and ChAT immunocytochemistry. *Brain Res.*, 479: 177–184.

A. Nordberg, K. Fuxe, B. Holmstedt and A. Sundwall (Eds.)
Progress in Brain Research, Vol. 79

CHAPTER 5

Mapping of brain nicotinic receptors by autoradiographic techniques and the effect of experimental lesions

Paul B.S. Clarke

Department of Pharmacology and Therapeutics, McGill University, 3655 Drummond St, Montreal, Canada H3G 1Y6

Introduction

Nicotinic binding sites have been detected using a number of radioligands in in vitro assays of mammalian brain. The most commonly used radioligands have been [³H]nicotine, [³H]acetylcholine and ^{125}I-labelled-α-bungarotoxin. All three bind with high affinity in a saturable and reversible manner, inhibited selectively by nicotinic agents (Romano and Goldstein, 1980; Marks and Collins, 1982; Schwartz et al., 1982). Tritiated nicotine and acetylcholine (ACh) produce virtually identical autoradiographic maps of high affinity binding when muscarinic receptors are occluded in advance (Clarke et al., 1985a), suggesting that these two agonists bind to the same population of nicotinic sites, a conclusion reinforced by more extensive neuropharmacological analysis (Marks et al., 1986; Martino-Barrows and Kellar, 1987). In contrast, ^{125}I-α-bungarotoxin labels a different population of molecules with a distinctly different neuroanatomical distribution (Marks and Collins, 1982; Schwartz et al., 1982; Clarke et al., 1985a; Wonnacott, 1986).

The properties of agonist binding have varied between studies. In addition to a high-affinity binding component (K_d 1 – 10 nM), some investigators have reported binding sites of lower affinity (typical K_d > 100 nM) (e.g. Marks and Collins, 1982; Sloan et al., 1984), whose pharmacological significance is not established (Lippiello and Fernandez, 1986; Martin et al., 1986).

Subconvulsive doses of nicotine, including those in the smoking range, appear to act at central nAChRs labelled with high affinity by [³H]nicotine and [³H]ACh. This is demonstrated, for example, by the upregulation of high-affinity agonist binding sites in rat brain which has been reported following chronic in vivo treatment with nicotine (Schwartz and Kellar, 1983; Marks et al., 1986). In contrast, upregulation of sites labelled by ^{125}I-α-bungarotoxin appears to require higher doses of nicotine than are encountered in cigarette smoking (Marks et al., 1986).

The neuroanatomical pattern of neuronal activation which is induced by acute treatment with nicotine has been visualized in rat brain using the 2-deoxyglucose technique (London et al., 1985a; Grunwald et al., 1987). These experiments suggest that the activational effects of nicotine are largely produced in brain areas having an appreciable density of high-affinity agonist binding sites. Hence, the radioligand binding and functional studies, described above, suggest that the brain receptors labelled with high affinity by [³H]nicotine are the principal pharmacological target for nicotine.

Neuroanatomy of high-affinity agonist binding sites in mammalian brain

High-affinity [³H]nicotine binding sites in brain have similar pharmacological characteristics in mouse (Marks and Collins, 1982), rat (Romano and Goldstein, 1980; Clarke et al., 1984; Lippiello and Fernandez, 1986), cow and monkey (Clarke, 1988), and also in human brain (Shimohama et al., 1985; Flynn and Mash, 1986).

There is little agreement as to the regional distribution of high-affinity agonist binding sites in microdissected rat brain tissue. In particular, varying amounts of binding have been attributed to the hippocampus and hypothalamus (Abood et al., 1981; Costa and Murphy, 1983; Benwell and Balfour, 1985; Larsson and Nordberg, 1985; Yamada et al., 1985; Marks et al., 1986; Martino-Barrows and Kellar, 1987). Although dissection techniques may vary from study to study, another source of variability between studies may be the presence of low-affinity binding sites. When saturation binding data is carefully analyzed, these sites are detected in certain assays but not in others. Most authors have assessed receptor density by determining binding at a single concentration of radioligand, rather than by Scatchard or other more sophisticated analysis.

A much greater degree of neuroanatomical detail can be achieved with receptor autoradiography. High-affinity nicotinic binding sites for [³H]nicotine, [³H]ACh, and to a lesser extent [³H]methylcarbachol have all been mapped autoradiographically in rat brain with virtually identical results (Clarke et al., 1984a, 1985; London et al., 1985b; Boksa and Quirion, 1987). The most detailed qualitative mapping analysis of [³H]nicotine sites has been provided by Clarke et al. (1984, 1985a), using an assay that detects only one population of high-affinity sites, as shown by highly linear Scatchard plots (Clarke, 1986) and Hill numbers typically near unity (Table I). It should be emphasized that the characteristics of [³H]nicotine binding reported here and previously (Clarke et al., 1984) were determined in rat brain

sections treated exactly as for autoradiography, except that radiolabelled sections were taken for scintillation counting rather than for film exposure. Because a close comparison of this sort was made, one can be confident that the autoradiographic images represent high-affinity binding to a single class of sites, with negligible low-affinity binding.

Dense autoradiographic labelling is observed in the medial habenula and interpeduncular nucleus; in the so-called specific motor and sensory nuclei of the thalamus, and in layers III and/or IV of cerebral cortex; in the substantia nigra pars compacta and ventral tegmental area; in the molecular layer of the dentate gyrus and the presubiculum; and in the superficial layers of the superior colliculus. Labelling is conspicuously sparse in the hippocampus and hypothalamus.

A more extensive anatomical analysis has been performed by Swanson et al. (1987) using a radioiodinated monoclonal antibody (mab 270), which reveals a neuroanatomical pattern that is very similar to that obtained with [³H]nicotine. The antibody differs from [³H]nicotine in that it labels

TABLE I

Hill numbers derived from binding of D,L-[³H]nicotine with and without cold displacer*

Inhibitor	n_H	r
Inhibition of D,L[³H]nicotine:		
L-Nicotine	0.94	0.998
D-Nicotine	1.09	0.998
Cytisine	1.34	0.998
Lobeline	1.45	0.983
DMPP	0.95	0.986
ACh/DFP	0.92	0.989
Anabasine	1.18	1.000
Carbachol	1.16	0.992
Piperidine	1.09	0.994
Coniiine	1.15	0.995
From Scatchard analysis:		
D,L-[³H]nicotine	1.00	0.998

* Raw data as in Clarke et al., 1984.

nAChRs in nerve fibre tracts, apparently in the process of axonal transport.

The distribution of high-affinity [³H]nicotine binding in rhesus monkey brain is reminiscent of that in the rat (Friedman et al., 1985; O'Neill et al., 1985). Thus, dense labelling occurs in the anterior thalamic nuclei, and in a band within cerebral cortex layer III; this band is most prominent in the primary sensory areas. Recently, the laminar distributions of cholinergic fibres and nicotinic receptors have been compared in monkey premotor cortex (Friedman et al., 1988). Sites labelled with [³H]nicotine were restricted to the lower part of layer IIIa, whereas cholinergic fibres were seen in almost all cortical layers, probably reflecting the more widespread distribution of muscarinic receptors. Despite clear parallels between the monkey and rat, there are also clear differences. Thus, in contrast to the rat, the medial habenula appears unlabelled in monkey. Although high-affinity agonist binding has not been mapped systematically in human brain, the autoradiographic distribution of sites within the thalamus (Adem et al., 1988) is similar to that reported for rat and monkey.

Central nAChRs: pre- vs. postsynaptic localization

In some brain areas, the release of ACh from nerve terminals appears to be enhanced by a presynaptic action of the transmitter on nicotinic "autoreceptors" (Rowell and Winkler, 1984; Beani et al., 1985; Araujo et al., 1988). Using electrophysiological techniques, Docherty and colleagues have detected nicotinic receptors which appear to be situated on nerve terminals of the fasciculus retroflexus (Brown et al., 1984). This mixed nerve conveys a massive cholinergic input to the interpeduncular nucleus, which possesses one of the highest densities of [³H]nicotine sites of any brain nucleus in the rat. In order to assess whether [³H]nicotine binding within this nucleus was actually located on retroflexus afferents, we removed this input by making bilateral electrolytic lesions some distance from the nucleus, at the level of the habenula (Clarke et al., 1986). Control subjects received sham surgery. Subjects were sacrificed at 3, 5 or 11 days postsurgery, and their brains were removed for histological examination and for [³H]nicotine and ¹²⁵I-α-bungarotoxin autoradiography, determined at a single concentration of radioligand. Significant decreases in [³H]nicotine binding were found as early as 3 days postlesion in several subnuclei of the interpeduncular nucleus, and these losses did not change up to 11 days. The decrease in labelling at 3 days was subsequently analyzed in a homogenate preparation of microdissected interpeduncular nucleus, and was found to reflect a reduction in the density (B_{max}) but not affinity (K_d) of high affinity binding sites (Table II). An additional low-affinity component of binding was unchanged by the lesion. Since transsynaptic degeneration does not appear to occur within this short postlesion period, and since axoaxonic contacts are not found within this nucleus, these experiments suggest that [³H]nicotine binding sites may be present on the fibres or terminals of this pathway. Thus, some high-affinity binding may be to nicotinic autoreceptors.

Clearly, not all nicotinic receptors are located on cholinergic terminals. For example, functional studies have shown that nicotine also acts on receptors located on noncholinergic nerve terminals to influence the release of neurotransmitters such as

TABLE II

Scatchard analysis of D,L-[³H]nicotine binding to interpedunculopontine nucleus tissue membranes (mean ± S.E.M.)*

	High-affinity site		Low-affinity site	
	B_{max} (fmol/mg prot)	K_d (nM)	B_{max} (fmol/mg prot)	K_d (nM)
Sham (n = 8)	146.9 ± 8.5	20 ± 1.7	596 ± 13	250 ± 22
Lesion (n = 8)	96.4 ± 5.3	23 ± 1.4	560 ± 17	290 ± 22

* Binding data were fitted to a two-site regression model (from Clarke et al., 1986).

dopamine, noradrenaline and serotonin (Chesselet, 1984). Consistent with these findings, lesion studies have shown that [3H]agonist binding sites are associated with noncholinergic nerve terminals in certain parts of the brain (Schwartz et al., 1983; Clarke and Pert, 1985; Prusky et al., 1987).

Electrophysiological studies have identified direct postsynaptic actions of nicotine on neuronal cell bodies or dendrites. This has been shown in the medial habenula, locus coeruleus and interpeduncular nucleus, which are all labelled by [3H]nicotine (Clarke et al., 1984). Nicotinic excitation in these areas is selectively blocked by C6-selective antagonists (Brown et al., 1983; Egan and North, 1986; McCormick and Prince, 1987), which contributes to the evidence that high-affinity agonist sites represent central nAChRs with a ganglion-like (C6) pharmacological profile (see Clarke, 1987a).

Nicotine and ascending dopaminergic pathways

Functional interactions between nicotine and the nigrostriatal dopaminergic system were first reported in the 1970s. Nicotine was found to release DA through local actions on DA terminals (see Chesselet, 1982), and Lichtensteiger et al. (1976, 1982) described excitatory effects of systemic and microiontophoretic nicotine on DA neurons located in the substantia nigra pars compacta (SNC). Neuronal excitation was inhibited by the nicotinic antagonist dihydro-beta-erythroidine (DHBE). Unfortunately, the electrophysiological criteria used to identify DA neurons were not sufficiently rigorous.

Consistent with these functional studies, our lesion experiments demonstrated that dopaminergic neurons possess [3H]nicotine binding sites not only at the level of somata, but also at the level of terminals (Clarke and Pert, 1985). These binding sites are present in both nigrostriatal and mesolimbic systems (Fig. 1)

We have examined the effects of systemic nicotine on SNC DA single cell firing rate (Clarke et al., 1985b). Subcutaneous and intravenous administrations of nicotine stimulated firing in a dose-dependent way, and the excitation was shown to be of central origin and prevented by the C6-selective antagonist mecamylamine. Mecamylamine alone reduced neuronal firing rate, pointing

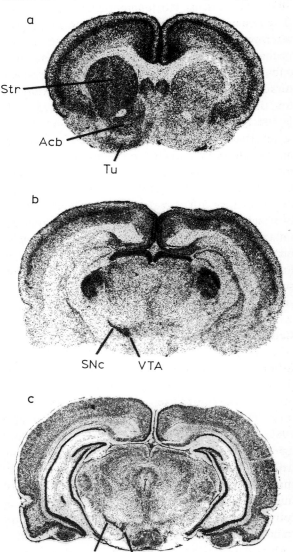

Fig. 1. Photomicrographs of [3H]nicotine receptor distribution in coronal sections of rat brain showing DA terminal regions (a) and cell body areas (b), together with underlying Nissl-stained tissue section (c). Near total unilateral destruction of DA neurons resulted in a marked reduction of receptor labelling. (From Clarke and Pert, 1985.)

to a possible endogenous cholinergic influence on SNC DA neurons. In subsequent work (Clarke et al., 1987), a probable source of cholinergic input to the SNC was traced to cell bodies in the pedunculopontine nucleus. An ipsilateral pathway, initially detected by neuroanatomical double-labelling procedures, was then stimulated by administering an excitatory amino acid (kainic acid) into the cholinergic cell body region, resulting in an immediate excitatory response among DA cells situated at a distance in the ipsilateral SNC. The remotely-triggered excitation of DA cells was prevented by administration of mecamylamine, indicating the involvement of nAChRs. These findings suggest that cholinergic neurons located in the pedunculopontine nucleus project directly onto and excite SNC DA neurons via postsynaptic nicotinic receptors. These data, although not definitive, appear to represent the best evidence to date of nicotinic cholinergic synaptic transmission in mammalian brain.

More recently, we have turned our attention to the mesolimbic DA system. As in the nigrostriatal system, nicotine stimulates neuronal firing and promotes DA release from terminals, at least in drug-naive animals (Imperato et al., 1986; Yoon et al., 1986; Rowell et al., 1987). These findings are of especial interest because the stimulant and reinforcing effects of certain drugs (e.g. d-amphetamine) appear to be mediated by activation of the mesolimbic DA system (Fibiger and Phillips, 1987). The reinforcing effects of nicotine are hard to measure in rats, whereas nicotine produces a robust locomotor stimulant effect if subjects have previously received several administrations of nicotine (see Clarke, 1987b). Rats were therefore pretreated for 2 weeks with daily injections of nicotine. A few days later, L-nicotine was shown to increase locomotor activity in a dose-dependent and stereoselective way. These behavioural effects were paralleled by dose-dependent and stereoselective effects on DA utilization, as measured by L-dopa accumulation following drug-induced inhibition of aromatic acid decarboxylase (Clarke et al., 1988). These effects of nicotine were restricted to

mesolimbic areas (nucleus accumbens, olfactory tubercle) and were not found in the neostriatum. In a subsequent experiment, rats were lesioned with 6-hydroxydopamine in order to produce selective depletion of mesolimbic DA. This lesion abolished the locomotor stimulant effect of nicotine. It therefore appears that the behavioural stimulant action of nicotine seen in tolerant rats occurs through autoradiographically-identified nicotinic receptors in the mesolimbic dopaminergic system.

References

Abood, L.G., Reynolds, D.T., Booth, H. and Bidlack, J.M. (1981) Sites and mechanisms for nicotine's actions in the brain. Neurosci. Biobehav. Rev., 5: 479–486.

Adem, A., Jossan, S.S., d'Argy, R., Brandt, R., Winblad, B. and Nordberg, A. (1988) Distribution of nicotinic receptors in human thalamus as visualized by ^{3}H-nicotine and ^{3}H-acetylcholine receptor autoradiography. J. Neural. Trans., 73: 77–83.

Araujo, D.M., Lapchak, P.A., Collier, B. and Quirion, R. (1988) Characterization of [^{3}H]N-methylcarbamylcholine binding sites and effect of N-methylcarbamylcholine on acetylcholine release in rat brain. J. Neurochem., 51: 292–299.

Beani, L., Bianchi, C., Nilsson, L., Nordberg, A., Romanelli, L. and Sivilotti, L. (1985) The effect of nicotine and cytisine on ^{3}H-acetylcholine release from cortical slices of guinea-pig brain. Naunyn-Schmiedebergs Arch. Pharmacol., 331: 293–296.

Benwell, M.E.M. and Balfour, D.J.K. (1985) Nicotine binding to brain tissue from drug-naive and nicotine-treated rats. J. Pharm. Pharmacol., 37: 405–407.

Boksa, P. and Quirion, R. (1987) [^{3}H]Methyl-carbachol, a new radioligand for nicotinic cholinergic receptors in brain. Eur. J. Pharmacol., 139: 323–333.

Brown, D.A., Docherty, R.J. and Halliwell, J.V. (1983) Chemical transmission in the rat interpeduncular nucleus in vitro. J. Physiol., 341: 655–670.

Brown, D.A., Docherty, R.J. and Halliwell, J.V. (1984) The action of cholinomimetic substances on impulse conduction in the habenulointerpeduncular pathway of the rat in vitro. J. Physiol., 353: 101–109.

Chesselet, M.F. (1984) Presynaptic regulation of neurotransmitter release in the brain: facts and a hypothesis. Neuroscience, 12: 347–375.

Clarke, P.B.S. (1986) Radioligand labelling of nicotinic receptors in mammalian brain. In A. Maelicke (Ed.), Mechanism of Action of the Nicotinic Acetylcholine Receptor, NATO ASI Series, Springer-Verlag, Berlin, pp. 345–358.

Clarke, P.B.S. (1987a) Recent progress in identifying nicotinic cholinoceptors in mammalian brain. *Trends Pharmacol. Sci.,* 8: 32 – 35.

Clarke, P.B.S. (1987b) Nicotine and smoking: a perspective from animal studies. *Psychopharmacology,* 92: 135 – 143.

Clarke, P.B.S. (1988) Autoradiographic mapping of putative nicotinic cholinoceptors in mammalian brain. In F. Clementi (Ed.), *Nicotinic Acetylcholine Receptors in the Nervous System,* NATO ASI Series, Springer-Verlag, Berlin, pp. 1 – 14.

Clarke, P.B.S. and Pert, A. (1985) Autoradiographic evidence for nicotine receptors on nigrostriatal and mesolimbic dopaminergic neurons. *Brain Res.,* 348: 355 – 358.

Clarke, P.B.S., Pert, C.B. and Pert, A. (1984) Autoradiographic distribution of nicotine receptors in rat brain. *Brain Res.,* 323: 390 – 395.

Clarke, P.B.S., Schwartz, R.D., Paul, S.M., Pert, C.B. and Pert, A. (1985a) Nicotinic binding in rat brain: autoradiographic comparison of ^{3}H-acetylcholine, ^{3}H-nicotine and ^{125}I-alpha-bungarotoxin. *J. Neurosci.,* 5: 1307 – 1315.

Clarke, P.B.S., Hommer, D.W., Pert, A. and Skirboll, L.R. (1985b) Electrophysiological actions of nicotine and substantia nigra single units. *Br. J. Pharmacol.,* 85: 827 – 835.

Clarke, P.B.S., Hamill, G.S., Nadi, N.S., Jacobowitz, D.M. and Pert, A. (1986) ^{3}H-nicotine and ^{125}I-alpha-bungarotoxin labelled receptors in the interpeduncular nucleus of rats. II. Effects of habenular deafferentation. *J. Comp. Neurol.,* 251: 407 – 413.

Clarke, P.B.S., Hommer, D.W., Pert, A. and Skirboll, L.R. (1987) Innervation of substantia nigra neurons by cholinergic afferents from the pedunculopontine nucleus in rats: neuroanatomical and electrophysiological evidence. *Neuroscience,* 23: 1011 – 1020.

Clarke, P.B.S., Fu, D.W., Jakubovic, A. and Fibiger, H.C. (1988) Evidence that mesolimbic dopaminergic activation underlies the locomotor stimulant action of nicotine in rats. *J. Pharmacol. Exp. Ther.,* 246: 701 – 709

Costa, L.G. and Murphy, S.D. (1983) ^{3}H-nicotine binding in rat brain: alteration after chronic acetylcholinesterase inhibition. *J. Pharmacol. Exp. Ther.,* 226: 392 – 397.

Egan, T.M. and North, R.A. (1986) Actions of acetylcholine and nicotine on rat locus coeruleus neurons in vitro. *Neuroscience,* 19: 565 – 571.

Fibiger, H.C. and Phillips, A.G. (1987) Role of catecholamine transmitters in brain reward systems: implications for the neurobiology of affect. In J. Engel and L. Oreland (Eds.), *Brain Reward Systems and Abuse,* Raven Press, New York, pp. 61 – 74.

Flynn, D.D. and Mash, D.C. (1986) Characterization of L-[^{3}H]nicotine binding in human cerebral cortex: comparison between Alzheimer's disease and the normal. *J. Neurochem.,* 47: 1948 – 1954.

Friedman, D.P., Clarke, P.B.S., O'Neill, J.B. and Pert, A. (1985) Distributions of nicotinic and muscarinic cholinergic receptors in monkey thalamus. *Soc. Neurosci. Abstr.,* 11: 307.3.

Friedman, D.P., Levey, A.E., O'Neill, J.B., Clarke, P.B.S., Price, D.L. and Kitt, C.A. (1988) Comparative laminar distributions of cholinergic fibres and receptors in monkey premotor cortex. *Soc. Neurosci. Abstr.,* 14: 257.7.

Grunwald, F., Schrock, H. and Kuschinsky, W. (1987) The effect of an acute nicotine infusion on the local cerebral glucose utilization of the awake rat. *Brain Res.,* 400: 232 – 238.

Imperato, A., Mulas, A. and di Chiara, G. (1986) Nicotine preferentially stimulates dopamine release in the limbic system of freely moving rats. *Eur. J. Pharmacol.,* 132: 337 – 338.

Larsson, C. and Nordberg, A. (1985) Comparative analysis of nicotine-like receptor-ligand interactions in rodent brain homogenate. *J. Neurochem.,* 45: 24 – 31.

Lippiello, P.M. and Fernandes, K.G. (1986) The binding of L-[^{3}H]nicotine to a single class of high affinity sites in rat brain membranes. *Mol. Pharmacol.,* 29: 448 – 454.

London, E.D., Connolly, R.J., Szikszay, M. and Wood, J.K. (1985a) Distribution of cerebral metabolic effects of nicotine in the rat. *Eur. J. Pharmacol.,* 110: 391 – 392.

London, E.D., Waller, S.B. and Wamsley, J.K. (1985b) Autoradiographic localization of [^{3}H]nicotine binding sites in the rat brain. *Neurosci. Lett.,* 53: 179 – 184.

Marks, M.J. and Collins, A.C. (1982) Characterization of nicotine binding in mouse brain and comparison with the binding of alpha-bungarotoxin and quinuclidinyl benzylate. *Mol. Pharmacol.,* 22: 554 – 564.

Marks, M.J., Stitzel, J.A., Romm, E. Wehner, J.M. and Collins, A.C. (1986) Nicotinic binding sites in rat and mouse brain: comparison of acetylcholine, nicotine and alpha-bungarotoxin. *Mol. Pharmacol.,* 30: 427 – 436.

Martino-Barrows, A.M. and Kellar, K.J. (1987) [^{3}H]Acetylcholine and [^{3}H](–)nicotine label the same recognition site in rat brain. *Mol. Pharmacol.,* 31: 169 – 174.

McCormick, D.A. and Prince, D.A. (1987) Acetylcholine causes rapid nicotinic excitation in the medial habenular nucleus of guinea pig, in vitro. *J. Neurosci.,* 7: 742 – 752.

O'Neil, J.B., Clarke, P.B.S., Friedmen, D.P. and Pert, A. (1985) Distributions of nicotinic and muscarinic receptors in monkey cerebral cortex. *Soc. Neurosci. Abstr.,* 11: 307.4.

Prusky, G.T., Shaw, C. and Cynader, M.S. (1987) Nicotine receptors are located on lateral geniculate nucleus terminals in cat visual cortex. *Brain Res.,* 412: 131 – 138.

Romano, C. and Goldstein, A. (1980) Stereospecific nicotine receptors on rat brain membranes. *Science,* 210: 647 – 649.

Rowell, P.P. and Winkler, D.L. (1984) Nicotinic stimulation of [^{3}H]acetylcholine release from mouse cerebral cortical synaptosomes. *J. Neurochem.,* 43: 1593 – 1598.

Rowell, P.P., Carr, L.A. and Garner, A.C. (1987) Stimulation of [^{3}H]dopamine release by nicotine in rat nucleus accumbens. *J. Neurochem.,* 49: 1449 – 1454.

Schwartz, R.D. (1986) Autoradiographic distribution of high affinity muscarinic and nicotinic cholinergic receptors labeled with [^{3}H]acetylcholine in rat brain. *Life Sci.,* 38: 2111 – 2119.

Schwartz, R.D. and Kellar, K.J. (1983) Nicotinic cholinergic receptor binding sites in brain: in vivo regulation. *Science,* 220: 214 – 216.

Schwartz, R.D., McGee, R. and Kellar, K.J. (1982) Nicotinic cholinergic receptors labeled by [^{3}H]acetylcholine in rat brain. *Mol. Pharmacol.,* 22: 56 – 62.

Schwartz, R.D., Lehmann, J. and Kellar, K.J. (1984) Presynaptic nicotinic cholinergic receptors labeled by [^{3}H]acetylcholine in rat brain. *Mol. Pharmacol.,* 22: 56 – 62.

Sloan, J.W., Todd, G.D. and Martin, W.R. (1984) Nature of nicotine binding to rat brain P$_2$ fraction. *Pharmacol.*

Biochem. Behav., 20: 899 – 909.

Swanson, L.W., Simmons, D.M, Whiting, P.J. and Lindstrom, J.M. (1987) Immunohistochemical localization of neuronal nicotinic receptors in the rodent central nervous system. *J. Neurosci.,* 7: 3334 – 3342.

Wonnacott, S. (1986) Alpha-bungarotoxin binds to low-affinity nicotine binding sites in rat brain. *J. Neurochem.,* 47: 1706 – 1712.

Yamada, S., Isogai, M., Kagawa, Y., Takayanagi, N., Hayashi, E., Tsuji, K. and Kosuge, T. (1985) Brain nicotinic acetylcholine receptors: biochemical characterization by neosurugatoxin. *Mol. Pharmacol.,* 28: 120 – 127.

Yoon, K.-W.P., Gessa, G.L., Boi, V., Naes, L., Mereu, G. and Westfall, T.C. (1986) Electrophysiological effects of nicotine on dopamine neurons. *Soc. Neurosci. Abstr.,* 12: 411.9.

A. Nordberg, K. Fuxe, B. Holmstedt and A. Sundwall (Eds.)
Progress in Brain Research, Vol. 79

CHAPTER 6

Single cell studies of the actions of agonists and antagonists on nicotinic receptors of the central nervous system

Terrance M. Egan

Department of Biochemistry, Brandeis University, Waltham, MA 02254-9110, U.S.A.

Introduction

Nicotine is one of the most widely used drugs in the world. Recent findings of a study initiated by the Office of the Surgeon General of the United States suggests that nicotine is addictive, and that the effects of cessation of nicotine intake in long-time users reflects both psychological and physiological components. While it is certain that the psychological component results from an action on the central nervous system (CNS), it is also likely that some of the desired physiological effects of this drug (e.g. the nicotine "high") stem from its CNS actions. This chapter, and the accompanying chapter by Westfall, reviews our understanding of the actions of nicotine on single cells of the CNS obtained from electrophysiological studies.

Nicotine acts after combining with membrane-bound receptors on nerve and muscle. These "nicotinic" receptors translate the chemical signal of the drug into an electrical impulse which spreads along defined pathways in the central and peripheral nervous systems to influence nerve and muscle activity. Nicotine is not, however, the natural ligand of the nicotinic receptor. Rather, the neurotransmitter acetylcholine (ACh) binds to nicotinic receptors and causes rapid excitations of effector tissues. Besides being the natural agonist, ACh is also often the preferred drug of choice in

the electrophysiological study of nicotinic receptors because it is rapidly inactivated by naturally occuring cholinesterases and therefore acts at the receptor for only a short period of time. Nicotine is not rapidly degraded, and its continued presence at the receptor results in receptor desensitization.

ACh is not a selective nicotinic agonist. It also activates cholinergic muscarinic receptors, and it is important that the effects resulting from activation of muscarinic receptors are clearly differentiated from those caused by nicotinic receptors. One way to separate the effects is on the basis of their time course following a nearly instantaneous application of ACh. Nicotinic effects are quick, having a fast onset of peak action and a rapid decay. For example, nicotinic fast excitatory postsynaptic potentials lasting a few milliseconds are recorded in skeletal muscle after stimulation of connecting motoneurons. When using less physiological methods to administer drugs, the exact time course of the nicotinic effect depends on the method used to apply ACh (e.g. application by iontophoresis or pressure ejection versus bath or systemic administration), making this variable a less reliable indicator of the nature of the receptor. Muscarinic effects are normally preceded by a delay between the moment of drug application and the onset of action, and the effect often lasts for seconds or even minutes.

The best way to classify the receptors is to test

the ability of selective nicotine and/or muscarinic antagonists to block the ACh effect. An action of ACh which is blocked by a reasonable concentration of a nicotinic antagonist can be assumed to be caused by activation of nicotinic receptors. The concentration of antagonist used to block the effect should be less than or equal to the lowest concentration of the drug shown in binding studies to occupy one-half of the total receptor population. At higher concentrations, many antagonists lose their specificity and will block many different kinds of receptors. An additional consideration when using nicotinic antagonists is that many of these drugs do not act at the receptor at all. Rather, they prevent the actions of ACh and nicotine by blocking the ionic channel activated by the receptor (see McCarthy et al., 1986). It is therefore wise to also look at the effects of putative nicotinic agonists. While agonists are in general less desirable classifiers of receptor identity than antagonists, there are some reasonably specific nicotinic agonists, including nicotine itself.

From both pharmacological and biochemical studies, it is apparent that ACh binds and activates subtypes of nicotinic receptors. Gross division splits the receptors into two classes. The first class is typical of neuromuscular junctions (n.m.j.). These receptors are blocked by a wide range of nicotinic antagonists including hexamethonium, d-tubocurarine, dihydro-β-erythroidine and α-bungarotoxin. The second class is typical of peripheral ganglia, where they transduce the signal between nerves of the peripheral nervous system (PNS). These receptors are also blocked by hexamethonium and dihydro-β-erythroidine, and by mecamylamine and certain snake toxins including toxin F (Lipton et al., 1987; Loring and Zigmond, 1988), but are not blocked by α-bungarotoxin. The exact identity (n.m.j.-like versus ganglion-like) of the nicotinic receptor of central neurons is unknown. Recent evidence suggests that central nicotinic receptors closely resemble their PNS counterparts in antagonist specificity (Egan and North, 1986; McCormick and Prince, 1987; Lipton et al., 1988). However, brain neurons do bind the n.m.j.-

specific nicotinic receptor antagonist α-bungarotoxin; it has yet to be determined whether these binding sites represent physiologically active receptors.

The CNS is obviously an important region in which to investigate the effects of nicotine for the reasons mentioned above. Unfortunately, much less is known about the actions of nicotine on single cells of the CNS than those of the PNS. This is because there is no good central preparation of neurons in which to study these effects. The ideal preparation for the study of single cell electrophysiology is one that presents a group of clean, accessible and mature neurons having a normal pattern of synaptic connectivity. The cells must have clean and smooth membranes if single nicotinic channels are to be studied using low resistance electrodes to clamp a patch of membrane. It must be possible to apply known concentrations of drugs to the cell under investigation to avoid the problem of non-specific actions of high drug doses. It is desirable that synaptic connections remain intact so that the effect of drugs at the synapse can be investigated. While some peripheral preparations do fulfill these requirements, to date no central preparation even comes close. In situ preparations suffer from accessibility problems, whereas in vitro preparations disrupt synaptic inputs. Cultures of neonate central neurons may present a viable alternative although they may not reliably represent the functional state of mature adult cells.

The focus of this chapter is the study of the pharmacology of nicotinic receptors of single CNS cells using electrophysiological means. Both in vivo and in vitro preparations are discussed.

Nicotine excites and inhibits central neurons

One of the earliest ways in which the actions of cholinergic agents on central neurons were studied was to record the change in electrical activity of a single in situ neuron caused by either systemic administration of nicotine, or by local application of ACh. The greatest advantage of the in situ extracellular method is that it is less invasive than the in vitro "brain slice" method described below. The normal pattern of connections between cells is left relatively unperturbed, permitting presumed cholinergic synaptic responses to be investigated, and thus it is the most useful technique for mapping the cholinoreceptive areas of the CNS. The method tells us that certain areas do or do not respond to a nicotinic agonist, and then whether or not this effect is blocked by nicotinic and/or muscarinic antagonists. It also delineates direct central actions of systemically applied nicotine from actions secondary to a peripheral effect (Clarke et al., 1985). However, it has the disadvantage of telling us little about the underlying ionic mechanism of the response. Also, in cases where drugs are applied by iontophoresis, the investigator does not have precise control over the final concentration of applied drug, making a firm pharmacological determination of receptor properties difficult. This problem is overcome if a suitable in vitro preparation is obtained.

The typical effect of ACh recorded with extracellular electrodes in situ or in vitro is to increase firing rate (or to increase the susceptibility to excitation by glutamate) (Table I). Investigators have attempted to separate these excitations into muscarinic and/or nicotinic receptor-mediated responses on the basis of response time and susceptibility to antagonists. In most areas of the brain (e.g. hippocampus, cerebral cortex, cerebellum and brain stem), the excitation by ACh is slow in onset and outlasts the period of drug application by several seconds. These slow excitations are mimicked by application of muscarinic, but not nicotinic, agonists, and are blocked by low concentrations (suggested by the low currents used to eject the drugs from the iontophoretic pipettes) of muscarinic antagonists. Nicotinic antagonists, even when applied at high ejection currents, typically have little or no effect. The conclusion therefore is that the predominant action of ACh in the brain is a muscarinic excitation. However, fast, and therefore presumed nicotinic, excitations have been recorded in the spinal cord (Curtis and Ec-

cles, 1958; Ryall, 1983), thalamus and geniculate (Phillis et al., 1967; Tebecis, 1970; McCormick and Prince, 1987), locus coeruleus (Egan and North, 1986) and cerebellum (de la Garza et al., 1987). These fast excitations are blocked largely by nicotinic antagonists. Regions showing rapid responses to ACh also respond to nicotinic agonists such as nicotine and cytisine.

In some regions of the CNS, ACh also inhibits the firing of a percentage of cells (see Table I), but the number of cells excited usually exceeds the number inhibited. One exception is the nucleus reticularis thalami were ACh has been reported to produce only inhibition (Ben-Ari et al., 1976a, b). Duggan and Hall (1975) have suggested that inhibition of central neurons by nicotinic antagonists may result from activation of inhibitory interneurons. An alternative explanation is offered by McCormick and Prince (1987); nicotinic inhibition may be secondary to excitatory effects of nicotine on cell action potential discharge (see section on Thalamus).

Currently, the most useful in vitro preparation is the "brain slice" (Dingledine, 1984). Here a thin ($300 - 600 \ \mu m$) slice of brain containing the area of interest can be studied under circumstances that allow precise application of drugs. For example, in many cases, the brain slice is completely submerged in a flowing superfusate to which can be added known concentrations of drugs. Cells within brain slices are also more amenable to voltage clamp using single electrodes and time-sharing amplifiers thus allowing nicotinic currents to be investigated.

From in vitro preparations, we known how ACh and nicotine excite muscle. The increase in cell excitability seen after nicotinic receptor activation results from a membrane depolarization (Takeuchi and Takeuchi, 1960). ACh (and nicotine) open a membrane ionic channel which is part of a receptor-channel protein complex. When open, the channel allows cations (mainly sodium and potassium ions) to enter the cell (Colquhoun and Sakmann, 1981, 1985), thus reducing the membrane potential and rendering the cell hyperexcitable. Active nicotinic channels similar to those

TABLE I

Single cell studies of central nicotinic receptors

CNS level	Cell type or location	Method	Nicotinic agonist effect	Effective nicotinic antagonists	α-BGT test	Presumed cholinergic pathway or cells of origin of ACh-containing fibers	References
Spinal cord	Renshaw cell	In situ (ext, int)	Excite	DH-β-E TEA, atropine mecamylamine	Yes	Recurrent collateral branches of motoraxons	Curtis and Eccles, 1958; Curtis and Ryall, 1966a, b; Duggan et al., 1976; Ryall, 1983
	Motoneurons	In vitro (int)	Excite	d-TC Hexamethonium	No	CAT-containing, small diameter neurons of the ventral horn	Jiang and Dun, 1987
	Clarke's column	In situ	Excite		No		Mysilinski and Randic, 1977
Cerebellum	Purkinje cells (A)	In situ (ext)	Excite	DH-β-E	No		Crawford et al., 1963; McCance and Phillis, 1964
	Purkinje cells (B)	In situ (ext)	Inhibit	Hexamethonium	No		De la Garza et al., 1987
	Interneurons	In situ (ext)	Excite	d-TC	No		De la Garza et al., 1987
Medulla	Medullary respiratory neurons	In situ (ext)	Excite, inhibit	DH-β-E Hexamethonium	No		Bradley and Lucy, 1983; Bohmer et al., 1987; Salmoiraghi and Steiner, 1963
	Unidentified	Culture (SC)	Excite		No		Arcava et al., 1987
Pons	Locus coeruleus	In situ (ext) In vitro (int)	Excite	DH-β-E Hexamethonium	Yes		Egan and North, 1985, 1986; Engberg and Svensson, 1980; Svensson and Engberg, 1980
Mesencephalon	Substantia nigra	In situ (ext)	Excite	DH-β-E			Clarke et al., 1985; Lichtensteiger et al., 1982
	Interpeduncular nucleus	In vitro (ext)	Excite	Hexamethonium d-TC Mecamylamine	Yes	Habenulo-interpeduncular tract	Brown et al., 1983, 1984

Region	Structure	Preparation	Effect	Antagonist	α-BGT Test	Comments	References
Diencephalon	Medial geniculate	In situ (ext) In vitro (int)	Excite		No	Pebunculopontine cells lateral dorsal tegmentar nuclei	Tebecis, 1970; McCormick and Prince, 1987a
	Lateral geniculate	In situ (ext) In vitro (int)	Excite		No	Pedunculopontine cells lateral dorsal tegmentar nuclei	McCormick and Prince, 1987a
	Medial habenular	In vitro (ext, int)	Excite Inhibit	Hexamethonium	No	Postcommissural septum	McCormick and Prince, 1987b
	Supraoptic nucleus	In situ (ext) In vitro (ext, int)	Excite	d-TC Hexamethonium Mecamylamine.	Yes		Bourque and Brown, 1987; Dreifuss and Kelly, 1972; Hatton et al., 1983
	Nucleus ventralis basalis thalami	In situ (ext)		amine	No	From neurons in the reticular formation	Anderson and Curtis, 1964
	Center Median nucleus	In situ (ext)	Inhibit	DH-β-E	No		Duggan and Hall, 1975
	Nucleus reticularis thalami	In situ (ext)	Inhibit	DH-β-E	No		Ben-Ari et al., 1976a,b
	Globus pallidus	In situ (ext)	Excite	Hexamethonium	No		Lamour et al., 1986
Rhinen-cepha-lon	Hippocampal pyramidal	In situ (ext) In vitro (sc)	Excite (?)		No		Arcava et al., 1987; Bird and Aghajanian, 1976; Cole and Nicoll, 1984; Rovira et al., 1978; Lipton et al., 1987
Retina	Retinal ganglion cells	In vitro (sc)	Excite	Toxin-F			

ext = Studies using extracellular electrodes; int = studies using intracellular electrodes; sc = studies of nicotinic single channels; α-BGT = alpha-bungarotoxin; DH-β-E = dihyro-β-erythroidine; d-TC = d-tubocurarine.

A "yes" or "no" entry under the heading α-BGT Test indicates whether or not α-BGT was tested. In all cases where it was tested, α-BGT had no effect.

just described have recently also been identified in neuronal preparations of the CNS (Arcava et al., 1987; Lipton et al., 1987). In the study of Arcava et al. (1987), central nicotinic channels of cultured embryonic rat brain neurons resembled their peripheral counterparts in mean channel conductance, mean open time, presence of substates, and tendency to open in bursts (e.g. multiple closed states) (Arcava et al., 1987). They differed in requiring 5 – 10 times larger agonist concentrations to activate the channel. The predominant substate had a slope conductance of about 20 pS, and therefore more closely resembled nicotinic channels of embryonic or denervated skeletal muscle preparations than those of adult tissues.

Pharmacology of nicotinic receptors at specific sites in CNS

Spinal cord. The actions of ACh on neurons of the spinal cord have been reviewed by Ryall (1983) and Jiang and Dun (1987). ACh is released from collateral branches of spinal motorneurons and excites Renshaw cells (Eccles et al., 1954, 1956). In addition, motorneurons themselves may be excited by ACh released from cholinergic interneurons (Jiang and Dun, 1987).

As noted by R.W. Ryall (1983), the best identified cholinergic synapse in the CNS is that between motor axon collaterals and the spinal interneurons known as Renshaw cells. ACh, applied either systemically or directly by iontophoresis, excites Renshaw cells of cats (Curtis and Ryall, 1966a, b) and rats (McLennan and Hicks, 1978). This action is mimicked by application of both muscarinic and nicotinic agonists, nicotine being a more potent agonist than ACh, and both muscarinic and nicotinic antagonists block a portion of the ACh response. In addition, the agonist acetyl-β-methylcholine produces a biphasic response in which the early phase is reduced by nicotinic antagonists and the later phase by muscarinic antagonists (Curtis and Ryall, 1966b; Ryall, 1983). These experiments point to the existence on Renshaw cells of both muscarinic and nicotinic receptors. The effects of nicotine are blocked by dihydro-β-erythroidine but not by α-bungarotoxin, and thus the receptors resemble ganglionic nicotinic receptors in their sensitivity to antagonists (Eccles et al., 1956; Duggan et al., 1976). However, based on the finding that the "selective" muscarinic antagonist atropine, administered in presumed small doses, is as effective an antagonist of nicotine as it is of muscarinic agonists (King and Ryall, 1979, 1981), Ryall (1983) has suggested that the nicotinic receptors on Renshaw cells belong in neither the ganglia-like or n.m.j.-like classes, but constitute a unique class of their own. The increase in Renshaw cell firing rate caused by ACh results from a membrane depolarization, which itself results from an increase in membrane ionic conductance (Zieglgansberger and Reiter, 1974).

Ventral horn motoneurons in spinal cord slices are also depolarized by ACh (Jiang and Dun, 1987). The depolarization is biphasic and has a rapid phase which is blocked by the nicotinic antagonists hexamethonium and d-tubocurarine, and a late phase which is blocked by muscarinic antagonists.

Locus coeruleus. This nucleus has been extensively studied both in situ using extracellular electrodes and in the in vitro brain slice preparation using extracellular and intracellular electrodes (Williams et al., 1984). A quick, brief application of ACh by ejection from a pressure-loaded micropipette to neurons of the in vitro locus coeruleus results in about 50% of the cases in a biphasic depolarization which has both a rapidly rising, quickly decaying component (lasting 1 – 2 s), and a slower, longer lasting component (lasting up to 60 s) (see Fig. 1 of Egan and North, 1986). The initial rapid component is blocked by the nicotinic channel blocker hexamethonium but is unaffected by muscarinic antagonists such as atropine or scopolamine. The slower component is also somewhat reduced by application of hexamethonium and other nicotinic antagonists. The portion of the slow phase which remains after nicotinic receptor antagonism is blocked by muscarinic antagonists

(Egan, 1984; Egan and North, 1985). The fact that nicotinic antagonists block a portion of the slow component may at first seem surprising, and may reflect a non-specific action of hexamethonium on muscarinic receptors. However, in our hands, hexamethonium (100 – 400 μM) did not block fast depolarizations caused by pressure-ejected glutamate, nor did it block slow depolarizations caused by muscarine, arguing against non-selective receptor antagonism. Rather, we reasoned that a portion of the slow ACh-excitation probably results from the continued presence of the drug in the vicinity of nicotinic receptors, an artifact of the pressure-ejection method of applying drug (Egan and North, 1986). That is to say that the pressure-ejection method cannot realistically be expected to reliably reproduce the rapid time course of cholinergic synaptic events. The pressure-ejection method does somewhat mimic the quick time-to-peak-response of nicotinic postsynaptic potentials because it produces a rapid drug application as a function of the high-ejection pressure. However, unlike synaptic release, where a high local concentration of ACh is rapidly degraded by acetylcholinesterase, the ejection method results in application of a high concentration of drug over a diffuse area, and the fade of the response is likely to be determined both by the action of cholinesterases (see Egan and North, 1986) and by a washout determined by the flow rate of the superfusate. Slow washout is apparently a significant determinant of the time course of action of ACh in the brain slice preparation and it seems to be responsible in part for the prolonged nicotinic action.

Under voltage-clamp at the resting membrane potential (about -55 mV), ACh causes a biphasic inward current to flow which has the same time course as the biphasic depolarization seen under current clamp (Egan and North, 1986). The fast, nicotinic portion of this inward current becomes larger in amplitude when the cell is held at potentials more negative than rest, and smaller at potentials positive to rest. The estimated reversal potential is about -16 mV. This reversal potential suggests that, as in the PNS, activation of nicotinic receptors in the CNS results in the flow of inward current through a channel which is relatively cation non-selective.

The ability of different types of nicotinic receptor antagonists to block that portion of the action of ACh remaining after elimination of muscarinic receptors (by preincubation with high doses of scopolamine (up to 1 μM)) was tested to determine whether nicotinic receptors on locus coeruleus neurons resemble ganglionic nicotinic receptors or those mediating cholinergic transmission at the neuromuscular junction (Egan and North, 1986). Dihydro-β-erythroidine, which blocks cholinergic nerve-nerve transmission in the PNS but not at the neuromuscular junction, reduces the nicotinic effect of ACh in the locus coeruleus by up to 70%. The neuromuscular nicotinic antagonist α-bungarotoxin has no effect on locus coeruleus neurons. These experiments suggest that receptors in the locus coeruleus more closely resemble nicotinic receptors of ganglion cells than those of the neuromuscular junction.

Superfusion of nicotine (1 – 100 μM) mimics the nicotinic effects of ACh (Egan and North, 1986). This is a direct effect of nicotine; indirect effects mediated through the PNS have also been reported (Svensson, this volume). It causes a membrane depolarization which is associated with an increased membrane conductance. The depolarization results from an inward current having a reversal potential identical to the inward current caused by ACh. The effect of nicotine is difficult to study as the cells show a marked desensitization to the action of this drug. A 1 min superfusion of 1 μM nicotine is often enough to cause a complete desensitization of the cell to this drug for up to 1 h. In addition, cells no longer able to respond to nicotine are also rendered insensitive to the nicotinic effects of ACh. While it is clear that nicotine desensitized these cells, it is not clear that the long time course of the desensitization is due to continued desensitization in the complete absence of agonist. As nicotine is very lipophilic, it may be that it takes a very long time to completely wash the nicotine out of the brain slice preparation.

Field stimulation of the area of the slice surrounding the locus coeruleus produces both a fast excitatory and a slow inhibitory postsynaptic potential in locus coeruleus neurons (Williams et al., 1984). The inhibitory postsynaptic potential results from release of noradrenaline (Egan et al., 1983). The transmitter responsible for the fast excitatory postsynaptic potential has not been determined, but it is unlikely to be ACh as hexamethonium does not affect it (Cherubini, Egan and Williams, unpublished observation).

Hippocampus. The hippocampus receives extensive cholinergic input from neurons originating in the septal nuclei (Swanson, 1978), and it therefore is a favorite target for both in situ and in vitro investigations of central cholinergic actions. Iontophoretic ACh excites hippocampal pyramidal cells in situ (Herz and Nacimiento, 1965). Muscarinic, but not nicotinic, agonists also excite these cells (but see Rovira et al., 1983), although both muscarinic and nicotinic antagonists block ACh-induced excitations (Bird and Aghajanian, 1976). However, in this same study, cholinergic antagonists also reduced amino acid-induced excitations, and the authors concluded that the block of effect of ACh by the nicotinic antagonist hexamethonium was mostly likely non-specific (Bird and Aghajanian, 1976). Other investigators have also failed to find a nicotinic component of the action of ACh (Cole and Nicoll, 1984). It is possible that the pharmacology of the ACh receptor population depends on the architecture of the pyramidal neuron. For example, Rovira et al. (1983) found different muscarinic/nicotinic effects of ACh when this drug was applied to the apical dendrites as opposed to the soma. Likewise, Arcava et al. (1987) found in cultured neurons that nicotinic receptor-channel complexes were concentrated in the region of the apical dendrite.

Interpeduncular nucleus. This nucleus receives an extensive cholinergic innervation from fibers travelling via the fasciculi retroflexi of Meynert and originating in the habenular nuclei and the septum. Cells within the nucleus have a high concentration of α-bungarotoxin binding sites, and are excited by ACh (Lake, 1973; Sastry, 1978; Ogato, 1979; Brown et al., 1983, 1984). Likewise, stimulation of the fasciculi retroflexus results in interpeduncular neuron excitation (Brown and Halliwell, 1981). The excitation by iontophoretically applied ACh is blocked by hexamethonium, but not by α-bungarotoxin (Brown et al., 1983). Interestingly, no nicotinic antagonist blocks the excitation caused by stimulation of the fiber tract, suggesting that while cells within the nucleus are cholinoceptive, they do not receive an active cholinergic input from afferents running in the fasciculi retroflexus (Brown et al., 1983). The excitation caused by stimulation of this tract is reduced by some amino acid antagonists.

The work of Brown et al. has also shown that ACh has a large presynaptic effect on fibers terminating in the interpeduncular nucleus. This idea was originally suggested by the finding that hexamethonium sometimes increased the frequency of discharge of interpeduncular neurons (Brown et al., 1983). Subsequent work showed that application of nicotinic agonists depressed the peak height of the compound action potential evoked by stimulation of afferent fibers in the fasciculi retroflexus (Brown et al., 1984) while increasing the latency to peak. The concentrations of agonists needed to produce these presynaptic effects were much lower than those needed to directly excite postsynaptic cells. These actions were blocked by hexamethonium, d-tubocurarine and mecamylamine, but α-bungarotoxin was without effect, keeping with the pattern of ganglion-like over n.m.j.-like nicotinic receptors on central neurons.

Thalamus. A very complete study of the effect of nicotinic agonists on the medial habenular nucleus contained in a thalamic slice has recently been reported by McCormick and Prince (1987). This nucleus receives a large presumed cholinergic input from the postcommissural septum, and itself sends cholinergic efferents to the interpeduncular nucleus (see above). Application of ACh to these neurons results in a rapid increase in action potential discharge followed by a longer-lasting decrease. Both the excitation and inhibition persist in the absence of synaptic input showing that the effects of ACh results from direct postsynaptic actions of this drug. The effect of ACh is mimicked by nicotinic agonists including nicotine, 1,1-dimethyl-4-phenylpiperazinium (DMPP) and cytisine. Muscarinic agonists have little or no effect. The relative ability to cause excitation/inhibition depends on the agonist; nicotine and cytisine cause large excitations but little inhibition. Both nicotinic agonist-induced excitations and inhibitions are blocked by hexamethonium.

The ionic mechanism of the dual response was studied with intracellular electrodes. A brief application of ACh causes a membrane depolarization followed by a membrane hyperpolarization. The depolarization is associated with an increased membrane conductance, and has an extrapolated reversal potential of -16 mV. The hyperpolarization is also associated with a rise in conductance, and has an estimated reversal potential (about -80 mV) consistent with the idea that it is caused by opening of potassium channels (see McCormick and Prince, 1987 for further discussion of the identity of the ionic current). The inhibition always follows the excitation, and is never seen in the absence of a preceding excitation. This finding suggests that the inhibition (e.g. hyperpolarization) results from the ionic changes taking place during the excitation (e.g. depolarization with its associated increase in action potential discharge). One possibility suggested by the authors is that calcium ion, which may enter the cell during the excitation, causes the activation of a calcium-dependent potassium conductance which hyperpolarizes the membrane.

Conclusions

There is ample evidence that ACh is a neurotransmitter at nicotine receptors of central neurons. However, exact characterization of these central nicotinic receptors has been hampered by the lack of a suitable preparation in which to study their pharmacology and biophysics. In situ studies using extracellular electrodes to monitor changes in discharge rate have been used to map the cholinoceptive areas of the brain. They have also shown that nicotinic agonists are capable of both exciting and inhibiting neurons in some brain areas. The excitation is a direct postsynaptic action of ACh; the inhibition may be direct, but most likely results either from indirect inhibition through synaptic connections, or as a result of activation of a membrane conductance as the result of the preceding excitation (for example, increase in a calcium-activated potassium conductance).

The in vitro brain slice preparation has allowed a more precise characterization of the receptor and currents underlying the nicotinic effects. In all areas of the CNS studied so far, central nicotinic receptors have been found to more closely resemble those of peripheral autonomic ganglia than those subserving nicotinic transmission at the n.m.j. Specifically, α-bungarotoxin, a neurotoxin which blocks nicotinic signal transmission between nerve and muscle, but not between peripheral nerves, also fails to block the effects of ACh and nicotine on central neurons. The ionic nature of the nicotinic effect resembles that of both peripheral nerve and muscle; ACh opens a membrane cation-selective channel which allows positive charge to enter the cell-producing depolarization.

What is needed now is a suitable preparation of adult central neurons amendable to patch clamp methods, which can be used to investigate the effects of nicotine at the single channel level. Alternatively, it is possible that molecular cloning techniques may in the near future supply a functional central nicotinic receptor expressed in an electrophysiological-accessible medium.

Acknowledgements

I thank Profs. R.A. North, D. Noble, H.D. Lux and I.B. Levitan for their support. Funded by N.I.H. grant 7F32 HL07269-03.

References

Anderson, P. and Curtis, D.R. (1964) The pharmacology of the synaptic and acetylcholine-induced excitations of ventrobasal thalamic neurons. *Acta. Physiol. Scand.,* 61: 100 – 120.

Aracava, Y., Deshpande, S.S., Swanson, K.L., Rapoport, H., Wonnacott, S., Lunt, G. and Albuquerque, E.X. (1987) Nicotinic acetylcholine receptors in cultured neurons from hippocampus and brain stem of the rat characterized by single channel recording. *FEBS Lett.,* 222: 63 – 70.

Ben-Ari, Y., Dingledine, R., Kanazawi, I. and Kelly, J.S. (1976a) Inhibitory effects of acetylcholine on neurones in the feline nucleus reticularis thalami. *J. Physiol. (Lond.),* 261: 647 – 671.

Ben-Ari, Y., Kanazawa, I. and Kelly, J.S. (1976b) Exclusively inhibitory action of iontophoretic acetylcholine on single neurons of feline thalamus. *Nature (Lond.),* 259: 327 – 330.

Bird, S.J. and Aghajanian, G.K. (1976) The cholinergic pharmacology of mammalian hippocampal pyramidal cells: a microiontophoretic study. *Neuropharmacology,* 15: 273 – 282.

Biscoe, T.J. and Straughan, D.W. (1966) Micro-electrophoretic studies of neurones in the cat hippocampus. *J. Physiol. (Lond.),* 183: 341 – 359.

Bloom, F.E., Costa, E. and Salmoiraghi, G.C. (1965) Anesthesia and the responsiveness of individual neurons of the caudate nucleus of the cat to acetylcholine, norepinephrine and dopamine. *J. Pharmacol. Exp. Ther.,* 150: 244 – 252.

Bohmer, G., Schmid, K., Schmidt, P. and Stehle, J. (1987) Cholinergic effects on spike-density and burst-duration of medullary respiration-related neurones in the rabbit: an iontophoretic study. *Neuropharmacology,* 26: 1561 – 1572.

Bourque, C.W. and Brown, D.A. (1987) Nicotinic receptor activation in rat supraoptic neurosecretory cells in vitro: effects of mecamylamine, hexamethonium and alpha-bungarotoxin. *Soc. Neurosci. Abstr.,* 13: 664.

Bradley, P.B. and Dray, A. (1972) Short-latency excitation of brain stem neurones in the rat by acetylcholine. *Br. J. Pharmacol.,* 45: 372 – 374.

Bradley, P.B. and Lucy, A.P. (1983) Cholinoceptive properties of respiratory neurones in the rat medulla. *Neuropharmacology,* 22: 853 – 858.

Bradley, P.B., Dhawan, B.N. and Wolstencroft, J.H. (1966) Pharmacological properties of cholinoceptive neurones in the medulla and pons of the cat. *J. Physiol. (Lond.),* 183: 665 – 674.

Brown, D.A. and Halliwell, J.V. (1981) An in vitro preparation of the diencephalic interpeduncular nucleus. In G.A. Kerkut and H.V. Wheal (Eds.), *Electrophysiology of Isolated CNS Preparations,* Academic Press, London, pp. 285 – 308.

Brown, D.A., Docherty, R.J. and Halliwell, J.V. (1983) Chemical transmission in the rat interpeduncular nucleus in vitro. *J. Physiol. (Lond.),* 341: 655 – 670.

Brown, D.A., Docherty, R.J. and Halliwell, J.V. (1984) The action of cholinomimetic substances on impulse conduction in the habenulointerpeduncular pathway of the rat in vitro. *J. Physiol. (Lond.),* 353: 101 – 109.

Clarke, P.B.S., Hommer, D.W., Pert, A. and Skirboll, L.R. (1985) Electrophysiological studies of nicotine on substantia nigra units. *Br. J. Pharmacol.,* 85: 827 – 836.

Cobbett, P., Mason, W.T. and Poulain, D.A. (1986) Intracellular control of supraoptic neurone (SON) activity in vitro by acetylcholine (ACh). *J. Physiol. (Lond.),* 371: 216P.

Cole, A. and Nicoll, R.A. (1984) The pharmacology of cholinergic excitatory responses in the hippocampal pyramidal cells. *Brain Res.,* 305: 283 – 290.

Colquhoun, D. and Sakmann, B. (1981) Fluctuations in the microsecond time range of the current through single acetylcholine receptor ion channels. *Nature (Lond.),* 294: 464 – 466.

Colquhoun, D. and Sakmann, B. (1983) Transmitter-activated bursts of openings. In B. Sakmann and E. Neher (Eds.), *Single-Channel Recording,* Plenum Press, New York, pp. 345 – 363.

Crawford, J.M., Curtis, D.R., Voorhoeve, P.E. and Wilson, V.J. (1963) Excitation of cerebellar neurones by acetylcholine. *Nature (Lond.),* 200: 579 – 580.

Curtis, D.R. and Eccles, R.M. (1958) The excitation of Renshaw cells by pharmacological agents applied electrophoretically. *J. Physiol. (Lond.),* 141: 433 – 445.

Curtis, D.R. and Ryall, R.W. (1966a) The excitation of Renshaw cells by cholinomimetics. *Exp. Brain Res.,* 2: 49 – 65.

Curtis, D.R. and Ryall, R.W. (1966b) The synaptic excitation of Renshaw cells. *Exp. Brain Res.,* 2: 81 – 96.

De la Garza, R., Bickford-Wimer, P.C., Hoffer, B.J. and Freedman, R. (1987) Heterogeneity of nicotine actions in the rat cerebellum: an in vivo electrophysiological study. *J. Pharmacol. Exp. Ther.,* 240: 689 – 695.

Dingledine, R. (1984) *Brain Slices,* Plenum Press, New York.

Dreifuss, J.J. and Kelly, J.S. (1972) The activity of identified supraoptic neurones and their response to acetylcholine by iontophoresis. *J. Physiol. (Lond.),* 220: 105 – 118.

Duggan, A.W. and Hall, J.G. (1975) Inhibition of thalamic neurones by acetylcholine. *Brain Res.,* 100: 445 – 449.

Duggan, A.W., Hall, J.G. and Lee, C.Y. (1976) Alpha bungarotoxin, cobra neurotoxin and excitation of Renshaw cells by acetylcholine. *Brain Res.,* 107: 166 – 170.

Eccles, J.C., Eccles, R.M. and Koketsu, K. (1954) Cholinergic and inhibitory synapses in a pathway from motor axon collaterals to motoneurons. *J. Physiol. (Lond.),* 126: 524 – 562.

Eccles, J.C., Eccles, R.M. and Fatt, P. (1956) Pharmacology of a central synapse operated by acetylcholine. *J. Physiol. (Lond.),* 131: 154 – 169.

Egan, T.M. (1984) The actions of acetylcholine and related drugs on neurons of the rat locus coeruleus. Ph.D. Thesis, M.I.T., Cambridge, MA.

Egan, T.M. and North, R.A. (1985) Acetylcholine acts on m2-

muscarinic receptors to excite rate locus coeruleus neurones. *Br. J. Pharmacol.,* 85: 733–735.

Egan, T.M. and North, R.A. (1986) Actions of acetylcholine and nicotine on rat locus coeruleus neurons in vitro. *Neuroscience,* 19: 565–571.

Egan, T.M., Henderson, G., North, R.A. and Williams, J.T. (1983) Noradrenaline-mediated synaptic inhibition in rat locus coeruleus neurones. *J. Physiol. (Lond.),* 345: 477–488.

Engberg, G. and Svensson, T.H. (1980) Pharmacological analysis of a cholinergic receptor-mediated regulation of brain norepinephrine neurons. *J. Neural Trans.,* 49: 137–150.

Guyenet, P.G. and Aghajanian, G.K. (1983) Acetylcholine, substance P and met-enkephalin in the locus coeruleus: pharmacological evidence for independent sites of action. *Eur. J. Pharmacol.,* 53: 319–328.

Hatton, G.I., Ho, Y.W. and Mason, W.T. (1983) Synaptic activation of phasic bursting in rat supraoptic nucleus neurones recorded in hypothalamic slices. *J. Physiol. (Lond.),* 345: 297–317.

Jiang, Z.G. and Dun, N.J. (1987) Actions of acetylcholine on spinal motoneurons. In N.J. Dun and R.L. Perlman (Eds.), *Neurobiology of Acetylcholine,* Plenum Press, New York, pp. 283–293.

King, K.T. and Ryall, R.W. (1981) A reevaluation of acetylcholine receptors on feline Renshaw cells. *Br. J. Pharmacol.,* 73: 455–460.

Lake, N. (1973) Studies of the habenulo-interpeduncular pathways in cats. *Expl. Neurol.,* 41: 113–132.

Lamour, Y., Dutar, P., Rascol, O. and Jobert, A. (1986) Basal forebrain neurons projecting to the rat frontoparietal cortex: electrophysiological and pharmacological properties. *Brain Res.,* 362: 122–131.

Lichtensteiger, W., Hefti, F., Felix, D., Huwyler, T., Melamed, E. and Schlumpf, M. (1982) Stimulation of nigrostriatal dopamine neurones by nicotine. *Neuropharmacology,* 21: 963–968.

Lipton, S.A., Aizerman, E. and Loring, R.H. (1987) Neural nicotinic acetylcholine responses in solitary mammalian retinal ganglia. *Pflugers Arch.,* 410: 37–43.

Loring, R.H. and Zigmond, R.E. (1988) Characterization of neuronal nicotinic receptors by snake venom neurotoxins. *Trends Neurosci.,* 11: 73–77.

McCance, I. and Phillis, J.W. (1964) Discharge patterns of elements in rat cerebellar cortex and their responses to iontophoretically applied drugs. *Nature (Lond.),* 204: 844–846.

McCarthy, M.P., Earnest, J.P., Young, E.F., Choe, S. and Stroud, R.M. (1986) The molecular neurobiology of the acetylcholine receptor. *Ann. Rev. Neurosci.,* 9: 383–414.

McCormick, D.A. and Prince, D.A (1987a) Actions of acetylcholine in the guinea-pig and cat medial and lateral geniculate nuclei, in vitro. *J. Physiol. (Lond.),* 392: 147–165.

McCormick, D.A. and Prince, D.A. (1987b) Acetylcholine causes a rapid nicotinic excitation in the medial habenular nucleus of guinea-pig, in vitro. *J. Neurosci.,* 7: 742–752.

McLennan, H. and Hicks, T.P. (1978) Pharmacological characterization of the excitatory cholinergic receptors of rat central neurones. *Neuropharmacology,* 17: 329–334.

McLennan, H. and York, D.H. (1966) Cholinergic mechanisms in the caudate nucleus. *J. Physiol. (Lond.),* 187: 163–175.

Mysilinski, N.R. and Randic, M. (1977) Responses of identified spinal neurons to acetylcholine applied by microelectrophoresis. *J. Physiol. (Lond.),* 269: 195–219.

Otago, N. (1979) Electrophysiology of mammalian hypothalamic and interpeduncular nucleus in vitro. *Experentia,* 35: 1202–1203.

Phillis, J.W., Tebecis, A.K. and York, D.H. (1967) A study of cholinoceptive cells in the lateral geniculate nucleus. *J. Physiol. (Lond.),* 192: 695–713.

Rovira, C., Cherubini, E and Ben-Ari, Y. (1983) Opposite actions of muscarinic and nicotinic agents on hippocampal dendritic negative fields recorded in rats. *Neuropharmacology,* 22: 239–243.

Ryall, R.W. (1983) Cholinergic transmission in the spinal cord. In R.A. Davidoff (Ed.), *Handbook of the Spinal Cord, Pharmacology,* Marcel Dekker, New York, pp. 203–239.

Salmoiraghi, G.C. and Steiner, F.A. (1963) Acetylcholine sensitivity of cat medullary neurons. *J. Neurophysiol.,* 26: 581–597.

Sastry, B.R. (1978) Effects of substance P, acetylcholine and stimulation of the habenula on rat interpeduncular neuronal activity. *Brain Res.,* 144: 404–410.

Segal, M. (1978) The acetylcholine receptor in the rat hippocampus: nicotinic, muscarinic, or both? *Neuropharmacology,* 17: 619–623.

Stollberg, J., Whiting, P.J., Lindstrom, J.M. and Berg, D.K. (1986) Functional blockade of neuronal acetylcholine receptors by antisera to a putative receptor from brain. *Brain Res.,* 378: 179–182.

Svensson, T.H. and Engberg, G. (1980) Effect of nicotine on single cell activity in the noradrenergic nucleus locus coeruleus. *Acta Physiol. Scand. Suppl.,* 479: 31–34.

Takegi, M. (1984) Actions of cholinergic drugs on cells in the interpeduncular nucleus. *Exp. Neurol.,* 84: 358–363.

Takeuchi, A. and Takeuchi, N. (1960) On the permeability of the end plate membrane during the action of transmitter. *J. Physiol. (Lond.),* 154: 52–67.

Tebecis, A.K. (1970) Properties of cholinoceptive neurons in the medial geniculate nucleus. *Br. J. Pharmacol.,* 38: 117–137.

Weiler, M.H., Misgeld, U. and Cheong, D.K. (1984) Presynaptic muscarinic modulation of nicotinic excitation in the rat neostriatum. *Brain Res.,* 296: 111–120.

Williams, J.T., Henderson, G. and North, R.A. (1984) Locus Coeruleus. In R. Dingledine (Ed.) *Brain Slices,* Plenum Press, New York, pp. 297–311.

Zieglgansberger, W. and Reiter, C. (1974) A cholinergic mechanism in the spinal cord of cats. *Neuropharmacology,* 13: 519–527.

A. Nordberg, K. Fuxe, B. Holmstedt and A. Sundwall (Eds.)
Progress in Brain Research, Vol. 79
© 1989 Elsevier Science Publishers B.V. (Biomedical Division)

CHAPTER 7

Galanin in the cholinergic basal forebrain: histochemical, autoradiographic and in vivo studies

T. Melander[a], T. Bartfai[b], N. Brynne[a], S. Consolo[c], G. Fisone[c], T. Hökfelt[a], C. Köhler[d], Ö. Nordström[b], E. Norheim-Theodorsson[e], A. Persson[f], G. Sedvall[f], W.A. Staines[a] and C.F. Wu[a]

[a] Department of Histology and Neurobiology, Karolinska Institutet, Stockholm, Sweden, [b] Department of Biochemistry, Arrhenius Laboratory University of Stockholm, Stockholm, Sweden, [c] Istituto di Ricerche Farmacologiche MarioNegri, 20157 Milan, Italy, [d] Department of Neurochemistry, ASTRA Pharmaceuticals, Södertälje, Sweden, [e] Department of Clinical Chemistry, Karolinska Hospital, Stockholm, Sweden, and [f] Department of Psychiatry, Karolinska Hospital, Stockholm, Sweden

Introduction

The peptide galanin (GAL) was isolated by Tatemoto, Mutt and collaborators (Tatemoto et al., 1983). GAL has, by the use of immunohisto-chemistry and radioimmunoassay, been mapped in the central nervous system of rat, pig and man (Rökaeus et al., 1984; Ch'ng et al., 1985; Skofitsch and Jacobowitz, 1985; Melander et al., 1986a). Among other areas, GAL immunoreactivity was found in the septum-basal forebrain complex, a region of interest due to its possible involvement in senile dementia of Alzheimer's type (SDAT) (Davies and Maloney, 1976; Whitehouse et al., 1982; Wolozin et al., 1986) and its relation to cognitive functions including memory and learning (Scoville and Milner, 1957; Stepien and Sierpinski, 1960; Mishkin, 1982). The chemical neuroanatomy of this area is rather well elucidated (for review see Fibiger, 1980 and Wenk et al., 1980). These nuclei have been shown to contain neurons staining for acetylcholinesterase (AChE) (e.g. Lewis and Shute, 1967; Jacobowitz and Palkovits, 1974;

Address for correspondence: Dr. T. Melander, Department of Histology and Neurobiology, Karolinska Institute, Box 60 400, 104 01 Stockholm, Sweden.

Lynch et al., 1977) as well as the acetylcholine synthesizing enzyme choline acetyltransferase (ChAT) (Kimura et al., 1980; Sofroniew et al., 1982; Armstrong et al., 1983; Houser et al., 1983; Mesulam et al., 1983), leaving little doubt as to their cholinergic nature.

It has been found that peptides in many instances occur in the same neurons as "classical transmitters", such as biogenic amines and/or amino acids, and may modulate the response to these messengers (see Cuello, 1982; Chan-Palay and Palay, 1983; Hökfelt et al., 1984). On the basis of these findings it has subsequently been shown that potent modulation of the effects of the classical transmitter can take place by co-stored peptides (e.g. Skirboll et al., 1981; Lundberg et al., 1982; Markstein and Hökfelt, 1984; Crawley et al., 1985). Also biochemical techniques studying interactions at the receptor level imply modulation relating to the costorage of peptides and classical transmitters (Abens et al., 1984; Agnati and Fuxe, 1984; Bartfai et al., 1984). In general, it appears as if the peptide actions are more of a modulatory nature with prolonged action, and can perhaps underlie many of the "slower" processes in the nervous system. Such "slow" events, influenced by peptidergic transmission, may include memory and learning, which are brain functions of relevance when studying the cholinergic systems.

The following concerns mainly the neuroanatomical features of GAL-immunoreactive (IR) neuronal systems in the basal forebrain system, including coexistence situations both at the cell body and fiber/varicosity level with acetylcholine. The study also includes analysis of $[^{125}I]$GAL binding sites in the hippocampal region and some interactions with cholinergic transmission.

Immunoreactive galanin in the septohippocampal pathway

The cholinergic afferents enter the hippocampal formation through three different pathways. The largest portion has been shown to project by the way of the dorsal system, mainly in the dorsal fornix/fimbria (e.g. Lewis and Shute, 1967; Storm-Mathisen, 1970). However, a minor subcortical ventral pathway has also been described (Gage et al., 1984).

With regard to peptidergic innervation of the hippocampal formation, little is known. However, Vincent and coworkers (1983) have shown that reticular neurons co-containing substance P-like immunoreactivity (LI) and ChAT-LI project to several cortical areas, and Everitt et al. (1984) found NPY-LI in some cells of the locus coeruleus.

Double-labelling immunohistochemistry for GAL and ChAT revealed that virtually all GAL-positive cell profiles in the medial septal nucleus and diagonal band nuclei co-contained ChAT-like immunoreactivity (Melander et al., 1985) and some projected to the hippocampal formation. It is of interest that the cholinergic neurons of the nucleus basalis, the major cholinergic subcortical projection to the neocortex, did not seem to contain GAL-like peptide in the rat.

In the monkey, GAL-immunohistochemistry combined with AChE histochemistry (Fig. 1) revealed that also here a GAL-like peptide is co-contained within putative cholinergic cell bodies (Fig. 1A – D). In this species, the peptide immunoreactivity was much more widespread, and extended into the various subregions of nucleus basalis (Fig. 1C, D). Thus, GAL-like immunoreactivity was seen in virtually all AChE-positive somata in this region of the monkey.

GAL-LI and AChE were co-contained within the hippocampus of the owl monkey (Melander and Staines, 1986) where an extensive GAL-IR fiber network was observed (Fig. 1E, F). Furthermore, neocortical areas in the monkey revealed a more extensive GAL-positive innervation, in agreement with the more widespread distribution in cholinergic cell bodies in nucleus basalis in this species. In contrast, the GAL-IR hippocampal fibers in the rat were not as extensive as the known cholinergic innervation but did decrease after fimbria lesions (Fig. 2A, B) (Melander et al., 1986b).

The apparent difference between coexistence at the cell body level versus the nerve fiber level in the rat could be due to several reasons. The peptide could be processed to a form less immunoreactive to the antibody, as it is transported from the cell body to the terminals. Alternatively, GAL-like peptide(s) could, in this system, simply be restricted to the cell body region or, more likely, the levels of peptide in the terminals are below the detection limit of our technique. The latter alternative may be related to the fact that our antibodies were raised towards naturally isolated pig GAL. Species differences in amino acid sequence could lower the affinity for the GAL-like peptide in the rat basal forebrain. In fact, such a structural difference has been suggested (Rökaeus et al., 1984).

Fig. 1. Immunofluorescence (a, c, e, f) and acetylcholinesterase (b, d) micrographs of the monkey medial septal nucleus (a, b) and basal nucleus (c, d) and dentate gyrus (e, f). Arrows indicate cells co-containing GAL-LI and AChE. (e) and (f) show GAL-IR fibers in the dentate gyrus. Scale bar in (a) and (e) indicates 50 μm; all micrographs have the same magnification except (e).

→

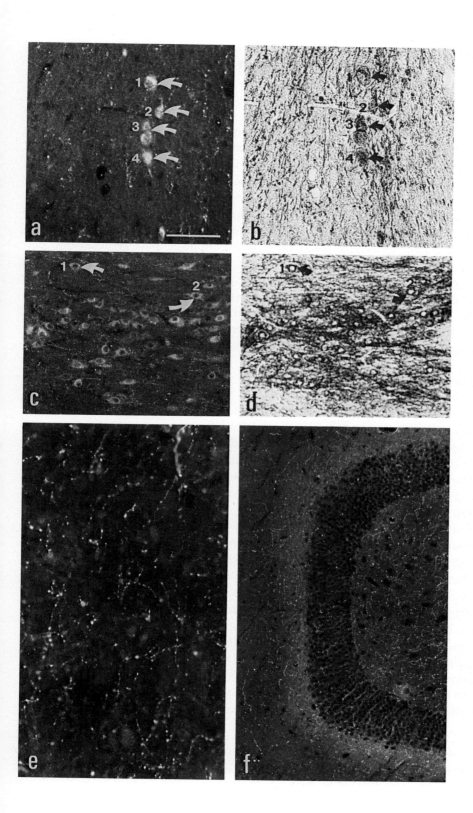

Autoradiographic localization of [^{125}I]GAL binding sites in the rat, human and monkey hippocampus

The distribution of [^{125}I]GAL binding sites has been analyzed in some detail in the rat CNS (Skofitsch et al., 1986; Melander et al., 1988). Another study (Servin et al., 1987) has shown that the GAL binding site in the rat brain consists of a saturable 50 kDa protein and that the binding is highly pH and ionic strength dependent with defined kilodalton and B_{max}. They also detected to our results comparable amounts of binding in discrete brain regions. Islands of higher labelling were seen in the outer three layers of the ventral entorhinal cortex. No binding could be detected in the somatosensory or cingulate cortices.

The hippocampal binding sites were in general located to the most ventral part, where the densest labelling was seen in the subiculum. Stratum oriens, most ventral pole, contained medium dense labelling in CA3 as well as in the CA1 and CA2 regions. Labelling over the pyramidal cell bodies was weaker than the surrounding layers. No detectable binding was seen in the dentate gyrus.

[^{125}I]GAL was bound to the dorsal hippocampal afferent fiber systems (supracallosal striae and fimbria/fornix) as well as to the areas containing possible cell bodies of origin (medial septum-diagonal band complex), indicating a probable presence of a GAL binding substance in hippocampal afferent neurons. Interestingly, in the fimbria the binding sites were concentrated to their lateral tip, suggesting that the receptor is transported only in those septohippocampal axons that are destined for the ventral hippocampus (Meibach and Siegel, 1977). In the human and monkey brain a much more extensive GAL binding was seen in the neocortex whereby in the human the highest density was seen in layer 6 and in the monkey in layers 4 and 5 (Köhler et al., 1989).

Effects of GAL on cholinergic transmission in the hippocampal formation

Experiments by Fisone and coworkers (1987) revealed that GAL can inhibit potassium and scopolamine-stimulated acetylcholine release in the ventral but not dorsal hippocampus in vitro and in vivo, respectively and that the [^{125}I]GAL binding in the hippocampal formation can be reduced by lesions of the afferent pathways to this region (Fig. 2C – F). Taken together, this indicates that the GAL binding sites, to some extent, may be presynaptic in the hippocampus. A behavioral correlate to the effects of GAL on cholinergic septohippocampal neurons has been presented by Sundström et al. (1988), who showed that intraventricular GAL has inhibiting effects on acquisition, but not retrieval of memory in a swim-maze paradigm.

Conclusions

The presence of GAL-like peptide(s) in the cholinergic forebrain neurons and the widespread occurrence of this coexistence situation in the monkey makes it relevant to discuss the concept of peptide co-transmission (see e.g. Hökfelt et al., 1984) as a means to increase diversity in chemical signalling in relation to brain functions such as memory and learning.

The finding that GAL inhibits scopolamine-stimulated acetylcholine release but does not affect basal release in this system could be a further example where the peptidergic mechanisms act only under high frequency discharge (see e.g. Lundberg et al., 1982).

These findings may also have implications in the human brain since distinct cortical binding pattern was observed concentrated to layer 6 as well as to the hippocampal formation. The findings may therefore also be pertinent to disease states such as

Fig. 2 (a, b) Immunofluorescence micrograph showing the depletion (b) of GAL-IR fibers from the hippocampal CA3 region of the rat after fimbria lesions. (c – f) Receptor autoradiograms showing the decrease in [^{125}I]GAL binding in the hippocampal formation after fimbria lesions (c, d) (horizontal section) and after ibotenic acid lesions of the septal complex (squared) (e, f). Note the more moderate effect of the more specifically cholinergic lesion in (f).

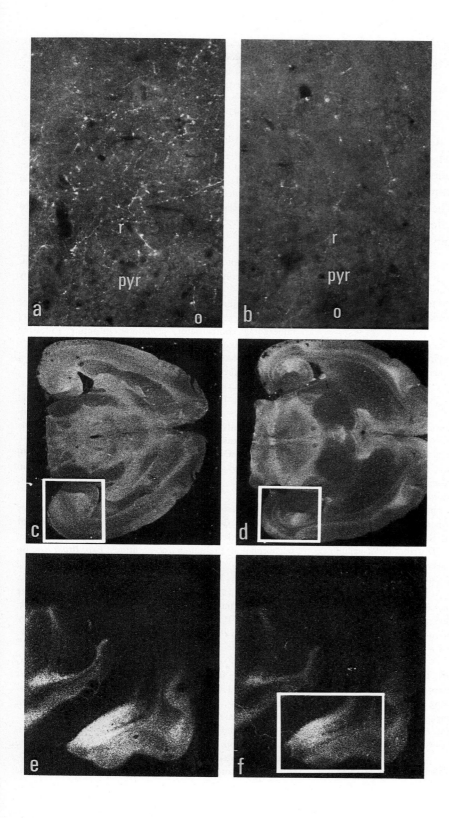

SDAT which have been thought to be associated with these forebrain structures (for references see Introduction).

Acknowledgements

This work was supported by the Swedish MRC (2887), National Institutes of Mental Health (MH 43230), Knut och Alice Wallenbergs Stiftelse, Konung Gustav V:s och Drottning Victorias Stiftelse and Thurings Stiftelse. The skilful technical assistance of Ms. A.C. Radesäter, Ms. W. Hiort and Ms. Siv Nilsson is gratefully acknowledged.

References

Abens, J., Westlind, A. and Bartfai, T. (1984) Chronic atropine treatment causes increase in VIP receptors in rat cerebral cortex. *Peptides,* 5: 375 – 377.

Agnati, L.F. and Fuxe, K. (1984) New concepts on the structure of the neuronal networks: the miniaturization and hierarchical organization of the central nervous system. *Biosci. Rep.,* 4: 93 – 98.

Armstrong, D.M., Saper, C.B., Levey, A.I. and Wainer, B.H. (1983) Distribution of cholinergic neurons in rat brain: demonstration by the immunohistochemical localization of choline acetyltransferase. *J. Comp. Neurol.,* 215: 53 – 61.

Bartfai, T., Westlind, A., Abens, J., Engström, C. and Alberts, P. (1984) On acetylcholine-vasoactive intestinal polypeptide coexistence and presynaptic interactions. In E.S. Vizi and K. Magyar (Eds.), *Regulation of Transmitter Function,* Proc. 5th Meeting Eur. Soc. Neurochem., pp. 497 – 500.

Ch'ng, J.L.C., Christofides, N.D., Anand, P., Gibson, S.J., Allen, Y.S., Su, H.C., Tatemoto, K., Morrison, J.F.B., Polak, J.M. and Bloom, S.R. (1985) Distribution of galanin immunoreactivity in the central nervous system and the responses of galanin-containing neuronal pathways to injury. *Neuroscience,* 16: 343 – 354.

Chan-Palay, V. and Palay, S.L. (Eds.) (1983) *Coexistence of Neuroactive Substances,* John Wiley and Sons, New York.

Coons, A.H. (1958) Fluorescent antibody methods. In J.F. Danielli (Ed.), *General Cytochemical Methods,* Academic Press, New York, pp. 399 – 422.

Crawley, J.N., Olschowka, L.A., Diz, D.I. and Jacobowitz, D.M. (1985) Behavioral investigation of the coexistence of substance P, corticotropin releasing factor, and acetylcholinesterase in lateral dorsal tegmental neurons projecting to the medial frontal cortex of the rat. *Peptides,* 6: 891 – 901.

Cuello, A.C. (Ed.) (1982) *Co-transmission,* MacMillan, London.

Davies, P. and Maloney, A.J. (1976) Selective loss of central cholinergic neurones in Alzheimer's disease. *Lancet,* ii: 1403.

Everitt, B.J., Hökfelt, T., Terenius, L., Tatemoto, K., Mutt, V. and Goldstein, M. (1984) Differential co-existence of neuropeptide Y (NPY)-like immunoreactivity with catecholamines in the central nervous system of the rat. *Neuroscience,* 11: 443 – 462.

Fibiger, H.C. (1982) The organization and some projections of cholinergic neurons of the mammalian forebrain. *Brain Res. Rev.,* 4: 327 – 288.

Fisone, G., Wu, F.C., Consolo, F., Nordström, Ö., Brynne, N., Bartfai, T., Melander, T. and Hökfelt, T. (1987) Galanin inhibits acetylcholine release in the ventral hippocampus of the rat: histochemical autoradiographic, in vivo and in vitro studies. *Proc. Natl. Acad. Sci. U.S.A.,* 88: 7339 – 7343.

Fuxe, K., Agnati, L.F., Benfenati, F., Celani, M., Zini, I., Zoli, M. and Mutt, V. (1983) Evidence for the existence of receptor-receptor interactions in the central nervous system. Studies on the regulation of monoamine receptors by neuropeptides. *J. Neural. Trans. Suppl.,* 18: 165 – 179.

Gage, F.H., Björklund, A. and Stenevi, U. (1984) Cells of origin of the ventral cholinergic septohippocampal pathway undergoing compensatory collateral sprouting following fimbria-fornix transection. *Neurosci. Lett.,* 44: 211 – 216.

Hökfelt, T., Johansson, O. and Goldstein, M. (1984) Chemical anatomy of the brain. *Science,* 225: 1326 – 1334.

Houser, C.R., Crawford, G.D., Barber, R.P., Salvaterra, P.M. and Vaughn, J.E. (1983) Organization and morphological characteristics of cholinergic neurons. An immunohistochemical study with a monclonal antibody to choline acetyltransferase. *Brain Res.,* 266: 97 – 119.

Kimura, H., McGeer, P.L., Peng, F. and McGeer, E.G. (1980) Choline acetyltransferase containing neurons in rodent brain demonstrated by immunohistochemistry. *Science,* 208: 1957 – 1959.

Köhler, C., Persson, A., Melander, T., Theodorsson-Norheim, E., Sedvall, G. and Hökfelt, T. (1988) Distribution of galanin binding sites in the monkey and human telencephalon. Preliminary observations. *Exp. Brain Res.* (in press).

Lundberg, J.M., Hedlund, B., Änggård, A., Fahrenkrug, J., Hökfelt, T., Tatemoto, K. and Bartfai, T. (1982) Costorage of peptides and classical transmitters in neurons. In S.R. Bloom, J.M. Polak and E. Lindenlaub (Eds.), *Systemic Role of Regulatory Peptides,* Schattauer, Stuttgart and New York, pp. 93 – 119.

Lynch, G., Rose, G. and Gall, C. (1977) Anatomical and functional aspects of the septo-hippocampal projections. In K. Elliott and J. Whelan (Eds.), *Functions of the Septo-Hippocampal System, Ciba Foundation Symposium,* Elsevier, Amsterdam, pp. 5 – 24.

Markstein, P. and Hökfelt, T. (1984) Effect of cholecystokinin-octapeptide on dopamine release from slices of cat caudate nucleus. *J. Neurosci.,* 4: 570 – 576.

Meibach, R.C. and Siegel, A. (1977) Efferent connection of the septal area in the rat: an analysis utilizing retrograde and anterograde transport methods. *Brain Res.,* 119: 1 – 20.

Melander, T. and Staines, Wm.A. (1986) A galanin-like peptide coexists in putative cholinergic somata of the septum-basal forebrain complex and in acetylcholinesterase containing fibers and varicosities within the hippocampus in the owl monkey (*Aoutus trivirgatus*). *Neurosci. Lett.,* 68: 17 – 22.

Melander, T., Staines, Wm.A., Hökfelt, T., Rökaeus, Å., Eckenstein, F., Salvaterra, P.M. and Weiner, B.H. (1985) Galanin-like immunoreactivity in cholinergic neurons of the septum-basal forebrain complex projecting to the hippocampus of the rat. *Brain Res.,* 360: 130 – 138.

Melander, T., Hökfelt, T. and Rökaeus, Å. (1986a) Distribution of galanin-like immunoreactivity in the rat CNS. *J. Comp. Neurol.,* 248: 475 – 517.

Melander, T., Staines, Wm.A. and Rökaeus, A. (1986b) Galanin-like immunoreactivity in hippocampal afferents in the rat, with special reference to cholinergic and noradrenergic inputs. *Neuroscience,* 19: 223 – 240.

Melander, T., Köhler, C., Nilsson, S., Hökfelt, T., Brodin, E., Theodorsson, E. and Bartfai, T. (1988) Autoradiographic quantitation and anatomical mapping of ^{125}I-galanin binding sites in the rat central nervous system. *J. Chem. Neuroanat.,* 1: 213 – 233.

Mesulam, M.M., Mufson, E.J., Wainer, B.H., and Levey, A.I. (1983) Central cholinergic pathways in the rat: an overview based on an alternative nomenclature (Ch1-Ch6). *Neuroscience,* 10: 1185 – 1201.

Mishkin, M. (1982) A memory system in the monkey. *Philos. Trans. R. Soc. Lond. Ser. B.,* 198: 85 – 95.

Scoville, W.B. and Milner, B. (1957) Loss of recent memory after bilateral hippocampal lesions. *J. Neurol. Neurosurg. Psychol.,* 20: 11 – 21.

Servin, A.L., Amiranoff, B., Rouyer-Fessard, C., Tatemoto, K. and Laburthe, M. (1987) Identification and molecular characterization of galanin receptor sites in rat brain. *Biochem. Biophys. Res. Commun.,* 144: 298 – 306.

Skirboll, L.R., Grace, A.A., Hommer, D.W., Rehfeld, J., Goldstein, M, Hökfelt, T. and Bunney, B.S. (1981) Peptide-monoamine coexistence: studies of the actions of cholecystokinin-like peptide on the electrical activity of midbrain dopamine neurons. *Neuroscience,* 6: 2114 – 2124.

Skofitsch, G. and Jacobowitz, D.M. (1985) Immunohistochemical mapping of galanin-like neurons in the rat central nervous system. *Peptides,* 6: 509 – 546.

Skofitsch, G., Sills, M.A. and Jacobowitz, G. (1986) Autoradiographic distribution of ^{125}I-galanin binding sites in the rat central nervous system. *Peptides,* 7: 1029 – 1042.

Sofroniew, M.V., Eckenstein, F., Thoenen, H. and Cuello, A.C. (1982) Topography of choline acetyltransferase-containing neurons in the forebrain of the rat. *Neurosci. Lett.,* 33: 7 – 12.

Stepien, L. and Sierpinski, S. (1960) The effect of local lesions of the brain upon auditory and visual recent memory in man. *J. Neurol. Neurosurg. Psychol.,* 23: 334 – 340.

Sundström, E., Archer, T. and Hökfelt, T. (1988) Galanin impairs acquisition but not retrieval of spatial memory in rats studied in the Morris swim maze. *Neurosci. Lett.,* 88: 331 – 335.

Tatemoto, K., Rökaeus, Å., Jörnvall, H., McDonald, T.J. and Mutt, V. (1983) Galanin − a novel biologically active peptide from porcine intestine. *FEBS Lett.,* 164: 124 – 128.

Vincent, S.R., Satoh, K., Armstrong, D.M. and Fibiger, H.C. (1983) Substance P in the ascending cholinergic reticular system. *Nature (Lond.),* 306: 688 – 691.

Wenk, H., Volker, B. and Meyer, U. (1980) Cholinergic projections from magnocellular nuclei of the basal forebrain to cortical areas in rats. *Brain Res. Rev.,* 2: 295 – 316.

Whitehouse, P.J., Price, D.L., Struble, R.G., Clarke, A.W., Coyle, J.T. and DeLong, M.R. (1982) Alzheimer's disease and senile dementia: loss of neurons in the basal forebrain. *Science,* 215: 1237 – 1239.

Wolozin, B.L., Pruchnicki, A., Dickson, D.W. and Davies, P. (1986) A neuronal antigen in the brains of Alzheimer patients. *Science,* 232: 648 – 649.

Young, III, W.S. and Kuhar, M.J. (1979) A new method for receptor autoradiography. ^{3}H-opioid receptor labeling in mounted tissue sections. *Brain Res.,* 179: 255 – 270.

Central Nicotinic Receptors – Radioligand Binding Studies

A. Nordberg, K. Fuxe, B. Holmstedt and A. Sundwall (Eds.)
Progress in Brain Research, Vol. 79
© 1989 Elsevier Science Publishers B.V. (Biomedical Division)

CHAPTER 8

Stereoselectivity of nicotinic receptors

Uli Hacksell and Charlotta Mellin

Department of Organic Pharmaceutical Chemistry, Uppsala Biomedical Center, University of Uppsala, S-75123 Uppsala, Sweden

Introduction

Nicotinic receptors have been isolated and their structures have been determined (Changeux et al., 1984). However, the loci of the sites where nicotinic ligands interact with the receptors (the "receptor sites") are still unknown. Thus, although much is known about the structure of the nicotinic receptor, we still have to rely on deduced models when attempting to rationalize structure-activity relationships for nicotinic agonists/antagonists and when designing novel ligands.

Acetylcholine, the endogenous transmitter at nicotinic and muscarinic receptors is a flexible and achiral molecule. Since various conformations of acetylcholine differ very little in energy, conformational studies of acetylcholine do not provide clues to the receptor-bound conformation and to the intramolecular distances between receptor-bound functional groups. A more fruitful approach when attempting to determine receptor-bound conformations/pharmacophores is to use chiral and semirigid or rigid derivatives. Such compounds have few low-energy conformations and the stereoselectivity that may be obtained is useful when discussing structural requirements for binding to or activation of the receptor. In the present study we have used (+)-anatoxin-a and (−)-cytisine − two semirigid and potent nicotinic-receptor agonists − to model nicotinic receptor activity. We have also included isoarecolone methiodide in

our model. The model is used to predict the stereoselectivity of a recently reported, potent but racemic bicyclic agonist called pyridol [3,4-b] homotropane (PHT) and to discuss the stereoselectivity of nicotine. The structures of the compounds under study are presented in Fig. 1.

Methods

We have used a method called "the active analogue approach", in which the energetically accessible conformational space for a number of bioactive compounds is considered when deducing the pharmacophore, i.e. the relative spatial positions of key functional groups (Marchall, 1987). The compounds may then be fitted to the pharmacophore in their pharmacophore conformations and the combination of their van der Waals volumes form a partial receptor-excluded volume (Sulfrin et al., 1981). Part of the volume outside the receptor-excluded volume belongs to the receptor-essential volume. This volume may be defined by use of inactive analogues.

The structural modelling was performed by use of the interactive computer graphics program MIMIC (Liljefors, 1983) and Chem-X (developed and distributed by Chemical Design Ltd, Oxford, England). Molecular mechanics (MMP2) calculations (Boyd and Lipkowitz, 1982) were performed using Allingers MMP2 1980 force field (Allinger, 1977) to which had been added amino group para-

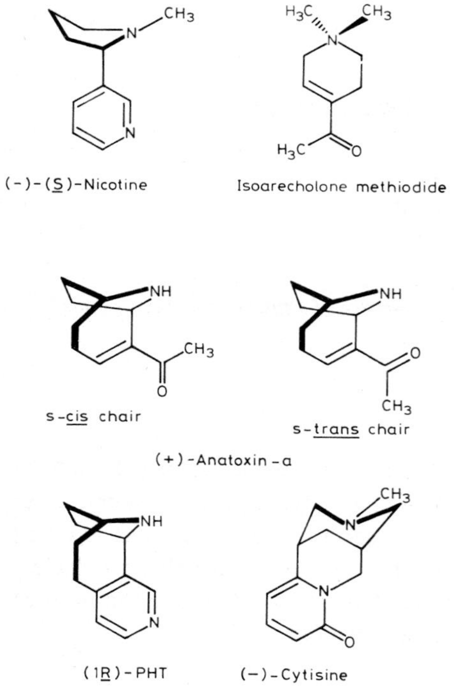

Fig. 1. Structures of the compounds under study.

meters (Profeta, Jr. and Allinger, 1985).

No parameters were available for the $N(sp^3)$-$C(sp^3)$-$C(sp^2)$ fragment. In these cases, calculations were performed using parameters for the corresponding carbon-containing moieties. Conformational geometries were obtained without restrictions in the minimization process. The receptor-excluded volume and excess volumes were constructed with the "Set Map" option in Chem-X using a grid size of 24^3 and contouring at the van der Waals level. All calculations were performed on a Microvax II computer.

Results and discussion

Background

Nicotine

Nicotine has a center of asymmetry at the carbon joining the pyrrolidine ring with the pyridine moiety. Thus, nicotine is chiral and there are two possible stereoisomers, two enantiomers, of nicotine. Natural nicotine, the (S)-enantiomer, has an $[\alpha]^{20}_D$ value of around $-169°$ as a base. However, several salts of (S)-nicotine, such as the hydrochloride or the sulfate, give positive $[\alpha]^{20}_D$ values. Therefore, the natural isomer of nicotine should be denoted $(-)$-(S)-nicotine to avoid unnecessary confusion. By analogy, the synthetic, non-natural enantiomer is $(+)$-(R)-nicotine.

A large number of investigations on the stereoselectivity of nicotine have been reported. The naturally occurring $(-)$, l- or $(-)$-(S)-enantiomer is the more potent agonist in vivo. However, in a pharmacodynamic sense, the potencies of the antipodes may be comparable since it has been reported that equimolar doses of the antipodes lead to higher brain levels of the $(-)$-(S)-enantiomer (Martin et al., 1983). Similarly, only $(+)$-(R)-nicotine is a substrate for the N-methyl transferase whereas the $(-)$-(S)-enantiomer is a competitive inhibitor (Cundy et al., 1985a, b). In most investigations, however, $(-)$-(S)-nicotine appears to have a greater affinity for both high- and low-affinity state nicotinic receptors than the $(+)$-(R)-enantiomer, e.g. see Ikushima et al. (1982), Martino-Barrows and Kellar (1987) and Reavill et al. (1988). Thus, a relevant model for nicotinic agonists should take into account that both nicotine enantiomers behave as nicotinic agonists.

Nicotine is a rather flexible molecule; there is a possibility of rotation around the bond joining the pyridine and pyrrolidine rings. In addition, the pyrrolidine nitrogen may undergo inversion and the pyrrolidine ring may adopt many different ring conformations. To evaluate the energy required for rotation around the pyridine-pyrrolidine bond, we performed molecular mechanic calculations (by use of the Rigid Rotation option in MIMIC) in which the two rings were rotated in increments of 10°. We identified two energy minima separated by an energy barrier slightly lower than 5 kcal/mol. The most favored conformations of nicotine in solution and in vacuo appear to be those in which the pyrrolidine and pyridine rings are perpendicular and the pyrrolidine ring is in an envelope conformation with the *N*-methyl and the pyridinyl substituents in a *trans*-relationship, compare e.g. Pullman, et al. (1971) and Pitner et al. (1978).

Cytisine

The nicotinic alkaloid (−)-cytisine is similar in potency to (−)-(*S*)-nicotine. Due to its high potency and considerable rigidity, (−)-cytisine is a very interesting nicotinic ligand. We have performed molecular mechanic calculations on cytisine and have found that the energetically most favorable geometries are those in which the piperidine ring adopts a chair conformation. Similar conclusions have been drawn by others (Beers and Reich, 1970).

Anatoxin-a

"Very Fast Death Factor" (VFDF), also known as anatoxin-a, is a highly interesting compound since it is fairly rigid and, more importantly, interacts stereoselectively with nicotine receptors (Swanson et al., 1986). (+)-Anatoxin-a binds with much higher potency to acetylcholine- and nicotine-labelled cholinergic receptors than does the (−)-enantiomer (Zhang et al., 1987). In addition, the (+)-enantiomer binds to the two nicotine-labelled high affinity sites whereas the (−)-enantiomer binds only to one low-affinity site.

Although anatoxin-a is a semirigid compound, it can adopt four distinctly different conformations. The chair conformations (Fig. 1) are more stable (around 1 kcal/mol) than the boat conformations. According to previous NMR- and molecular mechanic studies, the two chair conformations are populated at room temperature. In the solid state, anatoxin-a adopts the s-*trans* chair conformation (Koskinen and Rapoport, 1985).

Isoarecolone methiodide

The quaternary piperidinium derivative isoarecolone methiodide is a very potent nicotinic agonist when tested at the frog neuromuscular junction − it is, for example, 50 times more potent than carbachol (Spivak et al., 1986).

Isoarecholone methiodide can adopt conformations in which the enone system is *cis* or *trans*. Our molecular mechanic calculations indicate that the *trans*-conformation is of around 2 kcal/mol lower energy than the *cis*-conformation. These results agree well with those previously published (Spivak et al., 1986).

Pyrido[3,4-b]homotropane (PHT)

Recently, a number of racemic pyridohomotropane derivatives were reported (Kanne et al., 1986). These compounds are nicotinic agonists and can be considered as rigid derivatives of nicotine. (±)-PHT binds somewhat more potently to (−)-(*S*)-[^{3}H]nicotine-labelled receptors than nicotine itself, thereby demonstrating high affinity. In addition, PHT seems to be equipotent to nicotine in producing the prostration syndrome in rats (Kanne and Abood, 1986). PHT, which is a secondary amine, is the most potent of the reported pyridohomotropane derivatives. The *N*-methyl derivative is much less potent. In contrast, nornicotine is considerably less potent than nicotine. It is also noteworthy that 2-methyl PHT is much less active than PHT (Kanne and Abood, 1986).

The structure of (−)-(*S*)-nornicotine can be identified within the skeleton of (1*S*)-PHT. However, such a geometry of the nornicotine moiety in PHT corresponds to a high-energy conformation. Thus, biologically relevant structural comparisons between PHT and nornicotine/nicotine must be based on additional information.

The nicotinic pharmacophore

The pharmacophore for nicotinic activity is commonly accepted. In its simplest form, it consists of a positive charge (normally due to a protonated or quaternary nitrogen) which is believed to interact with the receptor via coulumbic forces, and a hydrogen-bond acceptor (oxygen or nitrogen of low basicity) which is thought to form a hydrogen bond with the receptor (Beers and Reich, 1970). These two points define a line but a pharmacophore plane is neccessary to enable a unified orientation of putative ligands. This is accomplished by including in the model both the nitrogen and its lone pair for pyridine-based molecules and the carbonyl oxygen and one of its lone pairs in ketones or carboxylic esters. Thus, the nicotine pharmacophore can be described by a triangle (Sheridan et al., 1986). Since the receptor protein is flexible, the lengths of the sides of the triangle (or the intramolecular distances within the pharmacophore) can vary within certain unknown limits.

Construction of a nicotine-receptor excluded volume

Anatoxin-a is one of surprisingly few stereoselective nicotine-receptor agonists. Thus, the structure of anatoxin should always be included when modelling nicotinic activity. It has been argued that the s-*cis* chair or boat conformation of (+)-anatoxin-a would correspond to the receptor-bound conformation (see, e.g. Koskinen and Rapoport, 1985). This is probably mainly due to the fact that the distance between the nitrogen and oxygen in this conformation is 6.04 Å, that is close to 5.9 Å, which is the nicotinic pharmacophore distance suggested by Beers and Reich. However, a superposition of N, O and carbonyl carbon of s-*cis*-(+)-anatoxin-a, (−)-cytisine, and isoarecholone methiodide produces a combined van der Waals volume which does not readily help in rationalizing the poor activity (inactivity) of (−)-anatoxin-a. In contrast, when the s-*trans* chair conformation of (+)-anatoxin-a was used when modelling the receptor-excluded volume, we arrived at a smaller volume that readily permits rationalization of the enantioselectivity of anatoxin-a (Fig. 2). When fitting (−)-cytisine to the pharmacophore, we used the carbonyl oxygen and not the nitrogen as the hydrogen bond donator; the distance between the two nitrogens would probably be too short. The energetically favored *trans*-conformation of isoarecolone methiodide was used in the construction of the receptor-excluded volume. However, the van der Waals

volume of the *cis*-conformation would not produce any excess volume.

Implications of the model

It has been indicated, based on structural comparisons of s-*cis*-anatoxin-a and PHT, that (1R)-PHT would be the active enantiomer (Kanne and Abood, 1988). If this were true, the C2-methyl would correspond to the acetyl methyl of s-*cis*

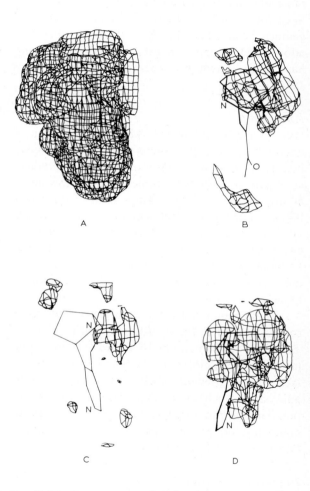

A

B

C

D

Fig. 2. Nicotine-receptor excluded volume constructed by superposition of the s-*trans* chair conformation of anatoxin-a, (−)-cytisine and isoarecholone methiodide (A), and excess volumes formed by fitting (−)-anatoxin-a (B), (1R)-PHT (C) and (1S)-PHT (D) to the receptor-excluded volume. In (A), the nicotinic pharmacophore is indicated by a solid line (the third point of the pharmacophore is oriented away from the viewer).

anatoxin-a. However, the low potency of 2-methyl PHT does not corroborate this hypothesis. In our model, in which an s-*trans* conformation is used as the bioactive conformation of (+)-anatoxin-a, the methyl groups of C2-methyl-(1*R*)-PHT and *N*-methyl-(1*R*)-PHT as well as (1*S*)-PHT and its derivatives produce excess volumes. Thus, our model also suggests that (1*R*)-PHT is the active PHT enantiomer (compare Fig. 2).

To fit optimally to the nicotinic pharmacophore, the nicotine enantiomers have to adopt different conformations. The two conformations used for the modelling studies correspond to the two identified low energy conformations of nicotine. When the selected conformations were fitted to our model, the nicotine enantiomers produced about the same amount of excess volume. These excess volumes are, however, not part of the nicotine-receptor essential volume since both enantiomers of nicotine possess nicotinic activity. Instead, these excess volumes may be used to extend the receptor-excluded volume.

Concluding remarks

Recently, a sophisticated model that includes electrostatic potentials was reported (Waters et al., 1988). However, the fairly simple model presented here is consistent with the high stereoselectivity of anatoxin-a and with the relatively low stereoselectivity of nicotine. The predictability of our model can be tested by determination of the pharmacological stereoselectivity of PHT. Thus, we consider it important to synthesize the enantiomers of PHT and to test them for nicotinic activity. In addition, these enantiomers may become important pharmacological tools for pharmacologists studying nicotinic receptors.

Acknowledgements

The financial support from the Swedish Natural Science Research Council, "Centrala försöksdjursnämnden" and 'Inga Britt and Arne Lundbergs forskningsstiftelse" is gratefully acknowledged.

References

Allinger, N.L. (1977) Conformational analysis. 130. MMP2. A hydrocarbon force field utilizing v_1 and v_2 torsional terms. *J. Am. Chem. Soc.*, 99: 8127–8134.

Beers, W.H. and Reich, E. (1970) Structure and activity of acetylcholine. *Nature (Lond.)*, 222: 917–922.

Boyd, D.B. and Lipkowitz, K.B. (1982) Molecular mechanics. The method and its underlying philosophy. *J. Chem. Educ.*, 59: 269–274.

Changeux, J.-P., Devillers-Thiery, A. and Chemouilli, P. (1984) Acetylcholine receptor: an allosteric protein. *Science*, 225: 1335–1345.

Cundy, K.C., Sato, M. and Crooks, P.A. (1985a) Stereospecific in vivo N-methylation of nicotine in the guinea pig. *Drug Met. Disp.*, 13: 175–185.

Cundy, K.C., Crooks, P.A. and Godin, C.S. (1985b) Remarkable substrate-inhibitor properties of nicotine enantiomers towards a guinea pig lung aromatic azaheterocycle N-methyltransferase. *Biochem. Biophys. Res. Commun.*, 128: 312–316.

Ikushima, S., Muramatsu, I., Sakakibara, Y., Yokotani, K. and Fujiwara, M. (1982) The effects of *d*-nicotine and *l*-isomer on nicotinic receptors. *J. Pharmacol. Exp. Ther.*, 22: 463–470.

Kanne, D.B. and Abood, L.G. (1988) Synthesis and biological characterization of pyridohomotropanes. Structure-activity relationships of conformationally restricted nicotinoids. *J. Med. Chem.*, 31: 506–509.

Kanne, D.B., Ashworth, D.J., Ashworth, M.J., Cheng, M.T., Mutter, L.C. and Abood, L.G. (1986) Synthesis of the first highly potent bridged nicotinoid. 9-Azabicyclo[4.2.1.] nona[2,3-c]pyridine (pyrido[3,4]homotropane) *J. Am. Chem. Soc.*, 108: 7864–7865.

Koskinen, A.M.P. and Rapoport, H. (1985) Synthetic and conformational studies on anatoxin-a: a potent acetylcholine agonist. *J. Med. Chem.*, 28: 1301–1309.

Liljefors, T. (1983) Molbuild — an interactive computer graphics interface to molecular mechanics. *Mol. Graphics*, 1: 111–117.

Marchall, G.R. (1987) Computer-aided drug design. *Ann. Rev. Pharmacol. Toxicol.*, 27: 193–213.

Martin, B.R., Tripathi, H.L., Aceto, M.D. and May, E.L. (1983) Relationship of the biodisposition of the stereoisomers of nicotine in the central nervous system to their pharmacological actions. *J. Pharmacol. Exp. Ther.*, 226: 157–163.

Martino-Barrows, A.M. and Kellar, K.J. (1987) [3H]Acetylcholine and [3H](−)nicotine label the same recognition site in rat brain. *Mol. Pharmacol.*, 31: 169–174.

Pitner, T.P., Edwards, W.B., III, Bassfield, R.L. and Whidby, J.F. (1978) The solution conformation of nicotine. A ^{1}H and ^{2}H nuclear magnetic resonance investigation. *J. Am. Chem. Soc.*, 100: 246–251.

Profeta, S.R., Jr. and Allinger, N.L. (1985) Molecular

mechanics calculations on aliphatic amines. *J. Am. Chem. Soc.*, 107: 1907 – 1918.

Pullman, B., Courriere, P. and Coubeils, J.L. (1971) Quantum mechanical study of the conformational and electronic properties of acetylcholine and its agonists muscarine and nicotine. *Mol. Pharmacol.*, 7: 397 – 405.

Reavill, C., Jenner, P., Kumar, R. and Stolerman, I.P. (1988) High affinity binding of [^{3}H](–)-nicotine to rat brain membranes and its inhibition by analogues of nicotine. *Neuropharmacology*, 27: 235 – 241.

Sheridan, R.P., Nilakantan, R., Dixon, J.S. and Venkataraghvan, R. (1986) The ensemble approach to distance geometry: application to the nicotinic pharmacophore. *J. Med. Chem.*, 29: 899 – 906.

Spivak, C.E., Gund, T.M., Liang, R.F. and Waters, J.A. (1986) Structural and electronic requirements for potent agonists at a nicotinic receptor. *Eur. J. Pharmacol.*, 120: 127 – 131.

Sulfrin, J.R., Dunn, D.A. and Marchall, G.R. (1981) Steric mapping of the L-methionine binding site of ATP:L-methionine S-adenosyltransferase. *Mol. Pharmacol.*, 19: 307 – 313.

Swanson, K.L., Allen, C.N., Aronstam, R.S., Rapoport, H. and Albuquerque, X. (1986) Molecular mechanisms of the potent and stereospecific nicotinic receptor agonist (+)-anatoxin-a. *Mol. Pharmacol.*, 29: 250 – 257.

Waters, J.A., Spivak, C.E., Hermsmeier, M., Yadav, J.S., Liang, R.F. and Gund, T.M. (1988) Synthesis, pharmacology, and molecular modelling studies of semirigid, nicotinic agonists. *J. Med. Chem.*, 31: 545 – 554.

Zhang, X., Stjärnlöf, P., Adem, A. and Nordberg, A. (1987) Anatoxin-a a potent ligand for nicotinic cholinergic receptors in the brain. *Eur. J. Pharmacol.*, 135: 457 – 458.

A. Nordberg, K. Fuxe, B. Holmstedt and A. Sundwall (Eds.)
Progress in Brain Research, Vol. 79
© 1989 Elsevier Science Publishers B.V. (Biomedical Division)

CHAPTER 9

α-Bungarotoxin receptors in the CNS

Barbara J. Morley

Research Division, Boys Town National Institute for Communication Disorders in Children, 555 North 30th Street, Omaha, NE 68131, U.S.A.

Introduction

Significant advances have been made in our understanding of the biochemistry and physiology of the nicotinic acetylcholine receptor (nAChR) present in skeletal muscle and in tissue derived from muscle, such as the electroplax of eel and electric fish. Acetylcholine (ACh) binds to a site on the nAChR which also binds neurotoxins, most notably α-bungarotoxin (BuTX). The skeletal muscle nAChR has long served as a model for receptor function, including the physiological effects of agonists and antagonists, receptor regulation, and studies of the receptor-ion channel complex.

BuTX also binds with high affinity to several neuronal tissues, including mammalian brain (Morley et al., 1977; Schmidt, 1977; Lukas and Bennett, 1978; Lukas et al., 1979). The receptor sites have a biochemistry and pharmacology similar to that of the nAChRs at the skeletal muscle synapse (Morley et al., 1977a; Schmidt, 1977; Lukas and Bennett, 1978; Lukas et al., 1979; Lukas and Bennett, 1980; Morley and Kemp, 1981; Kemp and Morley, 1985).

Immunological distinctions exist between classes of nAChRs. The data suggest that rat brain BuTX binding sites cross-react minimally or not at all with *Torpedo* or rat skeletal muscle nAChR (Morley et al., 1983a). Some cross-reactivity may exist between *Electrophorus* nAChR and rat brain BuTX binding sites (rev. in Lindstrom et al.,

1987). In general, however, it has been found that antibodies to skeletal muscle nAChR cannot be used to effectively purify, isolate, or label mammalian neuronal receptors that are recognized by BuTX (Lindstrom et al., 1987).

The equivalence of BuTX binding sites with functional nAChRs has been questioned on the basis of electrophysiological and release studies in a number of neural systems. BuTX does not block post-synaptic activation in chick sympathetic ganglion, rat superior cervical ganglion, the spinal Renshaw cell, and PC 12 cells (i.e. Duggan et al., 1976; Patrick and Stallcup, 1977; Carbonetto et al., 1978; Kouvelas et al., 1978) although blocking activity is found in frog sympathetic neurons (Marshall, 1981). BuTX does not antagonize the agonist-induced release of neurotransmitters from brain (rev. in Morley and Kemp, 1981). In rat brain, we found that BuTX altered the modulatory effects of carbachol on single unit recording in the inferior colliculus, but the effect was weaker than we would have expected of true neurotransmitter antagonism (Farley et al., 1983; Morley et al., 1983b).

Despite the disappointing electrophysiological data, there are biochemical data that support the supposition that the BuTX binding component is a bona fide nicotinic receptor. The 55 kDa subunit of the BuTX binding component can be labelled with [^{3}H]maleimidobenzyl trimethylammonium bromide (MBTA), a ligand that binds to a well-conserved region of the α-subunit identical to or near the agonist binding site in *Torpedo* nAChR. ACh blocks the binding of MBTA to the 55 kDa rat brain subunit, suggesting that the agonist binding site is on this subunit (Kemp et al., 1985).

Since agonists can apparently interact in vitro with the BuTX binding protein, it is paradoxical that little or no electrophysiological blocking of nicotinic actions can be detected with BuTX. Apparently this protein does not contribute significantly to nicotinic ion fluxes. One subunit of the chick brain BuTX binding protein shares sequence homology with *Torpedo* nAChR (Conti-Tronconi et al., 1985). This suggests that at least certain sequences of functional nAChRs have been conserved in the BuTX binding protein. Thus, the BuTX binding protein may serve a function in mammalian brain that is related to cholinergic neurotransmission.

In this chapter we will discuss efforts to (1) determine if cholinergic agents interact with the BuTX binding protein in vivo, and (2) investigate the relationship between the concentration of BuTX binding sites and the effect of nicotine on behaviors that are believed to be mediated by nicotinic cholinergic receptors.

The regulation of BuTX binding sites by cholinergic drugs

We have previously reported that the administration of choline chloride in the drinking water results in an increase in the number of BuTX binding sites in rat brain (Morley et al., 1977b; Morley and Garner, 1986; Morley and Fleck, 1987). The increase is apparent after 24 h of treatment, is dose-dependent, is age-dependent, and persists for several days after treatment is stopped. The effect is identical when an equivalent amount of choline as phosphadtidyl choline is administered in the drinking water.

The mechanism by which choline increases the concentration of toxin binding sites is not known. Choline may increase the release of ACh, act as an agonist at nicotinic receptor sites, uncover "hidden" receptor sites, or generally increase

membrane-bound proteins. Since we have no evidence that choline increases all membrane-bound proteins (Morley and Fleck, 1987), we are currently assuming that the effect of choline on BuTX binding sites is specific and related to its function as an nAChR.

In order to investigate the role of the presynaptic element in the up-regulation of toxin sites by choline, we administered ethylcholine aziridinium ion (AF64), a potentially specific neurotoxin which produces cholinergic hypoactivity (i.e. Mantione et al., 1981). Under certain circumstances, AF64 may also produce the degeneration of cholinergic axons (i.e. Millar et al., 1987). AF64 was infused into the lateral ventricles of rats as described in Fig. 1. Half of the rats were then administered 15 mg/ml of choline chloride in their drinking water.

After 7 days, the cortex, hippocampus, and hypothalamus of the choline-treated animals and

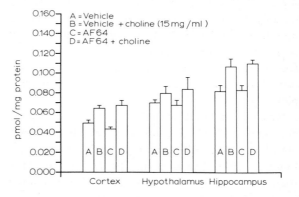

Fig. 1. AF64 was synthesized from acetylethylcholine mustard HCl as described by Mantione et al. (1981). The efficiency of synthesis was determined by the iodine titration method (Sandberg et al., 1986). Animals were infused with 2.5 nmol AF64 or artificial CSF in each lateral ventricle. Three days later groups B and D were administered 15 mg/ml choline chloride in their drinking water. After 4 days of choline treatment, the cortex, hippocampus, and hypothalamus were assayed for BuTX binding as previously described (Morley and Fleck, 1987). Groups B and D were significantly higher than the controls ($P < 0.01$) but not different from each other, indicating that choline chloride increases the number of toxin sites even when HACU is decreased by $\simeq 50\%$. The AF64-treated animals did not differ from controls.

controls were assayed for BuTX binding. As is shown in Fig. 1, choline chloride increased the number of toxin binding sites in both control and AF64-treated animals. AF64 did not affect the concentration of toxin sites nor did it prevent the up-regulation of BuTX sites by choline. These results suggest that the effect of choline on toxin sites may persist despite cholinergic hypoactivity and implies that choline may increase the density of toxin sites by a direct effect on the (presumed post-synaptic) receptors.

Apparent up-regulation of toxin sites is also produced by the chronic administration of nicotine (Morley and Fleck, 1988). The minimum dose of nicotine tartrate producing a significant increase in the number of toxin binding sites is about 3.0 mg/kg. The number of sites is increased within 24 h after treatment, but there appears to be an additional increase after 10 days of treatment (see Fig. 2). Upon the withdrawal of nicotine, the number of binding sites decreases to normal values within 1 – 4 days.

Treatment of rats with the acetylcholinesterase inhibitor, diisopropylfluorophosphate (DFP), results in a decrease in the number of toxin sites in the cortex, but not the hypothalamus or hippocampus (Fig. 3). After 4 days without treatment with DFP, the density of cortical toxin sites is no longer

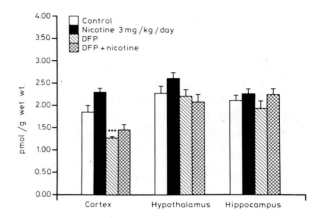

Fig. 3. Animals were injected with DFP saline for 10 days (1 mg/kg of DFP on day 1; 0.5 mg/kg on day 2 and 0.2 mg/kg on days 7 – 10). At the end of the 10 days, the cortex, hippocampus, and hypothalamus were assayed for BuTX binding. The concentration of binding sites in the cortex was significantly decreased in comparison to controls ($P < 0.001$), but the hypothalamus and hippocampus were not different from controls. Nicotine did not reverse the effect of DFP.

significantly decreased in comparison to controls (Fig. 4A). After 10 days of recovery, there is no difference at all between the groups (Fig. 4B). These data suggest that the density of toxin sites can be regulated by an agent that affects cholinergic activity, but that this effect is specific to the toxin sites found in the cortex.

As will be discussed in other chapters in this volume, the agonists $^3H(-)$nicotine and $[^3H]$ACh also bind with high affinity to mammalian brain, but they bind to a molecule that is distinct from the toxin binding component. Mecamylamine, a nicotinic ganglionic blocker that prevents the central effects of nicotine does not displace the binding of either agonists or BuTX to neural tissues. Mecamylamine is believed to be an effective antagonist in several areas of the brain rich in toxin binding sites, i.e. the inferior colliculus and hypothalamus, suggesting that mecamylamine could be interacting with nAChRs labelled by BuTX. In order to obtain some evidence that mecamylamine interacts in vivo with the toxin binding component, we investigated the effects of mecamylamine on the up-regulation

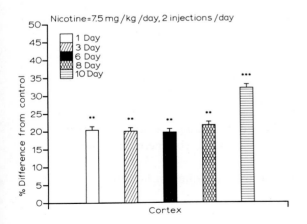

Fig. 2. Nicotine tartrate (7.5 mg/kg per day) was injected twice daily at 0900 and 1600 h for 1 – 10 days (** $P < 0.01$; *** $P < 0.001$ in comparison with controls).

of toxin sites by nicotine. While mecamylamine alone did not affect the concentration of toxin binding sites (Fig. 5A), the administration of mecamylamine 20 min before an injection of nicotine enhanced the effect of nicotine on toxin binding sites. The enhancement of up-regulation was a surprising finding, but mecamylamine has also been found to increase BuTX binding sites in cultured

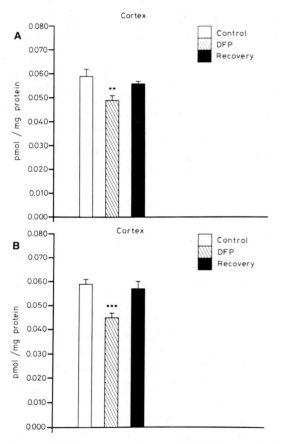

Fig. 4. (A) Thirty rats were injected with DFP or saline for 18 days in the following sequence: group 1, 18 days saline; group 2, 8 days saline followed by 10 days DFP; group 3, 4 days saline, 10 days DFP, 4 days saline. Following 4 days of recovery, the density of binding sites was not significantly different from controls. (B) Thirty rats were injected with saline or DFP in the following sequence: group 1, 30 days saline; group 2, 20 days saline, 10 days DFP; group 3, 10 days saline, 10 days DFP, 10 days saline. Following 10 days of recovery, the DFP-treated animals did not differ from controls (** $P < 0.01$; *** $P < 0.001$ in comparison with controls).

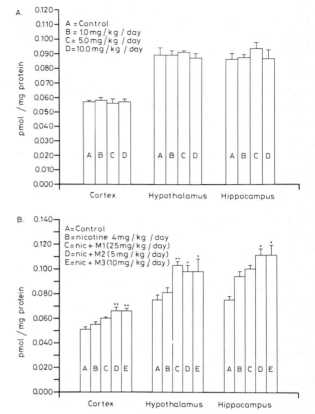

Fig. 5. Rats were injected with (A) mecamylamine or (B) mecamylamine 20 min prior to the injection of nicotine tartrate (3.0 mg/kg per day) at 0900 and 1600 h. The administration of mecamylamine alone had no significant effect on BuTX binding, but administered prior to nicotine injections, it enhanced the effect of nicotine (* $P < 0.05$; ** $P < 0.01$ in comparison with controls).

adrenal medulla chromaffin cells (Quik et al., 1987). These observations indicate that cholinergic drugs interact with BuTX binding sites in vivo, and that the BuTX binding protein may therefore be a functional nicotinic cholinergic receptor.

The effects of nicotine on behavior

Activity

In doses that produce the up-regulation of toxin binding, nicotine administered via osmotic minipumps results in a transient increase in behavioral activity as measured by telemetry (Fig. 6). Some of the effect of nicotine demonstrated in Fig. 6 relates to the absence of a depression in activity following surgery, which probably reflects the ability of nicotine to reverse the depressant effect of anesthesia.

Circadian rhythms

Rats maintained in a 12 h light-dark cycle demonstrate diurnal rhythms with a period of approximately 24 h (circadian rhythms). Circadian rhythms are under the control of an internal ''biological clock'' that is entrained by events in the environment. The primary entraining agent is the light-dark cycle (Turek, 1987).

Carbachol is effective in mimicking the phase-shifting effects of light on circadian rhythms of activity (Zatz and Herkenham, 1981) and gonadal

function (Earnest and Turek, 1983) and mecamylamine blocks the phase-shifting effects of carbachol (Keefe et al., 1987). In addition, the intraventricular injection of BuTX has been reported to block the effect of light on pineal activity (Zatz and Brownstein, 1981), but this effect could not be repeated (Miller and Billiar, 1986b).

The nuclei identified as oscillators of the circadian clock are located in the hypothalamus. The primary oscillator is the suprachiasmatic nucleus (SCN; Moore, 1982), an area rich in toxin binding sites (Miller and Billiar, 1986a) which are associated with axo-dendritic synapses (Miller et al., 1987). There is also electrophysiological evidence to support the idea that ACh is important in the entrainment of the circadian rhythm (Nishino et al., 1977).

In doses that increase motor activity and the concentration of toxin binding, nicotine failed to affect the circadian rhythm of activity either during the condition when the rhythm was entrained by a light-dark cycle (Fig. 7), or when animals were maintained in a condition of constant light for 2 weeks.

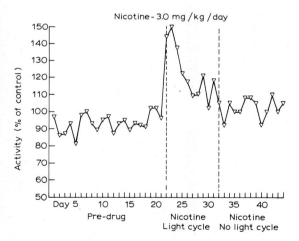

Fig. 6. Fourteen rats were implanted with battery-operated telemeters (Model VM-FH; Mini-Mitter Corp, Sunriver, OR) and baseline activity was recorded for 3 weeks. After 3 weeks, 21-day mini-pumps (Alza) were implanted into rats under ketamine/xylazine anesthesia and telemetry recordings continued. The mini-pumps delivered 3.0 mg/kg per day of nicotine or saline. Nicotine-treated animals had higher activity level than controls and this effect was accentuated because the activity of the control animals was depressed the first day following surgery, probably as a result of anesthesia. At the end of 2 weeks, the lighting was changed from a 12 h light-dark cycle to constant light.

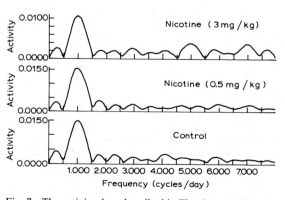

Fig. 7. The activity data described in Fig. 6 was analyzed using spectrum analysis (Data-Quest software, Mini-Mitter). Representative samples of the data for nicotine-treated and controls are shown. Both controls and nicotine-treated animals maintained 24 h rhythms (shown above as a peak of 1.0 cycles/day) in a 12 h light-dark cycle.

The effects of AF64 on the circadian rhythm and nAChRs

In order to further study the possibility that nAChRs might be involved in the generation or maintenance of the circadian clock, we attempted to produce cholinergic hypoactivity in the hypothalamus. We injected 5 nmol of AF64 into the third ventricle of rats and studied the effect of the drug on the circadian rhythm and the number of hypothalamic [^{125}I]BuTX and ^{3}H($-$)nicotine binding sites. Sodium-dependent, high-affinity choline uptake (HACU) was also measured to confirm the effectiveness of the neurotoxin. Animals injected with control solution all maintained circadian rhythms for activity with a period exactly 24 h, while all of the AF64-treated animals had disrupted or eliminated rhythms. Representative data from one experimental animal and one control animal are shown in Fig. 8.

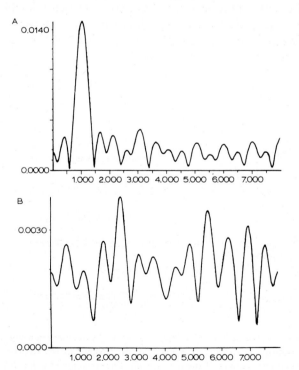

Fig. 8. An AF64-treated animal (bottom) had a disrupted rhythm in comparison to a control animal (top). Representative cases are shown here.

HACU was decreased by 54% in AF64-treated animals in comparison with controls (Fig. 9A). The concentration of binding sites for [^{125}I]BuTX in the hypothalamus of AF64-treated and control animals were not significantly different (see Fig. 9B). The number of binding sites for ^{3}H($-$) nicotine in the hypothalamus of AF64-treated and control animals were also not significantly different (Fig. 9C).

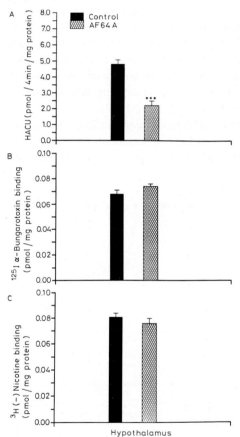

Fig. 9. (A) HACU was determined with a modification of the method described by Murrin and Kuhar (1976). HACU was decreased 54% in the hypothalamus 7 days after the intraventricular administration of 5 nmol AF64 (*** $P < 0.001$). (B) The concentration of BuTX binding sites was determined as previously described (Morley, 1981). BuTX binding in the hypothalamus was not significantly altered 7 days after the administration of AF64. (C) ^{3}H($-$)nicotine receptor assays were performed as described by Lippiello et al. (1986). The density of ^{3}H($-$)nicotine sites was not significantly altered 7 days after the administration of AF64.

Conclusions

The α-bungarotoxin binding component in rat brain is a well-characterized biochemical entity that has an agonist binding site. Despite a concerted effort, however, a role for this molecule in cholinergic neurotransmission has not been identified. Our studies of the in vivo regulation of the BuTX protein indicates quite clearly that physiologically relevant doses of cholinergic drugs can alter the density of receptors. This implies that the BuTX protein interacts with cholinergic drugs in vivo. In addition, the studies of the effects of DFP on toxin binding sites indicates that at least cortical receptors are probably present on neurons innervated by cholinergic axons.

The search for a physiologically relevant function for the BuTX binding protein has not yet been encouraging. Studies of physiological and behavioral characteristics under the control of the hypothalamus, however, may eventually lead to more promising results. The hypothalamus is associated with a wide variety of endocrine and homeostatic functions, many of which are believed to be mediated by nicotinic receptors. The hypothalamus is rich in BuTX binding sites and may make a good model with which to study the function of the BuTX binding protein.

Acknowledgements

This research was supported by Grants 85-36 and 86-36 from the State of Nebraska Department of Health to BJM. The author thanks Dr. Clement A. Stone of Merck Sharp & Dohme Research Laboratories for supplying the mecamylamine HCl and Ms. Charlotte Lieser for her assistance in preparing the manuscript.

References

Carbonetto, S.T., Fambrough, D.M. and Muller, K.J. (1978) Non-equivalence of α-bungarotoxin receptors and acetylcholine receptors in chick sympathetic neurons. *Proc. Natl. Acad. Sci. U.S.A.*, 75: 1016–1020.

Conti-Tronconi, B.A., Dunn, S.M.J., Barnard, E.A., Dolly, O.J., Lai, F.A., Ray, N. and Raftery, M.A. (1985) Brain and muscle nicotinic acetylcholine receptors are different but homologous proteins. *Proc. Natl. Acad. Sci. U.S.A.*, 82: 5208–5212.

Duggan, A.W., Hall, J.G. and Lee, C.Y. (1976) Alpha-bungarotoxin, cobra neurotoxin, and excitation of Renshaw cells by acetylcholine. *Brain Res.*, 107: 166–170.

Earnest, D.J. and Turek, F.W. (1983) Role for acetylcholine in mediating effects of light on reproduction. *Science*, 219: 77–79.

Farley, G.R., Morley, B.J., Javel, E. and Gorga, M.P. (1983) Single-unit responses to cholinergic agents in the rat inferior colliculus. *Hear. Res.*, 11: 73–91.

Kemp, G.E., Bentley, L., McNamee, M.G., and Morley, B.J. (1985) Purification and characterization of the α-bungarotoxin binding protein from rat brain. *Brain Res.*, 347: 274–283.

Keefe, D.L., Earnest, D.L., Nelson, D., Takahashi, J.S. and Turek, F.W. (1987) A cholinergic antagonist, mecamylamine, blocks the phase-shifting effects of light on the circadian rhythm of locomotor activity in the golden hamster. *Brain Res.*, 403: 308–312.

Kouvelas, E.D., Dichter, M.A. and Greene, L.A. (1978) Chick sympathetic neurons develop receptors for α-bungarotoxin in vitro but the toxin does not block nicotinic receptors. *Brain Res.*, 154: 83–93.

Lindstrom, J., Schoepfer, R. and Whiting, P. (1987) Molecular studies of the neuronal nicotinic acetylcholine receptor family. *Mol. Neurobiol.*, 1: 281–337.

Lippiello, P.M. and Fernandes, K.G. (1986) The binding of L-[³H]nicotine to a single class of high affinity sites in rat brain membranes. *Mol. Pharmacol.*, 29: 448–454.

Lukas, R.J. and Bennett, E.L. (1978) α-Bungarotoxin properties of a central nervous system nicotinic acetylcholine receptor. *Biochim. Biophys. Acta*, 554: 294–308.

Lukas, R.J. and Bennett, E.L. (1980) Interaction of nicotinic receptor affinity reagents with central nervous system α-bungarotoxin-binding receptor sulfhydryls and disulfides. *Mol. Pharmacol.*, 17: 149–155.

Lukas, R.J., Morimoto, H. and Bennett, E.L. (1979) Effects of thio-group modification and Ca^{2+} on agonist-specific state transitions of a central nicotinic acetylcholine receptor. *Biochemistry*, 18: 2384–2395.

Mantione, C.R., Fisher, A. and Hanin, I. (1981) The AF64A-treated mouse: possible model for central cholinergic hypofunction. *Science*, 213: 579–580.

Marshall, L.M. (1981) Synaptic localization of α-bungarotoxin binding which blocks nicotinic transmission at frog sympathetic neurons. *Proc. Natl. Acad. Sci. U.S.A.*, 78: 1948–1952.

Millar, T.J., Ishimotor, I., Boelen, M., Epstein, M.L., Johnson, C.D. and Morgan, I.G. (1987) The toxic effects of ethylcholine mustard aziridinium ion on cholinergic cells in the chicken retina. *J. Neurosci.*, 7: 343–356.

108

Miller, M.M. and Billiar, R.B. (1986a) A quantitative and morphometric evaluation of ^{125}I-alpha-bungarotoxin binding in the rat hypothalamus. *Brain Res. Bull.,* 16: 681 – 688.

Miller, M.M. and Billiar, R.B. (1986b) Relationship of putative nicotinic cholinergic receptors in the suprachiasmatic nucleus to levels of pineal serotonin N-acetyltransferase activity in the normally cycling female, the male, and the ovariectomized rat. *J. Pineal Res.,* 3: 159 – 168.

Miller, M.M., Billiar, R.B. and Beaudet, A. (1987) Ultrastructural distribution of alpha-bungarotoxin binding sites in the suprachiasmatic nucleus of the rat hypothalamus. *Cell Tissue Res.,* 250: 13 – 20.

Moore, R.Y. (1982) The suprachiasmatic nucleus and the organization of a circadian system. *Trends Neurosci.,* 5: 404 – 407.

Morley, B.J. (1981) The properties of brain nicotine receptors. *Pharmacol. Ther.,* 15: 111 – 122.

Morley, B.J. and Fleck, D.L. (1987) A time-course and dose-response study of the regulation of brain nicotinic receptors by dietary choline. *Brain Res.,* 421: 21 – 29.

Morley, B.J. and Fleck, D. (1988) The regulation of α-bungarotoxin binding sites by cholinergic agents. *J. Neurochem.,* (under revision).

Morley, B.J. and Garner, L.L. (1986) Increases in the concentration of brain α-bungarotoxin binding sites induced by dietary choline are age-dependent. *Brain Res.,* 378: 315 – 319.

Morley, B.J. and Kemp, G.E. (1981) Characterization of a putative nicotinic acetylcholine receptor in mammalian brain. *Brain Res. Rev.,* 3: 83 – 104.

Morley, B.J., Lorden, J.F., Bradley, R.J., Brown, G.B. and Kemp, G.E. (1977a) Regional distribution of nicotinic receptors in rat brain. *Brain Res.,* 134: 161 – 166.

Morley, B.J., Robinson, G.R., Brown, G.B., Kemp, G.E. and Bradley, R.J. (1977b) Effects of dietary choline on nicotinic acetylcholine in brain. *Nature (Lond.),* 266: 848 – 850.

Morley, B.J., Dwyer, D.S., Strang-Brown, P., Bradley, R.J. and Kemp, G.E. (1983a) Evidence that certain peripheral anti-acetylcholine receptor antibodies do not interact with brain BuTX binding sites. *Brain Res.,* 262: 109 – 116.

Morley, B.J., Farley, G.R. and Javel, E. (1983b) Nicotinic acetylcholine receptors in mammalian brain. *Trends Pharmacol. Sci.,* 4: 225 – 227.

Murrin, L.C. and Kuhar, M.J. (1976) *Mol. Pharmacol.,* 12: 1082 – 1090.

Nishino, H. and Koizumi, K. (1977) Responses of neurons in the suprachiasmatic nuclei of the hypothalamus to putative transmitters. *Brain Res.,* 120: 167 – 172.

Patrick, J. and Stallcup, W.B. (1977) Immunological distinction between acetylcholine receptor and the α-bungarotoxin binding component on sympathetic neurons. *Proc. Natl. Acad. Sci. U.S.A.,* 71: 4689 – 4692.

Quik, M., Geertsen, S. and Trifaro, J.M. (1987) Marked upregulation of the α-bungarotoxin site in adrenal chromaffin cells by specific nicotinic antagonists. *Mol. Pharmacol.,* 31: 385 – 391.

Sandberg, K., Schnaar, R.L. and Coyle, J.T. (1986) Method for the quantitation and characterization of the cholinergic neurotoxin, monoethylcholine mustard aziridinium ion (AF64A). *J. Neurosci. Methods,* 14: 143 – 148.

Schmidt, J. (1977) Drug binding properties of an α-bungarotoxin-binding component from rat brain. *Mol. Pharmacol.,* 13: 282 – 290.

Turek, F.W. (1987) Pharmacological probes of the mammalian circadian clock: use of the phase response curve approach. *Trends Pharmacol. Sci.,* 8: 212 – 217.

Zatz, M. and Brownstein, M.J. (1981) Injection of α-bungarotoxin near the suprachiasmatic nucleus blocks the effects of light on nocturnal pineal enzyme activity. *Brain Res.,* 213: 438 – 442.

Zatz, M. and Herkenham, M.A. (1981) Intraventricular carbachol mimics the phase-shifting effect of light on the circadian rhythm of wheel-running activity. *Brain Res.,* 212: 234 – 238.

A. Nordberg, K. Fuxe, B. Holmstedt and A. Sundwall (Eds.)
Progress in Brain Research, Vol. 79
© 1989 Elsevier Science Publishers B.V. (Biomedical Division)

CHAPTER 10

Characterization of neuronal nicotinic receptors using neuronal bungarotoxin

Ralph H. Loring*, David W. Schulz and Richard E. Zigmond**

Department of Biological Chemistry and Molecular Pharmacology, Harvard Medical School, Boston, MA 02115, U.S.A.

Snake venom neurotoxins, in particular α-bungarotoxin (ABT), have been extremely valuable in the study of nicotinic receptors in skeletal muscle and in electric organs. The high affinity of ABT for these receptors has aided in their quantitation and localization. It was natural, therefore, that ABT was used in many early attempts to study nicotinic receptors in autonomic ganglia and in the central nervous system, and high affinity binding sites for this toxin were indeed found in a variety of neuronal preparations (e.g. Greene et al., 1973; Fumagalli et al., 1976; Vogel and Nirenberg, 1976; Morley et al., 1977). Furthermore, competition for binding to these sites was seen with traditional nicotinic agonists and antagonists. Therefore, it was surprising that in most of the neuronal preparations tested, ABT did not block nicotinic transmission, as it does at the vertebrate neuromuscular junction (see Chiappinelli, 1985 for review). Such preparations include both autonomic ganglia (e.g. the rat superior cervical ganglion — Brown and Fumagalli, 1977) and areas of the central nervous system (e.g. the cat spinal cord — Duggan et al., 1976).

Current address: *Department of Pharmacology, Northeastern University, Boston, MA 02115, and **Center for Neuroscience, Case Western Reserve University School of Medicine, Cleveland, OH 44106, U.S.A.

Discovery of neuronal bungarotoxin — a neurotoxin that blocks ganglionic nicotinic receptors

In a study in our laboratory on the chick ciliary ganglion, it was found that certain lots of commercially available ABT *did* block nicotinic transmission, while others did not (Chiappinelli and Zigmond, 1978). When the different lots of toxin were examined by gel electrophoresis, those lots that *did not* block transmission showed a single band, with a molecular weight expected for ABT (8 kDa). However, the lots that did block transmission showed an additional band with a molecular weight of approximately 6.5 kDa (Chiappinelli et al., 1981).

Ravdin and Berg (1979) purified a toxin from *Bungarus multicinctus* venom that had a molecular weight smaller than ABT and that blocked the depolarizations produced by iontophoretically applied acetylcholine onto chick ciliary neurons in culture. They named this toxin bungarotoxin 3.1. When ABT, free of contamination by bungarotoxin 3.1, was tested in the same system, it had no effect. Our laboratory subsequently purified a toxin from the same venom, which had properties similar to that of bungarotoxin 3.1, but which seemed to behave differently on ion-exchange chromatography (Loring et al., 1983, 1984). We named this material toxin F, while raising the possibility that toxin F might be identical to bungarotoxin 3.1. Chiappinelli (1983) later reported purification of a toxin, which he named ϰ-bungarotoxin and which he reported differed from toxin F in its isoelectric point (Dryer and Chiappinelli, 1983). However, when all three materials were examined in the same isoelectric focussing gel, bungarotoxin 3.1, toxin F, and ϰ-bungarotoxin focussed identically (Loring et al., 1986). Further evidence for the similarity of these materials came from amino acid sequencing, which revealed that the sequences of ϰ-bungarotoxin (Grant and Chiappinelli, 1985) and toxin F (Loring et al., 1986) were identical. In addition, data on the first 16 amino acid residues of bungarotoxin 3.1 indicated that the N-terminal sequences of all three materials are identical (Loring et al., 1986). Finally, the amino acid composition of toxin F and bungarotoxin 3.1 are indistinguishable. Based on these data, we believe that the three toxins are actually the same molecule. Lindstrom et al. (1987) have recently suggested that this toxin be called "neuronal bungarotoxin" (NBT), and we and others have adopted this nomenclature.

NBT has 66 amino acids, including 10 half cystine residues. These characteristics indicate that the toxin is related to the class of "long" snake venom α-neurotoxins (Dufton and Hider, 1983). Around 30 long α-neurotoxins have been sequenced, and, thus, their sequences can be compared to that of NBT. While NBT shares approximately 50% sequence homology with these other toxins, it also contains a number of unique residues. Among these, one of the most interesting is the proline at position 36. This proline occurs immediately following an arginine-glycine pair. Since this arginine is thought to be important in the binding of α-neurotoxins to nicotinic receptors, the presence of a proline at position 36 may be partly responsible for the distinct pharmacological profile of NBT (Loring et al., 1986).

Characterization and localization of ganglionic nicotinic receptors using neuronal bungarotoxin

In addition to its action on chick ciliary neurons, NBT has now been found to block nicotinic recep-

tors in a number of other peripheral cholinoceptive cells, including rat (Sah et al., 1987) and chick (Chiappinelli and Dryer, 1984) sympathetic neurons and bovine adrenal chromaffin cells (Higgins and Berg, 1987). A recent study on neonatal rat sympathetic neurons in culture provides evidence for the specificity of the action of the toxin (Sah et al., 1987). When neurons from the superior cervical ganglion are maintained in tissue culture together with cardiac myocytes, the neurons form synaptic connections with the heart cells, with other sympathetic neurons, and often with themselves ("autapses"). The neurons also synthesize and release acetylcholine (Furshpan et al., 1976). Following electrical stimulation of one of these neurons, one can often record a muscarinically mediated hyperpolarization from a nearby cardiac cell and a nicotinically mediated depolarization from the sympathetic neuron itself. NBT (40 – 400 nM) produced a 50 – 100% decrease in the size of the nicotinic response but did not affect the muscarinic response (Sah et al., 1987). This blockade produced by NBT is long-lasting. Thus, the recovery from the blocking effect produced by perfusion with NBT (400 nM) for 10 – 15 min had a half-time of approximately 100 min (Sah et al., 1987). Interestingly, this long-lasting effect of NBT can be prevented by including in the medium the rapidly reversible nicotinic antagonist dihydro-β-erythroidine, suggesting that the two antagonists act at a common site. On the other hand, hexamethonium, also a rapidly reversible nicotinic antagonist, did not protect against the blocking action of NBT. Previous workers have suggested that hexamethonium is one of a number of nicotinic antagonists that act by blocking the ion channel of the nicotinic receptor rather than the agonist recognition site (Ascher et al., 1979).

Binding studies with [^{125}I]NBT in intact ciliary ganglia (Chiappinelli, 1983; Loring et al., 1984; Loring and Zigmond, 1987), in cultures of dissociated neurons from the ciliary ganglion (Halvorsen and Berg, 1986), and in cultures of sympathetic neurons (Loring et al., 1988) indicate that the toxin binds to two pharmacologically distinct sites. One of these sites is also recognized by ABT, while the other is not. Since ABT does not block nicotinic receptors in any of these preparations and since ABT does not interfere with the blocking action of NBT, it is the NBT-selective binding sites that most likely correspond to the nicotinic receptors in these systems. Autoradiographic studies in the sympathetic neuronal cultures indicate that there is a dramatic difference in the distribution of these two types of binding sites in the plasma membrane of these neurons (Loring et al., 1988). The NBT-selective sites are highly localized to membranes near synapses, while the sites recognized by both toxins appear to be excluded from these regions. The density of NBT-selective binding sites at synaptic membranes is quite high, approximately 5000 sites/μm^2, which is only about a factor of $3-4$ lower than that found for nicotinic receptors at the vertebrate neuromuscular junction (Loring and Zigmond, 1988).

Neuronal bungarotoxin blocks nicotinic receptors in rat and chick retina

The first evidence that NBT can block nicotinic receptors in the central nervous system was obtained in studies of the retina. Nicotinic transmission is believed to be important in stimulating ganglion cells in retina from a variety of species. For instance, in the rabbit, light stimulation causes release of acetylcholine, probably from amacrine cells (Masland and Livingstone, 1976; Massey and Neal, 1979), and nicotinic receptors are implicated in the resulting excitation of the ganglion cells (Masland and Ames, 1976; Ariel and Daw, 1982a, b). Similar findings of nicotinically driven ganglion cells have been reported in the retina of cat (Ikeda and Sheardown, 1982), carp (Negishi et al., 1978; Glickman et al., 1982) and mudpuppy (McReynolds and Miyachi, 1986).

Lipton et al. (1987) demonstrated, using patch-clamp recording techniques, that a subpopulation of isolated ganglion cells from rat retina possesses nicotinic receptors that are blocked by 200 nM NBT but not by 10 μM ABT. This blockade by NBT was partially reversible and, when agonist was added during the early phase of washout of the toxin, underlying single channel events appeared. These single channel events had a conductance of about 50 pS. These measurements provided the first documentation of single channel properties of a nicotinic receptor from an identified central nervous system neuron.

Although few electrophysiological studies have been done on the retina in the chick, other types of evidence suggest an important role for cholinergic innervation in this preparation as well. Certain lamina of the inner plexiform layer of the chick retina are richly innervated by fibers containing choline acetyltransferase-like immunoreactivity and acetylcholinesterase activity (Millar et al., 1985; Spira et al., 1987). Furthermore, the inner plexiform layer contains high-affinity choline uptake sites (Baughman and Bader, 1977) and muscarinic receptors (Sugiyama et al., 1977). A portion of the ABT binding found in chick retina is localized to these lamina of the inner plexiform layer (Vogel and Nirenberg, 1976; Vogel et al., 1977), although much of the ABT binding is localized elsewhere and there is no evidence that ABT blocks any functional receptors in this tissue.

We recently reported that NBT, but not ABT, blocks a hexamethonium-sensitive depolarization, recorded from chick retina, that is induced by either nicotinic or mixed cholinergic agonists (Loring et al., 1987). [^{125}I]NBT binds to two sites in homogenates of chick retina; one site that is shared by ABT (K_d = 15 nM) and another that is not (K_d = 3 nM). Since 10 μM ABT has no physiological effect in chick retina, it is most likely the NBT-selective sites that correspond to functional nicotinic receptors, as argued above for autonomic neurons. In the presence of unlabelled ABT (to prevent binding to the shared site), a variety of nicotinic ligands displaced [^{125}I]NBT, though no blockade was seen with hexamethonium, mecamylamine and pempidine, which are believed to act, in part, as channel blockers rather than competitive antagonists (Ascher et al., 1979). In addition, [^{125}I]NBT binding in the presence of unlabelled ABT is localized to the same sublayers of the inner plexiform layer that are richly innervated by fibers containing choline acetyltransferase-like immunoreactivity. These data indicate that the NBT binding sites are found in those regions of the chick retina where nicotinic transmission is expected to occur.

Neuronal bungarotoxin blocks nicotinic receptors in the corpus striatum

The first demonstration of functional nicotinic receptors in the mammalian central nervous system came from studies of Renshaw cells in the cat spinal cord (Curtis and Eccles, 1958). It was shown that the activation of these neurons caused by orthograde stimulation of motor neuron fibers was mimicked by the application of nicotinic agonists and that the excitation caused by either method of stimulation could be blocked by the ganglionic nicotinic antagonist dihydro-β-erythroidine (Curtis and Ryall, 1966a, b, c). Consistent with the pharmacological results in autonomic ganglia discussed above, ABT failed to antagonize nicotinic activation of Renshaw cells (Duggan et al., 1976).

Perhaps the functional effect mediated by central nicotinic receptors that has been most commonly demonstrated is the augmentation of neurotransmitter release. For example, following incubation of cortical tissue with [^{3}H]choline, it has been found that nicotinic agonists enhance the release of [^{3}H]acetylcholine from cortical slices (Beani et al., 1985) or synaptosomes (Rowell and Winkler, 1984). In addition, facilitation of the release of tritiated norepinephrine, dopamine, or serotonin has been documented in a variety of brain regions (reviewed by Chesselet, 1984; also, see below).

The importance of cholinergic neurons in the corpus striatum is indicated by the fact that choline acetyltransferase activity in this brain region is greater than that in almost any other region measured (Yamamura et al., 1974). The high den-

sity of muscarinic receptors ($B_{max} > 1000$ fmol/mg protein: Marchand et al., 1979; Salvaterra and Foders, 1979) relative to [^{125}I]ABT binding sites ($B_{max} < 20$ fmol/mg protein: Morley et al., 1977; Marchand et al., 1979; Salvaterra and Foders, 1979) raised the possibility that cholinergic synaptic transmission in this area is almost exclusively muscarinic in nature. More recent binding studies using ^{3}H-labelled nicotinic agonists have suggested that a somewhat higher level of nicotinic receptors may exist in the striatum ($B_{max} > 80$ fmol/mg protein: Martin and Aceto, 1981; Marks and Collins, 1982). Several laboratories have demonstrated that destruction of dopaminergic terminals in this region following administration of the neurotoxin 6-hydroxydopamine causes a significant decrease in the binding of radiolabelled nicotinic compounds (DeBelleroche et al., 1979; Schwartz et al., 1984; Clarke and Pert, 1985), suggesting that some of these nicotinic receptors are on dopaminergic nerve terminals.

We have chosen to use the nicotine-evoked release of dopamine from rat striatal slices as a model system in which to study the effect of NBT in the brain. A number of laboratories (Westfall, 1974; Giorguieff et al., 1976; Takano et al., 1983; Rapier et al., 1985) have demonstrated, using either synaptosomal or brain slice preparations from the striatum, that nicotinic agonists facilitate the release of [^{3}H]dopamine following preincubation with either [^{3}H]tyrosine or [^{3}H]dopamine. In our studies, we have measured changes in the release of endogenous dopamine, and, thus, we may be monitoring release from a different pool than that measured in these other studies. We first sought to establish that the stimulation of endogenous dopamine release following treatment with nicotine was, in fact, caused by a direct action on nicotinic receptors. This was of particular concern because of reports that this phenomenon is not always dependent on extracellular calcium, and is not always blocked by nicotinic antagonists (for a summary of these studies see Schulz and Zigmond, 1988). Using an experimental system designed to measure dopamine release over relatively short periods of time, we found that the enhancement of release during a 15-s period following exposure to 100 μM nicotine was completely calcium-dependent and was blocked by 1 mM hexamethonium. Furthermore, the increase in dopamine release was reduced from 150 to 25% above basal levels in the presence of 100 nM NBT. Finally, an equal concentration of ABT had no effect on nicotine-evoked dopamine release.

Future studies on nicotinic receptors in the central nervous system

It should be noted that, in the studies with 6-hydroxydopamine cited above, a majority of agonist binding sites are spared. These data suggest that, in addition to "presynaptic" nicotinic receptors, there are also postsynaptic nicotinic receptors in this brain region. Furthermore, Misgeld et al. (1980) have shown that locally evoked "fast excitatory" synaptic potentials in striatal slices are blocked by hexamethonium, mecamylamine, or d-tubocurarine, but not atropine. Because it is estimated that many of the neurons in the caudate-putamen are cholinoceptive (Bloom et al., 1965), an important area of future research will be the study of striatal nicotinic receptors that are not located on afferent terminals.

It is now generally recognized that nicotinic receptors play a larger role in synaptic transmission in the brain than previously assumed. For example, there is increasing evidence that nicotinic transmission may be important in many areas concerned with the processing of visual information. In an early study, Collier and Mitchell (1966) showed that either stimulation of the retina with light or electrical stimulation of the lateral geniculate nucleus resulted in increased release of acetylcholine in the visual cortex. More recently, it has been found that treatment with nicotine causes increased 2-deoxyglucose utilization in the optic chiasm, the superior colliculus, and the lateral geniculate nucleus (Pazdernik et al., 1982; Grunwald et al., 1987; London et al., 1988). Interesting-

ly, our own preliminary data indicate that among the areas of the rat brain in which NBT binding is highest are a number of regions involved in processing visual information, including the superior colliculus and the lateral geniculate nucleus (Schulz et al., 1988).

While, as stated above, ABT does not block nicotinic receptors in most neuronal preparations examined, exceptions do exist. For example, Marshall (1981) reported that ABT blocks transmission in a frog sympathetic ganglion and that ABT binds to synaptic sites in this preparation. Furthermore, de la Garza (1987, 1988) recently reported that, of two evoked potentials they examined in the rat cerebellum, one was blocked by ABT and the other was blocked by NBT, but not by ABT. Clearly, considerable work remains to sort out possible subtypes of neuronal nicotinic receptors. Evidence of the existence of such subtypes is now clearly suggested by molecular biological studies (Nef et al., 1988; Wada et al., 1988).

Over the past few years, several laboratories have accumulated evidence that central nicotinic receptors are decreased in Alzheimer's disease (Nordberg and Winblad, 1986; Shimohama et al., 1986; Whitehouse et al., 1986), suggesting a possible involvement of these receptors in the etiology of this disorder. The availability of NBT, together with the recent development of monoclonal antibodies (Lindstrom et al., 1987) and cDNA probes (Nef et al., 1988; Wada et al., 1988) for central nicotinic receptors, should greatly facilitate future studies in this area.

Acknowledgements

The research described in this chapter was supported by grants NS12651 and NS22472 from the National Institutes of Health. D.W.S was supported by a postdoctoral training grant from the National Institutes of Health (NS 07009) and R.E.Z., by a Research Scientist Award from the National Institutes of Mental Health (MH00162).

114

References

Ariel, M. and Daw, N.W. (1982a) Effects of cholinergic drugs on receptive field properties of rabbit retinal ganglion cells. *J. Physiol.,* 324: 135 – 160.

Ariel, M. and Daw, N.W. (1982b) Pharmacological analysis of directionally sensitive rabbit retinal ganglion cells. *J. Physiol.,* 324: 161 – 186.

Ascher, P., Large, W.A. and Rang, H.P. (1979) Studies on the mechanism of action of acetylcholine antagonists on rat parasympathetic ganglion cells. *J. Physiol.,* 295: 139 – 170.

Baughman, R.W. and Bader, C.R. (1977) Biochemical characterization and cellular localization of the cholinergic system in the chicken retina. *Brain Res.,* 138: 409 – 485.

Beani, L., Bianchi, C., Nilsson, L., Nordberg, A., Romanelle, L. and Sivilotti, L. (1985) The effect of nicotine and cytisine on ^{3}H-acetylcholine release from cortical slices of guinea-pig brain. *Naunyn. Schmiedebergs Arch. Pharmacol.,* 331: 293 – 296.

Bloom, F.E., Costa, E. and Salmoiraghi, G.C. (1965) Anesthesia and the responsiveness of individual neurons of the caudate nucleus of the cat to acetylcholine, norepinephrine, and dopamine administered by microelectrophoresis. *J. Pharmacol. Exp. Ther.,* 150: 244 – 252.

Brown, D.A. and Fumagalli, L. (1977) Dissociation of α-bungarotoxin binding and receptor block in the rat superior cervical ganglion. *Brain Res.,* 129: 165 – 168.

Chesselet, M.F. (1984) Presynaptic regulation of neurotransmitter release in the brain: facts and hypothesis. *Neuroscience,* 12: 347 – 376.

Chiappinelli, V.A. (1983) Kappa bungarotoxin: a probe for the neuronal nicotinic receptor in the avian ciliary ganglion. *Brain Res.,* 277: 9 – 21.

Chiappinelli, V.A. (1985) Actions of snake venom toxins on neuronal nicotinic receptors and other neuronal receptors. *Pharmac. Ther.,* 31: 1 – 32.

Chiappinelli, V.A. and Zigmond, R.E. (1978) α-Bungarotoxin blocks nicotinic transmission in the avian ciliary ganglion. *Proc. Natl. Acad. Sci. U.S.A.,* 75: 2999 – 3003.

Chiappinelli, V.A., Cohen, J.B. and Zigmond, R.E. (1981) Effects of α- and β-neurotoxins from the venoms of various snakes on transmission in autonomic ganglia. *Brain Res.,* 211: 107 – 126.

Chiappinelli, V.A. and Dryer, S.E. (1984) Nicotinic transmission in sympathetic ganglia: blockade by the snake venom neurotoxin kappa-bungarotoxin. *Neurosci. Lett.,* 50: 239 – 244.

Clarke, P.B.S. and Pert, A. (1985) Autoradiographic evidence for nicotine receptors on nigrostriatal and mesolimbic dopaminergic neurons. *Brain Res.,* 348: 355 – 358.

Collier, B. and Mitchell, J.F. (1966) The central release of acetylcholine during stimulation of the visual pathway. *J. Physiol. (Lond.),* 184: 239 – 254.

Curtis, D.R. and Eccles, R.M. (1958) The excitation of Ren-
shaw cells by pharmacological agents applied electrophoretically. *J. Physiol.,* 141: 435 – 445.

Curtis, D.R. and Ryall, R.W. (1966a) The excitation of Renshaw cells by cholinomimetics. *Exp. Brain Res.,* 2: 49 – 65.

Curtis, D.R. and Ryall, R.W. (1966b) The acetylcholine receptors of Renshaw cells. *Exp. Brain Res.,* 2: 66 – 80.

Curtis, D.R. and Ryall, R.W. (1966c) The synaptic excitation of Renshaw cells. *Exp. Brain Res.,* 2: 81 – 96.

DeBelleroche, J., Lugami, Y. and Bradford, H.F. (1979) Evidence for presynaptic cholinergic receptors on dopaminergic terminals: degeneration studies with 6-hydroxy-dopamine. *Neurosci. Lett.,* 11: 209 – 213.

De la Garza, R., McGuire, T.S., Freedman, R. and Hoffer, B.J. (1987) Selective antagonism of nicotine actions in the rat cerebellum with α-bungarotoxin. *Neuroscience,* 23: 887 – 891.

De la Garza, R., Hoffer, B.J. and Freedman, R. (1988) Selective antagonism of the electrophysiological actions of nicotine in the rat cerebellum using α-bungarotoxin and k-bungarotoxin. Abstracts, *NATO Advanced Research Workshop on Nicotinic Acetylcholine Receptors in the Nervous System,* p. 30.

Dryer, S.E. and Chiappinelli, V.A. (1983) Kappa-bungarotoxin: an intracellular study demonstrating blockade of neuronal nicotinic receptors by a snake venom neurotoxin. *Brain Res.,* 289: 317 – 321.

Dufton, M.J. and Hider, R.C. (1983) Conformational properties of the neurotoxins and cytotoxins isolated from elapid snake venom. *CRC Crit. Rev. Biochem.,* 14: 113 – 171.

Duggan, A.W., Hall, J.G. and Lee, C.Y. (1976) Alpha-bungarotoxin, cobra neurotoxin and excitation of Renshaw cells by acetylcholine. *Brain Res.,* 107: 166 – 170.

Fumagalli, L., DeRenzis, G. and Miani, N. (1976) Acetylcholine receptors: number and distribution in intact and deafferented superior cervical ganglion of the rat. *J. Neurochem.,* 27: 47 – 52.

Furshpan, E.J., MacLeish, P.R., O'Lague, P.H. and Potter, D.D. (1976) Chemical transmission between rat sympathetic neurons and cardiac myocytes developing in microcultures: evidence for cholinergic, adrenergic and dual-function neurons. *Proc. Natl. Acad. Sci. U.S.A.,* 73: 4225 – 4229.

Giorguieff, M.F., Le Floc'h, M.L., Westfall, T.C., Glowinski, J. and Besson, M.J. (1976) Nicotinic effect of acetylcholine on the release of newly synthesized ^{3}H-dopamine in rat striatal slices and cat caudate nucleus. *Brain Res.,* 106: 117 – 131.

Glickman, R.D., Adolf, A.R. and Dowling, J.E. (1982) Inner plexiform circuits in the carp retina: effects of cholinergic agonists, GABA and substance P on the ganglion cells. *Brain Res.,* 234: 81 – 99.

Grant, G.A. and Chiappinelli, V.A. (1985) Kappa-bungarotoxin: complete amino acid sequence of a neuronal nicotinic receptor probe. *Biochemistry,* 24: 1532 – 1537.

Greene, L.A., Sytkowski, A.J., Vogel, Z. and Nirenberg,

M.W. (1973) α-Bungarotoxin used as a probe for acetylcholine receptors of cultured neurones. *Nature (Lond.),* 243: 163 – 166.

Grunwald, F., Schrock, H. and Kuschinsky, W. (1987) The effect of an acute nicotinic infusion on the local cerebral glucose utilization of the awake rat. *Brain Res.,* 400: 232 – 238.

Halvorsen, S.W. and Berg, D.K. (1986) Identification of nicotinic acetylcholine receptors on neurons using an α-neurotoxin that blocks receptor function. *J. Neurosci.,* 7: 2547 – 2555.

Higgins, L.S. and Berg, D.K. (1987) Immunological identification of a nicotinic acetylcholine receptor on bovine chromaffin cells. *J. Neurosci.,* 7: 1792 – 1798.

Ikeda, H. and Sheardown, M.J. (1982) Acetylcholine may be an excitatory transmitter mediating visual excitation of "transient" cells with the periphery effect in the cat retina: iontophoretic studies in vivo. *Neuroscience,* 7: 1299 – 1308.

Lindstrom, J., Schoepfer, R. and Whiting, P. (1987) Molecular studies of the neuronal nicotinic acetylcholine receptor family. *Molec. Neurobiol.,* 1: 281 – 337.

Lipton, S.A., Aizenman, E. and Loring, R.H. (1987) Neural nicotinic acetylcholine responses in solitary mammalian retinal ganglion cells. *Pfluegers Arch.,* 410: 37 – 43.

London, E.D., Dam, M. and Fanelli, R.J. (1988) Nicotine enhances cerebral glucose utilization in central components of the rat visual system. *Brain Res. Bull.,* 20: 381 – 385.

Loring, R.H. and Zigmond, R.E. (1987) Ultrastructural distribution of [^{125}I]toxin F binding sites on chick ciliary neurons: synaptic distribution of a toxin that blocks ganglionic nicotinic receptors. *J. Neurosci.,* 7: 2153 – 2162.

Loring, R.H. and Zigmond, R.E. (1988) Characterization of neuronal nicotinic receptors by snake venom neurotoxins. *Trends Neurosci.,* 11: 73 – 78.

Loring, R.H., Chiappinelli, V.A., Zigmond, R.E. and Cohen, J.B. (1983) Characterization of a snake venom neurotoxin which blocks nicotinic transmission in autonomic ganglia. *Soc. Neurosci. Abstr.,* 9: 1143.

Loring, R.H., Chiappinelli, V.A., Zigmond, R.E. and Cohen, J.B. (1984) Characterization of a snake venom neurotoxin which blocks nicotinic transmission in the avian ciliary ganglion. *Neuroscience,* 11: 989 – 999.

Loring, R.H., Andrews, D., Lane, W. and Zigmond, R.E. (1986) Amino acid sequence of toxin F, a snake venom neurotoxin that blocks neuronal nicotinic receptors. *Brain Res.,* 38: 30 – 37.

Loring, R.H., Aizenman, E., Lipton, S.A. and Zigmond, R.E. (1987) Characterization of nicotinic receptors in chick retina. *Soc. Neurosci. Abstr.,* 13: 795.

Loring, R.H., Sah, D.W.Y., Landis, S.C. and Zigmond, R.E. (1988) The ultrastructural distribution of putative nicotinic receptors on cultured neurons from the rat superior cervical ganglion. *Neuroscience,* 24: 1071 – 1080.

Marchand, C.M.F., Hunt, S.P. and Schmidt, J. (1979) Putative acetylcholine receptors in hippocampus and corpus striatum of rat and mouse. *Brain Res.,* 160: 363 – 367.

Marks, M.J. and Collins, A.C. (1982) Characterization of nicotine binding in mouse brain and comparison with the binding of α-bungarotoxin and quinuclidinyl benzilate. *Molec. Pharmacol.,* 22: 554 – 564.

Marshall, L.M. (1981) Synaptic localization of α-bungarotoxin binding which blocks nicotinic transmission at frog sympathetic neurons. *Proc. Natl. Acad. Sci. U.S.A.,* 78: 1948 – 1952.

Martin, B.R. and Aceto, M.D. (1981) Nicotine binding sites and their localization in the central nervous system. *Neurosci. Biobehav. Res.,* 5: 473 – 478.

Masland, R.H. and Ames, A. (1976) Responses to acetylcholine of ganglion cells in an isolated mammalian retina. *J. Neurophysiol.,* 39: 1220 – 1235.

Masland, R.H. and Livingstone, C.J. (1976) Effect of stimulation with light on synthesis and release of acetylcholine by an isolated mammalian retina. *J. Neurophysiol.,* 39: 1210 – 1219.

Massey, S.C. and Neal, M.J. (1979) The light evoked release of acetylcholine from the rabbit retina in vivo and its inhibition by GABA. *J. Neurochem.,* 32: 1327 – 1329.

McReynolds, J.S. and Myachi, E.-I. (1986) The effect of cholinergic agonists and antagonists on ganglion cells in the mudpuppy retina. *Neurosci. Res. Suppl.,* 4: 5153 – 5161.

Millar, T., Ishimoto, I., Johnson, C.D., Epstein, M.L., Chubb, I.W. and Morgan, I.G. (1985) Cholinergic and acetylcholinesterase-containing neurons of the chicken retina. *Neurosci. Lett.,* 61: 311 – 316.

Misgeld, U., Weiler, M.H. and Bak, I.J. (1980) Intrinsic cholinergic excitation in the rat neostriatum: nicotinic and muscarinic receptors. *Exp. Brain. Res.,* 39: 401 – 409.

Morley, B.J., Lordon, J.F., Brown, G.B., Kemp, G.E. and Bradley, R.J. (1977) Regional distribution of nicotinic acetylcholine receptor in rat brain. *Brain Res.,* 134: 161 – 166.

Nef, B., Oneyser, C., Alliod, C., Couturier, S. and Ballivet, M. (1988) Genes expressed in the brain define three distinct neuronal nicotinic acetylcholine receptors. *EMBO J.,* 7: 595 – 601.

Negishi, K., Kato, S., Teranishi, T. and Laufer, M. (1978) An electrophysiological study on the cholinergic system in the retina. *Brain Res.,* 148: 85 – 93.

Nordberg, A. and Winblad, B. (1986) Reduced number of nicotine and [^{3}H]acetylcholine binding sites in the frontal cortex of Alzheimer brains. *Neurosci. Lett.,* 72: 115 – 119.

Pazdernik, T.L., Cross, R.S., Mewes, K., Samson, F. and Nelson, S.R. (1982) Superior colliculus activation by retinal nicotinic ganglion cells: a 2-deoxyglucose study. *Brain Res.,* 243: 197 – 200.

Rapier, C., Harrison, R., Lunt, G.C. and Wonnacott, S. (1985) Neosurugatoxin blocks nicotinic acetylcholine receptors in the brain. *Neurochem. Int.,* 7: 389 – 396.

Ravdin, P.M. and Berg, D.K. (1979) Inhibition of neuronal acetylcholine sensitivity by a toxin from Bungarus multicinctus venom. *Proc. Natl. Acad. Sci. U.S.A.*, 76: 2072 – 2076.

Rowell, P.P. and Winkler, D.L. (1984) Nicotinic stimulation of [³H]acetylcholine release from mouse cerebral cortical synaptosomes. *J. Neurochem.*, 43: 1593 – 1598.

Sah, D.W.Y., Loring, R.H. and Zigmond, R.E. (1987) Long term blockade by toxin F of nicotinic synaptic potentials in cultured sympathetic neurons. *Neuroscience*, 20: 867 – 874.

Salvaterra, P.M. and Foders, R.M. (1979) [¹²⁵I]₂α-Bungarotoxin and [³H]quinuclidinylbenzilate binding in central nervous systems of different species. *J. Neurochem.*, 32: 1509 – 1517.

Schulz, D.W. and Zigmond, R.E (1989) Neuronal bungarotoxin blocks the nicotinic stimulation of endogenous dopamine release from rat striatum. *Neurosci. Lett.* (in press).

Schulz, D.W., Aizenman, E., White, W.F. and Zigmond, R.E. (1988) Autoradiographic localization of nicotinic receptors in rat brain using ¹²⁵I-neuronal bungarotoxin. *Soc. Neurosci. Abstr.*, 14: 1328.

Schwartz, R.D., Lehmann, J. and Kellar, K.J. (1984) Presynaptic nicotinic cholinergic receptors labelled by [³H]acetylcholine on catecholamine and serotonin axons in brain. *J. Neurochem.*, 42: 1495 – 1498.

Shimohama, S., Taniguchi, T., Fujiwara, M. and Kameyama, M. (1986) Changes in nicotinic and muscarinic cholinergic receptors in Alzheimer-type dementia. *J. Neurochem.*, 46: 288 – 293.

Spira, A.W., Millar, T.J., Ishimoto, I., Epstein, M.L., Johnson, C.D., Dahl, J.L. and Morgan, I.G. (1987) Localization of choline acetyltransferase-like immunoreactivity in the embryonic chick retina. *J. Comp. Neurol.*, 260: 526 – 538.

Sugiyama, H., Daniels, M.P. and Nirenberg, M. (1977) Muscarinic acetylcholine receptors of the developing retina. *Proc. Natl. Acad. Sci. U.S.A.*, 74: 5524 – 5528.

Takano, Y., Sakurai, Y., Kowjimoto, Y., Honda, K. and Kamiya, H. (1983) Presynaptic modulation of the release of dopamine from striatal synaptosomes: differences in the effects of high K⁺ stimulation, methamphetamine and nicotinic drugs. *Brain Res.*, 279: 330 – 334.

Vogel, Z. and Nirenberg, M. (1976) Localization of acetylcholine receptors during synaptogenesis in retina. *Proc. Natl. Acad. Sci. U.S.A.*, 73: 1806 – 1810.

Vogel, Z., Maloney, G.J., Ling, A. and Daniels, M.P. (1977) Identification of synaptic acetylcholine receptor sites in retina with peroxidase-labelled α-bungarotoxin. *Proc. Natl. Acad. Sci. U.S.A.*, 74: 3268 – 3272.

Wada, K., Ballivet, M., Boulter, J., Connolly, J., Wada, E., Denaris, E.S., Swanson, L., Heineman, S. and Patrick, J. (1988) Functional expression of a new pharmacological subtype of brain nicotinic acetylcholine receptor. *Science*, 240: 330 – 334.

Westfall, T.C. (1974) Effect of nicotine and other drugs on the release of ³H-norepinephrine and ³H-dopamine from rat brain slices. *Neuropharmacology*, 13: 693 – 700.

Whitehouse, P.J., Martino, A.M., Antuono, P.G., Lowenstein, P.R., Coyle, J.T., Price, D.L. and Kellar, K.J. (1986) Nicotinic binding sites in Alzheimer's disease. *Brain Res.*, 371: 146 – 151.

Yamamura, H.I., Kuhar, M.J., Greenberg, D. and Snyder, S.H. (1974) Muscarinic cholinergic receptor binding: regional distribution in monkey brain. *Brain Res.*, 66: 541 – 546.

A. Nordberg, K. Fuxe, B. Holmstedt and A. Sundwall (Eds.)
Progress in Brain Research, Vol. 79
© 1989 Elsevier Science Publishers B.V. (Biomedical Division)

CHAPTER 11

Nicotinic acetylcholine receptor diversity: agonist binding and functional potency

Ronald J. Lukas

Division of Neurobiology, Barrow Neurological Institute, 350 West Thomas Road, Phoenix, AZ 85013, U.S.A.

Introduction

There is yet to emerge a consensus of scientific opinion regarding the relationships between sites (particularly those in the central nervous system; CNS) that interact with putative probes for nicotinic acetylcholine receptors (nAChR) and physiologically functional cholinergic receptors that are targets for the actions of nicotine (Barnard and Dolly, 1982; see Discussion). Among the first ligands used as probes for neuronal sites of nicotine's actions were the curaremimetic neurotoxins (Changeux et al., 1970), such as α-bungarotoxin (Bgt), which was isolated from the venom of the Formosan banded krait, *Bungarus multicinctus*. These peptide toxins are apparently the most "purely" competitive antagonists of nicotinic activation of acetylcholine (ACh) receptors at the neuromuscular junction (Katz and Miledi, 1978), and it was reasonable to assume that they might recognize the conserved features of binding sites for nicotine and ACh on other nicotinic receptors in the nervous system. However, coincident with the demonstration that high-affinity, specific, radiolabelled Bgt binding sites were present in the CNS (Eterovic and Bennett, 1974), had synaptic disposition (Lentz and Chester, 1977), and exhibited regional distributions that were consistent with the distribution of known cholinergic markers (Salvaterra et al., 1975), were observations that some functional responses to nicotine or

ACh at autonomic neuronal sites where toxin binding had been demonstrated were not sensitive to Bgt-mediated, high-affinity antagonism (reviewed in Chiappinelli, 1985; Loring and Zigmond, 1988). From these studies, the physiological relevance of neurotoxin binding sites was questioned, as was the global utility of those toxins as probes for authentic, functional nAChR not only in the autonomic nervous system, but in the CNS as well (Patrick and Stallcup, 1977a; Morley et al., 1979). Nevertheless, other neurotoxins were isolated from the venom of *Bungarus* species, were shown to exhibit primary sequence similarities with the class of Bgt-like neurotoxins, but were demonstrated to have unique quaternary structure, existing as dimers under physiological conditions (Chiappinelli, 1985; Chiappinelli et al., 1988; Loring and Zigmond, 1988). This class of toxins was shown to have functional antagonistic activity at some, but not all, neuronal sites where Bgt-like neurotoxins were functionally impotent, and these toxins are now being used to identify different classes of nAChR (Loring and Zigmond, 1988).

Similarly, some approaches involved the use of antibodies raised against nAChR from electric tissue or muscle as probes for neuronal nAChR, based on the rationale that these antibodies might recognize highly conserved epitopes that are characteristic of nAChR across tissues (reviewed in Lindstrom, 1986). However, immunogenicity of peripheral nAChR is dominated by an epitope called the main immunogenic region (Lindstrom, 1986); not all anti-electroplax nAChR antibodies are reactive across species and fewer yet have functional activity that would prove their specificity as probes for domains of physiologically relevant neuronal nAChR (but see Patrick and Stallcup, 1977b; Gomez et al., 1979; and review by Lindstrom, 1986).

In an attempt to address these issues, some of the work in our laboratory has involved studies of the interactions of small cholinergic drugs and anti-nAChR antibodies with binding sites for radiolabelled curaremimetic neurotoxins and studies of interactions of anti-nAChR antibodies with nAChR-like epitopes from a variety of neuronal and non-neuronal sources. The results of these studies (see Lukas, 1986a, 1988) suggested the existence of heterogeneity of toxin binding sites and nAChR-like antigens across and within tissues. Other results have shown that some neurotoxins and anti-nAChR antibodies or polyclonal, monospecific antisera exhibit functional potency at nAChR expressed in different clonal cell systems (Lukas, unpublished results).

In contrast to the controversy regarding the physiological relevance of neuronal sites that interact with neurotoxins or anti-nAChR antibodies and the global utility of those agents as probes for functional nAChR, little argument could be made with the operationally defined use of nicotine or ACh (under the appropriate conditions) as specific, physiologically relevant probes for neuronal nAChR. However, evidence suggested that concentrations of agonists in the high micromolar to low millimolar range were required to maximally activate nAChR function (Kasai and Changeux, 1971; Catterall, 1975; Patrick and Stallcup, 1977a), which was consistent with estimates of the vicinal concentration of ACh (about 300 μM) at the active neuromuscular synapse (Kuffler and Yoshikami, 1975). Therefore, even if radiolabelled agonists could be prepared to label high-affinity (nanomolar values of K_d) binding sites, the functional relevance of those sites to the presumptive low-affinity sites

that mediate the functional activities of nAChR would remain uncertain (see below). Undaunted by these considerations, workers in several laboratories used radiolabelled (−)nicotine to define stereoselective binding sites (Abood et al., 1980; Romano and Goldstein, 1980; Marks and Collins, 1982) or ^{3}H-labelled ACh ([^{3}H]ACh) in the presence of cholinesterase blockers and atropine (to block interactions with muscarinic receptors) to identify putative nAChR (Schwartz, et al., 1982) in neuronal tissues.

We now report on other studies in our laboratory that have concerned the properties of high-affinity binding sites for nicotinic agonists on membrane preparations from *Torpedo* electroplax, from rat brain, and from two model neuronal cell lines, the TE671 human medulloblastoma of neural tube origin (McAllister et al., 1977; Lukas, 1986b, 1988), and the PC12 rat pheochromocytoma of neural crest origin (Greene and Tischler, 1976; Lukas, 1986c, 1988a). Also presented are the results of ion efflux studies (Lukas and Cullen, 1989) that have been used to characterize the cellular ensemble of functional nAChR expressed by the PC12 or TE671 cell lines and to establish nAChR sensitivity to nicotine and other cholinergic agonists in order to address issues regarding the identity and functional relevance of high-affinity agonist binding sites. Preliminary accounts of this work have been presented (Lukas, 1986d − f, 1988; Lukas and Cullen, 1989).

Methods

[^{3}H]ACh was obtained from Amersham and used in binding assays (Lukas, 1986f) based on the procedure described by Schwartz et al. (1982). ^{86}Rb$^+$ efflux assays are according to Lukas and Cullen (1989). These assays of nAChR channel function take advantage of this receptor's permeability to monovalent cations as large as glucosamine (Yoshikami, 1981) and the cellular Na$^+$-K$^+$-ATPase's ability to concentrate ^{86}Rb$^+$ in TE671 and PC12 cells, and permits the assessment of nAChR-channel function by the measurement of

^{86}Rb$^+$ release during incubation of isotope-loaded cells in medium containing the appropriate cholinergic ligands. Membrane preparations from *Torpedo* or rat brain were as described (Lukas, 1984, 1986a) as were characterization, maintenance and handling of TE671 (Lukas, 1986b) and PC12 (Lukas, 1986c) cells.

Results

ACh binding assays. We have used the [^{3}H]ACh binding assay described by Schwartz et al. (1982) to study high-affinity binding sites for nicotinic cholinergic agonists on our model preparations. Membrane preparations from *Torpedo* electroplax, TE671 cells and rat brain express high-affinity [^{3}H]ACh binding sites (Table I) with ap-

parent macroscopic dissociation constants (obtained from ligand saturation isotherms and Scatchard analysis) of about 10 nM and Hill coefficients of about 1.0 (Fig. 1). In each case, binding is to a single class of apparently non-interacting sites at densities that are approximately equal to (*Torpedo* and rat brain) or one-half of (TE671 cells) the density of specific, high-affinity binding sites for radiolabelled Bgt (Table I), once the raw data displayed in Table I are corrected to account for the faster (relative to the very slow release of radiolabelled Bgt from binding sites) dissociation of [^{3}H]ACh from *Torpedo* membrane sites.

A small number of [^{3}H]ACh binding sites are also found on membrane preparations from PC12 cells. Scatchard plots of ligand binding data are best fit by a function that assumes the presence of two classes of sites on PC12 cells with macroscopic K_d values of 2 and 40 nM in the ratio of approximately 1 : 6, and less well by a function that assumes a single class of sites with a K_d of 9.8 nM (Table I). In the former case, the molar ratio of

TABLE I

[^{3}H]ACh binding parameters

Preparation	K_d	B_{max}	n_H	S_t/S_a
Torpedo electroplax	12.0	N.D.	1.05	1.34
TE671 cells	9.0	26.7	1.00	2.03
Rat brain	12.4	20.4	1.02	0.99
PC12 cells − one class	9.8	10.0	0.87*	20.5
PC12 cells − two classes	2.0	2.2	N.D.	13.3
	40.0	13.0		

Data from a single representative experiment are summarized for [^{3}H]ACh binding to membrane preparations derived from the indicated sources. Values for K_d (nM) and B_{max} (fmol/mg membrane protein except for *Torpedo* membranes) are derived from Scatchard analysis of saturation binding isotherms and assume the existence of one class of binding site (except where noted for PC12 cell membranes). Values for n_H are obtained from Hill plots (Fig. 1) of the same data over the entire range of ligand concentration used (0.1 – 100 nM) except where indicated*, where data are only for concentrations of [^{3}H]ACh from 0.1 – 30 nM. The parameter S_t/S_a represents the ratio of high-affinity binding sites at saturation for radiolabelled α-bungarotoxin to those for [^{3}H]ACh as obtained from Scatchard analysis of data taken with the same preparations. N.D. = not done.

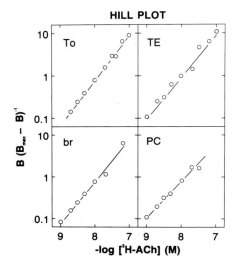

Fig. 1. Hill plots for [^{3}H]ACh binding to membrane preparations derived from *Torpedo* electroplax (To), TE671 cells (TE), PC12 cells (PC), or rat brain (br). Data from a single representative experiment (averages of two determinations at each ligand concentration) were fit by linear-regression analysis to yield the parameters shown in Table I.

sites with high affinity for radiolabelled Bgt to the total number of high- and lower-affinity sites for [3H]ACh is approximately 13 : 1, while in the latter case the ratio of high-affinity toxin sites to those for [3H]ACh is approximately 21 : 1 (Table I). The Hill coefficient for [3H]ACh binding is 0.87 for PC12 cells for data obtained at concentrations of radioligand between 0 and 30 nM (Fig. 1), but may be as high as 4.0 for data obtained at concentrations of [3H]ACh in excess of 30 nM.

We previously have reported that [3H]ACh binding to sites on *Torpedo* membranes or TE671 cells is fully sensitive to high-affinity (ca. 1 nM K_i) blockade by Bgt under the appropriate conditions (Lukas, 1988). By contrast, Bgt only partially (25 – 35%) blocks [3H]ACh binding to rat brain membranes, and then only with low affinity (ca. 1 μM), and at 1 μM has no effect on [3H]ACh binding to PC12 cell preparations. It should be noted that high-affinity binding sites for radiolabelled Bgt to each of these preparations are fully sensitive to blockade by ACh with apparent K_i values of 0.1 – 2 μM, which are between 10- and 700-fold higher than apparent K_i values for blockade of [3H]ACh binding by ACh to the same preparations (Lukas, 1988).

[3H]ACh binding sites from each of these preparations are also distinguishable on the basis of their characteristic sensitivities to blockade by other agonists or small antagonists (Table II). For example, nicotine and cytisine block high-affinity [3H]ACh binding to PC12 cell and rat brain preparations at much lower concentrations than those required to block [3H]ACh binding to TE671 cell or *Torpedo* membranes. By contrast, succinyldicholine and suberyldicholine are more effective blockers of [3H]ACh binding to TE671 cell and *Torpedo* membranes. Among the nicotinic antagonists tested, decamethonium, pancuronium and d-tubocurarine are more selective as blockers of [3H]ACh binding to TE671 and *Torpedo* preparations than to PC12 or brain membranes.

TABLE II

Drug competition toward [3H]ACh (10 nM) binding

Drug	IC$_{50}$ values (μM)			
	Torpedo	TE671	Brain	PC12
Agonists				
Acetylcholine	0.02	0.02	0.006	0.03
Nicotine	6.0	1.0	0.006	0.03
Cytisine	1.0	6.0	0.01	0.01
Suberyldicholine	0.02	0.01	0.3	0.1
Carbamylcholine	1.0	1.0	0.6	0.3
Succinyldicholine	1.0	0.3	60.0	500.0
Antagonists				
Decamethonium	10.0	0.3	30.0	300.0
d-Tubocurarine	1.0	0.1	100.0	10.0
Pancuronium	0.05	1.0	500.0	40.0
Alcuronium	0.3	5.0	600.0	1.0
Hexamethonium	100.0	400.0	500.0	> 1000.0
Mecamylamine	30.0	1000.0	> 1000.0	1000.0

Data from numerous ($n = 3 – 10$) replicate experiments are summarized for cholinergic ligand inhibition of high-affinity [3H]ACh binding to membranes prepared from *Torpedo* electroplax, TE671 cells, rat brain, or PC12 cells. IC$_{50}$ values were taken from inspection of ligand binding competition curves.

Ion efflux assays. We have used a $^{86}Rb^+$ efflux assay to characterize functional nAChR expressed by the PC12 and TE671 cell lines and to establish nAChR sensitivity to nicotine and other cholinergic agonists. Each agonist tested exhibits a characteristic dose-response profile for stimulation of nAChR function. This is illustrated in Fig. 2, which shows the nicotine concentration

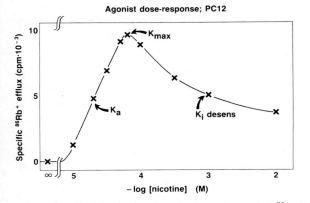

Fig. 2. Nicotine dose-response curve for activation of $^{86}Rb^+$ efflux from PC12 cells. PC12 cells were prepared as described (Lukas and Cullen, 1989) for ion efflux assay of nAChR function and subjected to exposure to nicotine for 5 min at the indicated concentration (abscissa, values in M, log scale) and assayed for the amount of liberated $^{86}Rb^+$ (ordinate, values as cpm $\cdot$ 10^{-3}). Curves similar to this were inspected to determine the parameters K_a, K_{max} and K_i as illustrated and E_{max} as described in the text for a variety of nicotinic agonists (Table III).

dependence of $^{86}Rb^+$ efflux from PC12 cells. Features of the dose-response profile may be described by four operationally defined parameters. K_a, or the activation constant, is the concentration of agonist at which the measured ion flux response is half of its maximum value. K_{max} is the lowest agonist concentration that promotes maximal ion efflux. The amount of ion flux at this agonist concentration is E_{max} (not illustrated on figure), and is presented in Table III as a percentage of the maximal ion flux response in the presence of 1 mM carbamylcholine under otherwise identical assay conditions. K_i, or the inactivation constant, is the concentration of agonist above K_{max} where, in some cases, ion flux falls again to half the maximal value, presumably due to an inhibitory process (see below) that is manifest at higher concentrations of some agonists.

From the data in Table III, it is clear that the relative efficacy of a given agonist and the characteristics of dose-response curves sometimes differ between PC12 and TE671 cells. In the most extreme case, succinyldicholine is a potent agonist of TE671 cellular nAChR responses, but is an antagonist when applied to PC12 cellular nAChR. Suberyldicholine and isoarecolone are more potent agonists on TE671 cells, while the selectivity is reversed for nicotine and cytisine, which are more potent agonists of nAChR on PC12 cells. General-

TABLE III

Agonist ion efflux activation parameters (values in μM)

Drug	TE671				PC12			
	K_a	K_{max}	K_i	E_{max}	K_a	K_{max}	K_i	E_{max}
Carbamylcholine	30	3000	> 10 000	100	200	2000	> 10 000	100
Nicotine	100	300	2000	45	20	100	1000	65
Cytisine	100	1000	10 000	100	10	300	> 10 000	65
Succinyldicholine	30	1000	> 10 000	95	antagonist			
Suberyldicholine	0.2	10	> 10 000	135	200	1000	5000	85
Isoarecolone	0.1	1	> 10 000	120	100	1000	> 10 000	100

Parameters describing dose-response characteristics for different nicotinic agonists activating ion efflux from TE671 or PC12 cells. Data are summarized from numerous replicate experiments ($n = 6-20$).

122

ly, the difference between K_a and K_{max} values is the smallest for nicotine, and values of E_{max} are depressed as a consequence.

Antagonist competition curves toward agonist activation can be used to determine the drug concentration that is effective in blocking one-half of the ion flux response, i.e. the apparent binding constant for antagonist interaction with functional nAChR (Lukas, 1986d, e; Lukas and Cullen, 1989). As functional antagonists, most curaremimetic neurotoxins are potent when acting on TE671 cell nAChR responses but are ineffective as blockers of PC12 cell nAChR (Lukas, 1988). Among the small antagonists, mecamylamine and hexamethonium exhibit relative, but not absolute, selectivity as blockers of nAChR responses on PC12 cells (Table IV).

Discussion

The results demonstrate that agonist binding subsites of putative nAChR detected by high-affinity [3H]ACh binding may be distinguished on the basis of their interactions with a diverse group of pharmacological agents. The simplest interpretation of these observations is that there are dif-

TABLE IV

Antagonist blockade of carbamylcholine (1 mM) activated ion efflux

Drug	IC$_{50}$ values (μM)	
	TE671	PC12
d-Tubocurarine	0.2	0.3
Pancuronium	0.2	0.3
Mecamylamine	30.0	0.2
Hexamethonium	3000.0	100.0

Data from numerous replicate experiments ($n = 3-18$) are summarized for nicotinic antagonist inhibition of nAChR-mediated ion efflux activated in the presence of 1 mM carbamylcholine. IC$_{50}$ values are taken from visual inspection of antagonist inhibition dose-response curves (Lukas and Cullen, 1989).

ferences in the molecular architecture of these ligand binding domains, and that the nAChR that carry these domains are a diverse group of macromolecules. Similarly, functional assays of nAChR expressed by TE671 or PC12 cells reveal pharmacological distinctions in agonist and antagonist binding efficacies. These distinctions may have structural bases, which would lead to the conclusion that functional nAChR are heterogeneous. These observations are consistent with the classical literature concerning differential pharmacology of nicotinic functional responses at the vertebrate neuromuscular synapse and on vertebrate autonomic ganglia and with earlier work from this laboratory showing diversity of curaremimetic neurotoxin binding sites and sites that interact with anti-nAChR antibodies (Lukas, 1986a, 1988), which also suggested the expression of nAChR heterogeneity. These results are also consistent with recent, convincing demonstrations of immunological, structural (Whiting and Lindstrom, 1987), and genetic (Boulter et al., 1986, 1987; Nef et al., 1986; Goldman et al., 1987; Deneris et al., 1988) diversity of putative nAChR expressed in muscle and in mammalian and vertebrate non-mammalian brain. Moreover, the data are concordant with reports that putative nAChR defined as high-affinity agonist and Bgt binding sites in the mammalian brain have different regional distributions (Clarke et al., 1985).

Nicotinic ligand binding sites. Some other provisional interpretations may be made regarding these results on the properties of high-affinity [3H]ACh binding sites on each preparation. On *Torpedo* membranes and at high-affinity [3H]ACh binding sites on TE671 cells, toxins, and at least some small drugs, compete for radiolabelled toxin binding and [3H]ACh binding with about equal efficacies. This implies that there is some degree of physical continuity between toxin, antagonist, and high-affinity agonist binding domains on these preparations and that differences in the apparent affinity of nicotine and ACh for high-affinity toxin and [3H]ACh binding sites may be explained if the agonist site is a subsite of the toxin binding domain. Slowly reversible toxin binding is thought to involve interactions between an extensive binding surface on the toxin (which includes three amino acid side chains in a spatial array that is presumed to mimic the charged and hydrophobic structure of ACh; reviewed in Babbit and Huang, 1985) with several polypeptide regions on the extracellular face of the nAChR ligand binding subunit (Mulac-Jericevic and Atassi, 1987). By contrast, readily reversible ACh binding presumably occurs only at one peptide sequence of the nAChR α-subunit (Kao et al., 1984). It is reasonable to contend that high levels of receptor occupancy would be required of a small, reversibly binding agonist to inhibit thermodynamically more stable toxin binding while homotypic inhibition of [^{3}H]ACh binding would be more efficient at relatively lower concentrations of competing agonist.

More complicated interpretations, which might invoke allosteric effects or agonist-induced state conversions to explain the results, would require extension of the experiments reported here (see Neubig et al., 1982). Another alternative explanation, i.e. that there are multiple (non-interconvertible?) ligand binding sites on *Torpedo* nAChR with different affinities for agonist (Conti-Tronconi et al., 1982; Dunn and Raftery, 1982), but only one class of these sites is capable of high-affinity interaction with Bgt, is a possibility that cannot be discounted by the current data.

On TE671 cells, there are two high-affinity toxin binding sites per high-affinity [^{3}H]ACh site, implying that one-half of the radiolabelled Bgt binding sites engage in lower affinity interaction with agonist. We have preliminary evidence that at least some of these agonist-insensitive sites are derived from cryptic plasma membranes or intracellular organelle membrane compartments (Lukas, unpublished observations). More rigorous tests would be required to determine the extent of physical overlap between these ligand binding domains and to critically evaluate models for multiple, non-interconvertible ligand binding sites on, or ligand-induced affinity state transitions of,

TE671 cellular nAChR.

In rat brain, there are at least two classes of high-affinity [^{3}H]ACh binding sites that can be distinguished on the basis of their interaction (with either low or no affinity) with Bgt. In addition, there would appear to be high-affinity radiotoxin binding sites that do not bind ACh with high affinity; K_i values for ACh blockade of toxin binding to these sites is about 700-fold higher than K_i values for homotypic inhibition by ACh of [^{3}H]ACh binding (Lukas, 1988). The simplest interpretation of these results is that the nearly unitary ratio of high-affinity toxin binding sites to high-affinity [^{3}H]ACh binding sites is accidental, and that about two-thirds of the high-affinity [^{3}H]ACh binding sites are completely distinct from toxin binding sites. This result would be consistent with autoradiographic comparisons of high-affinity toxin and ACh binding to rat brain tissue sections (Clarke et al., 1985). The other one-third of [^{3}H]ACh sites that are blocked by Bgt with low affinity either may have very limited physical overlap with high-affinity toxin binding domains or bear no relation at all to high-affinity toxin binding sites. Further work is necessary to test the alternative possibilities that there are multiple ligand binding sites on a limited number of classes of nAChR-like molecules or that ligand-induced affinity state transitions may account for the data.

On PC12 cells, there appear to be two classes of high-affinity agonist sites, neither of which interacts with neurotoxins with high or lower affinity. Nevertheless, the concentration of these sites is between 13- and 21-fold lower than the concentration of high-affinity binding sites for radiolabelled Bgt that are blocked by ACh with only low affinity. It is interesting to note that this result is also consistent with the use of PC12 cells as a sympathetic neuronal model, since the concentration on autonomic neurons of extra-synaptic sites for Bgt is estimated to be higher than that of junctional nAChR identified by other criteria (Jacob and Berg, 1983). The implication is that PC12 cells carry distinct classes of nAChR-like structures that bind either agonist or Bgt with high affinity.

The use of non-radiolabelled cholinergic drugs as competitors for high-affinity [³H]ACh binding to the model preparations permits distinctions to be made between these sites (see Table II). In general, the pharmacological profile of high-affinity [³H]ACh binding sites on PC12 cell or rat brain preparations is more "ganglionic" and that on TE671 cells or *Torpedo* membranes is more "neuromuscular junctional", based on the relative inhibition potencies of d-tubocurarine, decamethonium, and the *bis*-choline compounds as opposed to nicotine and cytisine.

nAChR functional properties. Qualitatively and quantitatively, the results of studies on nAChR function on TE671 and PC12 cells with both nicotinic agonists and antagonists demonstrate differences in these receptor types that are consistent with their designation as "neuromuscular junction-like" and "ganglionic-like" nAChR, respectively. In particular, nicotine and cytisine as agonists and mecamylamine and hexamethonium as antagonists exhibit selectivity for nAChR expressed by PC12 cells. These results are concordant with the observations described and discussed above that give similar classifications to the ligand binding sites that presumably are expressed by the functional nAChR on PC12 and TE671 cells.

The results also provide some insight into the possible mechanisms of action of nicotinic antagonists. Hexamethonium and, in particular, mecamylamine are functionally active at concentrations that are well below the concentrations needed to inhibit high-affinity [³H]ACh (or radiolabelled Bgt) binding to PC12 or TE671 cell preparations. This data might be interpreted in terms of the functional efficacy of these drugs as non-competitive inhibitors or channel blockers rather than as competitive inhibitors that act principally at agonist binding sites.

The relationships between functional and ligand binding sites. Debate continues regarding the relationships between ligand binding on nAChR expressed in peripheral tissues and sites that mediate activation of nAChR functional responses and desensitization (inactivation) of those responses on prolonged exposure to agonist (see Dunn and Raftery, 1982; Neubig et al., 1982; Conti-Tronconi and Raftery, 1986), largely due to technical limitations in the ability to directly measure very low affinity (high micromolar K_d values) binding of agonists, rather than indirectly through "reporter" groups that react with liganded or unliganded nAChR. For *Torpedo* electroplax nAChR, the high-affinity agonist binding state has been correlated with nAChR existing in a functionally "desensitized" form (Neubig et al., 1982). However, it remains an open question whether functional inactivation involves desensitization (Katz and Thesleff, 1957) or open channel block (Lapa et al., 1975; see Lukas and Cullen, 1989). Moreover, it is clear that different nAChR may undergo different state transitions and stages of inactivation (Epstein et al., 1980; Walker et al., 1981).

The data reported here cannot directly address these issues, since the methods employed lack the temporal resolution to detect and study the fast kinetics of functional activation, inactivation, or affinity state transitions. However, comparison of the effects of different drugs on ion flux and ligand binding might illuminate these issues. As is the case for studies of nAChR from electroplax, for both TE671 and PC12 cells the concentration of agonist necessary to produce a half-maximal ion efflux response generally is several orders of magnitude higher than the concentration needed to inhibit high-affinity radiolabelled Bgt binding (Lukas, 1986a, 1988), which, particularly for PC12 cells, also is higher than the concentration of agonist needed to block high affinity [³H]ACh binding. However, rank order potencies for agonist-mediated blockade of [³H]ACh binding are the same as for agonist activation (K_a values) of ion efflux responses for PC12 cells, or for TE671 cells, despite the fact that some agonists display selectivity in both assays for the class of nAChR expressed by one or the other types of cell. This lends circumstantial support to the notion that receptor functional activation and the expression of high-affinity agonist binding sites are linked. Drugs such as nicotine and cytisine, which are selective in either assay for ganglionic-like nAChR as expressed by PC12 cells, exhibit dose-response characteristics in ion efflux assays (illustrated by E_{max} values of less than 100) that suggest the expression of concentration-dependent (and perhaps time-dependent) inactivation of nAChR function that is not manifest by agonists such as carbamylcholine or isoarecolone. Comparative studies of the kinetics of the binding of these drugs to nAChR and phases of their activation and inactivation of receptor function by stopped-flow ion flux or by voltage or patch-clamp electrophysiological techniques might be pursued productively. By contrast, studies with toxins or antagonists would suffer interpretive uncertainties given the confounding influences of non-overlapping binding domains and of mechanisms of competitive vs. non-competitive inhibition discussed above.

Conclusions

In summary, the results of these studies support the concept of nAChR structural and functional diversity. The utility of clonal cell line models that express different nAChR isotypes in the studies of the properties of those receptors has been demonstrated. Further use of those cell lines promises to illuminate the relationships between ligand binding sites and sites that regulate activation and inactivation of nAChR functional responses, and to provide insight into the effects of nicotine at neuronal synapses.

Acknowledgements

This work was supported by National Institutes of Health Grant NS16821, the Council for Tobacco Research – U.S.A., Epi-Hab, Phoenix, and the Men's and Women's Boards of the Barrow Neurological Foundation. The author wishes to thank Mary Jane Cullen, Mercedeh Saba and Donna Kucharski for expert execution of many of these experiments, and Deirdre Anne Janus for excellent secretarial assistance.

References

Abood, L.G., Reynolds, D.T. and Bidlack, J.M. (1980) Stereospecific ^{3}H-nicotine binding to intact and solubilized rat brain membranes and evidence for its noncholinergic nature. *Life Sci.,* 27: 1307–1314.

Babbit, B. and Huang, L. (1985) Alpha-bungarotoxin immobilized and oriented on a lipid bilayer vesicle surface. *Biochemistry,* 24: 15–21.

Barnard, E.A. and Dolly, J.O. (1982) Peripheral and central nicotinic ACh receptors – how similar are they? *Trends Neurosci.,* 5: 325–327.

Boulter, J., Evans, K., Goldman, D., Martin, G., Treco, D., Heinemann, S. and Patrick, J. (1986) Isolation of a cDNA clone coding for a possible neuronal nicotinic receptor alpha-subunit. *Nature (Lond.),* 319: 368–374.

Boulter, J., Connolly, J., Deneris, E., Goldman, D., Heinemann, S. and Patrick, J. (1987) Functional expression of two neuronal nicotinic acetylcholine receptors from cDNA clones identifies a gene family. *Proc. Natl. Acad. Sci. U.S.A.,* 84: 7763–7767.

Catterall, W.A. (1975) Sodium transport by the acetylcholine receptor of cultured muscle cells. *J. Biol. Chem.,* 250: 1776–1781.

Changeux, J.-P., Kasai, M. and Lee, C.Y. (1970) Use of a snake venom toxin to characterize the cholinergic receptor protein. *Proc. Natl. Acad. Sci. U.S.A.,* 67: 1241–1247.

Chiappinelli, V.A. (1985) Actions of snake venom toxins on neuronal nicotinic receptors and other neuronal receptors. *Pharmac. Ther.,* 31: 1–32.

Chiappinelli, V.A., Dryer, S.E., Sorenson, E.M., Wolf, K.M., Grant, G.A., Chen, S.-J., Nooney, J.M., Lambert, J.J. and Hilder, R.C. (1988) Functional studies of neuronal nicotinic receptors utilizing kappa-neurotoxins. In F. Clementi, C. Gotti and E. Sher (Eds.), *Nicotinic Acetylcholine Receptors in the Nervous System, NATO ASI Series Vol. H25,* Springer-Verlag, Berlin, pp. 15–29.

Clarke, P.B.S., Schwartz, R.D., Paul, S.M., Pert, C.B. and Pert, A. (1985) Nicotinic binding in the rat brain: autoradiographic comparison of [^{3}H]-acetylcholine, [^{3}H]-nicotine, and [^{125}I]-alpha-bungarotoxin. *J. Neurosci.,* 5: 1307–1315.

Conti-Tronconi, B.M. and Raftery, M.A. (1986) Nicotinic acetylcholine receptor contains multiple binding sites: evidence from binding of alpha-dendrotoxin. *Proc. Natl. Acad. Sci. U.S.A.,* 83: 6646–6650.

Conti-Tronconi, B.M., Dunn, S.M.J. and Raftery, M.A. (1982) Independent sites of low and high affinity for agonists on *Torpedo californica* acetylcholine receptor. *Biochem.*

126

Biophys. Res. Commun., 107: 123 – 129.

Deneris, E.S., Connolly, J., Boulter, J., Wada, E., Wada, K., Swanson, L.W., Patrick, J. and Heinemann, S. (1988) Primary structure and expression of beta2: A novel subunit of neuronal nicotinic acetylcholine receptors. *Neuron,* 1: 45 – 54.

Dunn, S.M.J. and Raftery, M.A. (1982) Activation and desensitization of *Torpedo* acetylcholine receptor: evidence for separate binding sites. *Proc. Natl. Acad. Sci. U.S.A.,* 79: 6757 – 6761.

Epstein, N., Hess, G.P., Kim, P.S. and Noble, R.L. (1980) Inactivation (desensitization) of the acetylcholine receptor in *Electrophorus electricus* membrane vesicles by carbamylcholine: comparison between ion flux and alpha-bungarotoxin binding. *J. Memb. Biol.,* 56: 133 – 137.

Eterovic, V.A. and Bennett, E.L. (1974) Nicotinic cholinergic receptor in brain detected by binding of alpha-[^{3}H]bungarotoxin. *Biochim. Biophys. Acta,* 362: 346 – 355.

Goldman, D., Deneris, E., Luyten, W., Kochhar, A., Patrick, J. and Heinemann, S. (1987) Members of a nicotinic acetylcholine receptor gene family are expressed in different regions of the mammalian central nervous system. *Cell,* 48: 965 – 973.

Gomez, C.M., Richman, D.P., Berman, P.W., Burres, S.A., Arnason, B.G.W. and Fitch, F.W. (1979) Monoclonal antibodies against purified nicotinic acetylcholine receptor. *Biochem. Biophys. Res. Commun.,* 88: 575 – 582.

Greene, L.A. and Tischler, A.S. (1976) Establishment of a noradrenergic clonal line of rat adrenal pheochromocytoma cells which respond to nerve growth factor. *Proc. Natl. Acad. Sci. U.S.A.,* 73: 2424 – 2428.

Jacob, M.H. and Berg, D. (1983) The ultrastructural localization of alpha-bungarotoxin binding sites in relation to synapses on chick ciliary ganglion neurons. *J. Neurosci.,* 3: 260 – 271.

Kao, P.N., Dwork, A.J., Kaldany, R.-R.J., Silver, M.L., Wideman, J., Stein, S. and Karlin, A. (1984) Identification of the alpha subunit half-cystine specifically labeled by an affinity reagent for the acetylcholine receptor binding site. *J. Biol. Chem.,* 259: 11662 – 11665.

Kasai, M. and Changeux, J.-P. (1971) In vitro excitation of purified membrane fragments by cholinergic agonists. I. Pharmacological properties of the excitable membrane fragments. *J. Memb. Biol.,* 6: 1 – 23.

Katz, B. and Miledi, R. (1978) A re-examination of curare action at the motor end plate. *Proc. R. Soc. Lond. B,* 203: 119 – 133.

Katz, B. and Thesleff, S. (1957) A study of the ''desensitization'' produced by acetylcholine at the motor end-plate. *J. Physiol. (Lond.),* 138: 63 – 80.

Kuffler, S.W. and Yoshikami, D. (1975) The number of transmitter molecules in a quantum: an estimate from iontophoretic application of acetylcholine at the neuromuscular synapse. *J. Physiol. (Lond.),* 251: 465 – 482.

Lapa, A.J., Albuquerque, E.X., Sarvey, J.M., Daly, J. and Witkop, B. (1975) Effects of histrionicotoxin on the chemosensitive and electrical properties of skeletal muscle. *Exp. Neurol.,* 47: 558 – 580.

Lentz, T.L. and Chester, J. (1977) Localization of acetylcholine receptors in central synapses. *J. Cell Biol.,* 75: 258 – 267.

Lindstrom, J. (1986) Probing nicotinic acetylcholine receptors with monoclonal antibodies. *Trends Neurosci.,* 9: 401 – 407.

Loring, R.H. and Zigmond, R.E. (1988) Characterization of neuronal nicotinic receptors by snake venom neurotoxins. *Trends Neurosci.,* 11: 73 – 78.

Lukas, R.J. (1986a) Immunochemical and pharmacological distinctions between curaremimetic neurotoxin binding sites of central, autonomic, and peripheral origin. *Proc. Natl. Acad. Sci. U.S.A.,* 83: 5741 – 5745.

Lukas, R.J. (1986b) Characterization of curaremimetic neurotoxin binding sites on membrane fractions derived from the human medulloblastoma clonal line, TE671. *J. Neurochem.,* 46: 1936 – 1941.

Lukas, R.J. (1986c) Characterization of curaremimetic neurotoxin binding sites on cellular membrane fragments derived from the rat pheochromocytoma PC12. *J. Neurochem.,* 47: 1768 – 1773.

Lukas, R.J. (1986d) Heterogeneity of functional nicotinic acetylcholine receptors on clonal cell lines is revealed by the use of 86-rubidium ion efflux assays. *Biophys. J.,* 49: 3a.

Lukas, R.J. (1986e) Functional nicotinic acetylcholine receptors on the clonal line, TE671. *Trans. Am. Soc. Neurochem.,* 17: 286.

Lukas, R.J. (1986f) Relationships between high-affinity binding sites for nicotinic acetylcholine receptor agonists and alpha-bungarotoxin on membranes from rat brain, *Torpedo* electroplax, and the clonal cell lines, PC12 and TE671. *Soc. Neurosci. Abst.,* 12: 961.

Lukas, R.J. (1988) Evidence for functional and structural diversity of nicotinic acetylcholine receptors. In F. Clementi, C. Gotti and E. Sher (Eds.), *Nicotinic Acetylcholine Receptors in the Nervous System, NATO ASI Series Vol. H25,* Springer-Verlag, Berlin, pp. 15 – 29.

Lukas, R.J. and Cullen, M.J. (1988) An isotopic rubidium ion efflux assay for the functional characterization of nicotinic acetylcholine receptors on clonal cell lines. *Analyt. Biochem.,* 96: 207 – 212.

Marks, M.J. and Collins, A.C. (1982) Characterization of nicotine binding in mouse brain and comparison with the binding of alpha-bungarotoxin and quinuclidinyl benzilate. *Mol. Pharmacol.,* 22: 554 – 564.

McAllister, R.M., Issacs, H., Rougey, R., Peer, M., Au, W., Soukkup, S.W. and Gardner, M.B. (1977) Establishment of a human medulloblastoma cell line. *Int. J. Cancer,* 20: 206 – 212.

Morley, B.J., Kemp, G.E. and Salvaterra, P.M. (1979) Alpha-bungarotoxin binding sites in the CNS. *Life Sci.,* 24: 859 – 872.

Mulac-Jericevic, B. and Atassi, M.Z. (1987) Alpha-neurotoxin binding to acetylcholine receptor: localization of the full profile of the cobratoxin-binding regions on the alpha-chain of *Torpedo californica* acetylcholine receptor by a comprehensive synthetic strategy. *J. Prot. Chem.,* 6: 365 – 373.

Nef, P., Oneyser, C., Barkas, T. and Ballivet, M. (1986) Acetylcholine receptor related genes expressed in the nervous system. In A. Maelicke (Ed.), *Nicotinic Acetylcholine Receptor: Structure and Function,* Springer-Verlag, Berlin, pp. 417 – 422.

Neubig, R.R., Boyd, N.D. and Cohen, J.B. (1982) Conformations of *Torpedo* acetylcholine receptor associated with ion transport and desensitization. *Biochemistry,* 21: 3460 – 3467.

Patrick, J. and Stallcup, B. (1977a) Alpha-bungarotoxin binding and cholinergic receptor function on a rat sympathetic nerve line. *J. Biol. Chem.,* 252: 8629 – 8633.

Patrick, J. and Stallcup, W.B. (1977b) Immunological distinction between acetylcholine receptor and the alpha-bungarotoxin binding component on sympathetic neurons. *Proc. Natl. Acad. Sci. U.S.A.,* 74: 4689 – 4692.

Romano, C. and Goldstein, A. (1980) Stereospecific nicotinic receptors on rat brain membranes. *Science,* 210: 647 – 649.

Salvaterra, P.M., Mahler, H.R. and Moore, W.J. (1975) Subcellular and regional distribution of ^{125}I-labeled alpha-bungarotoxin binding in rat brain and its relationship to acetylcholinesterase and choline acetyltransferase. *J. Biol. Chem.,* 250: 6469 – 6475.

Schwartz, R.D., McGee, R., Jr. and Kellar, K.J. (1982) Nicotinic cholinergic receptors labeled by [^{3}H]acetylcholine in rat brain. *Mol. Pharmacol.,* 22: 56 – 62.

Walker, J.W., McNamee, M.G., Pasquale, E., Cash, D.J. and Hess, G.P. (1981) Acetylcholine receptor inactivation in *Torpedo californica* electroplax membrane vesicles: detection of two processes in the millisecond and second time regions. *Biochem. Biophys. Res. Commun.,* 100: 86 – 90.

Whiting, P. and Lindstrom, J. (1987) Purification and characterization of a nicotinic acetylcholine receptor from rat brain. *Proc. Natl. Acad. Sci. U.S.A.,* 84: 595 – 599.

Yoshikami, D. (1981) Transmitter sensitivity of neurons assayed by autoradiography. *Nature (Lond.),* 212: 929 – 930.

A. Nordberg, K. Fuxe, B. Holmstedt and A. Sundwall (Eds.)
Progress in Brain Research, Vol. 79
© 1989 Elsevier Science Publishers B.V. (Biomedical Division)

CHAPTER 12

Properties of putative nicotine receptors identified on cultured cortical neurons

Patrick M. Lippiello

Research and Development Department, Bowman Gray Technical Center, R.J. Reynolds Tobacco Company, Winston-Salem, NC 27102, U.S.A

Introduction

It has now been established, based on the work of a number of laboratories, that mammalian brain contains high-affinity receptors that bind nicotinic cholinergic agonists, but not the classical nicotinic antagonist, α-bungarotoxin (reviewed by Clarke, 1987). The binding properties of these sites have been characterized using [³H]nicotine (Marks and Collins, 1982; Lippiello and Fernandes, 1986) and [³H]acetylcholine (Schwartz et al., 1982). Several lines of evidence suggest that the sites recognized by these ligands represent a single clas of nicotinic cholinoceptors in the brain. For instance, the affinity and density, the pharmacological specificity, and the in vivo regulation of sites labelled by acetylcholine and nicotine are the same for both ligands (Marks et al., 1986; Martino-Barrows and Kellar, 1987). In addition, the regional distributions of sites labelled by [³H]nicotine and [³H]acetylcholine in brain tissue, based on autoradiographic evidence, are identical (Clarke et al., 1985). High-affinity agonist binding sites have in fact been solubilized and separated chromatographically from the α-bungarotoxin binding component in brain tissue (Sugiyama and Yamashita, 1986; Wonnacott, 1986) and have recently been purified using monoclonal antibodies directed against nicotinic cholinergic receptors (Whiting and Lindstrom, 1987) and anti-idiotypic

antibodies to nicotine (Abood et al., 1987). Apparently, α-toxin binding sites in the brain represent a separate class of potential cholinoceptors that bind nicotine with very low affinity (Marks and Collins, 1982; Wonnacott, 1986).

Although little is known about the functional significance of putative high-affinity nicotinic receptors in mammalian brain, it has been suggested that they represent functional nicotinic cholinergic receptors (Clarke, 1987). The anatomical localization of the sites coincides with known cholinergic systems in the brain (Clarke et al., 1984) and the sites can be down-regulated with acetylcholinesterase inhibitors (Schwartz and Kellar, 1983). Lesioning studies indicate that the sites are located pre-synaptically (Schwartz et al., 1984; Clarke and Pert, 1985) and the idea that these sites may play a neuromodulatory role is supported by studies demonstrating nicotine-evoked release of dopamine (Mills and Wonnacott, 1984), acetylcholine (Rowell and Winkler, 1984) and gamma-aminobutyric acid (Wonnacott et al., 1987) from synaptosomal preparations.

A number of laboratories have begun to study the localization and functional properties of nicotinic receptors using model neuronal cell culture systems. Putative nicotinic receptors have been identified on chick ganglion cells, using both neurotoxin and antibody probes (Halvorsen and Berg, 1986; Jacob et al., 1986), and a functional nicotinic receptor in a pheochromocytoma (PC12) cell line has been labelled with monoclonal antibodies to high-affinity receptors in rat brain (Whiting et al., 1987). The present studies describe a neuronal cell model based on primary monolayer cultures derived from fetal rat cortex. These cells are indistinguishable from cells in mature cortex (Kriegstein and Dichter, 1983) and we have found that they possess putative nicotinic receptors whose binding properties are essentially the same as in adult brain. Therefore, this cell system is a potentially useful model for studying the properties of neuronal nicotinic receptors in the brain.

Methods

Culture preparation

Cerebral cortical cultures were prepared from fetal rat brain tissue according to the methods of McCarthy and Harden (1981), with some modifications (Lippiello and Fernandes, 1988). Cells from 40 – 60 fetal cortices were dissociated mechanically, plated in poly-L-lysine-coated culture flasks in serum-containing medium (DMEM/Ham's F-12 (1 : 1) supplemented with glucose, glutamine, HEPES, antibiotics and 10% fetal bovine serum) for 2 days, and maintained in serum-free defined medium (Bottenstein and Sato, 1979) thereafter. Glial cell proliferation was arrested by inclusion of 5-fluorodeoxyuridine for the first 2 days in serum-free medium. Neuronal cultures prepared in this way contained approximately 95% neurons, as judged by indirect immunofluorescence using anti-neurofilament protein as a marker. Glial cell cultures were prepared by plating cortical cells in uncoated culture flasks and maintaining them in 10% serum-containing medium for 2 weeks, with a medium change every 2 – 3 days (McCarthy and deVellis, 1980). Glial cells were identified by indirect immunofluorescence, using glial fibrillary acidic protein (GFAP) as a marker.

Nicotine binding assays

Membranes were prepared from brain tissue (Lippiello and Fernandes, 1986) and from fetal cortical neurons (Lippiello and Fernandes, 1988) according to previously described methods, in Ca^{2+}/Mg^{2+}-free phosphate-buffered saline containing ethylenediaminetetraacetic acid (EDTA), iodoacetamide, and phenylmethylsulfonyl fluoride (PMSF). Equilibrium and kinetic binding assays were performed as described previously (Lippiello and Fernandes, 1986) by incubating membranes with L-[^{3}H]nicotine at 0°C, followed by rapid filtration through glass fiber filters under vacuum, and quantification of specifically bound radioactivity. Competition binding experiments were carried out as above with the addition of varying concentrations of competing compounds. Since the results obtained depended critically on the purity of the ligand, [^{3}H]nicotine was purified periodically by thin-layer chromatography, according to the methods of Marks et al. (1986).

Results and discussion

Cell cultures

All cultures were isolated by mechanical sieving techniques, rather than by trypsinization, inasmuch as it was found that inclusion of proteases resulted in erratic variations in the binding properties of the high-affinity nicotinic sites. Relevant to this, Abood et al. (1981) found that tryptic digestion of rat brain destroyed approximately 50% of the nicotine binding. The characteristics of primary cultures from rat cortex prepared by the present methods were the same as those previously described for both neurons (McCarthy and Harden, 1981; Kriegstein and Dichter, 1983) and glia (McCarthy and deVellis, 1980). Based on morphology, neuronal cultures contained pyramidal, fusiform and multipolar cell types. These cultures were composed primarily of neuronal cell types, as

131

judged by immunocytochemical staining with anti-neurofilament protein (Fig. 1A,B). Glial cultures, containing both large polygonal cells and fibrous astrocytes, were identified by staining with anti-GFAP (Fig. 1C,D). These cultures were devoid of neurons.

Properties of neuronal receptors

Equilibrium binding properties
The presence of high-affinity nicotinic receptors in cortical neurons was confirmed by measuring the binding of L-[³H]nicotine to cell membrane preparations. Nicotine binding to neuronal sites was saturable and specific (Fig. 2) and Scatchard and Hill analyses were consistent with a single class of non-interacting sites having a high affinity (K_d ca. 2 – 3 nM) for nicotine. These results were consistent with receptor properties determined for adult brain preparations (Lippiello and Fernandes, 1986). There was no evidence of additional, low-affinity sites. In general, the presence of low-

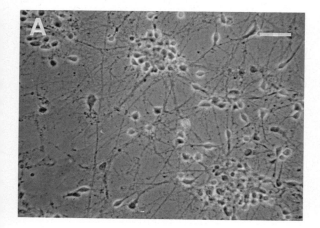
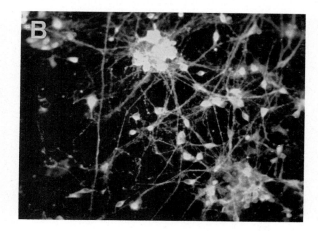
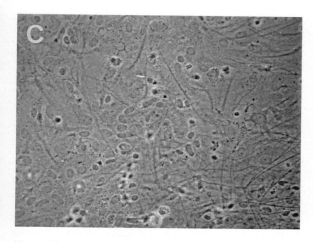
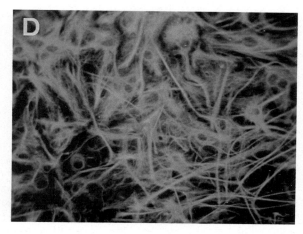

Fig. 1. Phase contrast and fluorescence photomicrographs of cultured neuronal and glial cells. Panels (A) and (B) represent phase contrast and fluorescent photomicrographs of cortical neurons. Cells were fixed with 4% paraformaldehyde and stained with anti-neurofilament protein, followed by rhodamine-conjugated goat anti-mouse IgG. The lower set of panels, (C) and (D), show phase contrast and fluorescent photomicrographs of glial cell cultures. Cells were fixed as above, and stained with anti-GFAP and FITC-conjugated anti-rabbit IgG, respectively. (From Lippiello and Fernandes, 1988.)

affinity sites has been reported (Sloan et al., 1984; Larsson and Nordberg, 1985), but not consistently detected in brain tissue using presently available binding assays. This is not unexpected since, based on kinetic considerations, the dissociation of ligand from low-affinity sites would be quite rapid. It has been suggested that additional low-affinity nicotine binding sites reported by some laboratories may result from proteolysis during membrane preparation (Lippiello and Fernandes, 1986), from low-temperature effects (Marks and Collins, 1982), or from impurities in the ligand (Lippiello

and Fernandes, 1986; Marks and Collins, personal communication).

The equilibrium binding properties of neuronal sites were compared with those in fetal, neonatal and adult cortical tissue (Table I). Although the apparent affinity of the sites for nicotine was the same in all tissues examined, the maximum number of sites (B_{max}) in fetal tissue and in cultured neurons was around $20-30\%$ of adult levels. This may reflect the absence of specific developmental signals under culture conditions or the absence of presynaptic sites located on cortical afferents which are normally present in vivo. There was no evidence of high-affinity binding sites on cultured glial cells, suggesting that putative nicotinic receptors are located only on neuronal cell types in the cortex.

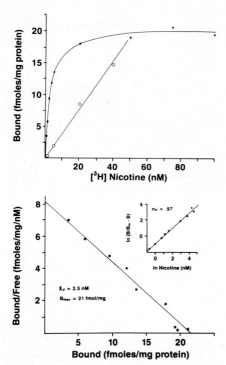

Fig. 2. Equilibrium binding properties of nicotinic sites in neuronal cell membrane preparations. Neuronal membranes (0.65 mg protein) were incubated with varying amounts of L-[^{3}H]nicotine ($1-100$ nM) for 2 h at $0°$. Specific binding was estimated, following rapid filtration, as the difference between total binding and blank incubations containing 100 μM unlabelled L-nicotine salicylate. Specific, saturable binding (●——●) and nonspecific binding (○——○) as a function of nicotine concentration are shown in the top panel. Scatchard and Hill (inset) analyses are shown in the bottom panel, together with estimates of the binding parameters, K_d and B_{max}. (From Lippiello and Fernandes, 1988.)

TABLE I

Equilibrium binding properties of high-affinity nicotinic sites in cortical tissue and cultured cells

Tissue	K_d (nM)	B_{max} (fmole/mg protein)
Whole cortex		
Fetal (18 days)	3.0 ± 0.5	20 ± 3
Neonatal (4 weeks)	4.0 ± 1.5	$110 \pm 10^{[a]}$
Adult	2.1 ± 0.6	$100 \pm 6^{[a]}$
Cultured cells		
Cortical neurons (2 weeks)	3.0 ± 0.8	25 ± 4
Cortical neurons (4 weeks)	4.2 ± 1.3	$31 \pm 5^{[b]}$
Cortical glia		< 1

[a] Significantly different from fetal ($p < 0.05$).
[b] Significantly different from fetal ($p < 0.01$).
Membranes prepared from whole cortex and cultured cells were incubated with varying concentrations of L-[^{3}H]nicotine ($1-100$ nM) for 2 h at $0°$. Blank incubations contained 100 μM L-nicotine salicylate. Specific binding was determined as the difference between total binding and blanks. Equilibrium binding parameters were estimated by Scatchard analysis. Values represent the Mean $\pm$ S.E.M. for at least three determinations.

Kinetic binding properties

The association kinetics were biphasic and the dissociation kinetics were first-order at all nicotine concentrations tested (Fig. 3). This agrees with the results of previous studies, using adult rat brain tissue (Lippiello and Fernandes, 1987), where it was shown that the complex binding kinetics exhibited by these sites are consistent with the two-state model originally proposed by Katz and Thesleff (1957). According to this model the binding sites, in the absence of ligand, exist as a mixture of low- and high-affinity forms, R and R' (Fig. 4). However, the ligand tends to stabilize the high-affinity conformer (R'N), resulting in the appearance of a single class of high-affinity sites at equilibrium. The slow-phase binding component seen in the association kinetics is a measure of the rate of isomerization from the low to the high-affinity conformation of the receptor. All of the kinetic rate constants determined for the neuronal sites agree well with those determined previously for adult brain (Lippiello and Fernandes, 1987).

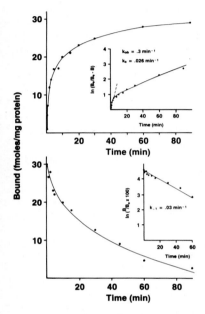

Fig. 3. Kinetics of L-[^{3}H]nicotine binding to neuronal membranes. Upper panel: membranes prepared from 2-week neuronal cultures (0.5 mg protein) were incubated for varying amounts of time (0 – 90 min) with 20 nM L-[^{3}H]nicotine at 0°. Specific binding was determined at successive time points following rapid filtration. Blank incubations contained 100 μM L-nicotine salicylate. The rates of association (k_{ob}) and isomerization (k_s) were estimated graphically (inset). Lower panel: dissociation kinetics were monitoried by incubating membranes with L-[^{3}H]nicotine for 2 h at 0°, followed by isotopic dilution with excess (1 mM) unlabelled L-nicotine salicylate and rapid filtration at successive times thereafter. The dissociation rate constant (k_{-1}) was determined graphically (inset). (From Lippiello and Fernandes, 1988.)

$$N + R \rightleftharpoons NR$$
$$\Updownarrow \qquad\qquad \Updownarrow$$
$$N + R' \rightleftharpoons NR'$$

Fig. 4. Adaptation of the two-state model of Katz and Thesleff (1957) to the binding of nicotine (N) to high (R') and low (R) affinity receptor conformations.

Pharmacological specificity

The pharmacological specificity of high-affinity sites identified in cultured neurons, assessed by inhibition of [^{3}H]nicotine binding, was the same as in adult brain (Table II). In general, nicotinic

TABLE II

Inhibition of [^{3}H]nicotine binding by nicotinic compounds

Compound	K_i (nM)		
	Cells	Brain[a]	n_H[b]
Cytisine	0.5	0.3	0.96
L-nicotine	3	3	0.87
Lobeline	17	9	0.90
DMPP	27	31	1.04
Carbachol	110	100	0.80
D-nicotine	100	125	0.90
Decamethonium	3506	3125	0.83

[a] Values for adult brain tissue, from Lippiello and Fernandes (1986).
[b] Hill coefficient for cell preparations.
Membranes (0.65 mg protein) prepared from cultured cortical neurons were incubated with 30 nM L-[^{3}H]nicotine for 2 h at 0° with varying concentrations of competing compounds. The K_i values shown represent the average of triplicate determinations.

134

agonists were the most potent competitive inhibitors of nicotine binding and antagonists were much less effective. The sites also exhibited modest stereoselectivity inasmuch as the D-isomer of nicotine was a significantly less effective competitive inhibitor of L-[³H]nicotine binding.

Summary

Cultured cortical neurons derived from fetal rat brain contain putative high-affinity nicotinic receptors. Cortical glial cells do not contain any high-affinity binding sites for nicotine. The affinity, kinetic binding properties and pharmacological specificity of neuronal sites are the same as those determined for adult brain tissue. However, the number of sites in cultured neurons is lower than in adult cortex. Cultured cortical neurons appear to be a suitable model system for defining the location(s) and functional properties of high-affinity nicotinic receptors in mammalian brain.

Acknowledgements

The author thanks Dr. John Wood for his kind gift of anti-neurofilament protein antibody, Dr. Ken McCarthy for sharing his expertise in immunocytochemical methodologies, and Ms. Regina W. Brim for preparation of the manuscript.

References

Abood, L.G., Reynolds, D.T., Booth, H. and Bidlack, J.M. (1981) Sites and mechanisms for nicotine's action in the brain. *Neurosci. Biobehav.*, 5: 479–486.

Abood, L.G., Langone, J.J., Bjercke, R., Lu, X. and Banerjee, S. (1987) Characterization of a purified nicotinic receptor from rat brain using idiotypic and anti-idiotypic antibodies. *Proc. Natl. Acad. Sci. U.S.A.*, 84: 6587–6590.

Bottenstein, J.E. and Sato, G.H. (1979) Growth of a rat neuroblastoma cell line in serum-free supplemented medium. *Proc. Natl. Acad. Sci. U.S.A.*, 76: 514–517.

Clarke, P.B.S. (1987) Recent progress in identifying nicotinic cholinoceptors in mammalian brain. *Trends Pharmacol. Sci.*, 8: 32–35.

Clarke, P.B.S. and Pert, A. (1985) Autoradiographic evidence for nicotine receptors on nigrostriatal and mesolimbic dopaminergic neurons. *Brain Res.*, 348: 355–358.

Clarke, P.B.S., Pert, C.B. and Pert, A. (1984) Autoradiographic distribution of nicotine receptors in rat brain. *Brain Res.*, 323: 390–395.

Clarke, P.B.S., Schwartz, R.D., Paul, S.M., Pert, C.B. and Pert, A. (1985) Nicotinic binding in rat brain: autoradiographic comparison of [³H]acetylcholine, [³H]nicotine, and [¹²⁵I]-alpha-bungarotoxin. *J. Neurosci.*, 5: 1307–1315.

Halvorsen, S.W. and Berg, D.K. (1986) Identification of a nicotinic acetylcholine receptor on neurons using an alpha-neurotoxin that blocks receptor function. *J. Neurosci.*, 6: 3405–3412.

Jacob, M.H., Lindstrom, J.M. and Berg, D.K. (1986) Surface and intracellular distribution of a putative neuronal nicotinic acetylcholine receptor. *J. Cell Biol.*, 103: 205–214.

Katz, B. and Thesleff, S. (1957) A study of the "desensitization" produced by acetylcholine at the motor end-plate. *J. Physiol. (Lond.)*, 138: 63–80.

Kriegstein, A.R. and Dichter, M.A. (1983) Morphological classification of rat cortical neurons in cell culture. *J. Neurosci.*, 3: 1634–1647.

Larsson, C. and Nordberg, A. (1985) Comparative analysis of nicotine-like receptor-ligand interactions in rodent brain homogenate. *J. Neurochem.*, 45: 24–31.

Lippiello, P.M. and Fernandes, K.G. (1986) The binding of L-[³H]nicotine to a single class of high affinity sites in rat brain membranes. *Mol. Pharmacol.*, 29: 448–454.

Lippiello, P.M. and Fernandes, K.G. (1988) Identification of putative high affinity nicotinic receptors on cultured cortical neurons. *J. Pharmacol. Exp. Ther.*, 246: 409–416.

Lippiello, P.M., Sears, S.B. and Fernandes, K.G. (1987) Kinetics and Mechanism of L-[³H]nicotine binding to putative high affinity receptor sites in rat brain. *Mol. Pharmacol.*, 31: 392–400.

Marks, M.J. and Collins, A.C. (1982) Characterization of nicotine binding in mouse brain and comparison with the binding of alpha-bungarotoxin and quinuclidinyl benzilate. *Mol. Pharmacol.*, 22: 554–564.

Marks, M.J., Stitzel, J.A., Romm, E., Wehner, J.M. and Collins, A.C. (1986) Nicotinic binding sites in rat and mouse brain: comparison of acetylcholine, nicotine, and alpha-bungarotoxin. *Mol. Pharmacol.*, 30: 427–436.

Martino-Barrows, A.M. and Kellar, K.J. (1987) [³H]Acetylcholine and [³H](-)nicotine label the same recognition site in rat brain. *Mol. Pharmacol.*, 31: 169–174.

McCarthy, K.D. and deVellis, J. (1980) Preparation of separate astroglial and oligodendroglial cell cultures from rat cerebral tissue. *J. Cell Biol.*, 85: 890–902.

McCarthy, K.D. and Harden, T.K. (1981) Identification of two benzodiazepine binding sites on cells cultured from rat cerebral cortex. *J. Pharmacol. Exp. Ther.*, 216: 183–191.

Mills, A. and Wonnacott, S. (1984) Antibodies to nicotinic acetylcholine receptors used to probe the structural and functional relationships between brain alpha-bungarotoxin binding sites and nicotinic receptors. *Neurochem. Int.*, 6: 249 – 257.

Rowell, P.P. and Winkler, D.L. (1984) Nicotinic stimulation of [³H]acetylcholine release from mouse cerebral cortical synaptosomes. *J. Neurochem.*, 43: 1593 – 1598.

Schwartz, R.D. and Kellar, K.D. (1983) Nicotinic cholinergic receptor binding sites in the brain: regulation in vivo. *Science,* 220: 214 – 216.

Schwartz, R.D., McGee, R. and Kellar, K.J. (1982) Nicotinic cholinergic receptors labeled by [³H]acetylcholine in rat brain. *Mol. Pharmacol.*, 22: 56 – 62.

Schwartz, R.D., Lehmann, J. and Kellar, K.J. (1984) Presynaptic nicotinic cholinergic receptors labeled by [³H]acetylcholine on catecholamine and serotonin axons in brain. *J. Neurochem.*, 42: 1495 – 1498.

Sloan, J.W., Todd, G.D. and Martin, W.R. (1984) Nature of nicotine binding to rat brain P2 fraction. *Pharmacol.*

Biochem. Behav., 20: 899 – 909.

Sugiyama, H. and Yukiko, Y. (1986) Characterization of putative nicotinic acetylcholine receptors solubilized from rat brains. *Brain Res.*, 373: 22 – 26.

Whiting, P. and Lindstrom, J. (1987) Purification and characterization of a nicotinic acetylcholine receptor from rat brain. *Proc. Natl. Acad. Sci.* U.S.A., 84: 595 – 599.

Whiting, P.J., Schoepfer, R., Swanson, L.W., Simmons, D.M. and Lindstrom, J.M. (1987) Functional acetylcholine receptor in PC12 cells reacts with a monoclonal antibody to brain nicotinic receptors. *Nature (Lond.),* 327: 515 – 518.

Wonnacott, S. (1986) Alpha-Bungarotoxin binds to low-affinity nicotine binding sites in rat brain. *J. Neurochem.*, 47: 1706 – 1712.

Wonnacott, S., Fryer, L. and Lunt, G.G. (1987) Nicotine-evoked [³H]GABA release from hippocampal synaptosomes. *J. Neurochem.*, 48: S72.

A. Nordberg, K. Fuxe, B. Holmstedt and A. Sundwall (Eds.)
Progress in Brain Research, Vol. 79
© 1989 Elsevier Science Publishers B.V. (Biomedical Division)

CHAPTER 13

Chronic nicotine exposure and brain nicotinic receptors – influence of genetic factors

Allan C. Collins and Michael J. Marks

School of Pharmacy and Institute for Behavioral Genetics, University of Colorado, Boulder, CO 80309, U.S.A.

Introduction

Several studies carried out over the last 30 years have suggested that genetic factors may be important in regulating the use of tobacco by humans. Fisher (1958a,b) was the first to suggest this possibility. He reported that concordance for smoking behavior (whether both of the twins in a pair were smokers or nonsmokers) was higher in monozygotic (identical or MZ) twins than in dizygotic (fraternal, DZ) twins. Some of the twins had not been reared together suggesting that the similarity between the twins was not due to very similar environmental influences. Because MZ twins are identical at all genetic loci whereas DZ twins exhibit the same degree of genetic identity as any other sibling pair, the greater concordance seen in the MZ twins suggested that genetic factors regulate smoking. Several other investigators (see, e.g. Todd and Mason, 1959; Raaschou-Nielsen, 1960; Dies and Reznikoff, 1969; Pedersen, 1981) have replicated Fisher's observations using the twin concordance method. Other studies have attempted to break smoking status in twins down into several groups: nonsmoker, "light", "heavy", and ex-smokers (Friberg et al., 1959; Kaprio et al., 1978; Crumpacker et al., 1979), and these studies concluded that genetic factors may regulate not only whether or not an individual smokes, but may also regulate how much.

Pedersen (1981) and Williams et al. (1981) also calculated heritability estimates for smoking in Swedish twins. Heritability is an estimate of the fraction of the total variance that may be ascribed to genetic factors, and may vary between 0 (no genetic influence) and 1.0 (total genetic control). Pedersen obtained a heritability of 0.62 in male twins and 0.49 in females. Williams et al. (1981) calculated heritability estimates in twins of varying age groups, and obtained values ranging from 0.55 in 18 – 20 year olds to 0.11 for those over 60. Heritability estimates for alcoholism are also considerably higher in younger people than in older people (Partanen et al., 1977) which suggests that smoking and alcohol abuse are both influenced by genetic and environmental factors, and that those individuals who smoke and drink at an early age are particularly susceptible to these agents because of genetic influences.

Although these investigations have suggested that genetic factors regulate smoking, these studies do not indicate precisely what may be inherited that regulates tobacco use. Nicotine is the most potent psychoactive agent in tobacco, and many studies indicate that people adjust their tobacco use in an attempt to carefully titrate their plasma nicotine levels (reviewed in, Gritz, 1980; Henningfield, 1984). Furthermore, pretreatment with the nicotinic receptor antagonist, mecamylamine, results in an increase in tobacco use (Stolerman et al., 1973; Pomerleau et al., 1987). In addition, several studies have demonstrated that people (smokers) differ in sensitivity to nicotine (Nesbitt, 1973; Jones, 1986). Whether individual differences in sensitivity to nicotine contribute to differences in smoking behavior has not been clearly established, but Kozlowsky and Herman (1984) have suggested this possibility based on the observation that different individuals seem to regulate their nicotine intake so as to fall within clearly defined boundaries. If plasma nicotine levels fall below a specified level, for each individual, smoking increases. If nicotine levels exceed a specified level, toxic symptoms appear and smoking is reduced. Although Kozlowski and Herman (1984) did not look for potential genetic regulation of this titration process, whenever individual differences are observed genetic regulation must be considered as a potential explanation. Thus, at this time, it seems likely that genetic factors influence smoking and that individual differences in sensitivity to nicotine exist. We do not know, however, whether these individual differences in sensitivity to nicotine are related to why people do, and do not, smoke.

Our studies have been designed to attempt to answer some of the questions concerning the relationship between individual differences in sensitivity to nicotine and the use of tobacco as well as to provide potential biochemical explanations for individual differences in sensitivity to nicotine. Because such studies cannot be done in humans, for ethical as well as practical reasons, we have studied nicotine response in the mouse. The mouse has been used because of the availability of a large number of inbred mouse strains. Our research has been geared towards answering two general questions:

(1) Do genetic factors regulate first dose sensitivity to nicotine and, if so, does first dose sensitivity correlate with the number or affinity of brain nicotinic receptors?

(2) Do genetic factors regulate the development of tolerance to nicotine, and, if so, is tolerance related to a change in the number or affinity of brain nicotine receptors?

These questions are being asked because it may be that those people who continue to use tobacco after initial experimentation are those people who are least affected by initial exposures to nicotine. This seems possible in view of the observation that young males from families with a positive history for alcoholism are less affected by alcohol, both behaviorally and physiologically (prolactin and cortisol release), than are young males from families that do not have a history of alcoholism (Schuckit and Gold, 1988). Alternatively, it may be that those individuals who develop tolerance to the actions of nicotine are more likely to persist in using tobacco; i.e. smokers are those individuals who develop tolerance to the noxious actions of nicotine thereby uncovering the reinforcing actions.

Methods

Animal subjects. Male mice of 19 inbred strains and two selected lines were used for these studies. Five of the strains were bred and raised in our mouse colony: A/J/Ibg, BALB/cByJ, C57BL/6J/Ibg, DBA/2J/Ibg, and C3H/2Ibg. Mice of the remaining 14 inbred strains were purchased from Jackson Laboratories, Bar Harbor, ME.: AKR/J, BUB/BnJ, CBA/J, C57BL/10J, C57/BRcdJ, C57L/J, C58/J, DBA/1J, LP/J, P/J, RIIIS/J, SJL/J, ST/bJ, and SWR/J. In addition, two selectively bred lines of mice, the LS and SS, were used. The LS and SS mice were selectively bred for differential sensitivity to ethanol (McClearn and Kakihana, 1981) and are also differentially sensitive to nicotine (de Fiebre et al., 1987). All mice were 60 – 90 days of age when used. They were maintained in cages with 3 – 5 like-sexed littermates and were provided with food (Wayne Sterilizable Lab Blox) and water.

Chronic infusions. Mice of seven of the strains (A, BUB, C3H, C57BL, DBA/2, LS, SS) were chronically infused with nicotine using the procedures developed previously (Marks et al., 1983).

With this procedure a cannula is implanted in the jugular vein and saline (control) or nicotine-containing saline is infused continuously at a rate of 35 μl/h. For the studies reported here, nicotine doses of 0 (saline), 0.5, 1.0, 2.0, 4.0, and 6.0 mg/kg/per h were infused into the animals for 7 days. For doses exceeding 2.0 mg/kg/per h, animals were started out at this dose, and dose was increased on a daily basis until the final dose was achieved. Animals were tested for tolerance to nicotine 2 h after nicotine infusion was stopped. This time is sufficient for the animals to eliminate the nicotine that had been infused (Marks et al., 1983).

Behavioral and physiological responses to nicotine. Nicotine administration affects a wide variety of behavioral and physiological parameters in both humans and animals. Consequently, we have developed a multifactorial test battery that monitors the effects of nicotine on respiratory rate, locomotor activity as measured in a Y-maze (the number of crosses into separate arms of the maze and the number of times that the animal rears up onto its hind legs, a form of exploratory activity, are measured), acoustic startle response, heart rate, and body temperature. These responses are measured following lower dose treatments. This test battery has been described in detail previously (Marks et al., 1985). Latency to and incidence of seizures is measured following treatment with higher doses of nicotine (Miner et al., 1984).

Dose-response curves were constructed for nicotine effects in each of the strains in naive animals and in seven of the strains following chronic treatment in an attempt to measure tolerance. Parameters such as ED_{50} values were calculated for each response. The results reported here will focus on the effects of nicotine on body temperature. For this effect, the ED_{-2} value was calculated. This value is the nicotine dose that results in a 2° decrease in body temperature.

Nicotinic receptor binding. Nicotinic receptors were measured in eight brain regions using L-[^{3}H]nicotine and α[^{125}I]bungarotoxin (BTX) as the ligands. These two ligands bind to two distinct sites with nicotinic properties and may represent high and low affinity binding sites for nicotine, respectively (Marks and Collins, 1982). The methods used to measure nicotine and BTX binding are described in detail elsewhere (Marks and Collins, 1982; Marks et al., 1986). Following dissection of the brains into eight regions (cortex, hindbrain, cerebellum, hypothalamus, striatum, hippocampus, superior and inferior colliculi, and midbrain) and homogenization in buffer (Krebs-Ringer's HEPES), membranes were prepared by centrifugation and frozen. On the day of assay, the pellet was resuspended in buffer and incubated for 10 min at 37°C after which it was centrifuged at 18 000 × g for 20 min. The resulting pellet was resuspended in buffer for use in the assays.

The binding of L-[^{3}H]nicotine was measured at 4°C by incubating samples (100 – 500 μg protein) for 2 h with 5 nM radiolabelled nicotine. Blanks contained 10 μM unlabelled nicotine. Incubations were terminated by diluting the samples with 3 ml of ice cold buffer followed immediately by filtration of the samples onto glass fiber filters that had been soaked overnight in buffer containing 0.5% polyethyleneimine which serves to reduce the nonspecific binding. The binding of [^{125}I]BTX was measured similarly except that incubations were carried out for 4 h at 37°C and the buffer contained 0.01% bovine serum albumin. Samples routinely contained 1 nM [^{125}I]BTX and blanks contained 1 mM nicotine. Protein was measured using the method of Lowry et al. (1951) and binding was calculated in terms of fmol/mg protein.

Results

The analysis of sensitivity to challenge with an acute dose of nicotine in the 19 inbred strains, as measured by the multifactorial test battery and seizures, produced a large data set. The complete results will be published elsewhere. The overall results will be summarized here so as to provide a rationale for the selection of the strains for the chronic treatment studies.

Correlation analysis indicated that the responses could be grouped into two major classes: a set characterized by Y-maze crosses, Y-maze rears and body temperature and another set characterized by the seizure measures. Principal component analysis indicated that nicotine's effects on respiratory rate and acoustic startle shared characteristics with both of these sets, while nicotine's effects on heart rate were fairly unique.

Figure 1 presents a summary of these results.

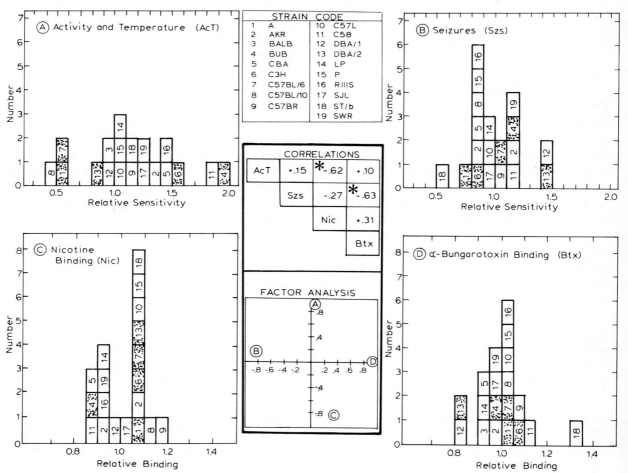

Fig. 1. Genetic influences on nicotinic receptors and response to nicotine. The histograms displayed in the four main panels represent the distribution of results obtained for the 19 inbred mouse strains. The upper left-hand panel presents the relative sensitivity of the strains to the effects of nicotine on Y-maze activity and body temperature. The upper right-hand panel presents the relative sensitivities of the strains to the seizure-inducing effects of nicotine. The lower left-hand panel presents the relative binding of L-[^{3}H]nicotine in the 19 strains, and the lower right-hand panel presents the relative binding of $\alpha - [^{125}I]$bungarotoxin (BTX). Values plotted for activity and body temperature are the arithmetic means of the ED_{50} values for the effects of nicotine on Y-maze crosses and rears and the ED_{-2} for nicotine effects on body temperature. The values plotted for the seizure measures represent the arithmetic mean of the ED_{50} values for nicotine-induced seizures following i.p. injection and latency to seizures following i.v. infusion. Binding data represent normalized averages obtained from ligand binding in six brain regions. The central portion of the figure presents a numerical code to identify the strains, correlation coefficients for relationships among the four variables, and a graphical representation of the results of a principal components factor analysis. Strains used in the chronic treatment studies are identified by stippling their corresponding data points.

The relative sensitivities of the strains were calculated for Y-maze crosses and rears (ED_{50} values for nicotine-induced decreases in activities were calculated for each strain) and body temperature (ED_{-2} values were calculated). Similarly, ED_{50} values for nicotine-induced seizures (the dose required to elicit seizures in 50% of the animals) and latency of seizures following i.v. infusion were estimated. Subsequently, the average of all 19 strains was calculated and given a value of 1.0. The strains were then rank ordered and these values are presented in Fig. 1, panel A, for the activity and temperature measures and in panel B for the seizure measures. The strains varied by approximately 4-fold for the activity and temperature measures and approximately 3-fold for the seizure measures. Activity and temperature-resistant strains included the BUB and C58 strains whereas sensitive strains included the A, C57BL/6 and C57BL/10 strains. Seizure-resistant strains included the DBA/1 and DBA/2 strains and a uniquely sensitive strain, the ST/b strain, was identified.

Panels C and D present the relative binding of nicotine and BTX in these 19 inbred strains. Relative binding was calculated by assigning the average binding, in terms of fmol/mg protein, for all of the brain regions in the 19 strains a value of 1.0. The mean binding for each of the strains was then calculated by averaging the binding in the various brain regions. This strain average was divided by the overall (19 strain) average to yield the relative binding. Binding values did not vary as widely among the strains as did the behavioral measures, but significant strain differences were seen. Nicotine binding appeared to be bimodal; high binding and low binding groups were seen. Strains at the highest end included the C57BL/10 and the C57BR while the lowest binding strains included BUB, CBA and C58 strains. BTX binding was trimodal with the ST/b strain showing very high binding and the DBA/1 and DBA/2 strains showing very low binding; all of the other strains were intermediate.

The middle panels in Fig. 1 present the strain

code along with the results of the correlational and factor analyses. Statistically significant correlations were seen between the activity-temperature measures and nicotine binding (-0.62) and between the seizure measures and BTX binding (-0.63) thereby suggesting that the nicotinic receptor that we measure with nicotine binding regulates the activity-temperature measures and that the BTX binding site regulates nicotine-induced seizures.

These studies allowed us to select strains that were maximally diverse with regard to their sensitivity to nicotine for use in the chronic treatment studies. The strains that we have selected include the C57BL/6, A, DBA/2, C3H and BUB strains. The C57BL/6 and A strains are very sensitive to the effects of nicotine on the activity-temperature measures whereas the DBA/2 strain is intermediate and the C3H and BUB strains are quite insensitive. We have added to these inbred strains, the selectively bred LS and SS lines. The LS are quite sensitive to nicotine and the SS are resistant (de Fiebre et al., 1987).

Figure 2 presents the effects of various challenge doses of nicotine on body temperature, measured 15 min after injection, in three of the mouse strains that had been infused chronically with saline or

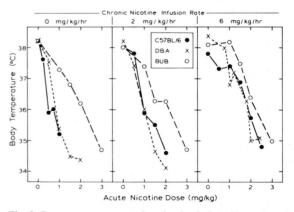

Fig. 2. Dose-response curves for nicotine-induced hypothermia. Dose-response curves were constructed with mice of the C57BL/6, DBA/2, and BUB strains after chronic infusion with saline (left), 2.0 mg/kg/per h nicotine (center) or 6.0 mg/kg/per h nicotine (right). Each point represents the mean value obtained from a minimum of six mice.

nicotine (2 or 6 mg/kg/per h). Note the marked strain differences in sensitivity to nicotine in the saline-infused control animals. Nicotine results in a shift to the right of the dose-response curves for nicotine effects on body temperature in the C57BL/6 and DBA/2 strains while the dose-response curves of the BUB mice are less affected by chronic nicotine infusion. These results suggest that mouse strains that are relatively sensitive to nicotine develop tolerance to nicotine in a dose-dependent fashion while resistant strains do not.

Figure 3 presents the effects of chronic infusion with varying doses of nicotine (0, 0.5, 1.0, 2.0, 4.0 and 6.0 mg/kg per h) on nicotine-induced

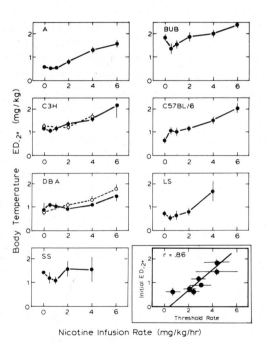

Fig. 3. Change in ED_{-2} values with chronic nicotine treatment. Seven different strains or lines of mice were chronically treated with the indicated doses of nicotine for 7 days and tested for the effects of challenge doses of nicotine 2 h after chronic treatment was stopped. The dose required to lower body temperature by 2°C was calculated from dose-response curves similar to those shown in Fig. 2. The data presented for C3H and DBA/2 mice were calculated from results previously published (Marks et al., 1986b). The panel in the lower right compares the basal ED_{-2} value for the seven mouse strains with the chronic infusion dose required to elicit a 0.3 mg/kg increase in the ED_{-2} values.

hypothermia as measured by the ED_{-2} value, in seven mouse strains. Consistent with the results presented in Fig. 2, some mouse strains and lines developed tolerance at lower doses than did others. The lower right-hand panel of Fig. 3 contrasts the dose required to elicit a 2°C decrease in body temperature, the ED_{-2} value, for control mice in each strain with the chronic nicotine infusion dose that appears to be at the threshold for tolerance induction. This value was estimated by calculating the infusion dose that elicits a 0.3 mg/kg change in the ED_{-2} value. The correlation between ED_{-2} in naive animals and the threshold dose for tolerance development is 0.86 indicating that the less sensitive mouse strains develop tolerance less readily than do the most sensitive strains.

Although not shown here, comparable results were obtained with the other tests that comprise the test battery. Thus, initial sensitivity to nicotine seems to predict the chronic treatment dose required to elicit tolerance to nicotine.

Figure 4 presents the effects of nicotine infusion on nicotine and BTX binding in the cortex of the BUB, DBA/2 and C57BL/6 mouse strains. Nicotine infusion resulted in a dose-dependent increase in nicotine binding in each of the strains, but subtle differences were evident. The BUB strain showed smaller increases in binding at the lower nicotine infusion doses, particularly the 1 and 2 mg/kg per h doses.

BTX binding was not affected by nicotine treatment until the higher infusion doses were achieved. The DBA strain exhibited significant increases in BTX binding at the 2 mg/kg/per h infusion dose. Higher doses were required in the C57BL/6 strain to elicit an increase in BTX binding and higher doses still were required for BUB mice.

Discussion

The results presented here clearly demonstrate that genetically regulated variability in sensitivity to a first dose of nicotine and in the ability to develop tolerance to nicotine exist in the mouse. These findings suggest that similar genetic influences in

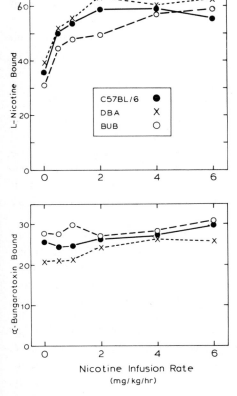

Fig. 4. Chronic treatment and nicotinic receptors. The effects of chronic nicotine infusion on the binding of L-[^{3}H]nicotine (4 nM) and $\alpha - $[^{125}I]bungarotoxin (1 nM) in cortical membranes are shown.

profound differences in receptor changes were not seen. This suggests that genetic influences on tolerance development involve regulation at another or additional levels.

A recent study from our laboratory (Miner and Collins, 1988) demonstrated that mouse strains differ in the development of acute tolerance or tachyphylaxis to high dose nicotine effects (seizures). DBA/2 mice show a marked reduction in sensitivity to nicotine-induced seizures if they are pretreated with subseizure doses of nicotine while C3H mice do not. We speculated that this difference might arise if nicotinic receptors from DBA/2 mice desensitize more readily, or resensitize at a slower rate, than do nicotinic receptors from C3H mice. Perhaps genetic factors regulate the functional status of mouse brain nicotinic receptors.

The chronic treatment studies reported here replicate previous studies from our laboratory (Marks et al., 1983, 1985, 1986b, 1987) that have demonstrated that tolerance to nicotine is dose-dependent and is accompanied by a dose-dependent increase in nicotine binding. Most of these earlier studies used DBA/2 mice. We have also demonstrated, using DBA/2 mice, that the time courses of tolerance acquisition and loss are paralleled by the increase and return to control, respectively, of brain nicotine binding (Marks et al., 1985). Tolerance was measured in these studies using our test battery. These results argue that receptor changes are a critical factor that regulates tolerance development. The recent observation that nicotine binding is greater in selected regions of autopsied brain from smokers, as compared to nonsmokers (Benwell et al., 1988) argues that these findings may be relevant to humans.

The results reported here do not agree entirely with our previous findings with DBA/2 mice in that some mouse strains, such as the BUB strain, do not develop marked tolerance even though chronic nicotine treatment elicits significant increases in nicotine binding at lower doses and BTX binding at higher doses. However, the present results are consistent with those obtained in a re-

humans may exist. Consequently, potential roles for initial sensitivity differences and in capacity for developing tolerance in the initiation and maintenance of tobacco use by humans must be considered. The observation that initial sensitivity differences among inbred mouse strains are due, in part, to differences in the number of brain nicotinic receptors with sensitive strains having greater numbers of receptors suggests the possibility that individual differences in sensitivity to nicotine in humans may also be regulated by differences in nicotinic receptor numbers. While more sensitive mouse strains developed tolerance to a greater degree than did the less sensitive strains,

144

cent comparison of tolerance development that used DBA/2 and C3H mice (Marks et al., 1986). In this study we confirmed the observation that DBA/2 mice develop a dose-dependent tolerance to nicotine. The C3H mice developed marginal tolerance even though they did respond to nicotine with a dose-dependent increase in nicotine binding. A comparison of DBA/2 and C3H receptor changes indicated that C3H mice receptor changes were subtly less than those observed in DBA/2 mice. The results reported here extend this observation. BUB and SS mice develop very little tolerance, yet these strains do exhibit an increase in nicotine binding. Such a result could be interpreted as suggesting that receptor changes are not the cause, or the only cause, of tolerance. However, before this conclusion is reached a more careful analysis is required.

We believe that a more careful analysis is required because of recent findings that were made using quantitative autoradiographic methods. We have recently completed an autoradiographic analysis of nicotine and BTX binding in several mouse strains. Interesting and provocative results are being obtained. For example, we have reported that LS and SS mice differ dramatically, approximately 2-fold, in their sensitivity to nicotine, but these two mouse lines do not differ in either nicotine or BTX binding as measured by our standard biochemical ("grind and bind") assays (de Fiebre et al., 1987). However, our autoradiographic analysis indicates that these mouse lines do, indeed, have different numbers of nicotinic receptors, but only in specific nuclei. Thus, the differential sensitivity to nicotine of these two mouse lines may be due to differences in receptor numbers in specific nuclei. Similar conclusions seem likely based on the results of a recent analysis of the effects of chronic nicotine infusion on nicotine and BTX binding in DBA/2 mice. Quantitative autoradiographic methods detected striking differences in the increase in nicotine and BTX binding in different brain nuclei; some nuclei show increases of 200–400% whereas other nuclei are virtually unaffected. If mouse strains differ in

receptor changes in critical brain nuclei, an explanation for differential tolerance may be at hand.

Our receptor binding assays detect a single, high affinity nicotine binding site. However, molecular genetic methods clearly demonstrate the existence of several nicotinic receptor subtypes that are differentially distributed in rat brain (Goldman et al.,1987). If one of these receptor subtypes is critically involved in regulating the behavioral responses to nicotine that we measure, and mouse strains differ in the number or regional distribution of these receptor subtypes, the differential tolerance development that we have detected may be explained.

In summary, our results suggest that genetic factors regulate tolerance development. Mouse strains that are less sensitive to a first dose of nicotine develop less tolerance to nicotine than do strains that are more sensitive. These strain differences are explained, at least in part, by differences in the number of brain nicotine and BTX binding sites. All mouse strains exhibit increases in brain nicotinic ligand binding following chronic treatment, but strains that differ in tolerance have only subtle differences in receptor up-regulation. More careful analyses using autoradiographic methods are in progress. These studies may be useful in assessing fully the role of receptor changes in tolerance development. In addition, it may be that the newly discovered receptor subtypes are differentially affected by chronic nicotine treatment in the various mouse strains and that this leads to differential tolerance development. Lastly, it may be that receptor changes are more related to the development of dependence on nicotine. Analysis of this potential role will require the development of adequate methods of quantifying a nicotine withdrawal syndrome, if it exists, in the mouse.

Acknowledgements

The technical assistance of Elena Romm, Steven Campbell and Jerry Stitzel is gratefully acknowledged. This work was supported by a grant from the National Institute on Drug Abuse (DA-03194) and the National Institute on Alcohol Abuse and Alcoholism (AA-06391).

References

Benwell, M.E.M., Balfour, D.J.K. and Anderson, J.M. (1988) Evidence that tobacco smoking increases the density of (−)-[^{3}H]nicotine binding sites in human brain. *J. Neurochem.,* 50: 1243 – 1247.

Crumpacker, D.W., Cederlof, R., Friberg, L., Kimberling, W.J., Sorenson, S., Vandenberg, S.G., Williams, J.S., McClearn, G.E., Grever, B., Iyer, H., Krier, M.J., Pedersen, N.L., Price, R.A. and Roulette, I. (1979) A twin methodology for the study of genetic and environmental control of variation in human smoking behavior. *Acta Gen. Med. Gemell.,* 28: 173 – 195.

De Fiebre, C.M., Medhurst, L.J. and Collins, A.C. (1987) Nicotine response and nicotinic receptors in long-sleep and short-sleep mice. *Alcohol,* 4: 493 – 501.

Dies, R. and Reznikoff, M. (1969) Personality and smoking patterns in a twin population. *J. Proj. Tech. Pers. Assess.,* 33: 457 – 463.

Fisher, R.A. (1958a) Lung cancer and cigarettes? *Nature (Lond.),* 182: 180.

Fisher, R.A. (1958b) Cancer and smoking. *Nature (Lond.),* 182: 596.

Friberg, L., Kaij, L., Dencka, S.J. and Jonsson, E. (1959) Smoking habits of monozygotic and dizygotic twins. *Br. Med. J.,* 1: 1090 – 1092.

Goldman, D., Deneris, E., Luyten, W., Kochhar, A., Patrick, J. and Heinemann, S. (1987) Members of a nicotinic acetylcholine receptor gene family are expressed in different regions of mammalian central nervous system. *Cell,* 48: 965 – 973.

Gritz, E.R. (1980) Smoking behavior and tobacco abuse. In N.K. Mello (Ed.), *Advances in Substance Abuse,* JAI Press, Greenwich, CT, pp. 91 – 158.

Henningfield, J.E. (1984) Behavioral pharmacology of cigarette smoking. *Adv. Behav. Pharmacol.,* 4: 131 – 210.

Jones, R.A. (1986) Individual differences in nicotine sensitivity. *Add. Behav.,* 11: 435 – 438.

Kaprio, J., Sarna, S., Koshenvuo, M. and Rantasalo, I. (1978) *The Finnish Twin Registry: Baseline Characteristics, Section II,* University of Helsinki Press, Helsinki.

Kozlowski, L.T. and Herman, C.P. (1984) The interaction of psychosocial and biological determinants of tobacco use: more on the boundary model. *J. Appl. Soc. Psychol.,* 14: 244 – 256.

Lowry. O.H., Rosebrough, N.L., Farr, A.L. and Randall, R.J. (1951) Protein measurement with the Folin phenol reagent. *J. Biol. Chem.,* 193: 265 – 275.

McClearn, G.E. and Kakihana, R. (1981) Selective breeding for ethanol sensitivity: short-sleep and long-sleep mice. In G.E. McClearn, R.A. Deitrich and V.G. Erwin (Eds.), *Development of Animal Models as Pharmacogenetic Tools,* DHHS Publication No. (ADM) 81 – 1133, Washington D.C., pp. 147 – 159.

Marks, M.J. and Collins, A.C. (1982) Characterization of nicotine binding in mouse brain and comparison with the binding of α-bungarotoxin and quinuclidinyl benzilate. *Mol. Pharmacol.,* 22: 554 – 564.

Marks, M.J. and Collins, A.C. (1985) Tolerance, cross-tolerance, and receptors after chronic nicotine or oxotremorine. *Pharmacol. Biochem. Behav.,* 22: 283 – 291.

Marks, M.J., Burch, J.B. and Collins, A.C. (1983) Effects of chronic nicotine infusion on tolerance and cholinergic receptors. *J. Pharmacol. Exp. Ther.,* 226: 806 – 816.

Marks, M.J., Romm, E., Bealer, S. and Collins, A.C. (1985a) A test battery for measuring nicotine effects in mice. *Pharmacol. Biochem. Behav.,* 23: 325 – 330.

Marks, M.J. Stitzel, J.A. and Collins, A.C. (1985b) Time course study of the effects of chronic nicotine infusion on drug response and brain receptors. *J. Pharmacol. Exp. Ther.,* 235: 619 – 628.

Marks, M.J., Romm, E. Gaffney, D.K. and Collins, A.C. (1986a) Nicotine-induced tolerance and receptor changes in four mouse strains. *J. Pharmacol. Exp. Ther.,* 237: 809 – 819.

Marks, M.J., Stitzel, J.A. and Collins, A.C. (1986b) A dose-response analysis of nicotine tolerance and receptor changes in two inbred mouse strains. *J. Pharmacol. Exp. Ther.,* 239: 358 – 364.

Marks, M.J., Stitzel, J.A., Romm, E., Wehner, J.M. and Collins, A.C. (1986c) Nicotinic binding sites in rat and mouse brain: comparison of acetylcholine, nicotine and α-bungarotoxin. *Mol. Pharmacol.,* 30: 427 – 436.

Marks, M.J., Stitzel, J.A. and Collins, A.C. (1987) Influence of kinetics of nicotine administration on tolerance development and receptor levels. *Pharmacol. Biochem. Behav.,* 27: 505 – 512.

Miner, L.L. and Collins, A.C. (1988) Effect of nicotine pretreatment on nicotine-induced seizures. *Pharmacol. Biochem. Behav.,* 29: 375 – 380.

Miner, L.L., Marks, M.J. and Collins, A.C. (1984) Classical genetic analysis of nicotine-induced seizures and nicotinic receptors. *J. Pharmacol. Exp. Ther.,* 231: 545 – 554.

Nesbitt, P.D. (1973) Smoking, physiological arousal, and emotional response. *J. Personal. Soc. Psychol.,* 25: 137 – 144.

Partanen, J., Bruun, K. and Markkanen, T. (1977) Inheritance of drinking behavior: a study of intelligence, personality, and use of alcohol of adult twins. In E.M. Pattison, M.B. Sobell and L.C. Sobell (Eds.), *Emerging Concepts of Alcohol Dependence.* Springer Publishing Co., Inc., New York, pp. 331 – 343.

Pedersen, N.L. (1981) Twin similarity for usage of common drugs. In L. Gedda et al. (Eds.), *Twin Research, Vol. 3:*

Epidemiological and Clinical Studies, Alan Liss, N.Y., pp. 53 – 59.

Pomerleau, C.S., Pomerleau, O.F. and Majchrzak, M.J. (1987) Mecamylamine pretreatment increases subsequent nicotine self-administration as indicated by changes in plasma nicotine level. *Psychopharmacology,* 91: 391 – 393.

Raaschou-Nielsen, E. (1960) Smoking habits in twins. *Dan. Med. Bull.,* 7: 82 – 88.

Schuckit, M.A. and Gold, E.O. (1988) A simultaneous evaluation of multiple markers of ethanol/placebo challenges in sons of alcoholics and controls. *Arch. Gen. Psychiatr.,* 45: 211 – 216.

Stolerman, I.P., Goldfarb, T., Fink, R. and Jarvik, M.E. (1973) Influencing cigarette smoking with nicotine antagonists. *Psychopharmacologia,* 28: 247 – 259.

Todd, G.F. and Mason, J.I. (1959) Concordance of smoking habits in monozygotic and dizygotic twins. *Heredity,* 13: 417 – 444.

Williams, J.S., Crumpacker, D.W. and Krier, M. (1981) Genetic and environmental variance fractions and correlation estimates for smoking behavior in a Swedish population (cited by Pedersen, N.) In L. Gedda et al. (Eds.), *Twin Research, Vol. 3: Epidemiological and Clinical Studies,* Alan Liss, N.Y., pp. 53 – 59.

Release of Acetylcholine and
other Transmitters by Nicotine

A. Nordberg, K. Fuxe, B. Holmstedt and A. Sundwall (Eds.)
Progress in Brain Research, Vol. 79

CHAPTER 14

Effect of nicotine on the release of acetylcholine and amino acids in the brain

Lorenzo Beani[a], Clementina Bianchi[a], Luca Ferraro[a], Lena Nilsson[b], Agneta Nordberg[b], Luciana Romanelli[a], Piero Spalluto[a], Anders Sundwall[c] and Sergio Tanganelli[a]

[a] Department of Pharmacology, University of Ferrara, Italy, [b] Department of Pharmacology, University of Uppsala, Sweden and [c] AB Leo Helsingborg, Sweden

Introduction

The central effects of nicotine are mainly due to its interaction with specific brain receptors (Morley, 1981; Aceto and Martin, 1982; De La Garza et al., 1987). Their stimulation increases the firing rate of many neurones and, consequently, the release of the respective transmitters (Balfour, 1982). Since these receptors are also present on the nerve endings, including cholinergic nerve endings (Giorguieff et al., 1979; Yoshida and Imura, 1979; Nordberg and Larsson, 1980; Rowell and Winkler, 1984; Schwartz et al., 1984; Beani et al., 1985; Moss and Wonnacott, 1985), their stimulation does enhance the neurosecretory process, independently of the axonal impulse flow.

The renewed interest in nicotinic transmission within the brain is due to the recognized links between Alzheimer disease and Meynert nucleus lesions, determining a consistent loss of corticipetal cholinergic input and of nicotinic receptors in various brain areas (Nordberg and Winblad, 1986; Shimohama et al., 1986; Kellar et al., 1987). Thus, facilitation of synaptic transmission by nicotinic agonists might be considered one of the possible therapeutic approaches to senile dementia (Nilsson et al., 1987). On the other hand, our knowledge about nicotine effects on neurotransmitter systems are mainly centered on brain monoamines (Balfour, 1982), whereas very little is known about nicotine influence on putative amino acid transmitters. Furthermore, data on possible changes in the neurosecretory responses following chronic nicotine treatment are lacking, although numerous reports show consistent adaptive changes at the receptor level (Schwartz and Kellar, 1983; Larsson et al., 1986; Romanelli et al., 1988).

The aim of this study was (1) to determine the influence of nicotine on D-[^{3}H]aspartate, endogenous GABA and acetylcholine (ACh) release in brain slices and in freely moving rats and guinea-pigs and (2) to check for possible alterations in the cholinergic secretory response and in nicotinic receptors following chronic nicotine treatment.

Materials and methods

Guinea-pig brain slices were prepared, set up in small superfusion chambers and kept at rest or electrically stimulated as previously described (Beani et al., 1978) so as to obtain a calcium- and tetrodotoxin-sensitive release. Striatal slices preloaded with D-[^{3}H]aspartate 0.3 μM for 30 min were used to study the effect of nicotine on stimulus-triggered tritium efflux from the glutamatergic structures. Cortical slices kept at rest were employed to examine the drug's influence on the spontaneous efflux of endogenous GABA from the gabaergic cortical neurones (Beani et al., 1986). The outflow of D-[³H]aspartate and of endogenous GABA and ACh from the parietal cortex of freely moving guinea-pigs and rats was studied according to the epidural cup method as described by Beani and Bianchi (1970, see also Antonelli et al., 1986).

The amount of transmitters present in the samples (physiological saline solution superfusing the slices or kept in contact for fixed times with the parietal cortex of the animals) was determined with (1) radiometric assay for D-[³H]aspartate, according to the procedure already followed for [³H]ACh assay (Beani et al., 1984); (2) GCMS analysis for GABA (see Beani et al., 1986); (3) luminescence method (Blomquist et al., 1987) or bioassay on the guinea-pig ileum (Beani et al., 1978) for ACh. Cortical nicotine receptors were measured using [³H]nicotine (7 nM) or [³H]ACh (10 nM) in the presence of atropine (Zhang et al., 1987). (−)Nicotine was administered s.c. or i.c.v. Its chronic treatment was performed with 0.45 mg · kg⁻¹, given twice daily for 16 days and the cortical ACh release was checked 24 h after.

Results

In vitro studies. Nicotine 6.2 − 62 μM increased the electrically-evoked release of D-[³H]aspartate from striatal slices stimulated at 20 Hz for 2 min. The increase reached +56% after nicotine 62 μM (Fig. 1). No drug effect was found on the spontaneous D-[³H]aspartate release. Except for this last point, these results agree with those showing nicotine facilitation of [³H]ACh release from guinea-pig cortical slices (Beani et al., 1985). Thus, nicotine receptors appear to positively modulate the glutamatergic cortico-striatal projections as well. This conclusion is similar to the one reached by Gongwer et al. (1987) who found increased release of endogenous glutamate by nicotine in potassium-depolarized slices of rat nucleus accumbens.

Nicotine 62 − 620 μM proved also to increase the spontaneous GABA efflux from guinea-pig cortical slices (Fig. 2). At variance with the facilitation on D-[³H]aspartate and [³H]ACh (prevented by D-

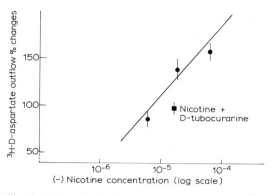

Fig. 1. Effect of nicotine on electrically-evoked D-[³H]aspartate efflux from guinea-pig caudatal slices. The parameters of the stimulation (repeated twice at the 60th and 90th min of superfusion) were: frequency 20 Hz, pulse strength 30 mA · cm⁻² and duration 5 ms, train duration 2 min. The facilitatory effect of nicotine, added during the second stimulation period (S₂) was calculated as percent increase of the S₂/S₁ ratio. Mean ± S.E.M. of 5 − 6 experiments. Note the antagonism by d-tubocurarine 4.5 μM.

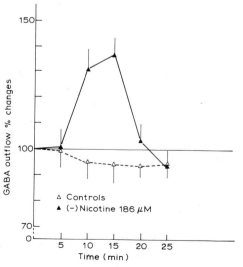

Fig. 2. Effect of nicotine 186 μM on the spontaneous efflux of endogenous GABA from guinea-pig cortical slices. The facilitatory effect of nicotine, added from the arrow onwards, was calculated as percent of the initial value. Mean ± S.E.M. of 5 − 6 experiments. Note the disappearance of drug effect some minutes after addition.

tubocurarine) the effect of nicotine on GABA was antagonized by mecamylamine 50 μM and by tetrodotoxin 2 μM. These findings agree with the recent reports of Limberger et al. (1986) and Wonnacott et al. (1987) who described facilitation of GABA release in hippocampal synaptosomes and caudatal slices. The only difference is represented by the ineffectiveness of tetrodotoxin in rat caudate slices (Limberger et al., 1986) suggesting a different localization and function of the nicotinic receptors in this structure with respect to guinea-pig cerebral cortex.

In vivo studies. At present, no reports are available regarding the effects of nicotine on the release of putative amino acid transmitters from the cortex of freely moving animals. To study the response of the cortical glutamatergic structures, it was worth considering preloading the cortex underlying the cup with D-[^{3}H]aspartate (0.6 μM for 5 h), then carefully washing, collecting and renewing at fixed intervals the saline solution in the cup and measuring its tritium content. This approach overcomes the troubles and errors deriving from the efflux of non-neuronal glutamate which may mask the actual changes of neuronal glutamate outflow. This simple procedure makes it possible to detect an increase D-[^{3}H]aspartate caused, e.g. by a CNS stimulant such as pentylenetetrazole, administered at subconvulsant doses. A consistent peak of increased tritium outflow has been found 30 – 45 min after nicotine 0.9 mg · kg^{-1} s.c. (Fig. 3), able to enhance the endogenous ACh release as well (see below). Preliminary trials indicate that the facilitatory effect of nicotine is prevented by mecamylamine 2 mg · kg^{-1} i.p.

A more intriguing situation arises when examining nicotine influence on cortical GABA outflow. The drug is apparently ineffective up to 0.9 mg · kg^{-1}, but it inhibits the amino acid outflow at higher doses. The GABA reduction reaches a value of 73% with respect to pretreatment levels after nicotine 3.6 mg · kg^{-1} (Fig. 4). This effect is prevented by pretreating the animals with

mecamylamine 8 mg · kg^{-1} and is no longer evident when the spontaneous outflow of GABA is reduced by adding tetrodotoxin 0.5 μM to the cup solution. Since the GABA responses to nicotine obtained in the animal are opposite to those found in brain slices, it is tenable that the drug can, in vivo, stimulate a neuronal system able to inhibit the GABA cells. Previous work showed that

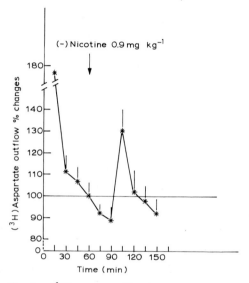

Fig. 3. D-[^{3}H]aspartate efflux from the parietal cortex of freely moving guinea-pig. Nicotine 0.9 mg · kg^{-1} s.c. consistently increased the outflow of tritium in 4 typical experiments.

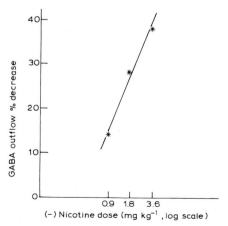

Fig. 4. Relationship between nicotine dose s.c. and percent reduction of cortical GABA outflow from the parietal cortex of freely moving guinea-pigs. Mean of 5 – 8 experiments.

opiates inhibit cortical GABA outflow (Antonelli et al., 1986). Thus, the influence of an opiate antagonist on nicotine-induced inhibition of GABA was tested. The initial trials have shown that naloxone 3 mg · kg^{-1}, administered 10 min before nicotine, prevents any reduction in GABA outflow. Therefore, it seems that the endogenous opioid system, stimulated by nicotine (Eiden et al., 1984; Molinero and Del Rio, 1987), in turn inhibits GABA release (Brennan et al., 1980).

Analysis of nicotine's effect on the cholinergic system has been carried out in naive and chronically treated (see setion on Methods) guinea-pigs and rats. Nicotine dose-dependently enhanced ACh release in both animal species (Fig. 5), not only by s.c. administration (0.45 – 0.9 mg · kg^{-1} s.c., see also Armitage et al., 1969; Erickson et al., 1973), but also by i.c.v. injection (1 – 5 μg), through mecamylamine-sensitive receptors. The chronic treatment has brought to light interesting species differences. Twenty-four hours after the last drug injection, the rats showed a basal ACh release significantly higher than that of naive animals, while the guinea-pigs behaved in the opposite manner. However, nicotine 0.45 mg · kg^{-1} still enhanced transmitter outflow (Table I) giving rise to a similar percent increase. The effect, as absolute increase of release, was, however, reduced with respect to normal animals in guinea-pigs while it was slightly enhanced in rats. Interestingly, the number of the cortical nicotinic receptors increased by 25 – 30% in the rat but it was unchanged in the guinea-pig. Consequently, the contribution of each receptor in determining the increase in ACh release (in terms of absolute transmitter amounts) was reduced in chronically treated animals, no matter what changes were found in the receptors.

Concluding remarks

From a general point of view, our findings confirm and extend previous reports on ACh release and fit well with the findings on myenteric plexus synaptosomes (Table II). They also agree with the old hypothesis (Beani et al., 1964), in which new interest is being taken (Wessler et al., 1986; Vizi et al., 1987; Bowman et al., 1988; Matzner et al., 1988) of nicotinic receptors increasing ACh release at the n.m.j. Clearly, nicotine facilitates the efflux of this transmitter from brain slices and from the cerebral cortex of naive and subchronically treated guinea-pigs and rats. In addition, evidence has been presented for the first time that this drug not only facilitates D-[^{3}H]aspartate and GABA outflow from brain slices, but increases D-[^{3}H]aspartate and reduces GABA outflow from the cortex of the unanaesthetized guinea-pigs.

The receptors involved belong to the n.m.j. subtype in the cortical cholinergic and striatal glutamatergic nerve endings. Conversely, the stimulation elicited by nicotine, both on the GABAergic interneurones in vitro and on ACh, GABA and glutamate cells in vivo is abolished by mecamylamine. This finding can be explained by assuming that nicotine receptors of the n.m.j. subtype are located on some axons and nerve terminals. On the other hand, receptors of the ganglionic subtype appear responsible for the increased ACh and amino acid release from the whole animal (as well as from slices containing intact GABA neurones). Possibly, species characteristics might account for differences in nicotine

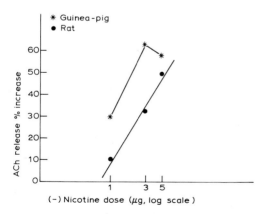

Fig. 5. Relationship between nicotine dose i.c.v. and percent increase of cortical acetylcholine release from the parietal cortex of freely moving guinea-pigs and rats. Mean of 5 – 8 experiments. Note the higher sensitivity of the guinea-pig to nicotine.

153

receptor subtypes and in their cellular localization (cell bodies or nerve endings).

Paying attention to the tolerance phenomenon of the cholinergic system, it is noteworthy that the response to nicotine after 16 days of treatment assumes peculiar aspects depending on the species examined. Thus, the basal ACh release 24 h after ceasing drug treatment is higher in subchronically treated rats than in naive animals, whereas the opposite happens in guinea-pigs. The reason for these differences is not clear. It is of interest, however, that the drug response is proportionally maintained in both species. Since the high affinity binding sites are increased in the rat and unaffected in the

TABLE I

Effect of (−)nicotine 0.45 mg · kg^{-1} on cortical release of ACh in naive and chronic nicotine-treated guinea-pigs and rats

Experimental conditions	No. of experiments	ACh release ng · cm^2/30 min		% Increase
		Basal	30 min after (−)nicotine	
Guinea-pigs				
Naive	10	23.9 ± 3.1	36.0 ± 4.4*	150 ± 5.9
Nicotine-treated	10	19.0 ± 1.9	25.2 ± 2.5*	132 ± 5.7**
Rats				
Naive	8	26.0 ± 6.0	40.0 ± 6.0*	153 ± 11.0
Nicotine-treated	8	44.0 ± 6.0**	62.0 ± 10.0*	140 ± 16.0

Significantly different from pretreatment values, * $p < 0.01$; significantly different from naive animals, ** $p < 0.05$.

TABLE II

Nicotinic facilitation of ACh release in the CNS

Experimental conditions and 1st author	Brain region and species	Type of preparation	Active doses or concentration	Antagonists
In vitro				
Rowell, 1984	Cerebral cortex mouse	Synaptosomes	100 µM	Hexamethonium
Beani, 1985	Cerebral cortex guinea-pig	Slices	1.8 – 180 µM	d-tubocurarine tetrodotoxin
Moss, 1987	Hippocampus rat	Synaptosomes	1 – 10 µM	Mecamylamine (not d-tubocurarine)
Meyer, 1987	Cerebral cortex rat	Minces Synaptosomes	10 – 100 µM inactive	Not tested
Araujo, 1987, 1988	Cerebral cortex rat	Slices	1 µM	d-Tubocurarine (not tetrodotoxin)
In vivo				
Armitage, 1969	Cerebral cortex rat	Cup technique	2 µg/kg i.v. every 30 s	Not tested
Erickson, 1973	Cerebral cortex rat	Cup technique	0.4 mg/kg i.p.	Not tested

guinea-pig, the conclusion can be drawn that, in both cases, partial tolerance develops. The persistence of a neurochemical response in terms of ACh release may account for the maintained reward found in the animals submitted to nicotine self-administration tests (Goldberg et al., 1981).

Parallel to the increased activity of the cholinergic structures, nicotine stimulates the glutamatergic neurones in vivo and in vitro. Furthermore, nicotine enhances the endogenous GABA release in vitro, even if at concentrations slightly higher than those active on ACh release. On the contrary, nicotine inhibits GABA outflow in freely moving animals. This reduction is prevented not only by mecamylamine but also by the topical application of tetrodotoxin and by naloxone administration. Thus, a reduced firing rate of the intracortical GABA interneurones is responsible for the reduction in the outflow: the opioid system appears to switch nicotine influence on GABA from excitation (as found in vitro) to inhibition.

In conclusion, nicotine evokes in unanaesthetized animals a series of complex neuronal responses represented by increased ACh release (remaining, even if reduced, after subchronic treatment), increased D-[^{3}H]aspartate and reduced GABA outflow. Such an imbalance between excitatory and inhibitory signals certainly increases cortical excitability, possibly even leading to toxicological consequences. In fact, the suppression of GABA control, together with increased glutamatergic activity, might promote cellular damage when concomitant metabolic or blood supply troubles are present. On the other hand, the increased efficacy of the excitatory inputs may explain the favourable effects of nicotine in maintaining the awake state and in facilitating the memory processes.

Acknowledgements

This work was supported by grants from Consiglio Nazionale delle Ricerche and Ministero della Pubblica Istruzione, and from grants of the Swedish Medical Research Council.

References

Aceto, M.D. and Martin, B.R. (1982) Central actions of nicotine. Med. Res. Rev., 2: 43 – 62.

Antonelli, T., Beani, L., Bianchi, C., Rando, S., Simonato, M. and Tanganelli, S. (1986) Cortical acetylcholine release is increased and τ-aminobutyric acid outflow is reduced during morphine withdrawal. Br. J. Pharmacol., 89: 853 – 860.

Araujo, D.M., Lapchak, P.A., Collier B. and Quirion, R. (1987) N-methylcarbamyl choline, a novel nicotinic agonist, enhances acetylcholine release from rat brain slices. Soc. Neurosci., 13: 1191.

Armitage, A.G., Hall, G.H. and Sellars, C.M. (1969) Effects of nicotine on electrocortical activity and acetylcholine release from cat cerebral cortex. Br. J. Pharmacol., 35: 152 – 160.

Balfour, D.J.K. (1982) The effects of nicotine on brain neurotransmitter systems. Pharmacol. Ther., 16: 269 – 282.

Beani, L. and Bianchi, C. (1970) Effects of adrenergic blocking and antiadrenergic drugs on the acetylcholine release from the exposed cerebral cortex of the conscious animal. In E. Heilbronn and A. Winter (Eds.), Drugs and Cholinergic Mechanisms in CNS, Research Institute of National Defence, Stockholm, pp. 369 – 386.

Beani, L., Bianchi, C. and Ledda, F. (1964) The effect of tubocurarine on acetylcholine release from the motor nerve terminals. J. Physiol., 174: 172 – 183.

Beani, L., Bianchi, C., Giacomelli, A. and Tamberi, F. (1978) Noradrenaline inhibition of acetylcholine release from guinea-pig brain. Eur. J. Pharmacol., 48: 89 – 93.

Beani, L., Bianchi, C., Siniscalchi, A., Sivilotti, L., Tanganelli, S. and Veratti, E. (1984) Different approaches to study acetylcholine release: endogenous ACh versus tritium efflux. Naunyn-Schmiedeberg's Arch. Pharmacol., 328: 119 – 126.

Beani, L., Bianchi, C., Nilsson, L., Nordberg, A., Romanelli, L. and Sivilotti, L. (1985) The effect of nicotine and cytisine on ^{3}H-acetylcholine release from cortical slices of guinea-pig brain. Naunyn-Schmiedeberg's Arch. Pharmacol., 331: 293 – 296.

Beani, L., Tanganelli, S., Antonelli, T. and Bianchi, C. (1986) Noradrenergic modulation of cortical acetylcholine release is both direct and γ-aminobutyric acid-mediated. J. Pharmacol. Exp. Ther., 236: 230 – 236.

Blomquist, I., Lundin, A., Nordberg, A. and Sundwall, A. (1987) Development of an automatic luminometric assay of choline and acetylcholine for studying acetylcholine metabolism in brain tissue in vitro. In M. Dowdall and J. Hawthorne (Eds.), Cellular and Molecular Basis of Cholinergic Function, Ellis Horvard Ltd., Chichester, Sussex U.K., pp. 368 – 378.

Bowman, W.C., Marshall, I.G., Gibb, A.J. and Harborne, A.J. (1988) Feedback control of transmitter release at the neuromuscular junction. TIPS, 9: 16 – 20.

Brennan, M.J.W., Cantril, R.C. and Wylie, B.A. (1980)

Modulation of synaptosomal GABA release by enkephalin. *Life Sci.,* 27: 1097 – 1101.

Briggs, C.A. and Cooper, J. (1982) Cholinergic modulation of the release of [^{3}H]-acetylcholine from synaptosomes of the myentheric plexus. *J. Neurochem.,* 38: 501 – 508.

De La Garza, R., Bickford-Wimer, P.C., Hoffer, B.J. and Freedman, R. (1987) Heterogeneity of nicotinic actions in the rat cerebellum: an in vivo electrophysiologic study. *J. Pharmacol. Exp. Ther.,* 240: 689 – 695.

Eiden, L.E., Giraud, P., Dave, J.R., Hotchkiss, A.J. and Affolter, H.V. (1984) Nicotine receptor stimulation activates enkephalin release and biosynthesis in adrenal chromaffin cells. *Nature (Lond.),* 312: 661 – 663.

Erickson, C.K., Graham, D.T. and U'Prichard, T. (1973) Cortical cups for collecting free acetylcholine in awake rats. *Pharmacol. Biochem. Behav.,* 1: 743 – 746.

Giorguieff-Chesselet, M.F., Kemel, M.L., Wandsheer, D. and Glowinsky, J. (1979) Regulation of dopamine release by presynaptic nicotinic receptors in rat striatal slices: effect of nicotine in a low concentration. *Life Sci.,* 25: 1257 – 1262.

Goldberg, S.R., Spealman, R.D. and Goldberg, D.M. (1981) Persistent behaviour at high rates maintained by intravenous self-administration of nicotine. *Science,* 214: 573 – 575.

Gongwer, M., Robison, P. and McBride, W.J. (1987) Cholinergic control of amino acid and monoamine transmitters in slices of nucleus accumbens. *Soc. Neurosci.,* 13: 703.

Kellar, K.J., Whitehouse, P.J., Martino-Barrows, A.M., Marcus, K. and Price, D.L. (1987) Muscarinic and nicotinic cholinergic binding sites in Alzheimer's disease cerebral cortex. *Brain Res.,* 439: 62 – 68.

Larsson, C., Nilsson, L., Halen, A. and Nordberg, A. (1986) Subchronic treatment of rats with nicotine: effects on tolerance and on [^{3}H]-acetylcholine and [^{3}H]-nicotine binding in the brain. *Drug Alcohol Depend.,* 17: 37 – 45.

Limberger, N., Spath, L. and Starke, K. (1986) A search for receptors modulating the release of γ-[^{3}H]-aminobutyric acid in rabbit caudate nucleus slices. *J. Neurochem.,* 46: 1109 – 1117.

Matzner, H., Parnas, H. and Parnas, I. (1988) Presynaptic effects of d-tubocurarine on neurotransmitter release at the neuromuscular junction of the frog. *J. Physiol.,* 398: 109 – 121.

Meyer, E.M., Arendash, G.W., Judkins, J.H., Ying, L., Wade, C. and Kem, W.R. (1987) Effects of nucleus basalis lesions on the muscarinic and nicotinic modulation of [^{3}H]-acetylcholine release in the rat cerebral cortex. *J. Neurochem.,* 49: 1758 – 1762.

Morley, B.J. (1981) The properties of brain nicotine receptors. *Pharmacol. Ther.,* 15: 111 – 122.

Molinero, M.T. and Del Rio, J. (1987) Substance P, nicotinic-acetylcholine receptors and antinociception in the rat. *Neuropharmacology,* 26: 1715 – 1720.

Moss, S.J. and Wonnacott, S. (1985) Presynaptic nicotinic autoreceptor in rat hippocampus. *Biochem. Soc. Trans.,* 13: 164 – 165.

Nilsson, L., Adem, A., Hardy, J., Winblad, B. and Nordberg, A. (1987) Do tetrahydroaminoacridine (THA) and physostigmine restore acetylcholine release in Alzheimer brain via nicotinic receptors? *J. Neural Trans.,* 70: 357 – 368.

Nordberg, A. and Larsson, C. (1980) Studies on muscarinic and nicotinic binding sites in brain. *Acta Physiol. Scand. Suppl.* 479: 19 – 23.

Nordberg, A. and Winblad, B. (1986) Reduced number of [^{3}H]-nicotine and [^{3}H]-acetylcholine binding sites in the frontal cortex of Alzheimer brains. *Neurosci. Lett.,* 72: 115 – 119.

Reese, J.H. and Cooper, J.R. (1984) Noradrenergic inhibition of the nicotinically-stimulated release of acetylcholine from guinea-pig ileal synaptosomes. *Biochem. Pharmacol.,* 33: 1144 – 1147.

Romanelli, L., Ohman, B., Adem, A. and Nordberg, A. (1988) Subchronic treatment of rats with nicotine: interconnection of nicotinic receptor subtypes in brain. *Eur. J. Pharmacol.,* 148: 289 – 291.

Rowell, P.P. and Winkler, D.L. (1984) Nicotinic stimulation of [^{3}H]acetylcholine release from mouse cerebral cortical synaptosomes. *J. Neurochem.,* 43: 1593 – 1598.

Schwartz, R.D. and Kellar, K.J. (1983) Nicotinic cholinergic receptor binding sites in the brain: regulation in vivo. *Science,* 220: 214 – 216.

Schwartz, R.D., Lehmann, J. and Kellar, K.J. (1984) Presynaptic nicotinic cholinergic receptors labeled by [^{3}H]acetylcholine on catecholamine and serotonin axons in brain. *J. Neurochem.,* 42: 1495 – 1498.

Shimohama, S., Taniguchi, T., Motohatsu, F. and Kameyama, M. (1986) Changes in nicotinic and muscarinic cholinergic receptors in Alzheimer-type dementia. *J. Neurochem.,* 46: 288 – 293.

Yoshida, K. and Imura, H. (1979) Nicotinic cholinergic receptors in brain synaptosomes. *Brain Res.,* 172: 453 – 459.

Wessler, I., Halank, M., Rasbach, J. and Kilbinger, H. (1986) Presynaptic nicotine receptors mediating a positive feed-back on transmitter release from the rat phrenic nerve. *Naunyn-Schmiedeberg's Arch. Pharmacol.,* 334: 365 – 372.

Wonnacott, S., Fryer, L., Lunt, G.G., Freund, R.K., Jungschaffer, D.A. and Collins A.C. (1987) Nicotinic modulation of gabaergic transmission in the hippocampus. *Soc. Neurosci.* 13: 1352.

Vizi, E.S., Somogyi, G.T., Nagashima, H., Duncalf, D., Chaudhry, I.A., Kabayashi, O., Goldiner, P.L. and Foldes, F.F. (1987) Tubocurarine and pancuronium inhibit evoked release of acetylcholine from the mouse hemidiaphragm preparation. *Br. J. Anaesth.,* 59: 226 – 231.

Zhang, X., Stjernlöf, P., Adem, A. and Nordberg, A. (1987) Anatoxin-a, a potent ligand for nicotinic cholinergic receptors in rat brain. *Eur. J. Pharmacol.,* 135: 457 – 458.

A. Nordberg, K. Fuxe, B. Holmstedt and A. Sundwall (Eds.)
Progress in Brain Research, Vol. 79
© 1989 Elsevier Science Publishers B.V. (Biomedical Division)

CHAPTER 15

Presynaptic modulation of transmitter release by nicotinic receptors

Susan Wonnacott, Jane Irons, Catherine Rapier, Beverley Thorne and George G. Lunt

Department of Biochemistry, University of Bath, Bath BA2 7AY, U.K.

Introduction

Nicotine acts on many transmitter systems in different parts of the brain to promote transmitter release, and these modulatory actions may underlie some of the psychopharmacological and behavioural effects of nicotine (reviewed by Balfour, 1982). Although many of these early studies were ascribed to a direct presynaptic action of nicotine, mediated by nicotinic acetylcholine receptors on the nerve terminals, high concentrations of nicotine were commonly used and the nicotinic pharmacology of the effect was often poorly established. With increasing evidence in favour of multiple classes of putative nicotinic receptors in the brain from ligand binding studies (Wonnacott, 1987), protein chemistry (Whiting and Lindstrom, 1987) and molecular biology (Wada et al., 1988), we have re-examined the presynaptic actions of nicotine. Using isolated nerve terminals (synaptosomes) we can be confident that nicotine is acting presynaptically, and we have developed a superfusion technique permitting the repetitive stimulation of the preparation (Rapier et al., 1988). Perhaps the best characterised presynaptic action of nicotine concerns the enhancement of dopamine release in the striatum (reviewed by Chesselet, 1984), so our initial studies focussed on this system. Subsequently we have extended this research to the hippocampus. The phar-macological profile of the presynaptic nicotinic receptor suggests a correlation with high affinity binding sites for [^{3}H]nicotine in the brain, and leads us to propose a model for the presynaptic modulation of transmitter release by nicotinic receptors.

Nicotinic modulation of dopamine release from striatal synaptosomes

Nicotine provoked the release of [^{3}H]dopamine from synaptosomes isolated from rat striata and preincubated with radiolabelled transmitter (Rapier et al., 1985, 1987, 1988). Nicotine-evoked release was concentration-dependent over the range 10^{-8} to 10^{-3} M, and the half maximal response was observed with 3.8 μM nicotine (Rapier et al., 1988). Although a dose-response relationship has not previously been reported for nicotine-evoked dopamine release in the striatum, Giorguieff-Chesselet et al. (1979a) demonstrated that 1 μM nicotine caused the release of [^{3}H]dopamine (newly synthesised from [^{3}H]tyrosine) from striatal slices. Nicotine-evoked dopamine release from nucleus accumbens (minced tissue) had an EC$_{50}$ of 0.5 μM (Rowell et al., 1987). Moreover, low micromolar concentrations are likely to be in the range of smoking doses of nicotine. Clearly, studies using high (> 10^{-4} M) concentrations of nicotine are of dubious

158

physiological significance, and such concentrations may promote transmitter release by other mechanisms (Arqueros et al., 1978).

In Fig. 1, nicotine is compared with other agonists for their ability to stimulate [³H]dopamine release from striatal synaptosomes. At 1 μM (Fig. 1a), cytisine is marginally more potent while dimethylphenylpiperazinium (DMPP) is slightly weaker than (−)nicotine. The action of nicotine is markedly stereoselective. Indeed, dose-response curves for (−) and (+)nicotine indicate that a 100-

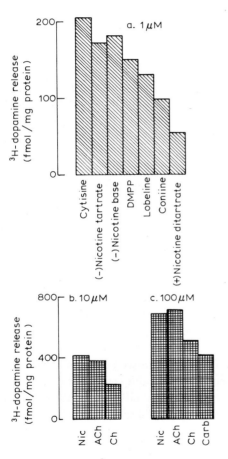

Fig. 1. Release of [³H]dopamine from striatal synaptosomes by nicotinic agonists. Agonists were compared at a single concentration of (a) 1 μM (b) 10 μM or (c) 100 μM for their ability to release [³H]dopamine from perfused synaptosomes (Rapier et al., 1988). Basal release was subtracted and stimulated release converted to fmol/mg protein by reference to the specific activity of the [³H]dopamine.

fold higher concentration of (+)nicotine is necessary to elicit the half maximal response (Rapier et al., 1988). The lack of stereoselectivity reported by Connelly and Littleton (1983) for the nicotine-induced release of [³H]dopamine from whole brain synaptosomes may again reflect the high concentrations ($10^{-4} - 10^{-2}$ M) employed; taken together with the slight Ca^{2+} dependence and meagre inhibition by pempidine seen in this study it seems probable that non-specific mechanisms were contributing to the observed effects.

Nicotine and acetylcholine were equipotent when compared at both 10 μM (Fig. 1b) and 100 μM (Fig. 1c) concentrations, and choline proved to be effective in releasing [³H]dopamine. The efficacy of nicotinic agonists favours a nicotinic receptor mechanism, and this is supported by the sensitivity of agonist-evoked [³H]dopamine release to nicotinic antagonists (Fig. 2). Thus release elicited by nicotine and DMPP could be inhibited by the ganglionic blocking drugs mecamylamine and pempidine, by dihydro-β-erythroidine which is effective at both ganglionic and neuromuscular synapses, and by the novel marine toxin neosurugatoxin (Rapier et al., 1985). The specificity of these agents was exemplified by their inability to inhibit transmitter release stimulated by a depolarising concentration of potassium (Fig. 2). Neosurugatoxin is a very potent ganglionic nicotinic antagonist (Hayashi et al., 1984) devoid of activity at the neuromuscular junction. In contrast, α-bungarotoxin failed to attenuate agonist-evoked [³H]dopamine release (Fig. 2; Rapier et al., 1985). De Belleroche and Bradford (1978) previously reported a small inhibition by α-bungarotoxin (0.19 μM) of acetylcholine (0.3 mM) evoked release of [³H]dopamine from striatal synaptosomes, but this effect was not significant and possible contamination of the toxin by neuronal bungarotoxin was not excluded. Intrinsic nicotinic excitation demonstrated electrophysiologically in rat striatal slices (Misgeld et al., 1980) was suppressed by mecamylamine and d-tubocurarine, but not by α-bungarotoxin. In agreement with the data from synaptosomes (Fig. 2), nicotine- or acetylcholine-induced release of [³H]dopamine from rat striatal slices was partially blocked by mecamylamine and pempidine (Giorguieff et al., 1976, 1977; Giorguieff-Chesselet et al., 1979a), consistent with a presynaptic nicotinic receptor of the ganglionic (C6) type.

The stereoselectivity for (−)nicotine displayed by this presynaptic receptor and its sensitivity to neosurugatoxin but not α-bungarotoxin are also characteristic of the high affinity binding sites for [³H]nicotine in rat brain (Rapier et al., 1985; Wonnacott, 1986) and distinguish this ligand bin-

ding site from that for α-[¹²⁵I]bungarotoxin. On this evidence we can propose that the presynaptic nicotinic receptor on dopaminergic nerve terminals and [³H]nicotine binding sites may be equivalent. There are moderate numbers of [³H]nicotine binding sites in the rat striatum compared with low levels of α-[¹²⁵I]bungarotoxin binding sites (Clarke et al., 1985; Marks et al., 1986). Lesion experiments with 6-hydroxydopamine support the presence of both types of binding sites on dopaminergic nerve terminals in the striatum (de Belleroche et al., 1979; McGeer et al., 1979; Schwartz et al., 1984; Clarke and Pert, 1985). However, the functional significance of α-bungarotoxin binding sites in mammalian brain is presently unclear, whereas there is very good evidence for the nicotinic receptor status of [³H]nicotine binding sites. Immunoaffinity purification of the [³H]nicotine binding protein (Whiting and Lindstrom, 1987) followed by N-terminal amino acid sequencing of the agonist binding subunit (Whiting et al., 1987) demonstrates that this subunit is identical to the protein product of the α_4-gene (coding for a nicotinic receptor α-subunit) cloned from rat brain (Goldman et al., 1987). Functional expression in *Xenopus* oocytes of α_4 in combination with the β_2-gene product (Boulter et al., 1987) results in depolarising responses to acetylcholine or nicotine (1 μM) that are blocked by neuronal bungarotoxin but not by α-bungarotoxin. These results are consistent with the agonist sensitivity of the presynaptic nicotinic receptor modulating dopamine release (Fig. 1) and its insensitivity to α-bungarotoxin (Fig. 2). Moreover, neuronal bungarotoxin inhibits nicotine-evoked dopamine release from striatal slices (Zigmond et al., this volume).

The characteristic nicotinic depolarisations observed in the oocyte expression system imply that the receptor protein includes an integral ion transduction mechanism. Nicotine-evoked [³H]dopamine release is inhibited by histrionicotoxin (Rapier et al., 1987), a non-competitive antagonist that acts at the ion channel of the muscle nicotinic receptor (Albuquerque et al., 1973). The

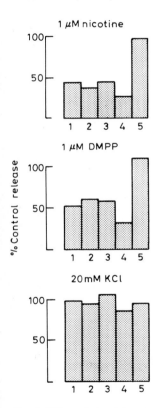

Fig. 2. Effect of nicotinic antagonists on evoked [³H]dopamine release. Synaptosomes loaded with [³H]dopamine were perfused in the presence or absence of antagonist for 20 min before stimulation with nicotinic agonists (1 μM) or KCl (20 mM). Evoked release in the presence of antagonist is presented as a percentage of the control response in the absence of antagonist. 1: Mecamylamine (5 μM); 2: pempidine (5 μM); 3: neosurugatoxin (0.05 μM); 4: dihydro-β-erythroidine (0.5 μM); 5: α-bungarotoxin (0.25 μM).

concentration of histrionicotoxin producing 50% blockade of striatal [³H]dopamine release was 5 μM, in close agreement with the sensitivity of muscle responses (Spivak et al., 1982). These data suggest that the presynaptic receptor operates via a cation channel that may closely resemble that of the muscle receptor. This is also supported by their mutual sensitivity to ketamine (Rapier et al., 1987). A disparity between the presynaptic nicotinic receptor mediating dopamine release and high affinity binding sites for [³H]nicotine is the sensitivity of the former to antagonists such as mecamylamine and pempidine (Fig. 2) that fail to inhibit ligand binding (MacAllan et al., 1988). It is increasingly recognised however that such blocking drugs are likely to be non-competitive antagonists that act at the level of the ion channel (Varanda et al., 1985).

The in vitro demonstration of nicotine-evoked dopamine release from striatal nerve terminals does not reveal the physiological significance of such a mechanism. There are cholinergic interneurones in the striatum (McGeer et al., 1975) in intimate association with dopaminergic terminals that could provide an endogenous source of nicotinic agonist. Acetylcholine-evoked release of [³H]dopamine from cat caudate nucleus in vivo

has been demonstrated (Giorguieff et al., 1976) and was partially blocked by hexamethonium and mecamylamine. These researchers subsequently reported that the presynaptic enhancement of release is seen only in the absence of nigral activation (Giorguieff-Chesselet et al., 1979b). However, in this study substance P was employed to stimulate nigro-striatal neurones; substance P can itself inhibit nicotinic receptor activation (Eardley and McGee, 1985) and this may be an alternative explanation of the absence of response to acetylcholine under these conditions. The effects of nicotine in the presence of depolarising stimuli have not been assessed in striatal preparations in vitro.

Nicotinic modulation of transmitter release from hippocampal synaptosomes

To address the question of how widespread presynaptic nicotinic receptors are in the mammalian brain, we have commenced a study of neurotransmitter release in the hippocampus. Micromolar concentrations of nicotine stimulate the release of both [³H]acetylcholine and [³H]GABA (Fig. 3) and the dose-response data for [³H]GABA release indicate that 5 μM nicotine

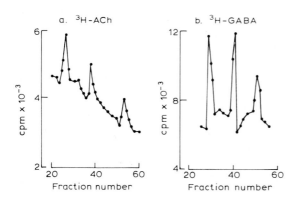

Fig. 3. Nicotine-evoked transmitter release from hippocampal synaptosomes. Hippocampal synaptosomes preloaded with (a) [³H]choline or (b) [³H]GABA were stimulated with successive pulses (50 μl) of 10 μM (−)nicotine. Transmitter release was monitored in successive fractions (350 μl) collected from the perfused preparation.

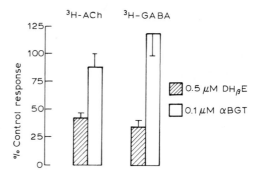

Fig. 4. Inhibition of nicotine-evoked transmitter release from hippocampal synaptosomes. Hippocampal synaptosomes preloaded with [³H]choline or [³H]GABA were perfused in the presence or absence of antagonist as described in the legend to Fig. 2.

produces the half maximal response. The nicotinic stimulation of both transmitters is insensitive to α-bungarotoxin but inhibited by dihydro-β-erythroidine (Fig. 4). Histrionicotoxin also inhibits nicotine-evoked [³H]acetylcholine release (Rapier et al., 1987) and [³H]GABA release (Fig. 5) from hippocampal synaptosomes. Thus far the presynaptic nicotinic receptors on cholinergic and GABAergic terminals in the hippocampus appear identical to those responsible for the modulation of striatal dopamine release.

Nicotinic autoreceptors on cholinergic terminals may subserve a physiological role in the feedback regulation of acetylcholine release. Hexamethonium sensitive nicotine- and DMPP-evoked release of acetylcholine from cortical synaptosomes has been reported (Rowell and Winkler, 1984), and a similar phenomenon has been demonstrated in synaptosomes of the myenteric plexus (Briggs and Cooper, 1982). The specific nicotinic agonist methylcarbamylcholine elicits [³H]acetylcholine release from hippocampal and cortical slices but not from striatal slices (Araujo et al., 1988) suggesting a regional specificity in the distribution of nicotinic autoreceptors. The presence of nicotinic receptors corresponding to [³H]nicotine binding

sites on cholinergic terminals in cortical and limbic regions may explain the deficits in these ligand binding sites in the brains of Alzheimer patients (Nordberg and Winblad, 1986; Whitehouse et al., 1986).

A model for the mechanism of nicotine-evoked transmitter release

From the data available concerning the presynaptic modulation of transmitter release by nicotine and other agonists, we can propose a model to account for this phenomenon (Fig. 6). Thus the receptor is shown schematically as a pentameric transmembrane protein as in the case of the muscle nicotinic receptor, except that the neuronal protein is likely to consist of only two types of subunits (Whiting and Lindstrom, 1987; Wada et al., 1988). The recognition site on the α-subunits will bind nicotine and other agonists, and the competitive antagonist dihydro-β-erythroidine, but unlike the corresponding subunit in muscle, the neuronal site has lost the ability to bind α-bungarotoxin. Instead, neosurugatoxin and neuronal bungarotoxin are likely to act in the vicinity of this site in neuronal but not muscle receptors.

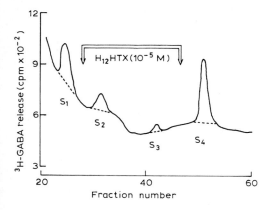

Fig. 5. Inhibition of nicotine-evoked [³H]GABA release from hippocampal synaptosomes by perhydrohistrionicotoxin. Hippocampal synaptosomes preloaded with [³H]GABA were stimulated with successive pulses of (−)nicotine (10 μM). Perhydrohistrionicotoxin (H₁₂HTX; 10 μM) was introduced into the perfusion buffer for the period indicated. The response to nicotine is seen to recover after removal of the toxin.

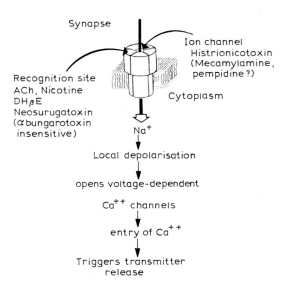

Fig. 6. Schematic model for the presynaptic modulation of transmitter release by nicotinic receptors.

162

By analogy with the muscle nicotinic receptor, agonist binding is presumed to promote an allosteric change in the protein that opens an integral ion channel, permitting the influx of cations into the nerve terminal. This is supported by the histrionicotoxin sensitivity, and it is plausible that the non-competitive antagonists mecamylamine and pempidine also act at the channel. Nicotine-evoked dopamine release in the striatum (Giorguieff-Chesselet et al., 1979a; Rapier et al., 1988) and methylcarbamylcholine-evoked acetylcholine release in the hippocampus (Araujo et al., 1988) is not inhibited by tetrodotoxin which indicates that the voltage-dependent Na^+ channel is not involved in the response. The local depolarisation arising from cation flux through the nicotinic channel results in Ca^{2+}-dependent transmitter release. Omission of Ca^{2+} from the perfusion buffer resulted in a 60% inhibition of nicotine-evoked release of [^{3}H]dopamine (Rapier et al., 1988), [^{3}H]GABA and [^{3}H]acetylcholine. Similar Ca^{2+}-dependence has been reported by other groups (e.g. Giorguieff-Chesselet et al., 1979a; Rowell et al., 1984, 1987) and accords with the Ca^{2+}-dependence exhibited by K^+- and veratridine-stimulated transmitter release (Rapier et al., 1988). Thus it is envisaged that neurotransmitter release triggered by nicotine occurs by the same Ca^{2+}-dependent process as that induced by nerve stimulation. It remains to be established what contribution presynaptic nicotinic stimulation may make in the presence of cell firing and in concert with other presynaptic receptors, both inhibitory and stimulatory, that may be present on the same nerve terminal. Nevertheless, the widespread distribution of presynaptic nicotinic receptors in the brain makes these likely targets for nicotine, whereby it may subtly influence the synaptic output of many neurotransmitters.

Acknowledgements

This research was supported by grants from The Tobacco Advisory Council, R J Reynolds Tobacco Co., and The Mental Health Foundation, and postgraduate studentships (to C.R. and B.T.) from The Science and Engineering Research Council.

References

Albuquerque, E.X., Barnard, E.A., Chiu, T.H., Lapa, A.J., Dolly, J.O., Jansson, S.E., Daly, J.W. and Witkop, B. (1973) Acetylcholine receptor and ion conductance modulator sites at the murine neuromuscular junction: evidence from specific toxin reactions. Proc. Natl. Acad. Sci. U.S.A., 70: 949–953.

Araujo, D.M., Lapchak, P.A., Collier, B. and Quirion, R. (1988) Characterisation of [^{3}H]N-methylcarbamylcholine binding sites and effect of N-methylcarbamylcholine on acetylcholine release in rat brain. J. Neurochem., 51: 292–299.

Arqueros, L., Naquira, D. and Zunino, E. (1978) Nicotine induced release of catecholamines from rat hippocampus and striatum. Biochem. Pharmacol., 27: 2667–2674.

Balfour, D.J.K. (1982) The effects of nicotine on brain neurotransmitter systems. Pharmacol. Ther., 16: 269–282.

Boulter, J., Connolly, J., Denens, E., Goldman, D., Heinemann, S. and Patrick, J. (1987) Functional expression of two neuronal nicotinic acetylcholine receptors from cDNA clones identifies a gene family. Proc. Natl. Acad. Sci. U.S.A., 84: 7763–7769.

Briggs, C.A. and Cooper, J.R. (1982) Cholinergic modulation of the release of [^{3}H]acetylcholine from synaptosomes of the myenteric plexus. J. Neurochem., 38: 501–508.

Chesselet, M.F. (1984) Presynaptic regulation of neurotransmitter release in the brain. Neuroscience, 12: 347–375.

Clarke, P.B.S. and Pert, A. (1985) Autoradiographic evidence for nicotine receptors on nigrostriatal and mesolimbic dopaminergic neurons. Brain Res., 348: 355–358.

Clarke, P.B.S., Schwartz, R.D., Paul, S.M., Pert, C.B. and Pert, A. (1985) Nicotinic binding in rat brain: autoradiographic comparison of [^{3}H]acetylcholine, [^{3}H]nicotine and [^{125}I]αbungarotoxin. J. Neurosci., 5: 1307–1315.

Connelly, M.S. and Littleton, J.M. (1983) Lack of stereoselectivity in ability of nicotine to release dopamine from rat synaptosomal preparations. J. Neurochem., 41: 1297–1302.

De Belleroche, J. and Bradford, H.F. (1978) Biochemical evidence for the presence of presynaptic receptors on dopaminergic nerve terminals. Brain Res., 142: 53–68.

De Belleroche, J., Lugmani, Y. and Bradford, H.F. (1979) Evidence for presynaptic cholinergic receptors on dopaminergic terminals: degeneration studies with 6-hydroxydopamine. Neurosci. Lett., 11: 209–213.

Eardley, D. and McGee, R. (1985) Both substance P agonists and antagonists inhibit ion conductance through nicotinic acetylcholine receptors on PC12 cells. Eur. J. Pharmacol., 114: 101–104.

Giorguieff, M.F., le Floc'h, M.L., Westfall, T.C., Glowinski, J. and Besson, M.J. (1976) Nicotinic effect of acetylcholine on the release of newly synthesised [^{3}H]dopamine in rat striatal slices and cat caudate nucleus. Brain Res., 106: 117–131.

Giorguieff, M.F., le Floc'h, M.L., Glowinski, J. and Besson, M.J. (1977) Involvement of cholinergic presynaptic receptors of nicotinic and muscarinic types in the control of the spontaneous release of dopamine from striatal dopaminergic terminals in the rat. J. Pharmacol. Exp. Ther., 200: 535–544.

Giorguieff-Chesselet, M.F., Kemel, M.L., Wandscheer, D. and Glowinski, J. (1979a) Regulation of dopamine release by presynaptic nicotinic receptors in rat striatal slices: effect of nicotine in a low concentration. Life Sci., 25: 1257–1262.

Giorguieff-Chesselet, M.F., Kemel, M.F. and Glowinski, J. (1979b) The presynaptic stimulating effect of acetylcholine on dopamine release is suppressed during activation of nigrostriatal dopaminergic neurons in the cat. Neurosci. Lett., 14: 177–182.

Goldman, D., Deneris, E., Luyten, W., Kochhar, A., Patrick, J. and Heinemann, S. (1987) Members of a nicotinic acetylcholine receptor gene family are expressed in different regions of the mammalian central nervous system. Cell, 48: 965–973.

Hayashi, E., Isogai, M., Kagawa, Y., Takayanagi, N. and Yamada, S. (1984) Neosurugatoxin, a specific antagonist of nicotinic acetylcholine receptors. J. Neurochem., 42: 1491–1494.

MacAllan, D., Lunt, G.G., Wonnacott, S., Swanson, K., Rapoport, J. and Albuquerque, E.X. (1988) Methyllycaconitine and anatoxin-a differentiate between nicotinic receptors in vertebrate and invertebrate nervous systems. FEBS Lett., 226: 357–363.

Marks, M.J., Slitzel, J.A., Romm, E., Wehner, J.M. and Collins, A.C. (1986) Nicotinic binding sites in rat and mouse brain: comparison of acetylcholine, nicotine and αbungarotoxin. Mol. Pharmacol., 30: 427–436.

McGeer, E.G., McGeer, P.L., Grewaal, D.S. and Singh, V.K. (1975) Striatal cholinergic interneurones and their relation to dopaminergic nerve endings. J. Pharmacol. (Paris), 6: 143–152.

McGeer, P.L., McGeer, E.G. and Innanen, V.T. (1979) Dendro axonic transmission. I. Evidence from receptor binding of dopaminergic and cholinergic agents. Brain Res., 169: 433–441.

Misgeld, V., Weiler, M.H. and Bak, I.J. (1980) Intrinsic cholinergic excitation in the rat neostriatum: nicotinic and muscarinic receptors. Exp. Brain Res., 39: 401–409.

Nordberg, A. and Winblad, B. (1986) Reduced number of [^{3}H]nicotine and [^{3}H]acetylcholine binding sites in the frontal cortex of Alzheimer brains. Neurosci. Lett., 72: 115–121.

Rapier, C., Harrison, R., Lunt, G.G. and Wonnacott, S. (1985) Neosurugatoxin blocks nicotinic acetylcholine receptors in the brain. Neurochem. Int., 7: 389–396.

Rapier, C., Wonnacott, S., Lunt, G.G. and Albuquerque, E.X. (1987) The neurotoxin histrionicotoxin interacts with the putative ion channel of the nicotinic acetylcholine receptors in the central nervous system. FEBS Lett., 212: 292–296.

Rapier, C., Lunt, G.G. and Wonnacott, S. (1988) Stereoselective nicotine-induced release of dopamine from striatal synaptosomes: concentration dependence and repetitive stimulation. J. Neurochem., 50: 1123–1130.

Rowell, P.P. and Winkler, D.L. (1984) Nicotinic stimulation of [^{3}H]acetylcholine release from mouse cerebral cortical synaptosomes. J. Neurochem., 43: 1593–1598.

Rowell, P.P., Carr, L.A. and Garner, A.C. (1987) Stimulation of [^{3}H]dopamine release by nicotine in rat nucleus accumbens. J. Neurochem., 49: 1449–1454.

Schwartz, R.D., Lehmann, J. and Kellar, K.J. (1984) Presynaptic nicotinic cholinergic receptors labelled by [^{3}H]acetylcholine on catecholamine and serotonin axons in brain. J. Neurochem., 42: 1495–1498.

Spivak, C.E., Maleque, M.A., Oliveira, A.C., Masukawa, L.M., Tokuyama, T., Daly, J.W. and Albuquerque, E.X. (1982) Actions of the histrionicotoxins at the ion channel of the nicotinic acetylcholine receptor and at the voltage-sensitive ion channels of muscle membranes. Mol. Pharmacol., 21: 351–361.

Varanda, W.A., Aracava, Y., Sherby, S.M., VanMeter, W.G., Eldefrawi, M.E. and Albuquerque, E.X. (1985) The acetylcholine receptor of the neuromuscular junction recognises mecamylamine as a noncompetitive antagonist. Mol. Pharmacol., 28: 128–137.

Wada, K., Ballivet, M., Boulter, J., Connolly, J., Wada, E., Deneris, E.S., Swanson, L.W., Heinemann, S. and Patrick, J. (1988) Functional expression of a new pharmacological subtype of brain nicotinic acetylcholine receptor. Science, 240: 330–332.

Whitehouse, P.J., Martino, A.M., Antuono, P.G., Lowenstein, P.R., Coyle, J.T., Price, D.L. and Kellar, K.J. (1986) Nicotinic acetylcholine binding in Alzheimer's disease. Brain Res., 371: 146–151.

Whiting, P. and Lindstrom, J. (1987) Purification and characterisation of a nicotinic acetylcholine receptor from rat brain. Proc. Natl. Acad. Sci. U.S.A., 84: 595–599.

Whiting, P., Esch, F., Schimasaki, S. and Lindstrom, J. (1987) Neuronal nicotinic acetylcholine receptor β-subunit is coded for by the cDNA clone α$_4$. FEBS Lett., 219: 459–463.

Wonnacott, S. (1986) αBungarotoxin binds to low affinity nicotine binding sites in rat brain. J. Neurochem., 47: 1706–1712.

Wonnacott, S. (1987) Brain nicotine binding sites. Human Toxicol., 6: 343–353.

A. Nordberg, K. Fuxe, B. Holmstedt and A. Sundwall (Eds.)
Progress in Brain Research, Vol. 79
© 1989 Elsevier Science Publishers B.V. (Biomedical Division)

CHAPTER 16

Influence of nicotine on the release of monoamines in the brain

David J.K. Balfour

Department of Pharmacology and Clinical Pharmacology, University Medical School, Ninewells Hospital, Dundee, Scotland, U.K.

Introduction

The effects of nicotine on brain monoamine systems have been the subject of study in a number of laboratories using both in vivo and in vitro procedures (Balfour, 1984a, 1989). This chapter will focus on the evidence from in vivo studies which support the conclusion that nicotine influences the secretion of catecholamines and 5-hydroxy-tryptamine (5-HT) in the brain and will consider particularly the results of recent experiments designed to investigate the role of the mesolimbic dopamine (DA) and hippocampal 5-HT systems in the psychopharmacological responses to the drug.

The effects of nicotine on brain noradrenaline

A majority of the early studies on the effects of nicotine on catecholamine turnover in the brain used a postmortem approach in which the concentrations of noradrenaline (NA) or DA and their metabolites were measured in discrete regions of brain. These experiments provided reasonably consistent evidence that the acute and chronic administration of nicotine stimulated NA turnover in the rodent brain and that the responses were antagonised by ganglionic blocking drugs, such as mecamylamine, which could cross the blood-brain barrier (Bhagat, 1970; Morgan and Pfeil, 1979). However, a more recent study by Kirch et al.

(1987) has shown that, if subcutaneous minipumps are used to infuse nicotine chronically for 21 days, NA turnover in the frontal cortex is reduced. The reason for the difference between this and the earlier studies remains to be established with certainty, although the data appear consistent with the hypothesis that stimulation of central nicotinic receptors evoked by the administration of nicotine boli is associated with increased NA turnover, whereas the chronic infusions cause desensitisation of the receptors which results in decreased NA turnover.

The effects of nicotine on brain dopamine systems

Turnover studies

The administration of nicotine is also reported to stimulate DA turnover in rodent brain (Andersson et al., 1981). Interestingly this study showed that DA turnover was increased in both the nucleus accumbens and medial striatum following the intraperitoneal injection of 1 mg/kg nicotine whereas the administration of a lower dose (0.3 mg/kg) exerted a significant effect only in the nucleus accumbens, data which suggested that the mesolimbic DA system is more sensitive to nicotine than the nigrostriatal system. A more recent study by Fuxe and his colleagues (1986) has shown that rats exposed intermittently to cigarette smoke also exhibit increased DA turnover in the mesolimbic system of the brain. This effect was blocked by the prior administration of mecamylamine and, therefore, the authors concluded that the response was caused by stimulation of central nicotinic receptors by the nicotine present in the smoke. As in the case of NA, chronic infusions of nicotine appeared to decrease DA turnover (Kirch et al., 1987).

Electrophysiological and microdialysis studies

The effects of acute nicotine on DA turnover suggested that the drug stimulated DA secretion in the brain and that the mesolimbic system was more sensitive to the drug than the nigrostriatal system. The electrophysiological study of Clarke et al. (1985) provided clear evidence that the systemic administration of nicotine could stimulate the rate of firing of the DA-secreting neurones in the substantia nigra. These data were supported by the results of experiments of Mereu et al. (1987) who, in addition, found that the cells in the ventral tegmental area were stimulated by lower doses of nicotine than those needed to stimulate the neurones in the substantia nigra. Both groups reported that the effects of nicotine on DA cell firing could be antagonised by the prior administration of mecamylamine. However, Clarke et al. (1985) found that systemic injections of chlorisondamine, a quaternary ganglion blocking drug which does not readily enter the brain, also antagonised the effects of nicotine on the DA-secreting cells of the pars reticulata of the substantia nigra. In contrast, chlorisondamine did not antagonise the effects of nicotine on the DA neurones in pars compacta. These authors concluded that the increased firing observed in the pars compacta of rats given nicotine was mediated by central nicotinic receptors whereas its somewhat greater effects on the neurons of the pars reticulata reflected a central response to the effects of nicotine on peripheral systems.

Imperato and her colleagues (1986) have examined the effects of nicotine on DA secretion in freely moving rats by using the technique of in vivo microdialysis. Their experiments also showed that nicotine stimulated the secretion of DA in both the mesolimbic and nigrostriatal DA systems. In addition, they provided further evidence that the mesolimbic system was more sensitive to nicotine than the nigrostriatal system since the ED_{50} for the effect on DA secretion in the nucleus accumbens was lower (0.32 mg/kg s.c.) than the ED_{50} (1.85 mg/kg s.c.) for its effects on striatal DA release. The responses were antagonised by mecamylamine but not hexamethonium, indicating that they were probably mediated by stimulation of central rather than peripheral nicotinic receptors.

Behavioural studies

In some of their experiments, Imperato et al. (1986) also recorded the locomotor activity of the rats. When this was done, there was a clear temporal relationship between the effects of nicotine on DA secretion in the nucleus accumbens and the increase in the proportion of the trial time spent exhibiting locomotor activity. Acute administration of the dose of nicotine used for the behavioural experiments (0.6 mg/kg s.c.) does not evoke increased locomotor activity in many test situations (Clarke, 1987) and it is possible, therefore, that the changes in mesolimbic DA secretion correlated with an attenuation of freezing behaviour rather than an increase in locomotor activity per se. Recent studies in this laboratory, which have shown that acute nicotine appears to ameliorate stress-induced reductions in locomotor activity (Vale and Balfour, 1988), provide some support for this hypothesis. In these experiments, the locomotor activity of rats was measured on an elevated open platform and compared with the activity measured on a less aversive platform of the same dimensions enclosed with 25 cm high sides. Saline-treated rats tested on the open platform were significantly less active ($p < 0.01$) than those tested on the enclosed platform (Fig. 1). The acute

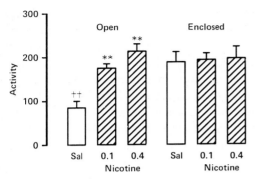

Fig. 1. The effects of environmental stress on the psychomotor response to nicotine. Rats (n = 7 per group) were given subcutaneous injections of saline or nicotine (0.1 or 0.4 mg/kg) 3 min before being placed on an elevated platform (40 cm square − open environment) or in an enclosed activity box (40 cm square with 25 cm high sides − enclosed environment). The activity was measured for 10 min using 4 infrared photobeams placed at equal intervals along two adjacent sides of the platform or the activity box. The results are expressed as means ± S.E.M. Analysis of variance indicated that the response to nicotine was influenced significantly by the test environment [F drug × environment $(2,36)$ = 3.9; $p < 0.05$]. ††, Significantly different ($p < 0.01$) from rats tested in the enclosed environment; **, significantly different ($p < 0.01$) from rats treated with saline (Duncan's test). (Data taken from Vale and Balfour, 1988).

administration of nicotine (0.1 or 0.4 mg/kg s.c.) increased the activity ($p < 0.01$) of the rats tested on the open platform to levels which were not significantly different from those recorded for saline-treated rats tested on the enclosed platform. The nicotine injections had no significant effects on the higher levels of locomotor activity measured on the enclosed platform (Fig. 1). The chronic administration of 0.1 mg/kg nicotine also failed to stimulate the activity of rats tested on the enclosed platform whereas the chronic administration of the higher dose (0.4 mg/kg) caused increased activity on both the open and enclosed platforms. In our view, our data support the conclusion that nicotine can attenuate stress-induced falls in activity and that this response to the drug contributes significantly to its psychostimulant properties, particularly when it is given in low doses. In contrast to the results with nicotine, the acute administra-

tion of d-amphetamine (0.5 mg/kg) stimulated the activity of rats tested on both the open and enclosed platforms to the same extent and, thus, the response to this drug appeared to be unaffected by the difference in the aversive properties of the two test environments (Fig. 2). The stimulant effects of cocaine on locomotor activity also appear to be unaffected by the aversive properties of the test environment used (Graham and Balfour, unpublished results). There is good evidence to suggest that the psychostimulant properties of d-amphetamine are related to its ability to stimulate DA secretion in the mesolimbic system (Kelly et al., 1975). Therefore, since our studies indicated that the responses to acute nicotine were only apparent when the animals were tested on the open platform, the data suggest that the effects of the drug on locomotor activity cannot be explained simply in terms of increased DA secretion in the nucleus accumbens.

In a series of studies currently in progress in my laboratory, the psychomotor responses to nicotine

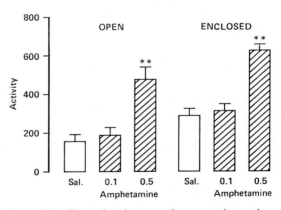

Fig. 2. The effects of environmental stress on the psychomotor response to d-amphetamine. Rats ($n \doteq 6$ per group) were given subcutaneous injections of saline or d-amphetamine (0.1 or 0.5 mg/kg) 30 min prior to testing on the elevated open platform (open environment) or in the enclosed activity box (enclosed environment). The activity was measured for 10 min. The results are expressed as means ± S.E.M. The effects of d-amphetamine and of test environment were significant [F $(2,30)$ = 35; $p < 0.001$ and $F(1,30)$ = 27; $p < 0.001$, respectively]. The interaction between the effects of the drug and the test environment was not significant. **, Significantly different ($p < 0.01$) from rats treated with saline (Duncan's test).

are being investigated in animals in which the mesolimbic DA system has been lesioned by the bilateral injection of 6-hydroxydopamine into the nucleus accumbens. The preliminary results of these experiments (Vale and Balfour, unpublished results) suggest that the lesions reduce the activity of rats treated acutely with nicotine whereas the stimulant response observed in animals treated chronically with the drug remains unaffected (Fig. 3). The reduction in the spontaneous activity of lesioned rats given saline was not significant. In agreement with previous studies (Kelly et al., 1975), the lesions abolished ($p < 0.01$) the

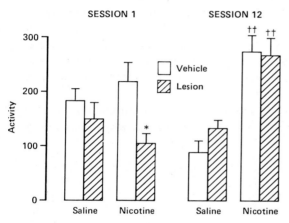

Fig. 3. The effects of nucleus accumbens lesions on the locomotor response to nicotine. Rats ($n = 10$ per group) were lesioned in the nucleus accumbens septi with 6-hydroxydopamine using the procedure described by Kelly et al. (1975). Control animals received intra-accumbens injections of the vehicle (0.1% ascorbic acid in saline) in place of the neurotoxin. Starting 10 days after surgery, half the rats from each group were given daily injections of nicotine (0.4 mg/kg s.c.) for 12 days, the remaining rats in each group being given saline. Three minutes after each injection, the rats were placed in an activity box (40 cm square with 25 cm high sides) for 10 min. This figure shows the activity measured during the first and last session of the study for the lesioned (hatched columns) and vehicle-treated (open columns) rats. Analysis of variance indicated that the effects of nicotine [F drug $\times$ sessions $(1,15) = 27$; $p < 0.001$] and of the lesion [F lesion $\times$ sessions $(1,15) = 8.5$; $p < 0.05$] on locomotor activity were influenced by the number of days of treatment. *, Significantly different ($p < 0.05$) from rats given intracerebral injections of the vehicle; ††, significantly different ($p < 0.01$) from rats treated subcutaneously with saline (Duncan's test).

locomotor response to d-amphetamine (0.5 mg/kg s.c.). Although there are a number of explanations for these preliminary data which remain to be resolved by more experiments, the results clearly suggest that the stimulant response to nicotine is not invariably associated with increased DA secretion in the nucleus accumbens.

Many of the drugs which have been shown to stimulate or potentiate DA secretion in the brain, particularly in the mesolimbic system, have also been shown to act as substrates in self-administration schedules with experimental animals (Wise and Bozarth, 1981, 1987). This is thought to be the case because the DA-secreting pathways form part of the reward system of the brain. Experimental animals can also be trained to self-administer nicotine although the drug appears to be significantly less rewarding in this paradigm than the more established psychostimulant drugs such as amphetamine or cocaine (Balfour, 1984b; Clarke, 1987). In most instances in which the self-administration of nicotine has been demonstrated successfully, the schedule has incorporated additional features which appear to be necessary to elicit the response. Thus, for example, Lang et al. (1977) used rats which were maintained at 80% of their normal body weight and a schedule in which the correct response delivered a food reward as well as an injection of nicotine. Using this schedule, rats can be trained to press for nicotine at a moderate rate. However, if the the mesolimbic DA system is lesioned with 6-hydroxydopamine at the level of the nucleus accumbens, acquisition of the response is greatly impaired (Singer et al., 1982), data which clearly support the conclusion that this system mediates nicotine self-administration in this schedule.

The effects of nicotine on hippocampal 5-HT

Postmortem studies in rats

The administration of nicotine to experimental rats causes a regionally-selective reduction in the concentrations of 5-HT and its principal metabolite, 5-

hydroxyindoleacetic acid (5-HIAA), in the hippocampus (Benwell and Balfour, 1979). The withdrawal of nicotine for 24 h does not reverse these effects. Studies with synaptosomes derived from the brains of treated rats indicated that both the acute and chronic administration of nicotine depressed 5-HT biosynthesis in the hippocampus and that, in animals treated chronically with the drug, there was evidence for down-regulation of the amino acid transporter system for L-tryptophan present in the synaptosome fraction prepared from hippocampus (Benwell and Balfour, 1982a).

Postmortem studies in humans

In a more recent study, we have found that tobacco smoking also appears to be associated with regionally-selective reductions in the concentrations of 5-HT and 5-HIAA in the hippocampus (Benwell, Anderson and Balfour, in preparation). In tissue taken at postmortem from the brains of subjects who smoked cigarettes (7 to 15 per day during the 2 years prior to death) the concentration of 5-HT in the hippocampal formation (including Ammon's horn and the subiculum) was reduced

($p < 0.05$) from 0.091 ± 0.008 $\mu g/g$ of tissue to 0.060 ± 0.008 $\mu g/g$ when compared with the levels found in age-matched non-smoking controls. The concentrations of 5-HIAA in the hippocampal formation and the hippocampal neocortex (Brodmann area 27) were reduced from 0.449 ± 0.032 to 0.307 ± 0.031 $\mu g/g$ ($p < 0.05$) and from 0.366 ± 0.020 to 0.253 ± 0.19 $\mu g/g$ ($p < 0.01$) respectively. Smoking was also associated with a significant reduction ($p < 0.05$) in the concentration of 5-HIAA in the median raphe nuclei, the area of the brain which provides the principal serotonergic innervation to the hippocampus, from 15.41 ± 0.71 to 11.93 ± 1.28 $\mu g/g$. Smoking did not reduce the concentrations of 5-HT or 5-HIAA in three other regions of the brain, the gyrus rectus (Brodmann area 11), the cerebellar cortex and samples taken from the rostral part of the medulla oblongata. In the hippocampal formation and hippocampal neocortex cigarette smoking was also associated with reduced [3H]5-HT and [3H]-8-hydroxy-2-(di-*n*-propylamino)-tetralin ([3H]-8-OH-DPAT) binding to 5-HT$_1$ and 5-HT$_{1A}$ receptors respectively whereas no changes in [3H]5-HT or [3H]-8-OH-DPAT binding were observed in the four other brain regions studied

TABLE I

Effects of smoking on radioligand binding to 5-HT$_1$ and 5-HT$_{1A}$ receptors in human brain

Brain region	Radioligand binding (fmol/mg protein)			
	Non-smokers [3H]5-HT	[3H]-8-OH-DPAT	Smokers [3H]5-HT	[3H]-8-OH-DPAT
Hippocampal formation	353 ± 25	125 ± 11	438 ± 34	183 ± 15*
Hippocampal neocortex	252 ± 24	66 ± 11	386 ± 34**	121 ± 11**
Gyrus rectus	367 ± 28	87 ± 9	412 ± 36	98 ± 8
Cerebellar cortex	98 ± 18	ndb	83 ± 12	ndb
Medulla oblongata	24 ± 4	ndb	14 ± 3	ndb
Median Raphe nuclei	123 ± 13	32 ± 5	120 ± 15	34 ± 6

The radioligand binding studies were performed with membranes prepared from the brains of 18 non-smokers (7 males and 11 females) and 12 smokers (6 males and 6 females) using the method of Middlemiss and Fozard (1983) with [3H]5-HT (7 nM) and [3H]-8-OH-DPAT (1 nM) as the radioligands for the 5-HT$_1$ and and 5-HT$_{1A}$ receptors, respectively. Significantly different from non-smokers * $p < 0.05$; ** $p < 0.01$.
ndb = No detectable binding.

(Table I). Cigarette smoking had no effects on [^{3}H]ketanserin binding to 5-HT$_2$ receptors in any of the brain regions assayed. These data provide clear support for the conclusion that tobacco smoking is associated with reduced 5-HT turnover in the hippocampus. Since the changes in hippocampal 5-HT and 5-HIAA are very similar to those found in rats treated with nicotine, it seems reasonable to suggest that the effects of tobacco smoke on brain 5-HT systems are mediated by the nicotine present in the smoke. The results also provide further support for the hypothesis that the effects of nicotine on hippocampal 5-HT may mediate psychopharmacological responses to the drug which are important to its role in maintaining the tobacco smoking habit.

Studies on the possible functional significance of the effects of nicotine on hippocampal 5-HT

The psychopharmacological consequences of reduced 5-HT secretion in the hippocampus remain to be established with certainty. Indeed, since reduced 5-HT turnover in the hippocampus appears to occur in both rats treated with nicotine and nicotine-withdrawn rats it is not clear whether the effect mediates responses to nicotine per se or the expression of the symptoms of nicotine withdrawal. Studies completed some time ago in our laboratory showed that the acute administration of nicotine to unstressed rats caused an increase in plasma corticosterone but that rats rapidly develop tolerance to this effect (Benwell and Balfour, 1979). These experiments also showed that if nicotine is withdrawn from rats following a period of chronic treatment a modest but significant increase in the plasma corticosterone level is observed, a change which is consistent with, but by no means clear evidence for, an anxiogenic response to nicotine withdrawal.

Using the elevated X-maze test for anxiety, Critchley and Handley (1987) have shown that stimulation of 5-HT$_{1A}$ autoreceptors causes changes in rat behaviour similar to those found in animals treated with an anxiogenic drug. In addi-

tion, File et al. (1987) have used the technique of in vivo microdialysis to show that there is a positive correlation between reduced 5-HT secretion in the hippocampus and the expression of increased anxiety in the elevated X-maze test. Since there are many reports that the inhalation of cigarette smoke exerts a "transquillising" effect (Gilbert, 1979; Balfour, 1989), it seemed reasonable to suggest that nicotine might have the properties of an anxiolytic drug and that the reduced 5-HT turnover in the hippocampus of nicotine-withdrawn rats could be associated with the expression of the symptoms of anxiety. Initial studies with the elevated X-maze, however, indicated that neither the acute nor the subchronic administration of nicotine (0.4 mg/kg s.c.) exerted anxiolytic or anxiogenic-like activity in this test when the rats were tested shortly (3 min) after the injection (Balfour et al., 1986b). However, in a more recent study in which rats were tested 90 min after injection of the drug, both acute and subchronic nicotine (0.4 mg/kg s.c.) were found to evoke changes in the behaviour of rats in the maze which were characteristic of an anxiolytic drug (Graham and Balfour, unpublished results). The full significance of these results remains to be clarified although analysis of the data using the entries into the enclosed runways of the maze as a covariant indicates that the anxiolytic-like activity is independent of the psychostimulant response to the drug.

In a study specifically designed to investigate the role of 5-HT secretion in the hippocampus in the behavioural responses to nicotine, Balfour et al. (1986a) examined the effects of the drug on the behaviour of rats with lesions of the serotonergic fibres which innervate the hippocampus, again using the elevated X-maze test for anxiety. Destruction of the 5-HT-secreting nerve terminals in the hippocampus evoked a change in the behaviour of the rats similar to that seen in animals given an anxiogenic drug to the extent that it decreased ($p < 0.05$) the relative number of entries made into the open runways of the maze. Nicotine did not attenuate this effect. Indeed, the reduction in open runway entries appeared to be greater in the rats treated acutely with the drug, although this was not statistically significant when compared with the effects observed in rats given saline or treated subchronically with nicotine. The lesions also failed to influence the increase in total activity, measured as the total number of entries made into all four runways of the maze, observed in rats treated subchronically (7 daily injections) with nicotine.

The results of the studies with nicotine suggest that it causes changes in 5-HT secretion in the hippocampus which, in the absence of the drug, evoke changes in behaviour which are characteristic of the response to an anxiogenic stimulus. However, the experiments completed to date have failed to provide any evidence to support the conclusion that nicotine exerts effects on rat behaviour similar to those of a classical anxiolytic drug such as diazepam (Balfour et al., 1986b). In addition, the withdrawal of nicotine, following a period of chronic treatment, does not alter the behaviour of rats, in the elevated X-maze test at least, in a way which mimics the effects of an anxiogenic drug (Vale and Balfour, 1987). However, other studies in our laboratory (Benwell and Balfour, 1982b), using plasma corticosterone as the measure of the response, have shown that nicotine does interact with the process by which rats, chronically exposed to an aversive stimulus, habituate to the stimulus. Clearly, tests other than those used to date are required to detect the behavioural consequences of this effect.

Concluding remarks

The data summarised in this chapter have provided clear evidence that nicotine stimulates DA secretion in the brain, particularly in the mesolimbic system, and decreases 5-HT secretion in the hippocampus. The functional significance of these effects remains to be established with certainty although experiments with animals treated acutely with nicotine suggest that that mesolimbic DA systems may be implicated in the psychomotor response to the drug. However, the preliminary results with the animals with lesions of the mesolimbic DA system suggested that this lesion does not invariably impair the psychostimulant response to the drug and, therefore, the role of this system as a mediator of the locomotor responses to nicotine requires further investigation. The data available support the conclusion that the mesolimbic DA system also mediates nicotine self-administration although the role of DA in this specific aspect of nicotine psychopharmacology appears to have been the subject of only one study. The possible behavioural consequences of the reduction in hippocampal 5-HT evoked by nicotine remain even less clear than for its effects on DA secretion and, indeed, there is no direct evidence to associate them with any of the responses to the drug. Nevertheless, the fact that smokers appear to inhale sufficient nicotine to influence the turnover of 5-HT in the hippocampus encourages the belief that this system mediates an effect of nicotine which could be related to its role in the tobacco smoking habit. If this proves to be the case then the data currently available appear most consistent with the hypothesis that they mediate one or more of the effects of nicotine withdrawal.

Acknowledgements

The studies reported here from my own laboratory were performed in collaboration with my colleagues Maureen Benwell, John Anderson and Alan Vale and with the financial support of the Scottish Home and Health Department and the Medical Research Council.

References

Andersson, K., Fuxe, K. and Agnati, L.F. (1981) Effects of single injections of nicotine on the ascending dopamine pathways in the rat: evidence for increases of dopamine turnover in the mesostriatal and mesolimbic dopamine neurons. *Acta Pysiol. Scand.*, 112: 345 – 347.

Balfour, D.J.K. (1984a) The effects of nicotine on brain neurotransmitter systems. In D.J.K. Balfour (Ed.), *Nicotine and the Tobacco Smoking Habit, International Encyclopedia* of Pharmacology and Therapeutics, Section 114, Pergamon Press, New York, pp. 61 – 74.

Balfour, D.J.K. (1984b) The pharmacology of nicotine dependence: a working hypothesis. In D.J.K. Balfour (Ed.), Nicotine and the Tobacco Smoking Habit, International Encyclopedia of Pharmacology and Therapeutics, Section 114, Pergamon Press, New York, pp. 101 – 112.

Balfour, D.J.K. (1989) Nicotine as the basis of the tobacco smoking habit. Pharmacol. Ther. (in press).

Balfour, D.J.K., Benwell, M.E.M. Graham, C.A. and Vale, A.L. (1986a) Behavioural and adrenocortical responses to nicotine measured in rats with selective lesions of the 5-hydroxytrypaminergic fibres innervating the hippocampus. Br. J. Pharmacol., 89: 341 – 347.

Balfour, D.J.K., Graham, C.A. and Vale, A.L. (1986b) Studies on the possible role of brain 5-HT systems and adrenocortical activity in behavioural responses to nicotine and diazepam in an elevated X-maze. Psychopharmacology, 90: 528 – 532.

Benwell, M.E.M. and Balfour, D.J.K. (1979) Effects of nicotine and its withdrawal on plasma corticosterone and brain 5-hydroxyindoles. Psychopharmacology, 63: 7 – 11.

Benwell, M.E.M. and Balfour, D.J.K. (1982a) Effects of nicotine administration on the uptake and biosynthesis of 5-HT in rat brain synaptosomes. Eur. J. Pharmacol., 84: 71 – 77.

Benwell, M.E.M. and Balfour, D.J.K. (1982b) Effects of chronic nicotine administration on the response and adaptation to stress. Psychopharmacology, 76: 160 – 162.

Bhagat, B. (1970) Influence of chronic administration of nicotine on the turnover and metabolism of noradrenaline in the rat brain. Psychopharmacologia (Berl.), 18: 325 – 332.

Clarke, P.B.S. (1987) Nicotine and smoking: a perspective from animal studies. Psychopharmacology, 92: 135 – 143.

Clarke, P.B.S., Homer, D.W., Pert, A. and Skirboll, L.R. (1985) Electrophysiological actions of nicotine on substantia nigra single units. Br. J. Pharmacol., 85: 827 – 835.

Critchley, M.A.E. and Handley, S.L. (1987) Effects in the X-maze anxiety model of agents acting on 5-HT$_1$ and 5-HT$_2$ receptors. Psychopharmacology, 93: 502 – 506.

File, S.E., Curle, P.F., Baldwin, H.A. and Neal, M.J. (1987) Anxiety in the rat is associated with decreased release of 5-HT and glycine from the hippocampus. Neurosci. Lett., 83: 318 – 322.

Fuxe, K., Andersson, K., Harfstrand, A. and Agnati, L.F. (1986) Increases in dopamine utilization in certain limbic terminal populations after a short period of intermittent exposure of male rats to cigarette smoke. J. Neural Trans., 15 – 29.

Gilbert, D.G. (1979) Paradoxical tranquilizing and emotion-reducing effects of nicotine. Psychol. Bull., 86: 643 – 661.

Imperato, A., Mulas, A. and Di Chiara, G. (1986) Nicotine preferentially stimulates dopamine release in the limbic system of freely moving rats. Eur. J. Pharmacol., 132: 337 – 338.

Kirch, D.G., Gerhardt, G.A., Shelton, R.C., Freedman, R. and Wyatt, R.J. (1987) effect of chronic nicotine administration on monoamine and monoamine metabolite concentrations in rat brain. Clin. Neuropharmacol., 10: 376 – 383.

Kelly, P.H., Seviour, P.W. and Iversen, S.D. (1975) Amphetamine and apomorphine response in the rat following 6-OHDA lesions of the nucleus accumbens septi and corpus striatum. Brain Res., 94: 507 – 522.

Lang, W.J., Latiff, A.A., McQueen, A. and Singer, G. (1977) Self administration of nicotine with and without a food delivery schedule. Pharmacol. Biochem. Behav., 7: 65 – 70.

Mereu, G., Yoon, K-W,P., Boi, V., Gessa, G.L., Naes, L. and Westfall, T.C. (1987) Preferential stimulation of ventral tegmental area dopamine neurons by nicotine. Eur. J. Pharmacol., 141: 395 – 399.

Middlemiss, D.N. and Fozard, J.R. (1983) 8-Hydroxy-2-(di-n-propylamino)tetralin discriminates between subtypes of 5-HT recognition site. Eur. J. Pharmacol., 90: 151 – 153.

Morgan, W.W. and Pfeil, K.A. (1979) Mecamylamine blockade of nicotine enhanced noradrenaline turnover in rat brain. Life Sci., 24: 417 – 420.

Singer, G., Wallace, M. and Hall, R. (1982) Effects of dopaminergic nucleus accumbens lesions on the acquisition of schedule induced self injection of nicotine in the rat. Pharmacol. Biochem. Behav., 17: 579 – 581.

Vale, A.L. and Balfour, D.J.K. (1987) The role of hippocampal 5-HT in the effects of nicotine on habituation to an elevated X-maze. Eur. J. Pharmacol., 41: 313 – 317.

Vale, A.L. and Balfour, D.J.K. (1988) The influence of aversive stimuli on behavioural responses to nicotine. In M.J. Rand (Ed.), The Pharmacology of Nicotine, IRL Press Ltd., Eynsham, Oxford, pp. 354 – 355.

Wise, R.A. and Bozarth, M.A. (1981) Brain substrates for reinforcement and drug self-administration. Prog. Neuro-Psychopharmacol., 5: 467 – 474.

Wise, R.A. and Bozarth, M.A. (1987) A psychomotor stimulant theory of addiction. Psychol. Bull., 94: 469 – 492.

A. Nordberg, K. Fuxe, B. Holmstedt and A. Sundwall (Eds.)
Progress in Brain Research, Vol. 79
© 1989 Elsevier Science Publishers B.V. (Biomedical Division)

CHAPTER 17

Regulation by nicotine of midbrain dopamine neurons

Thomas C. Westfall[a], Giampaolo Mereu[a], Lillian Vickery[a], Holly Perry[a], Linda Naes[a], and Kong-Woo P. Yoon[b]

Departments of [a] Pharmacology and [b] Neurosurgery, St. Louis University School of Medicine, 1402 South Grand Boulevard, St. Louis, MO 63104, U.S.A.

Introduction

The purpose this chapter is to summarize the results of recent studies in our laboratory where we have utilized two approaches to study the action of nicotine on midbrain dopamine neurons. The first was a neurochemical approach where we have examined the effect of nicotine on the release of dopamine in vitro using isolated slices and in vivo utilizing push-pull perfusion of rat striatum in freely moving unanesthetized rats. The second approach was electrophysiological where we have measured the firing rates of dopamine and non-dopamine cells in the substantia nigra and ventral tegmental area of Tsai following systemic injection of nicotine   following injection of the alkaloid into the terminal regions of these dopamine neurons, namely the caudate nucleus and nucleus accumbens.

Materials and methods

Neurochemical experiments – slices

Male Sprague-Dawley rats weighing between 160 and 300 g obtained from SASCO (St. Louis, MO) were used in all experiments. Animals were maintained under standard conditions with food and water ad libitum. Animals were sacrificed by decapitation and striata rapidly removed in a cold room (4°C) using glass manipulators. Striata were chopped into slices 0.5 mm thick with a McIlwain tissue chopper and placed in vials containing ice-cold Krebs-Ringer bicarbonate buffer, containing in mM: NaCl, 118; KCl, 4.85; $CaCl_2$, 2.5; $MgSO_4$, 1.15; KH_2PO_4, 1.15; glucose, 11.1; and $NaHCO_3$, 25.0. Release of dopamine in response to nicotine was determined by three separate procedures and experiments: (1) release of [^{3}H]dopamine from slices preincubated for 30 min with the labelled transmitter (newly taken up dopamine); (2) release of endogenous transmitter as measured by HPLC-EC (endogenous dopamine); (3) release of [^{3}H]dopamine continuously formed by superfusion of slices with [^{3}H]tyrosine (newly synthesized dopamine).

Release of newly taken up [^{3}H]dopamine

Following preparation of the slices, they were placed in glass minivials containing 750 μl of Krebs-Ringer bicarbonate, bubbled with 95% $O_2 - 5\%$ CO_2 and maintained at 37°C in a water bath. The slices were preincubated for 10 min and then transferred to a vial and incubated for 30 min in the presence of [^{3}H]dopamine final concentration 5×10^{-8} M. The slices were placed in vials containing normal buffer at 15 or 30 min intervals. They were then exposed to various concentrations of nicotine for 5 min. In experiments in which the effect of various perturbations or manipulations on the nicotine-induced release of dopamine were examined, the slices were placed in a modified buffer for 15 or 30 min prior to exposure to nicotine. Following the 5 min stimulation period, the slices were placed in another vial containing normal buffer and washed for 1 min before analysis. In addition to examining the effect of various concentrations of nicotine ranging from 10 to 1000 μM, the effect of removing extracellular calcium and the effect of mecamylamine, d-tubocurarine or tetrodotoxin on the release of [^{3}H]dopamine produced by 0.5 and 1.0 mM nicotine was examined. The amount of [^{3}H]dopamine released into the media was determined by liquid scintillation spectrometry following absorption and elution from alumina.

[^{3}H]dopamine was determined by counting 1 ml of sample in 10 ml of Scinti Verse E (Fisher) in a Packard Tri Carb 300 C or 4000 Series liquid scintillation counter. The nicotine-induced [^{3}H]dopamine release was calculated as the difference between stimulated [^{3}H]dopamine release and an approximation of spontaneous release during the period of stimulation based on prestimulation levels of basal release. This difference was divided by the total [^{3}H]dopamine released in superfusates plus the [^{3}H]dopamine remaining in the tissue at the end of the experiment and expressed as a percent fractional release. The amount of [^{3}H]dopamine in the tissue at the start of any time interval was calculated by adding cumulatively the amount of [^{3}H]dopamine released into the medium to the amount of [^{3}H]dopamine in the tissue at the end of the experiment.

Endogenous dopamine
Following the placement of the striata into glass vials, they were vigorously vortexed to separate the slices. Two slices from each striatum were selected and placed in a wire-mesh basket in mini-vials containing 750 μl buffer and incubated at 37°C in a shaking water bath and bubbled with 95% $O_2 - 5\%$ CO_2. The slices were incubated for at least 15 min before being challenged with various concentrations of nicotine for 5 min. The slices were then transferred to vials containing either a modified or normal buffer and incubated for another 15 or 30 min. The slices were then placed

in a buffer containing nicotine for a second 5 min stimulation period. In some experiments following the second 5 min stimulation period, the slices were placed in another vial containing normal buffer for an additional 15 or 30 min followed by exposure for a third time to nicotine. Following the third 5 min stimulation period, the slices were placed in a vial containing normal buffer and washed for 1 min before analysis. Release was studied under several sets of conditions. The release of dopamine was determined by measuring the amount in the media and tissue. Four fractions from each experiment were assayed for dopamine; the two 5 min stimulating fractions and the two 5 min fractions immediately preceding them. The media fractions were concentrated by an alumina batch technique using [^{3}H]dopamine as an internal standard. The eluate from the alumina was then analyzed for released dopamine by HPLC-EC. The solvent used routinely for analysis was 10% methanol; 125 mM acetic acid, 2 mM EDTA and 2 mM HSA. The pH was adjusted to 3.8 with 13.3 N NaOH. Dopamine was quantified by comparison of peak heights with a known amount of authentic dopamine.

The release of endogenous dopamine was expressed as the nanograms released per 5 min or as the percent fractional release. The nicotine-induced dopamine release or percent fractional release was calculated as the difference between stimulated dopamine release and an approximation of spontaneous release during the period of stimulation based on prestimulation levels of basal release in a fashion analogous to [^{3}H]dopamine release already mentioned.

Release of newly synthesized [^{3}H]dopamine
Slices were placed in individual superfusion chambers and continuously superfused at a flow rate of 200 μl/min with [^{3}H]tyrosine. [^{3}H]tyrosine was purified by successive passage over alumina and Dowex columns and maintained under an atmosphere of nitrogen. The [^{3}H]tyrosine was reconstituted in buffer before each experiment and used at a final concentration of $40 - 50$ μCi/ml. A steady-state formation of [^{3}H]dopamine occurs

following 25–30 min of superfusion. Once a steady state of [^{3}H]dopamine was obtained, the effect of nicotine in the absence or presence of d-tubocurarine, mecamylamine, the removal of extracellular Ca^{2+} or the presence of tetrodotoxin was assessed. Superfusate effluents were continuously collected in 2.5 min fractions by means of a fraction collector. [^{3}H]dopamine in the superfusate effluents was determined following passage over amberlite followed by alumina columns according to previously published techniques. In experiments in which perturbations were assessed, the drugs were present 15 min prior to stimulation by nicotine. Calcium was removed from the superfusion buffer 15 or 30 min prior to stimulation with nicotine. The manner in which the data were calculated was based on the method described by Nieoullan et al. (1977) and utilized by us previously (Voigt et al., 1985). The amount of [^{3}H]dopamine released during the steady state obviously varies from animal to animal. Therefore, each animal served as his own control.

The control (100%) value for each animal was calculated by taking the mean value of dopamine released from the three fractions preceding the drug treatment. Release of dopamine in each successive fraction was expressed as a percentage of the control value. Data from various animals were pooled and the mean ± standard error of the mean calculated.

Neurochemical experiments – push-pull perfusion

Animals were anesthetized with a mixture of ketamine/acepromazine and a 21-gauge guide cannula stereotaxically placed in the striatum A +1.0, L +3.0, V −6.0 with respect to bregma (Paxinos and Watson, 1982). Five to seven days later, the rats were lightly anesthetized with metofane and a 28-gauge push-pull cannula assembly introduced into the guide cannula to perfuse the striatum with artificial CSF (18 μl/min). Rats were allowed to recover from the anesthesia and CSF effluent collected in 10 min fractions for analysis of dopamine by HPLC-EC detection. In other experiments,

newly synthesized dopamine was measured following perfusion with [^{3}H]tyrosine (40–50 μCi/ml). Drug treatments were carried out 120 min after the start of perfusion to allow the release of [^{3}H]dopamine to reach a steady state. Samples were passed over amberlite and alumina columns and analyzed by liquid scintillation spectrometry as described above for the in vitro experiments.

Electrophysiological experiments

Systemic injections
Male Sprague-Dawley rats were anesthetized with 3% halothane in air, tracheotomized and intubated. A femoral vein was cannulated with silastic tubing for subsequent drug administrations, then the head restrained in a David Kopf stereotaxic apparatus. Microelectrodes were lowered into the substantia nigra 3–3.5 mm rostral and 1.5–2.0 mm lateral to the lambda point, according to the coordinates of Paxinos and Watson (1982). Nigral cells were usually encountered 7–8 mm ventral to the cortical surface. The spontaneous extracellular action potentials from A_9 and A_{10} dopamine cells were recorded and processed using conventional methodology. Dopamine cells were recognized on the basis of their well-established electrophysiological, pharmacological and anatomical characteristics described by us and others (Bunney et al., 1973; Mereu et al., 1984, 1987).

All surgery was performed under general anesthesia (2.0–3.5% halothane in air) and the wound edge and pressure points were infiltrated with a long-acting local anesthetic (2% mepivocaine hydrochloride). Diazepam 0.250 mg/kg was administered intravenously (i.v.) in order to reduce the animal stress. At this dosage, diazepam was shown not to have any significant effect on the dopaminergic neuronal activity (Ross et al., 1982). After these procedures, d-tubocurarine (5 mg/kg i.v.) was injected, halothane discontinued and the tracheal catheter connected to an artificial respirator. During the experiments, additional doses of local anesthetic, benzodiazepine and paralyzing agent were administered as needed. At the end of each experiment, the animal was sacrificed with an overdose (1 g/kg i.v.) of chloral hydrate (CH) slowly injected until complete cessation of cardiac activity. Rats were then perfused transcardially with 0.9% saline followed by 10% buffered formalin. Brains were removed for histological analysis of recording sites. Nicotine tartrate, mecamylamine HCl and apomorphine HCl (Sigma, St. Louis) were freshly dissolved in 0.9% saline. Body temperature, heart rate and cortical EEG were monitored throughout the experiment. Haloperidol was diluted from commercially available solution. Zetidoline (a gift from Dow-Lepetit, Milano) was diluted in saline and stored at 2°C in the dark. All drugs were neutralized to pH 6.0 – 7.0 and injected i.v. at a fixed volume of 1 ml/kg body weight.

Injections into target neurons

Animals were prepared as described above and recording electrodes placed into the substantia nigra (see above) or ventral tegmental area of Tsai. In these latter studies, microelectrodes were lowered into VTA 3 – 3.5 mm rostral and 0.5 – 1.2 mm lateral to the lambda point according to the coordinates of Paxinos and Watson (1982). Dopaminergic neurons were usually encountered 7.5 – 8.2 mm ventral to the cortical surface immediately following a flourish of neuronal activity for 1.0 min signifying a passage of electrode through the red nucleus. Their spontaneous extracellular active potentials were recorded and processed using conventional methodology.

Drugs were injected into the caudate nucleus or nucleus accumbens through an assembly of three 1.5 mm glass pipettes glued together and pulled (Fig. 1). The tip of such three-barrel micropipettes was broken back up to 25 μm (outside diameter). Ejection was obtained by connecting each pipette to a micropressure (nitrogen) pump set at 1.5 atm. Injection volumes were estimated by monitoring the lemniscus shift in each precalibrated pipette during the pressure application through a stereomicroscope. A volume of 0.5 μl was delivered in

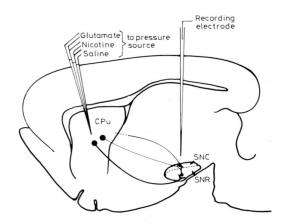

Fig. 1. A schema depicting the placement of the recording electrode in the substantia nigra compacta or substantia nigra reticulata and the three-barrel injection pipette in the striatum. A similar arrangement was used for recording from A_{10} or nondopamine VTA cells with the injection pipette in the nucleus accumbens.

1 – 3 s for each single injection. For caudal injections, the center of the injected area was 1.5 mm rostral and 2.0 mm lateral to bregma and 4.5 mm ventral to the cortical surface. It was approached by an angle of 16° from the sagittal plane to avoid the possible drug diffusion into the lateral ventricle.

For accumbens injections, the center of the injected area was 2.0 mm rostral and 1.2 mm lateral to bregma point, 7.2 mm ventral to cortex. It was also approached at an angle of 16°. At the end of the experiments, the injection sites were marked by the administration of 0.5 μl of 1% pontamine sky blue in 0.15 M NaCl and examined histologically.

Results

Release – in vitro

Nicotine in concentrations of 0.1 to 10 mM produced a concentration dependent release of newly taken up [³H]dopamine (Fig. 2) or endogenous dopamine (Fig. 3) from incubated striatal slices.

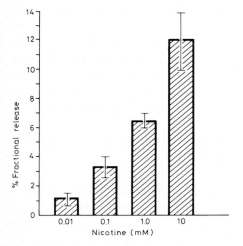

Fig. 2. The effect of nicotine on the release of dopamine from striatal slices previously incubated with [³H]dopamine to label the endogenous stores. Data are plotted as percent fractional release of [³H]dopamine against the concentration of nicotine in mM. Each bar is the mean ± S.E.M. of 5 – 7 experiments. A significant release was observed by 0.1, 1.0 and 10 mM ($p < 0.05$, $p < 0.01$ and $p < 0.001$, respectively).

Simultaneous administration of the nicotinic-cholinergic antagonist mecamylamine or tetrodotoxin failed to alter the nicotine-induced release of [³H]dopamine (Fig. 4) or endogenous dopamine (Fig. 3). Similarly removal of extracellular Ca^{2+} from the incubation buffer failed to alter the nicotine-induced release of [³H]dopamine (data not shown) or dopamine (Fig. 3). The use of the third technique to analyze the release of dopamine, namely [³H]dopamine newly synthesized from [³H]tyrosine in superfused slices resulted in dramatically different results. It was possible to detect a nicotine-induced release of dopamine with concentrations of nicotine as low as 0.01 μM (Westfall et al., 1987). This was not possible with either of the other ways of assessing dopamine release. In addition, the administration of mecamylamine and d-tubocurarine (Fig. 5), tetrodotoxin (data not shown) or removal of extracellular Ca^{2+} attenuated the nicotine-induced release of newly synthesized dopamine. Table I represents a summary of these various experiments.

Release – in vivo

The addition of nicotine to the perfusion buffer of push-pull cannulae in freely-moving unanesthetiz-

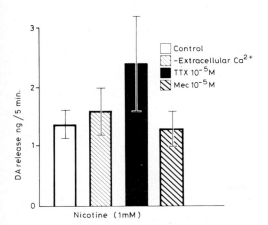

Fig. 3. The effect of nicotine (1 mM) to release endogenous dopamine from incubated striatal slices in the absence or presence of various perturbations. Removal of extracellular calcium or presence of tetrodotoxin (10^{-5} M) or mecamylamine (10^{-5} M). Data are plotted as the dopamine released in nanograms per 5 min and each bar represents the mean ± S.E.M. of 5 – 8 experiments. None of the perturbations altered the nicotine-induced response.

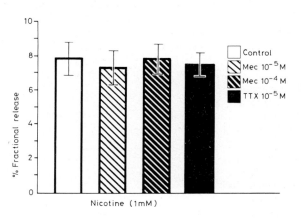

Fig. 4. Effect of mecamylamine (10^{-5} M; 10^{-4} M) or tetrodotoxin (10^{-5} M) on the nicotine-induced release of [³H]dopamine from striatal slices as described in Fig. 2. Data are plotted in a similar fashion as Fig. 2. Neither mecamylamine or tetrodotoxin (TTX) altered the nicotine-induced release.

178

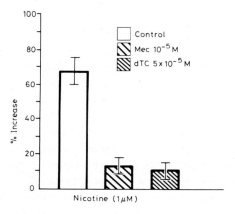

Fig. 5. The effect of mecamylamine (10^{-5} M) or d-tubocurarine (5×10^{-5} M) on the nicotine-evoked (1 μM) release of [^{3}H]dopamine in slices continuously superfused with [^{3}H]tyrosine. Data are plotted as the percent increase of [^{3}H]dopamine over the spontaneous release. Each bar represents the mean ± S.E.M. of 4–6 experiments. Both mecamylamine and d-tubocurarine produced a significant attenuation ($p < 0.01$) of the nicotine-evoked response.

TABLE I

Comparison of low and high concentrations of nicotine in releasing dopamine from rat striatal slices

Item	Low concentrations	High concentrations
Method of assessing	Newly synthesized [^{3}H]dopamine	Newly taken up dopamine Endogenous dopamine
Concentration range to see effect	1–100 μM	100–1000 μM
Ca^{2+}-dependent	Yes	No
Tetrodotoxin-dependent	Yes	No
Antagonist block d-Tubocurarine	Yes	No
Mecamylamine	Yes	No
Amine pump blocker	–	No

ed rats resulted in a release of endogenous dopamine (Fig. 6) or newly synthesized [^{3}H]dopamine (Fig. 7). A similar release could be seen following the subcutaneous administration of nicotine (Fig. 6). As was seen in vitro, the release of newly synthesized dopamine was a more sensitive way of assessing the effect of nicotine.

Preferential stimulation of A_{10} neurons after intravenous nicotine

The intravenous administration of increasing doses of nicotine (50–500 μg/kg) resulted in an increase in the firing rate of A_9 and A_{10} dopamine neurons (Fig. 8). The increase in firing rate could be prevented or reversed by subsequent administration of mecamylamine (data not shown). While i.v. nicotine increased the firing rate of A_9 neurons some 30% above base line, the same range of doses of nicotine was more potent as well as effective in stimulating the A_{10} cell up to 75% above the basal rate. After nicotine-induced stimulation, low doses of apomorphine (10–50 μg/kg) inhibited the firing

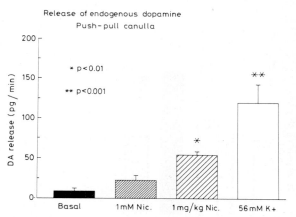

Fig. 6. The effect of nicotine administered by local administration in the push-pull cannula (1 mM) or by subcutaneous administration (1 mg/kg) or potassium in the push-pull cannula on the release of endogenous dopamine into the perfusate during push-pull perfusion of the striatum. Data are plotted as dopamine released in picograms per min versus drug administration. Each bar is the mean ± S.E.M. of 4–6 experiments. Both nicotine and K$^+$ produced a significant release of dopamine.

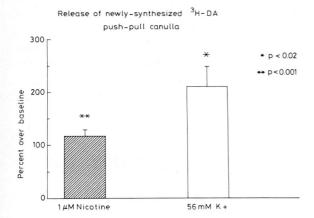

Fig. 7. The effect of the local administration into the push-pull cannula of nicotine or potassium on the release of [³H]dopamine continuously synthesized from [³H]tyrosine. Data are plotted as percent above basal vs. drug. Each bar is the mean ± S.E.M. of 5–7 observations. Both nicotine and K^+ produced a significant release of dopamine.

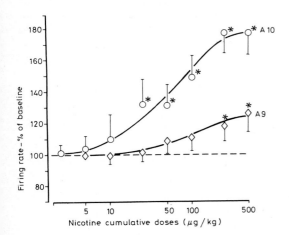

Fig. 8. Effect of cumulative i.v. doses of nicotine on the firing rate of A_9 (diamonds) or A_{10} (circles) dopaminergic cells in rats maintained on local anesthesia. Each point represents the averaged change (± S.E.M.) in the firing rate obtained from 12–19 cells and expressed in percent in respect to basal rates. Each cell (one per rat) was recorded for at least 10 mm in order to have a stable basal rate. The drug effect after a single cumulative dose was estimated at the peak of the firing change. Asterisk indicates $p < 0.05$ (Student's t-test with respect to basal rate).

rate activity in a fashion similar to that seen in control rats (data not shown).

Stimulation of dopamine and nondopamine neurons in the substantia nigra and VTA following injection of nicotine into target regions

The intrastriatal injection of nicotine resulted in a stimulation of 8 out of 11 A_9 cells examined. A representative stimulation is depicted in Fig. 9. There was a mean increase of 89% above the base line (Table II). Two cells were excited after a period of inhibition and one cell had no response. A similar response was seen following the intrastriatal administration of glutamate (Table II) where 12 of 14 cells were stimulated for a period of 2 min or more.

Similar results were observed in the VTA dopamine cells where 8 of 13 cells were excited following the microinjection of nicotine into the nucleus accumbens (Fig. 10 and Table III). The extent of excitation was 46.9% of the base line of firing rate. The microinjection of glutamate into the nucleus accumbens excited 19 out of 20 A_{10} cells.

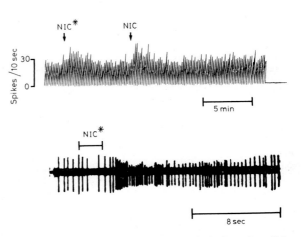

Fig. 9. Firing rate histograms (spikes/10 s) of an A_9 cell (top panel) following two consecutive injections of nicotine into the striatum. Asterisk represents the time of the single sway oscilloscope picture (bottom panel). Nicotine was shown to increase the firing rate with a characteristic decrease in action potential amplitude and increase in frequency (bottom panel).

The cells were stimulated by 10% of the basal rate. Stimulation ranged from 5 to 30 s. The increase in the firing rate was invariably accompanied by diminished amplitude of the action potential.

In contrast to what was seen for the A_9 cells following the intrastriatal injection of nicotine, the nondopamine cells of the substantia nigra reticulata were transiently inhibited by both nicotine and glutamate (Fig. 11 and Table II). The effect of nicotine was biphasic in which inhibition was followed by a rebound excitation in 6 out of 10 cells. Glutamate produced almost complete inhibition of the cells (90%). Following this temporary inhibition, the cells showed complete recovery in a few minutes.

The response of the nondopamine cells in the VTA region following injections of nicotine into the nucleus accumbens were inconsistent. Out of 10 cells, four were inhibited, four were stimulated and two had no response. The degree of inhibition was 57.15% of the basal firing rate, while the cells which were excited, increased by 32.2% of the basal firing rate. In contrast, the injection of glutamate into the nucleus accumbens decreased the firing rate in 9 of 10 cells. The decrease was

71.8% of the basal firing rate. The duration of the inhibition ranged from 10 to 160 s.

The control injection of saline into the striatum or nucleus accumbens failed to alter the firing rate of any neurons examined in A_9, A_{10}, SNR or nondopamine VTA cells. A summary of the elec-

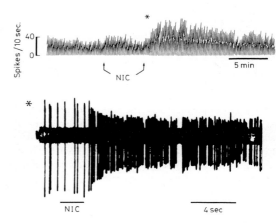

Fig. 10. Firing rate histograms (spikes/10 s) of an A_{10} cell (top panel) and single sweep oscilloscope depicted by the asterisk (bottom panel) following two consecutive injections of nicotine into the nucleus accumbens. A similar pattern as seen in A_9 (Fig. 9) was observed.

TABLE II

The response of SNC A_9 and SNR Daergic neurons to intrastriatal injection of glutamate and nicotine

	Caudate injection	Response	% Increase above base	% Decrease below base
SNC (A_9) Dopamine Neurons	Glutamate (14)	12 + 2 −	87.4 (8.2 S.E.)	
	Nicotine (11)	8 + 2 − 1 no response	89.0 (25.2 S.E.)	
SNR nonDA Neurons	Glutamate (10)	9 − 1 no response		92.2 (3.6 S.E.)
	Nicotine (9)	7 − 2 no response		75.5 (11.6 S.E.)

+ = Excitation; − = inhibition; S.E. = standard error. Base refers to baseline firing rate obtained from a trace obtained for 10 min before the injections.

trophysiological experiments is depicted in Table IV.

Discussion

The present experiments have shown that nicotine has several effects on midbrain dopamine neurons, including a release from rat striatal slices following in vitro or in vivo administration and an increase in the firing rate of dopamine neurons following injection systemically or into target neurons. The effects on release are consistent with the results of numerous investigators including ourselves who have observed a similar nicotine-induced release of dopamine (Westfall, 1974; Giorguieff et al., 1977; Arqueros et al., 1978; DeBelleroche and Bradford, 1978; Westfall et al., 1983; 1987). There now appears to be good agreement that nicotine releases dopamine from dopaminergic neurons in the central nervous system. The mechanism by which nicotine exerts this effect has been more controversial. Some investigators have reported that the effect of nicotine or acetylcholine on dopamine release can be attenuated by classical nicotinic an-

TABLE III

The response of VTA A_{10} and nonDaergic neurons to intrastriatal (nucleus accumbens) injections of glutamate and nicotine

	NAc injection	Response	% Increase above base	% Decrease below base
VTA (A_{10}) Dopamine Neurons	Glutamate (20)	19 + 1 −	80.15 (15.4 S.E.)	
	Nicotine (13)	8 + 3 − 2 no response	46.87 (9.0 S.E.)	
VTA nonDA Neurons	Glutamate (10)	9 − 1 no response		71.8 (6.4 S.E.)
	Nicotine (10)	4 − 4 + 2 no response	32.25 (10.5 S.E.)	57.5 (24.0 S.E.)

TABLE IV

Summary: Electrophysiological experiments

Drug	Site of administration	Response (firing rate)	
Nicotine	Striatum	Substantia nigra reticulata	↓
Nicotine	Striatum	A_9 DA cells	↑
Nicotine	I.V.	A_9 DA cells	↑
Glutamate	Striatum	Substantia nigra reticulata	↓
Glutamate	Striatum	A_9 DA cells	↑
Nicotine	Nucleus accumbens	VTA non-DA cells	↓ ↑
Nicotine	Nucleus accumbens	VTA DA cells	↑
Glutamate	Nucleus accumbens	VTA non-DA cells	↓
Glutamate	Nucleus accumbens	VTA DA cells	↑

182

tagonists (Westfall, 1974; Giorguieff et al., 1976; DeBelleroche and Bradford, 1978), while others have failed to observe such a reduction by similar drugs (Connelly and Littleton, 1983; Marien et al., 1983). Similar discrepancies have been reported for the importance of extracellular calcium with some authors observing that extracellular calcium is required (Giorguieff-Chesselet et al., 1979; Connelly and Littleton, 1983), while others suggest it is not a necessity (Arqueros et al., 1978; Marien et al.,

1983; Takano et al., 1983) for the nicotine-induced release of dopamine.

The results of the present experiments may provide an explanation for some of the discordant results. In previous studies, high concentrations of nicotine or acetylcholine were used to induce the release of dopamine from striatal slices. In the present studies, we were able to monitor the release of newly synthesized dopamine in continuously superfused striatal slices. Using this method of assessing the release of dopamine, we observed that very low concentrations of nicotine produced such an effect. This release was blocked by ganglionic nicotinic antagonists, and required the presence of extracellular calcium. In contrast, much larger concentrations of nicotine were necessary to induce a release of newly taken up [^{3}H]dopamine or endogenous dopamine. The higher concentrations also followed a concentration-effect relationship, but the mechanism of inducing release was different since the same antagonists (d-tubocurarine and mecamylamine) failed to antagonize the nicotine effect. In addition, removal of extracellular calcium had no effect on the release of dopamine produced by the high concentrations of nicotine.

A summary of the comparison between low and high concentrations of nicotine is depicted in Table

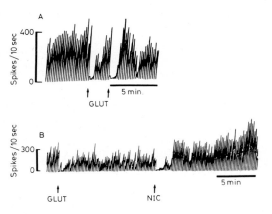

Fig. 11. Firing rate histograms (spikes/10 s) of two substantia nigra reticulata neurons to successive intrastriatal injections of glutamate (panel A) or the injection of glutamate followed by nicotine (panel B). The effect of nicotine was biphasic inhibition followed by excitation.

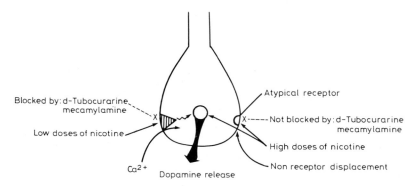

Fig. 12. A scheme showing the dual mechanism of action of nicotine in releasing dopamine from the rat striatum. Low concentrations of nicotine evoke release following activation of a nicotine receptor. The response can be blocked by classical nicotinic antagonists (d-tubocurarine, mecamylamine) and requires extracellular calcium. High concentrations release dopamine either by a displacement mechanism similar to sympathomimetic amines or following activation of an atypical nicotinic receptor. This response is not blocked or irregularly blocked by classical nicotinic antagonists and does not require extracellular calcium.

I. These data are highly suggestive of the idea that nicotine can release dopamine from striatal slices by at least two mechanisms. The releasing effect of low concentrations appears to require the activation of nicotinic-cholinergic receptors and requires extracellular calcium, while the effect of high concentrations may be independent of nicotinic-cholinergic receptors and does not require extracellular calcium. A schema depicting these two mechanisms of action is shown by Fig. 12.

In addition to being able to evoke a release of dopamine presumably by activation of receptors located on dopamine terminals, nicotine also increases the firing rate of dopamine cell bodies. This was observed for dopamine neurons located in both the substantia nigra and ventral tegmental areas. This increased firing rate was seen following the systemic administration of nicotine or following application to the target regions for these dopaminergic neurons, namely the striatum and nucleus accumbens.

An important aspect of our studies was the observation that i.v. nicotine appears more potent and efficacious in activating A_{10} than A_9 neurons. This finding is similar to several behavioral, biochemical and electrophysiological effects of several known drugs of abuse including morphine, cocaine, heroin and ethanol in the nigrostriatal and mesolimbic, mesocortical dopaminergic pathways (Lyness et al., 1979; Bozarth and Wise, 1981; Wise and Bozarth, 1982; Roberts and Koob, 1983; Matthews and German, 1984; Gessa et al., 1985). It appears that nicotine now joins this list of drugs as well as electrical brain stimulation in having in common the fact that they produce rewarding effects followed by dependence as well as preferentially stimulating mesolimbic, mesocortical dopamine neurotransmission. It is possible that such dopaminergic pathways may constitute the biological substrate for producing the rewarding effects of these drugs.

The mechanisms by which the systemic administration of nicotine increases the firing rate of dopamine cell bodies in the A_{10} and A_9 regions are not clear. There is evidence for direct stimulation

of the cell bodies (Lichtensteiger et al., 1982) presumably through its action on nicotinic-cholinergic receptors. The present study provides evidence that an additional site of action may be the target region of these dopamine projections. A large number of anatomical, physiological and biochemical studies have, in fact, well characterized the nigro-striato-nigral loop (Moore, 1971; Aghajanian and Bunney, 1974; Grace and Bunney, 1979; Dray, 1980; Collilngridge and Davis, 1981; Deniau and Chevaliler, 1985), suggesting that nigral afferents include terminals from striatum releasing gamma-aminobutyric acid (GABA) but also enkephalin and substance P (Dray, 1980; Collilngridge and Davis, 1981). The inhibitory effect of the striatonigral projections has been demonstrated electrophysiologically by extracellular (Precht and Yashida, 1971a, b) and intracellular (Grace and Bunney, 1985) techniques. The inhibitory influence of the striatonigral projection, however, affects the neuronal populations both in the pars compacta (A_9) and pars reticulata (SNR). Even though the monosynaptic inhibitory striatonigral projection onto the dopaminergic neuron has been well demonstrated (Precht and Yoshida, 1971), the anatomical studies have shown that the majority of the striatonigral projection is destined for SNR (Rinvik and Grofova, 1970; Dray, 1980; Szabo, 1980; Somogyi et al., 1981).

There is growing evidence that the so-called nigro-striato-nigral loop consists not only of the striatonigral projection and dopaminergic neurons which project back to the striatum, but also involves the nondopaminergic SNR neurons which have been demonstrated to influence other brainstem and thalamic nuclei (Deniau et al., 1978; Deniau and Chevaliler, 1985) with a tonic inhibition. The existence of the connection between the SNR and the A_9 neurons have been suggested by many investigators (Grace and Bunney, 1979; Dray, 1980; Waszczak et al., 1980; Grace and Bunney, 1985). Even though the details of the circuitry within the substantia nigra are not known, there is evidence to support the fact that the A_9 cells are under a tonic inhibitory control (Grace and Bun-

184

ney, 1979; Waszczak, 1980).

The precise mechanism by which the intrastriatal or intraaccumbens injection of nicotine results in stimulation of the dopamine cells in the A_9 and A_{10} regions is unclear. However, some evidence for the mechanism has been obtained in the present experiments. First, we have observed that the injection of glutamate, which would result in the pharmacological stimulation of the striatal output neurons, results in the inhibition of substantia nigra reticulata cells and VTA nondopamine neurons and stimulation of the A_9 and A_{10} cells via disinhibition. A similar effect was observed for nicotine in the substantia nigra although the effect in the VTA was less clear. We suggest that the effect of the striatal nicotinic effect on the dopamine cells is mediated through the SNR neurons producing inhibition of these neurons and stimulation of the dopamine neurons by disinhibition. Our findings suggest that nicotine can stimulate the activity of the dopaminergic neurons not only by its effect on the dopamine cell bodies but also by its effects on the target regions of the dopaminergic projections.

The precise mechanism of nicotine's effect in the central nervous system is still unknown. The behavioral effects of nicotine are probably mediated not only by the dopamine system but also on multiple neuronal circuitries. Nevertheless, it is clear that activation of the dopamine system plays an important role in the action of nicotine in the central nervous system.

Acknowledgements

This work was supported in part by USPHS grant NIDA 02668 (Thomas C. Westfall) and St. Louis University Biomedical Research Support Grant (Kong-Woo P. Yoon).

References

Aghajanian, G.K. and Bunney, B.S. (1974) Pre- and post-synaptic feedback mechanism in central dopaminergic neurons. In P. Seeman and G.M. Brown (Eds.), *Neurotrans-mitters and Brain Function,* University of Toronto Press, Toronto, pp. 4–11.

Arqueros, L., Naquira, D. and Zunino, E. (1978) Nicotine-induced release of catecholamines from rat hippocampus and striatum. *Biochem. Pharmacol.,* 27: 2667–2674.

Bozarth, M.A. and Wise, R.A. (1981) Intracranial self administration of morphine into the ventral tegmental area in rats. *Life Sci.,* 26: 551.

Bunney, B.S., Aghajanian, G.K. and Roth, R.H. (1973) Comparison of effects of L-Dopa, amphetamine, and apomorphine on firing rate of rat dopaminergic neurons. *Nature New Biol.,* 245: 123.

Collilngridge, G. and Davis, J. (1981) The influence of striatal stimulation and putative neurotransmitters on identified neurons in the rat SN. *Brain Res.,* 212: 345–359.

Connelly, M.S. and Littleton, J.M. (1983) Lack of stereoselectivity in ability of nicotine to release dopamine from rat synaptosomal preparations. *J. Neurochem.,* 41: 1297–1302.

Deniau, J.M. and Chevaliler, G. (1985) Disinhibition as the basic process in expression of striatal functions. II. The striato-nigral influence on thalamocortical cells of the ventromedial thalamic nucleus. *Brain Res.,* 334: 227–233.

Deniau, J.M., Hammond, C., Risak, A. and Feger, J. (1978) Electrophysiological properties of identified neurons of the rat substantia nigra (pars compacta and pars reticulata): evidence for the existence of branched neurons. *Exp. Brain Res.,* 32: 490–422.

DeBelleroche, J. and Bradford, H.F. (1978) Biochemical evidence for the presence of presynaptic receptors on dopaminergic nerve terminals. *Brain Res.,* 142: 53–68.

Dray, A. (1980) The physiology and pharmacology of mammalian basal ganglia. *Prog. Neurobiol.,* 14: 221–335.

Gessa, G.L., Muntoni, F., Collu, M., Vrgin, L. and Mereu, G. (1985) Low doses of ethanol activate dopaminergic neurons in the ventral tegmental area. *Brain Res.,* 348: 201.

Giorguieff, M.F., LeFloc'h, M.L., Glowinski, J. and Besson, M.J. (1977) Involvement of cholinergic presynaptic receptors of nicotinic and muscarinic types in the control of the spontaneous release of dopamine from striatal dopaminergic terminals in the rat. *J. Pharmacol. Exp. Ther.,* 200: 535–544.

Giorguieff-Chesselet, M.F., Kemel, M.L., Wandscheer, D. and Glowinski, J. (1979) Regulation of dopamine release by presynaptic nicotinic receptors in rat striatal slices: effect of nicotine in a low concentration. *Life Sci.,* 25: 1257–1262.

Grace, A.A. and Bunney, B.S. (1979) Paradoxical GABA excitation of nigral dopaminergic cells: indirect mediate through reticulate inhibitory neurons. *Eur. J. Pharmacol.,* 59: 211–218.

Grace, A.A. and Bunney, B.S. (1985) Opposing effects of striatonigral feedback pathways on midbrain dopamine cell activity. *Brain Res.,* 333: 271–284.

Lichtensteiger, W., Hefti, F., Felix, D., Huwyler, T., Melamud, E. and Schlumpf, M. (1982) Stimulation of nigrostriatal dopamine neurons by nicotine. *Neurophar-*

macology, 21: 963 – 968.

Lyness, W.H., Friedli, W.N. and Moore, K.E. (1979) Destruction of dopaminergic nerve terminals in nucleus accumbens: effects of D-amphetamine self-administration. *Pharmacol. Biochem. Behav.,* 11: 553 – 556.

Matthews, R.T. and German, D.C. (1984) Electrophysiological evidence for excitation of rat ventral tegmental area dopamine neurons by morphine. *Neuroscience,* 11: 617.

Marien, M., Brien, J. and Jhamandas, K. (1983) Regional release of ^{3}H-dopamine from rat brain in vitro: effects of opioids on release induced by potassium, nicotine and α-glutamic acid. *Can. J. Physiol. Pharmacol.,* 61: 43 – 60.

Mereu, G.P., Fanni, B. and Gessa, G.L. (1984) General anesthetic prevent dopaminergic neuron stimulation by neuroleptics. In E. Usdin, A. Carlsson and J. Engel (Eds.), *Catecholamines: Neuropharmacology and Central Nervous Systems: Neurology and Neurobiology, Vol. 8,* Alan R. Liss, Inc., New York, pp. 353 – 358.

Mereu, G.P., Yoon, K.P., Boi, V., Gessa, G.L., Naes, L. and Westfall, T.C. (1987) Preferential stimulation of ventral tegmental area dopaminergic neurons by nicotine. *Eur. J. Pharmacol.,* 141: 395 – 399.

Moore, R.Y., Bhatnagar, R.B. and Heller, A. (1971) Anatomical and chemical studies of a nigroneostriatal projection in the cat. *Brain Res.,* 30: 119 – 135.

Nieoullon, A., Cheramy, A. and Glowinski, J. (1977) An adaptation of the push-pull cannula method to study the in vivo release of ^{3}H-dopamine synthesized from ^{3}H-tyrosine in the cat caudate nucleus: effects of physical and pharmacological treatments. *J. Neurochem.,* 28: 819 – 828.

Paxinos, G. and Watson, C. (1982) *The Rat Brain in Stereotaxic Coordinates,* Academic Press, New York.

Precht, W. and Yoshida, M. (1971) Monosynaptic inhibition of neurons of the substantia nigra by picrotoxin. *Brain Res.,* 32: 225 – 228.

Precht, W. and Yoshida, M. (1971) Blockage of caudate-evoked inhibition of neurons in the substantia nigra by picrotoxin. *Brain Res.,* 32: 229 – 233.

Ribak, C.E., Vaughn, J.E. and Roberts, E. (1979) The GABA neurons and their axon terminals in rat corpus striatum as demonstrated by GAD immunohistochemistry. *J. Comp.*

Neurol., 186: 261 – 284.

Rinvik, E. and Grofova, I. (1970) Observation on the fine structure of the substantia nigra in the cat. *Exp. Brain Res.,* 11: 229 – 248.

Roberts, D.C.S. and Koob, G.F. (1983) Disruption of cocaine self-administration following β-hydroxydopamine lesion of the ventral tegmental area in rats. *Pharmacol Biochem. Behav.,* 17: 901.

Ross, R.J., Waszczak, B.L., Lee, E.K. and Walters, J.R. (1982) Effects of benzodiazepines on single unit activity in the substantia nigra pars reticulata. *Life Sci.,* 31: 1025.

Szabo, J. (1980) Distribution of striatal afferent from the mesencephalon in the cat. *Brain Res.,* 188: 3 – 21.

Takano, Y., Sakurai, Y., Kohjimoto, Y., Honda, K. and Kamiya, H.O. (1983) Presynaptic modulation of the release of dopamine from striatal synaptosomes: Differences in the effects of high K$^+$ stimulation, methamphetamine and nicotinic drugs. *Brain Res.,* 279: 330 – 334.

Voigt, M.M., Wang, R.Y. and Westfall, T.C. (1985) The effects of cholecystokinin on the in vivo release of newly synthesized ^{3}H-dopamine from the nucleus accumbens of the rat. *J. Neurosci.,* 5: 2744 – 2749.

Waszczak, B.L., Eng, N. and Walters, J.R. (1980) Effects of muscimol and picrotoxin on single unit activity of substantia nigra neurons. *Brain Res.,* 188: 185 – 197.

Westfall, T.C. (1974) Effect of nicotine and other drugs on the release of ^{3}H-norepinephrine and ^{3}H-dopamine from rat brain slices. *Neuropharmacology,* 13: 693 – 700.

Westfall, T.C., Grant, H. and Perry, H. (1983) Release of dopamine and 5-hydroxytryptamine from the rat striatal slices following activation of nicotinic-cholinergic receptors. *Gen. Pharmacol.,* 14: 321 – 325.

Westfall, T.C., Perry, H. and Vickery, L. (1987) Dual mechanism of action of nicotine on dopamine release in the rat striatum. In W.R. Martin, G.R. Van Loon, E.T. Iwamoto and L. Davis (Eds.), *Tobacco Smoking and Nicotine,* Plenum Publishing Corporation, New York, pp. 209 – 223.

Wise, R.A. and Bozarth, M.A. (1982) Action of drugs of abuse on brain reward systems: An update with specific attention to opiates. *Pharmacol. Biochem. Behav.,* 19: 239.

A. Nordberg, K. Fuxe, B. Holmstedt and A. Sundwall (Eds.)
Progress in Brain Research, Vol. 79
© 1989 Elsevier Science Publishers B.V. (Biomedical Division)

CHAPTER 18

Nicotine and opioid peptides

Volker Höllt and Gabriele Horn

Physiologisches Institut der Universität München, Pettenkoferstraße 12, D-8000 München, F.R.G.

Introduction

Several lines of evidence suggest a relationship between nicotinic cholinergic receptors and opioid peptides. Thus, Karras and Kane (1980) have reported that the opioid antagonist naloxone reduces cigarette smoking behavior. More specifically, Pomerleau et al. (1983) have found elevated levels of the opioid peptide β-endorphin in the plasma of smokers. Similarly, acute administration of nicotine into rats resulted in a release of β-endorphin and of ACTH from the anterior pituitary into the blood stream (Cam et al., 1979; Conte-Devolx et al., 1981; Andersson et al., 1983). Nicotine appears to exert its stimulatory effect most likely at the level of the hypothalamus by causing a release of corticotropin releasing factor (Calogero et al., 1988).

In contrast to the acute treatment, normal levels of ACTH and/or glucocorticoids have been observed after chronic administration of nicotine (Cam and Bassett, 1984; Andersson et al., 1985; Sharp and Beyer, 1986) indicating that tolerance to the stimulatory effect of nicotine on the hypo-thalamo-pituitary-adrenal axis can develop. Moreover, chronic administration of nicotine into mice did not change the levels of immunoreactive β-endorphin in the pituitary (Rosecrans et al., 1985). The same authors also investigated the effect of chronic nicotine treatment on the levels of β-endorphin, Met-enkephalin and other neuropeptides in various areas of the mouse brain. With the exception of the levels of β-endorphin in the

hypothalamus, no significant changes in the brain concentrations of the peptides have been found following chronic nicotine administration.

Peptide levels, however, reflect the balance between biosynthesis and release of peptides in a tissue. Therefore, an altered activity of a peptidergic neuron might not be detectable, if a change in release is associated by a compensatory change in the biosynthesis of the peptide. The biosynthetic activity of a peptidergic neuron is reflected by the level of mRNA coding for the respective peptide precursor. Using RNA-hybridization techniques we have investigated the effect of chronic nicotine treatment of rats on the tissue levels of mRNAs coding for the three opioid peptide precursors proopiomelanocortin (POMC), pro-enkephalin (PENK) and prodynorphin (PDYN)).

Methods

Male Wistar rats (200 – 220 g) were chronically treated with nicotine (3 mg/kg per day) or saline for 7 days by s.c. implanted osmotic minipumps. The rats were killed by decapitation and the trunk blood collected for determination of nicotine serum levels. Nicotine was measured in the sera by the Forschungschungslabor Prof. F. Adlkofer (Hamburg, F.R.G.) using gas chromatography. After removing the brain from the skull, the hypothalamus, the striatum and the hippocampus were dissected out; the pituitary was divided in situ into anterior and neurointermediate lobe and the medulla was isolated from the adrenal gland. The tissues were weighed and immediately frozen on dry ice. Total RNA was extracted from the tissues using the LiCl-method (Auffray and Rougeon, 1980). RNA samples were denatured with glyoxal, electrophoresed on a 1.2% agarose gel and transferred to nylon filter sheets. The filters were hybridized at 60°C to radiolabelled cRNA probes complementary to rat PDYN, PENK and/or mouse POMC mRNAs. The single stranded RNA probes were synthesized with RNA polymerases and α-^{32}P-UTP. The PDYN probe contains 74 bases complementary to nucleotide

513–587 of rat PDYN mRNA (Höllt et al., 1987). For preparation of the PENK probe a 938 kb SmaI/SacI restriction fragment of plasmid pRPE2 (a gift from Dr. Sabol, Bethesda) was subcloned into plasmid pBS (blue scribe, Genofit). Likewise, for preparation of the POMC probe a 150 bp HindIII restriction fragment of plasmid ME-150 (a gift from Dr. Roberts, New York) was subcloned into pBS. After hybridization, the filters were washed, dried and exposed to X-ray film at −70°C.

For the determination of peptide levels, the tissues were homogenized in 1 M acetic acid and

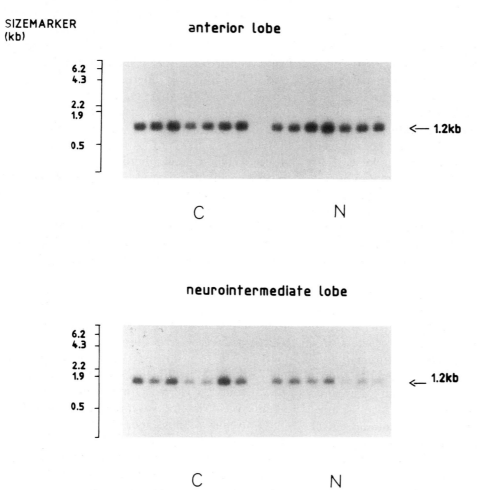

Fig. 1. 'Northern' blot analysis of RNA from anterior and neurointermediate pituitary lobes of control rats (C) and of rats chronically treated with nicotine (N). Each lane represents 3 μg of total RNA from anterior lobes or 0.2 μg RNA from neurointermediate lobes obtained from individual rats. Lamba HindIII/EcoRI restriction fragments were used as size markers.

than centrifuged. The supernatants were lyophiliz-
ed, redissolved in RIA buffer and measured for
immunoreactive (ir) peptides using antisera against
Met-enkephalin (a gift from Dr. K.H. Voight,
Marburg, F.R.G.), dynorphin(1-8) (a gift from Dr.
E. Weber, Portland, U.S.A.) and β-endorphin
'DA'. The characteristics of the antisera have been
described (Weber et al., 1982; Bommer et al.,
1987; Höllt et al., 1979).

Results

Infusion of nicotine (3 mg/kg per day) into rats for
7 days resulted in serum levels of nicotine between
50 and 70 μg/ml similar to that seen in heavy
smokers (Pomerleau et al., 1983). Figure 1 shows
an RNA blot analysis of RNA samples isolated

from pituitary lobes of control rats and of the
chronically nicotine-treated rats. A single mRNA
species of 1.2 kb hybridized to the POMC cRNA
probe – identical in size to that of mouse and rat
POMC mRNA (Civelli et al., 1982). When the
staining of the bands was quantified by densitometry there was no evidence for nicotine-
induced alterations in the level of mRNA in the
anterior lobe of the pituitary (Fig. 2a). In contrast,
there was a significant decrease in the POMC
mRNA levels in the neurointermediate pituitary
lobes after chronic nicotine treatment by about
50% (Figs. 1 and 2a). There was also a concomitant decrease in the tissue levels of ir-β-endorphin
in the neurointermediate pituitary in the nicotine-
treated rats, whereas the levels of the ir-peptide in
the anterior pituitary lobes were not significantly

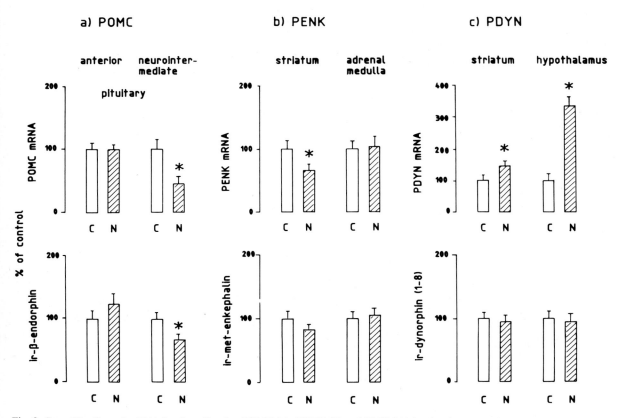

Fig. 2. Quantification of mRNA levels coding for POMC (a), PENK (b) and PDYN (c) by densitometry of the RNA blots. The corresponding peptide levels were measured by radioimmunoassay. Open bars represent mean ± S.E.M. of control rats (C); hatched
bars give the respective values for nicotine-treated rats (N) (n = 6 – 12).

altered. There was even a tendency for an increase in ir-β-endorphin levels in the adenohypophysis (Fig. 2a).

Figure 3 shows a 'Northern' blot analysis of RNA isolated from striatal tissue. The blot was sequentially hybridized with cRNA probes complementary to rat PDYN and PENK mRNAs. PDYN mRNA migrates as a single band with an apparent molecular size of 2.6 kb, whereas PENK mRNA migrates as a single species of about 1.45 kb. Densitometric analysis of the bands revealed a significant increase in the levels of PDYN mRNA in the striatum of about 50%, whereas the levels of PENK mRNA in this structure were slightly decreased (about 30%) after chronic nicotine ad-

ministration (Fig. 2b,c). No significant alterations in the level of the PENK-derived peptide Met-enkephalin or of the PDYN-derived peptide dynorphin (1-8) were found in the striata of nicotine-treated rats (Fig. 2b).

In the adrenal medulla no change in the PENK mRNA levels and of ir-Met-enkephalin was found in nicotine-treated rats as compared to control rats (Fig. 2b). In contrast, there was a marked increase (3 to 4-fold) in the PDYN mRNA levels in the hypothalamus of nicotine-treated animals. Despite this increase in PDYN mRNA levels, no change in the level of ir-dynorphin(1-8) could be observed (Fig. 2c).

SIZEMARKER
(kb)

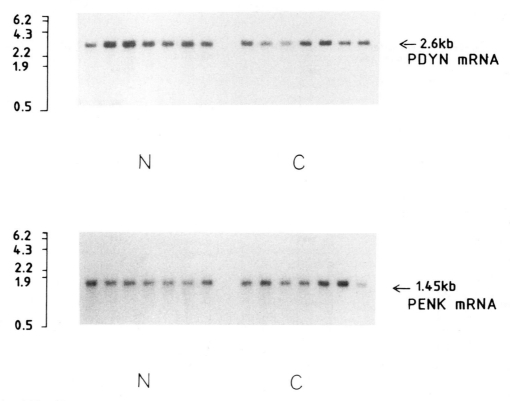

Fig. 3. 'Northern' blot of RNA from striatum of control rats (C) and of rats chronically treated with nicotine (N). Each lane represents 5 μg of total RNA obtained from individual rats. The blot was first hybridized to the PDYN probe washed and then hybridized to the PENK probe. Lambda HindIII/EcoRI restriction fragments were used as size markers.

Discussion

Our results demonstrate that the levels of mRNAs coding for the three opioid precursors POMC, PENK and PDYN are altered in some rat tissues after chronic nicotine treatment for 7 days. This indicates that some of the nicotinic receptors mediating this response are not, or not completely, desensitized during the course of nicotine treatment.

Tolerance, however, appears to develop to other effects. For instance, acute administration of nicotine has been shown to activate the hypothalamus-anterior pituitary-adrenal axis causing elevated levels of ACTH/β-endorphin and corticosteroids, whereas unchanged levels of ACTH and corticosteroids were found after chronic nicotine treatment (Cam and Bassett, 1984; Andersson et al., 1985; Sharp and Beyer, 1986). Our inability to find any significant change in the POMC mRNA or ir-β-endorphin levels in the anterior lobe of chronically nicotine-treated rats supports these results.

In contrast to its stimulatory effect on the secretion of ACTH-β-endorphin from the anterior pituitary, acute nicotine application has been shown to inhibit the release of the POMC-derived pepide α-MSH from the intermediate pituitary (Conte-Devolx et al., 1981). Our findings that the POMC-mRNA levels and the levels of ir-β-endorphin in the neurointermediate lobes of chronically nicotine-treated rats are decreased indicate that the receptors mediating the inhibition of α-MSH release and POMC-biosynthesis in the intermediate lobe are not, or not completely, desensitized during the chronic nicotine treatment. There is general agreement that the release and biosynthesis of POMC peptides in the intermediate pituitary are under the inhibitory control of the tubero-infundibular dopaminergic system. Acute nicotine treatment has been shown to increase the dopamine turnover in the turbero-infundibular dopaminergic neurons (Andersson et al., 1983). A maintained activation of the tubero-infundibular dopamine systems which has been found after chronic exposure to nicotine or to smoking (Dominiak et al., 1984; Andersson et al., 1985) appears to be responsible for the decreased gene expression of POMC in the intermediate lobe after chronic nicotine treatment.

Acute nicotine treatment has also been shown to activate the mesostriatal dopaminergic system. The cell bodies of the neurons are localized in the pars compacta of the substantia nigra; their axons project to the striatum. Both cell bodies (post-synaptic) and terminals (pre-synaptic) contain nicotinic acetylcholine receptors. Nicotine binds to these receptors to stimulate dopamine release and turnover in the striatum (Lichtensteiger et al., 1982; Rapier et al., 1988). In the striatum, dopamine has been reported to inhibit PENK biosynthesis by decreasing the levels of PENK mRNA (Tang et al., 1983; Morris et al., 1988). The decrease of PENK mRNA levels in the striatum might be explained by an increased activity of the mesostriatal dopaminergic system which is maintained during chronic nicotine treatment.

The stimulatory effect of chronic nicotine on the PDYN-mRNA levels in the striatum are more difficult to explain, since the mesostriatal dopaminergic system has no influence on PDYN neurons in the striatum (Morris et al., 1988a,b). A raphe-striatal serotonergic pathway has been shown to enhance levels of PDYN mRNA in the striatum (Morris et al., 1988b). Moreover, an increased serotonin turnover has been observed in the brain of rats after chronic nicotine treatment (Dominiak et al., 1984). An activation of serotonin pathways by chronic nicotine might therefore increase the PDYN levels in the striatum. Alternatively, the direct activation of the cholinergic interneurons in the striatum by nicotine might be responsible for this effect. In our experiments, chronic nicotine resulted in a marked increase in the PDYN levels in the hypothalamus. In this structure, PDYN is predominantly synthesized by the same magnocellular neurons of the nucleus supraopticus and the nucleus paraventricularis which produce vasopressin (Watson et al., 1982). Moreover, PDYN-derived peptides have been shown to be concomitantly released with vaso-pressin from the neurohypophysis of rats (Höllt et al., 1981). Acute nicotine treatment has long been known to cause an increase in the release of vasopressin from the posterior pituitary (Castro de Souza and Rocha E Silva, 1977). It is thus possible that the continuous activation of the vasopres-sin/dynorphin neurons in the hypothalamus by nicotine results in the increase in the PDYN mRNA in the hypothalamus. In situ hybridization experiments using probes complementary to vaso-pressin and dynorphin are necessary to verify this hypothesis.

A well characterized nicotinic cholinergic synapse is localized on the chromaffin cells of the adrenal medulla. In fact, acute administration of nicotine in rats has been shown to release high amounts of adrenaline and noradrenaline from the adrenal medulla (e.g. Dominiak et al., 1984). Moreover, nicotine has been shown to stimulate the concomitant release of catecholamines and of Met-enkephalin from cultured bovine adrenal medullary chromaffin cells in vitro (Eiden et al., 1984). Furthermore, chronic exposure of chromaf-fin cells to nicotine increases the mRNA levels coding for PENK and for tyrosine hydroxylase in vitro (Eiden et al., 1984; Kley et al., 1987). Surpris-ingly, there was no evidence from our experiments for an alteration in the PENK-mRNA levels in the adrenal medulla after chronic infusion of nicotine in rats. It thus appears that the nicotinic acetyl-choline receptors on the rat adrenal medullary cells in vivo desensitize more rapidly than the receptors on bovine adrenal medullary cells in vitro after prolonged exposure to nicotine. It is possible that the desensitization of the nicotinic receptors in vivo is modified by other peptides (Boksa and Livett, 1984; J.P. Changeux, personal communi-cation). Our results are in line with those of Fuchs (1985) who found no change in the levels of cate-cholamines in plasma, nor in the activity of tyro-sine hydroxylase and the levels of catecholamines in the adrenal medulla after chronic nicotine infu-sion (5 mg/kg per day for 7 days) in rats.

In conclusion, our findings show that chronic nicotine treatment of rats is associated with an altered biosynthetic activity of some, but not all, cells synthesizing the opioid peptide precursors POMC, PENK and PDYN. It is assumed that some nicotinic receptors, such as those in the adrenal medulla and those in the hypothalamus which mediate the stimulation of the hypothala-mus-pituitary-adrenal axis, rapidly desensitize in response to prolonged nicotine treatment. In con-trast, no or much less tolerance developed in those nicotinic receptors which activate the tubero-in-fundibular dopaminergic neurons or to those which mediate the increase of PDYN mRNA in the hypo-thalamus. The reason for this differential desen-sitization is not known. Evidence for the existence of multiple nicotinic receptors has recently been provided (Goldman et al., 1987). These receptors may differ in their desensitization characteristics. Alternatively, neuropeptides costored with acetyl-choline in the presynaptic terminals may differen-tially modify the desensitization characteristics of the nicotinic receptors. In any case, our findings provide evidence for an altered state of activity of the three opioid peptide systems under chronic ni-cotine treatment and suggest an intimate relation-ship between nicotinic receptors and endogenous opioid peptides.

Acknowledgement

Supported by the Deutscher Forschungsrat Rauchen und Gesundheit.

References

Andersson, K., Siegel, R., Fuxe, K. and Eneroth, P. (1983) In-travenous injection of nicotine induce very rapid and discrete reductions of hypothalamic catecholamine levels associated with increases of ACTH, vasopressin and prolactin secretion. *Acta Physiol. Scand.*, 118: 35–40.

Andersson, K., Eneroth, P., Fuxe, K., Mascagni, F. and Agnati, L.F. (1985) Effect of chronic exposure to cigarette smoke on amine levels and turnover in various hypothalamic catecholamine nerve terminal systems and on the secretion of pituitary hormones in the male rat. *Neuroendocrinology*, 41: 462–466.

Auffray, C. and Rougeon, F. (1980) Purification of mouse immunoglobin heavy chain mRNAs from total myeloma tumor RNA. *Eur. J. Biochem.,* 107: 303 – 312.

Boksa, P. and Livett, B.G. (1984) Desensitization of the nicotinic response in isolated adrenal chromaffin cells. *J. Neurochem.,* 42: 607 – 617.

Bommer, M., Liebisch, D., Kley, N., Herz, A. and Noble, E. (1987) Histamine affects release and biosynthesis of opioid peptides primarily via H_1-receptors in bovine chromaffin cells. *J. Neurochem.,* 49: 1688 – 1696.

Calogero, A.E., Galucci, W.T., Bernardini, R., Saoutis, C., Gold, P.W. and Chrousos, G.P. (1988) Effect of cholinergic agonists and antagonists on rat hypothalamic corticotropin-releasing hormone secretion in vitro. *Neuroendocrinology,* 47: 303 – 308.

Cam, G.R. and Bassett, J.R. (1984) Effect of prolonged exposure to nicotine and stress on the pituitary-adrenocortical response; the possibility of cross-adaptation. *Pharmacol. Biochem. Behav.,* 20: 221 – 226.

Cam, G.R., Bassett, J.R. and Cairncross, K.D. (1979) The action of nicotine on the pituitary-adrenal-cortical axis. *Arch. Intern. Pharmacodyn. Ther.,* 237: 49 – 66.

Castro de Souza, B. and Rocha E Silva, M. (1977) The release of vasopressin by nicotine: further studies on its site of action. *J. Physiol.,* 265: 297 – 311.

Conte-Devolx, B., Oliver, C., Giraud, P., Gillioz, P., Castanas, E., Lissitzky J.-C., Boudouresque, F. and Millet, Y. (1981) Effect of nicotine on in vivo secretion of melanocorticotrophic hormones in the rat. *Life Sci.,* 28: 1067 – 1073.

Civelli, O., Birnberg, N. and Herbert, E. (1982) Detection and quantitation of proopiomelanocortin mRNA in pituitary and brain tissues from different species. *J. Biol. Chem.,* 257: 6783 – 6787.

Dominiak, P., Kees, F., and Grobecker, H. (1984) Changes in peripheral and central catecholaminergic and serotoninergic neurons of rats after acute and subacute administration of nicotine. *Klin. Wochenschr.,* 62 (Suppl. II): 76 – 80.

Eiden, L.E., Giraud, P., Dave, J.R., Hotchkiss, A.J. and Affolter, H.-U. (1984) Nicotinic receptor stimulation activates enkephalin release and biosynthesis in adrenal chromaffin cells. *Nature (Lond.),* 312: 661 – 663.

Fuchs, G. (1985) Wirkungen von Nikotin und Kotinin auf periphere und zentrale katecholaminhaltige Neuronen der Ratte. Thesis, University of Regensburg, F.R.G.

Goldmann, D., Deneris, E., Luyten, W., Kochhar, A., Patrick, J. and Heinemann, S. (1987) Members of a nicotinic acetylcholine receptor gene family are expressed in different regions of the mammalian central nervous system. *Cell,* 48: 965 – 973.

Höllt, V., Gramsch, C. and Herz, A. (1979) Immunoassay of β-endorphin. In A. Albertini, M. da Prada and B.A. Peskar (Eds.), *Radioimmunoassay in Cardiovascular Medicine,* Elsevier/North-Holland Biomedical Press, Amsterdam, pp. 293 – 307.

Höllt, V., Haarmann, I., Seizinger, B.R. and Herz, A. (1981) Levels of dynorphin-(1-13)-immunoreactivity in rat neurointermediate pituitaries are concomitantly altered with those of leucine enkephalin and vasopressin in response to various endocrine manipulations. *Neuroendocrinology,* 33: 333 – 339.

Höllt, V., Haarmann, I., Millan, M.J. and Herz, A. (1987) Prodynorphin gene expression is enhanced in the spinal cord of chronic arthritic rats. *Neurosci. Lett.,* 73: 90 – 94.

Karras, A. and Kane, J. (1980) Naloxone reduces cigarette smoking. *Life Sci.,* 27: 1541 – 1545.

Kley, N., Loeffler, J.P., Pittius, C.W. and Höllt, V. (1987) Involvement of ion channels in the induction of proenkephalin A gene expression by nicotine and cAMP in bovine chromaffin cells. *J. Biol. Chem.,* 262: 4083 – 4089.

Lichtensteiger, W., Hefti, F., Felix, D., Huwyler, T., Melamed, E. and Schlumpf, M. (1982) Stimulation of nigrostriatal dopamine neurones by nicotine. *Neuropharmacology,* 21: 963 – 968.

Morris, B.J., Haarmann, I., Kempter, B., Höllt, V. and Herz, A. (1986) Localization of prodynorphin mRNA by in situ hybridization using a synthetic oligonucleotide probe. *Neurosci. Lett.,* 69: 104 – 108.

Morris, B.J., Höllt, V. and Herz, A. (1988a) Dopaminergic regulation of striatal proenkephalin mRNA and prodynorphin mRNA: contrasting effects of D_1 and D_2 antagonists. *Neuroscience,* 25: 525 – 532.

Morris, B.J., Reimer, S., Höllt, V. and Herz, A. (1988b) Regulation of striatal prodynorphin mRNA levels by the raphe-striatal pathway. *Mol. Brain Res.,* (in press).

Pomerleau, O.F., Fertig, J.B., Seyler, L.E. and Jaffe, J. (1983) Neuroendocrine reactivity to nicotine in smokers. *Psychopharmacology,* 81: 61 – 67.

Rapier, C., Lunt, G.G. and Wonnacott, S. (1988) Stereoselective nicotine-induced release of dopamine from striatal synaptosomes: concentration dependence and repetitive stimulation. *J. Neurochem.,* 50: 1123 – 1130.

Rosecrans, J.A., Hendry, J.S. and Hong, J.-S. (1985) Biphasic effects of chronic nicotine treatment on hypothalamic immunoreactive β-endorphin in the mouse. *Pharmacol. Biochem. Behav.,* 23: 141 – 143.

Sharp, B.M. and Beyer, H.S. (1986) Rapid desensitization of the acute stimulatory effects of nicotine on rat plasma adrenocorticotropin and prolactin. *J. Pharmacol. Exp. Ther.,* 238: 486 – 491.

Tang, F.G., Costa, E. and Schwartz, J. (1983) Increase of proenkephalin mRNA and enkephalin content of rat striatum after daily injection of haloperidol for 2 to 3 weeks. *Proc. Natl. Acad. Sci. U.S.A.,* 80: 3841 – 3844.

Watson, S.J., Akil, H., Fischli, W., Goldstein, A., Zimmermann, E., Nilaver, G. and von Wimmersma Greidanus, T.B. (1982) Dynorphin and vasopressin: common localization in magnocellular neurons. *Science,* 216: 85 – 87.

Weber, E., Evans, C.J. and Barchas, J.D. (1982) Predominance of the amino-terminal octapeptide fragment of dynorphin in rat brain regions. *Nature (Lond.),* 299: 77 – 79.

SECTION VI

Endocrine Actions of Nicotine

A. Nordberg, K. Fuxe, B. Holmstedt and A. Sundwall (Eds.)
Progress in Brain Research, Vol. 79
© 1989 Elsevier Science Publishers B.V. (Biomedical Division)

CHAPTER 19

Neurochemical mechanisms underlying the neuroendocrine actions of nicotine: focus on the plasticity of central cholinergic nicotinic receptors

K. Fuxe[a], K. Andersson[a], P. Eneroth[b], A. Jansson[a], G. von Euler[a], B. Tinner[a], B. Bjelke[a] and L.F. Agnati[c]

[a] Department of Histology and Neurobiology, Karolinska Institutet, S-104 01 Stockholm, [b] Department of Applied Biochemistry, Huddinge University Hospital, Huddinge, Sweden and [c] Department of Human Physiology, University of Modena, Modena, Italy

Introduction

We have in recent review articles summarized our work on the neuroendocrine actions of nicotine and exposure to cigarette smoke as well as the effects of these treatments on central monoamine neurons (Fuxe et al., 1987, 1988a, b). This chapter will focus on two aspects. The first aspect deals with the existence of dopamine/acetylcholine costoring tubero-infundibular neurons (Tinner et al., 1988) and that tanycytes (modified glial cells in the median eminence) containing D-1 dopamine (DA) receptors play a role in mediating some of the neuroendocrine actions on nicotine and exposure to cigarette smoke. The second aspect deals with the plasticity of central cholinergic nicotinic receptors of the ganglionic type including their regulation via receptor-receptor interactions (see Fuxe et al., 1988c) and permanent alteration by combined pre- and postnatal exposure to nicotine (Jansson et al., 1988). It must be emphasized that with regard to the neuroendocrine actions of nicotine, they appear to be mediated via two types of ganglionic central cholinergic nicotinic receptors which mainly differ via their rate of desensitization upon repeated nicotine treatment.

On the existence of tubero-infundibular dopamine/acetylcholine costoring neurons

Large numbers of choline acetylase (CAT) immunoreactive (IR) pericarya have been demonstrated in the preoptic area, in the hypothalamus and especially within the arcuate nucleus (Tago et al., 1987; Rao et al., 1987). In a recent study we have obtained evidence for the existence of a subpopulation of tubero-infundibular neurons, costoring CAT and tyrosine hydroxylase (TH) immunoreactivities by using combined immunoperoxidase and immunofluorescence histochemistry on the same section (Fuxe et al., 1988b; Tinner et al., 1988). As seen in Figs. 1 and 2, the TH and CAT IR nerve cell bodies were present both within the dorsomedial and ventrolateral part of the arcuate nucleus. In the dorsomedial part of the arcuate nucleus TH IR is a marker for DA pericarya projecting to the external layer of the median eminence (Agnati et al., 1988; Meister et al., 1988a). Instead, within the ventromedial part of the arcuate nucleus and the periarcuate region TH IR is predominantly found within the growth hormone releasing factor (GRF) IR nerve cell bodies (Everitt et al., 1986; Meister et al., 1986). Thus,

198

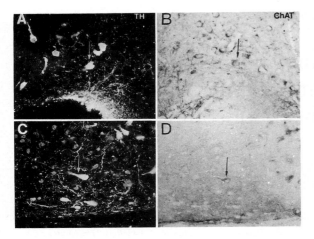

Fig. 1. Coronal sections of the arcuate nucleus, double-stained for TH (immunofluorescence histochemistry) and choline acetylase (ChAT) IR (ABC kit). (A,B) Ventrolateral part of the arcuate nucleus and lateral part of the median eminence containing TH IR nerve cells and nerve terminals, respectively. Medial direction is to the right and the ventral surface is below. Arrow indicates one TH IR nerve cell body (A) showing also choline acetylase IR (B). (C,D) TH (C) and ChAT (D) IR nerve cells and processes in the periarcuate nucleus. Arrow indicates a nerve cell showing both TH and choline acetylase IR. Bar length indicates 50 μm. Bregma level: −2.60 mm.

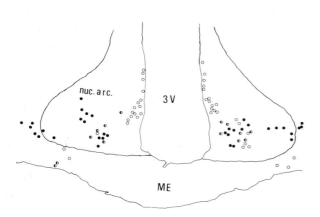

Fig. 2. Schematic drawing of one coronal section through the mediobasal hypothalamus, showing distribution of ChAT and TH IR nerve cells. The level is located at bregma −2.50 mm. Open circles indicate TH IR neurons and filled circles indicate ChAT IR neurons. Nerve cells showing coexistence are marked by half-filled circles. The arcuate nucleus (nuc. arc.), the median eminence (ME) and the third ventricle (3V) are outlined.

GRF and acetylcholine costoring tubero-infundibular neurons projecting to the median eminence may also exist. The results strongly indicate important intraneuronal interactions between DA and acetylcholine and acetylcholine and GRF in neuroendocrine regulation at the median eminence level. It remains to be shown, if nicotine preferentially acts on the DA/acetylcholine costoring neurons and on the acetylcholine/GRF costoring neurons innervating the median eminence versus those DA and GRF IR tubero-infundibular neurons not containing acetylcholine.

Studies on the DARPP-32 immunoreactive tanycytes of the median eminence and their relationship to luteinizing hormone releasing hormone and tyrosine hydroxylase immunoreactive nerve terminals in the external layer of the median eminence

Previous work has demonstrated the codistribution of DA and cyclic AMP regulated phosphoprotein-32 (DARPP-32) IR tanycytes and TH IR

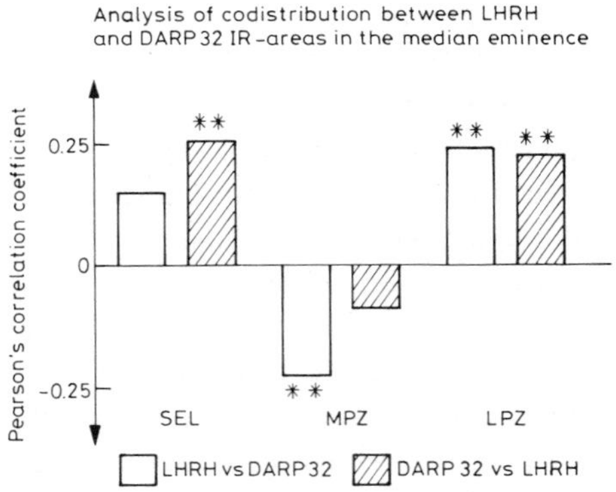

Fig. 3. Correlation between density values of corresponding unitary squares in the median eminence (see Fuxe et al., 1988d). Pearson's correlation coefficient has been used. LHRH versus all DARPP-32 positive squares (□) and all DARPP-32 versus all LHRH positive squares (▨) are illustrated in the following three parts of the median eminence: subependymal layer (SEL), medial palisade zone (MPZ) and lateral palisade zone (LPZ). The codistribution is significant for DARPP-32 versus LHRH in the SEL and LPZ region and for LHRH versus DARPP-32 in the LPZ. Instead, LHRH versus DARPP-32 in the MPZ is shown to be significantly inversely correlated.

nerve terminals within the external layer of the median eminence (Fuxe et al., 1988d, e; Meister et al.,
1988b). As seen in Fig. 3, the LHRH and DARPP-
32 IR profiles in the subependymal layer and in the
lateral palisade zone significantly codistribute with
each other, considering all the LHRH positive
squares in the respective regions (see Fuxe et al.,
1988d, e). These results open up the possibility that
the DARPP-32 IR tanycytes, probably possessing
D-1 DA receptors, participate in the inhibitory
dopaminergic regulation of LHRH release from
the median eminence.

Neuroendocrine actions of nicotine indicate the existence of two types of central ganglionic cholinergic nicotinic receptors

As seen in Fig. 4, nicotine in a low dose produces
rapid increases of prolactin, vasopressin and
ACTH serum levels, which are associated with a
delayed increase of serum corticosterone levels
(Andersson et al., 1983). In the doses used,
mecamylamine (1 mg/kg) but not α-bungarotoxin
(1 μg) given into the lateral ventricle appears
capable of counteracting the nicotine-induced increase of the serum luteinizing hormone (LH)
levels in the male rat (Fig. 5). These results indicate
that a ganglionic cholinergic nicotinic receptor is
involved in mediating these excitatory neuroendocrine actions of nicotine. Studies by other
authors (Sharp and Beyer, 1986; Kellar, this
volume) indicate that these central nicotinic receptors of the ganglionic type rapidly desensitize.

Repeated high doses of nicotine will instead produce marked inhibitory effects on serum TSH,
prolactin and LH levels associated with an increase
of serum corticosterone levels (Andersson et al.,
1985a) (see Fig. 6). This treatment with nicotine
does not lead to any significant effects on serum
FSH levels and vasopressin levels at the time interval analyzed. The inhibitory effects of nicotine
seen on TSH, prolactin and LH secretion are all
blocked by previous treatment with mecamylamine
(see Andersson et al., 1980, Andersson, 1985).
However, the excitatory effects of acute intermit-

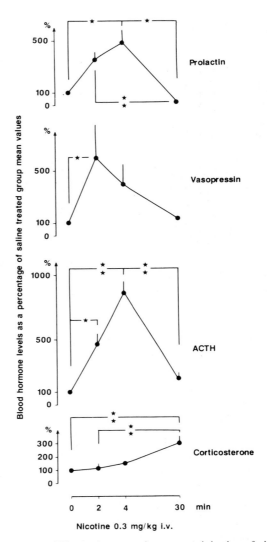

Fig. 4. Effects of an acute intravenous injection of nicotine on
serum levels of prolactin, vasopressin, ACTH and corticosterone of the normal male rat. Saline or nicotine (0.3
mg/kg) was given i.v. into the jugular vein 2 min, 4 min or 20
min before decapitation. The control group consists of a pooled
value from the respective 2 and 30 min control groups.
Mean ± S.E. in per cent of respective 2, 4 or 20 min control
group mean value. $n = 3$ in nicotine 2 min vasopressin group;
$n = 4$ in nicotine 30 min corticosterone group; $n = 9$ in control groups except for vasopressin control group; $n = 6$ in all
other groups. The absolute hormone concentrations (equivalent
with 100%) are: prolactin in 19.0 ± 2.5 ng/ml, vasopressin
2.5 ± 0.15 pg/ml, ACTH 91 ± 6 pg/ml and corticosterone
183 ± 38 nmol/l. Statistical analysis according to Wilcoxon;
one-way classification, comparing all possible pairs of
treatments. * $p < 0.05$; ** $p < 0.01$.

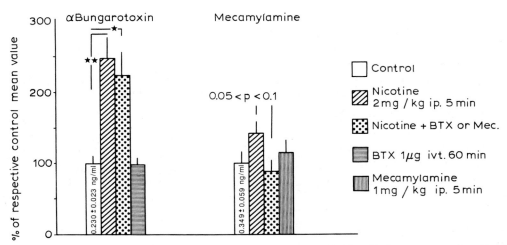

Fig. 5. Effects of α-bungarotoxin (60 min pretreatment) and mecamylamine (5 min pretreatment) on the nicotine-induced increase of serum LH levels in the male rat. Means ± S.E.M. ($n = 8$) are shown. LSD test for multiple comparisons. * $p < 0.05$; ** $p < 0.01$.

tent treatment with nicotine on serum corticosterone levels are only, in part, counteracted by mecamylamine. In contrast to the excitatory effects of nicotine on serum LH and prolactin levels, the inhibitory effects of acute intermittent nicotine treatment on serum LH and prolactin but not TSH levels are maintained upon chronic exposure to nicotine or to cigarette smoke (see Fig. 7) (Andersson et al., 1985b). In contrast, the excitatory effects of nicotine on serum corticosterone secretion have largely disappeared upon a previous chronic exposure to cigarette smoke (see Fig. 7). These results imply that there exist two types of central cholinergic nicotinic receptors of the ganglionic type involved in neuroendocrine regulation, their major difference being the degree to which they desensitize to repeated nicotine treatment.

On the role of tubero-infundibular dopamine neurons in the mediation of the inhibitory effects of nicotine on prolactin and LH secretion

Acute intermittent treatment with high doses of nicotine in the male and female rat in the same way as performed in the neuroendocrine experiments

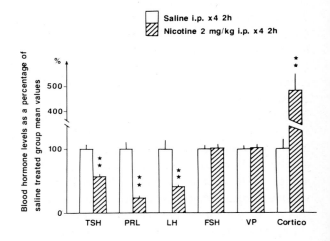

Fig. 6. Effects of nicotine on serum levels of TSH, prolactin, LH, FSH, vasopressin and corticosterone in the normal male rat. Saline or nicotine (2 mg/kg, i.p.) was given 4 times with 30-min intervals, the animals being decapitated 30 min after the last injection. Means ± S.E.M. are given in per cent of control and saline-treated group mean values; $n = 12 - 18$. The serum levels of the respective hormone were (equivalent with 100%): prolactin 7.8 ± 0.9 ng/ml, LH 11.3 ± 1.6 ng/ml and FSH 600 ± 36 ng/ml (using RP-1 standards); TSH 683 ± 41 pg/ml, vasopressin 2.7 ± 0.2 pg/ml and corticosterone 93 ± 15 nmol/l (using RP-2 standards). Statistical analysis according to the Wilcoxon test, one-way classification, comparing all possible pairs of treatments: * $p < 0.05$; ** $p < 0.01$.

leads to an increase in the depletion of the noradrenaline (NA) and DA stores in various catecholamine nerve terminal systems of the hypothalamus after TH inhibition, indicating that this type of nicotine treatment increases NA and DA utilization in the median eminence and NA utilization in other parts of the hypothalamus (see reviews by Fuxe et al., 1987, 1988a, b). These effects are also counteracted by mecamylamine pretreatment, suggesting the involvement of a ganglionic cholinergic nicotinic receptor which may be located on the nerve cell membrane of the catecholamine neuron systems (Schwartz et al., 1984, Härfstrand et al., 1987). In view of the inhibitory role of the tubero-infundibular DA neurons in the control of prolactin and LH secretion we have therefore suggested that the in-

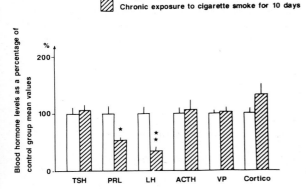

Fig. 7. Effects of chronic exposure to unfiltered cigarette smoke on prolactin, LH, FSH, ACTH, vasopressin and corticosterone levels in the male rat peripheral blood. The animals were treated chronically with two cigarettes per day for 10 days and the last day the animals were exposed for one cigarette 4 times with 30-min intervals and decapitated 30 min after the last smoke exposure. Means ± S.E.M. are expressed in percentage of the means obtained in the respective control group. $n = 12$ in the air-treated group and $n = 9$ in the cigarette smoke exposure group. The absolute hormone levels (equivalent with 100%) are: TSH 287 ± 29 ng/ml, prolactin 7.3 ± 1 ng/ml, LH 21.5 ± 3.0 ng/ml, ACTH 96 ± 10 pg/ml, vasopressin 3.5 ± 0.10 pg/ml and corticosterone 187 ± 15 nmol/l (using RP-1 standards). Statistical analysis according to the Wilcoxon test, one-way classification, comparing all possible pairs of treatments. * $p < 0.05$; ** $p < 0.01$.

hibitory effects of nicotine on LH and prolactin secretion are at least in part mediated via an increased release of DA in the median eminence (see Fuxe et al., 1987, 1988a, b). In agreement with this view we have also been able to observe (see Fig. 8) that the increases in DA utilization in the medial and lateral palisade zone are maintained following chronic exposure to cigarette smoke and as stated above also the inhibitory neuroendocrine effects of nicotine on LH and prolactin serum levels are maintained following such a chronic exposure to cigarette smoke (Andersson et al., 1985b). It should also be noted that there exists a differential tolerance development in the nicotinic cholinergic receptors controlling the NA nerve terminal systems of the hypothalamus, since the periventricular NA nerve terminal systems still show an increase in NA utilization to an acute nicotine treatment following chronic exposure to cigarette smoke, which is not true for the paraventricular and median eminence NA nerve terminal systems.

The DA nerve terminals within the medial palisade zone probably release DA as a prolactin inhibitory factor into the primary capillary plexus (see Fuxe et al., 1978; Andersson et al., 1981). Instead, the DA nerve terminals in the lateral palisade zone are probably involved in mediating the inhibitory effects of nicotine on LHRH release and subsequently on LH secretion. Thus, these DA nerve terminals are associated with LHRH nerve terminals (see above), an interaction which also may involve the tanycyte processes (see above). In agreement with these views we have recently obtained evidence that the D-1 receptor antagonist SCH-23390 (Fig. 9) but not the D-2 receptor antagonist raclopride can counteract the inhibitory effects of exposure to cigarette smoke on serum LH levels (see Andersson et al., 1988a; Fuxe et al., 1988d, e). In view of the fact that D-1 DA receptors are frequently associated with DARPP-32 IR it seems likely, based on these pharmacological experiments, that the tanycytes may be involved in the DA regulation of LHRH release (see also Meister et al., 1988b).

As outlined in Fig. 10, the involvement of D-1

202

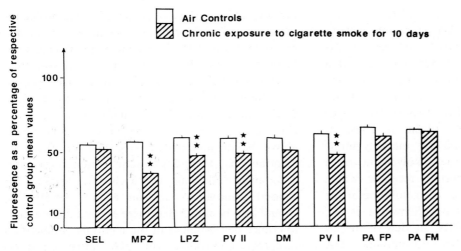

Fig. 8. Effect of chronic exposure to unfiltered cigarette smoke on αMT-induced CA fluorescence disappearance in various hypothalamic DA and NA nerve terminals of the male rat. Means ± S.E.M. are expressed in percent of means obtained in the respective control groups. $n = 11 - 12$ in air-treated groups; $n = 9$ in the groups exposed to unfiltered smoke. The absolute concentrations of CAs have been expressed in nmol/g tissue wet weight. SEL 162 ± 5, MPZ 194 ± 4, LPZ 541 ± 11, PV2 116 ± 2, DM 250 ± 5, PV1 137 ± 3, PAFP 227 ± 5 and PAFM 315 ± 6. Statistical analysis according to the Wilcoxon test, one-way classification, comparing all possible pairs of treatments. * $p < 0.05$; ** $p < 0.01$. SEL = subependymal layer of the median eminence; MPZ = medial palisade zone of the median eminence; LPZ = lateral palisade zone of the median eminence; PV II = posterior periventricular hypothalamic region; DM = dorsomedial hypothalamic nucleus; PVI = anterior periventricular hypothalamic region; PA FP = parvocellular part of the paraventricular hypothalamic nucleus; PA FM = magnocellular part of the paraventricular hypothalamic nucleus.

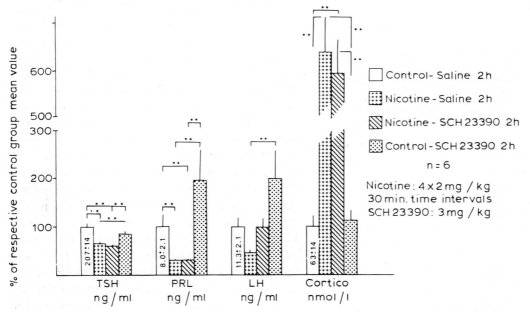

Fig. 9. Effects of SCH-23390 and nicotine on serum levels of TSH, prolactin, LH, FSH, vasopressin, corticosterone and testosterone in the normal male rat. Saline or nicotine (2 mg/kg, i.p.) was given 4 times with 30-min intervals, the animals being decapitated 30 min after the last injection. Saline or the D-1 DA antagonist SCH-23390 (3.0 mg/kg, i.p.) was given 5 min before the onset of the first repetitive saline or nicotine injection. Means ± S.E.M. are given in per cent of saline-treated group mean values. The serum levels of respective hormone in the control groups are given in ng/ml (using RP-1 standards). Statistical analysis according to the Wilcoxon test; one-way classification, comparing all possible pairs of treatment. * $p < 0.05$; ** $p < 0.01$.

DA receptor containing tanycytes may also explain the early excitatory and the delayed inhibitory effects of nicotine on LHRH release. It seems possible that nicotine can directly affect DA and DA/acetylcholine costoring neurons innervating the median eminence, originating from cell bodies in the arcuate nucleus. The activation of these ganglionic cholinergic nicotinic receptors may lead to an increase in the release of DA in part due to an increase in firing rate. DA may act in part on D-1 DA receptors located on the tanycyte processes. The activation of these D-1 DA receptors may lead to morphological changes in the tanycytes, so that they will withdraw and unmask DA receptors located on the LHRH axon terminals. These receptors can then be reached by the released DA. In this way the delayed inhibitory effects by nicotine on LHRH release may be in part

explained. This type of nicotinic receptor of the ganglionic type located on the tubero-infundibular DA and DA/acetylcholine costoring neurons is different from the one located possibly directly located on the LHRH immunoreactive neurons contributing to the excitatory effects on LH release (see Fig. 10), since unlike the latter type it is resistant to desensitization.

Studies on withdrawal from chronic exposure to cigarette smoke have also provided further evidence for an involvement of DA in the medial palisade zone in the inhibitory control of prolactin secretion. Thus, at the 48 h time interval following withdrawal there is a reduction of serum prolactin level, which is associated with a selective increase of DA utilization in the medial palisade zone (Andersson et al., 1988b).

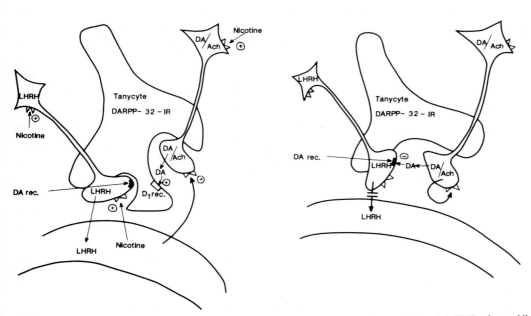

Changes induced by nicotine

Median eminence

Fig. 10. Schematic illustration on the role of tanycytes in the action of nicotine on DA and LHRH release. Nicotine may directly affect DA and acetylcholine costoring neurons leading to an increase of the release of DA, which *inter alia* acts on D-1 DA receptors located on the tanycyte processes. The activation of these D-1 DA receptors may lead to a morphological change in the tanycytes (right), so that DA receptors located on LHRH axon terminals can be reached by released DA. In this way, the delayed inhibitory effect on LHRH release might be explained. The excitatory effects of nicotine on LHRH release can instead be mediated via nicotinic receptors directly located on the LHRH neurons and which may rapidly desensitize.

Modulation of ganglionic cholinergic nicotinic receptors via receptor-receptor interactions

In recent work (Andersson et al., 1988a) we have been able to show that not only does a D-1 receptor antagonist such as SCH-23390 counteract the inhibitory effects of nicotine on LH secretion but it also inhibits the nicotine-induced depletion of NA stores in the hypothalamus and in the preoptic area. Thus, D-1 DA receptors may be located on the NA nerve terminal systems of the hypothalamus and the preoptic area, where they may facilitate the ability of cholinergic nicotinic receptors of the ganglionic type to increase the release of NA. These results open up the possibility for the existence of an interaction between D-1 DA receptors and cholinergic nicotinic receptors of the ganglionic type in the NA nerve terminal membrane. To illustrate the possible existence of such a receptor-receptor interaction in the NA nerve terminal membranes we have analyzed the effects of the D-1 agonist SKF-38393 on the binding of [^{3}H]nicotine (2 nM) in membrane preparations from the frontoparietal cortex of the male rat. As seen in Fig.

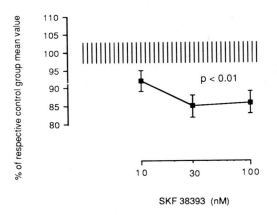

Fig. 11. Effects of SKF-38393 (10 – 100 nM) on the binding of 2 nM [^{3}H]nicotine in membrane preparations of the rat frontoparietal cortex. The values (means ± S.E.M.) are expressed as a percentage of the respective control group mean value. The striped area represents the variability of control groups (S.E.M.). The experiments involved 16 (control), 8 (10 nM SKF-38393), 8 (30 nM SKF-38393) and 8 (100 nM SKF-38393) observations. The significances refer to the Jonckhere-Terpstra test for ordered alterations.

11, SKF-38393 is capable in a concentration-related way to inhibit the binding of [^{3}H]nicotine in this membrane preparation. These results open up the possibility that the facilitatory effects of D-1 receptor activation on NA release (Andersson et al., 1988a) are associated with a reduction of the facilitatory effects of nicotine on NA release due to the existence of a receptor-receptor interaction, leading to a reduced affinity of the high affinity [^{3}H]nicotine binding site. D-1 DA receptors have been shown to exist within the frontoparietal cortex (see Fuxe et al., 1988f).

We have also recently obtained evidence that α_2-adrenoreceptors may participate in the regulation of high affinity but not low affinity [^{3}H]nicotine binding sites in membrane preparations of the frontoparietal cortex (Fuxe et al., 1988c). Thus, clonidine in low concentrations was capable of reducing the high affinity [^{3}H]nicotine binding without influencing the low affinity [^{3}H]nicotine binding in membrane preparations from this region. These results open up the possibility that activation of α_2-autoreceptors, inhibiting the release of NA, also reduces the influence of cholinergic nicotinic receptors on NA release by reducing the affinity of these [^{3}H]nicotine binding sites via a receptor-receptor interaction within the membrane (see Fig. 12). In contrast, nicotine (10 nM) has not been found capable of modulating the binding characteristics of [^{3}H]paraminoclonidine binding sites in membrane preparations of the frontoparietal cortex as revealed in Scatchard analysis (see Fig. 13). These latter results indicate the possibility that nicotinic cholinergic receptors of the ganglionic type may facilitate NA release without turning off the α_2-adrenoreceptor function, at least with regard to the α_2-recognition site.

Possible permanent alterations in central cholinergic nicotinic receptor function following pre- and postnatal nicotine treatment

In recent experiments, pregnant Sprague-Dawley rats have received (±) nicotine in the drinking water (50 mg base/l) during the entire prenatal and during the first 3 postnatal weeks giving rise to a moderate dose of approximately 4 mg/kg per 24 h

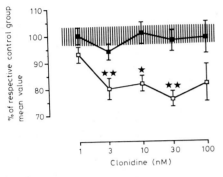

Fig. 12. Effects of clonidine (1 – 100 nM) on the binding of 2 nM (□) and 100 nM (■) [³H]nicotine in membrane preparations of the rat frontoparietal cortex. The values (means ± S.E.M.) are expressed as a percentage of the respective control group mean value. The striped area represents the variability of the control groups (S.E.M.). The experiments on low-affinity [³H]nicotine binding sites involved 16 (control), 16 (1 nM clonidine), 8 (3 nM clonidine), 16 (10 nM clonidine), 8 (30 nM clonidine) and 8 (100 nM clonidine) observations in duplicates. The corresponding number of observations for high-affinity [³H]nicotine binding was 24, 8, 23, 24, 23 and 8, respectively. * $p < 0.05$; ** $p < 0.01$. The significances refer to the control group using Dunnet's multiple comparison test.

of nicotine. The effects of such combined perinatal treatment on neuroendocrine function and on DA utilization in the medial and lateral palisade zone of the median eminence was evaluated in the off-spring. Some of the results are shown in Fig. 14 (Jansson et al., 1988). One week after withdrawal from nicotine in the drinking water the offspring of nicotine-treated rats demonstrated a highly significant lowering of prolactin serum levels and an increase of DA utilization in the medial and lateral palisade zone, which does not take place after 3 weeks of postnatal treatment alone (data not shown). Of substantial interest is the observation that the increase of DA utilization in the median eminence is maintained even 6 months later. At this time, however, the serum prolactin levels are no longer reduced. These results indicate that tubero-infundibular DA neurons may have become permanently altered by the pre- and postnatal exposure to nicotine. However, the inhibition of prolactin secretion is no longer observed probably due to the compensatory activation of other transmitter systems involved in enhancing the release of prolactin from the anterior pituitary gland. Such a permanent alteration in discrete regions of the brain may be brought about by unknown genomic actions of the prenatal nicotine treatment, which becomes very important to in-

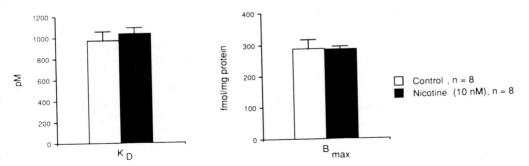

Fig. 13. Effects of nicotine (10 nM) on the binding characteristics of [³H]paraminoclonidine in membrane preparations of the rat frontoparietal cortex as evaluated by Scatchard analysis. The values are expressed as means ± S.E.M. There was no significant difference according to the Wilcoxon two-samples test; $n = 8$.

206

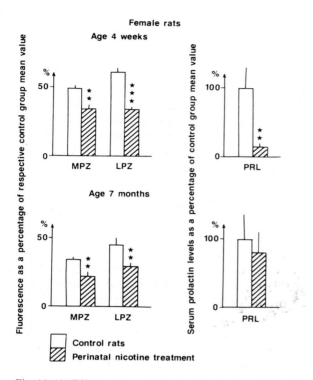

Fig. 14. (A) Effects of pre- and postnatal treatment of nicotine (50 mg nicotine/11 drinking water) on serum prolactin levels and on α-MT-induced CA disappearance in MPZ and LPZ of the 4-week-old female rat one week after cessation of nicotine treatment and in the 7-month-old diestrous rat, 6 months following cessation of nicotine treatment. α-MT (250 mg/kg) was given i.p. 2 h before killing. Means ± S.E.M. are given as a percentage of respective control group mean value; $n = 5 - 7$ rats in the control groups and $n = 6 - 9$ in the nicotine-treated groups. The absolute concentrations of CA (referred to as 100% in the control group) are given in nmol/g tissue wet weight (4-week-old: MPZ 310 ± 12, LPZ 722 ± 32; 7-month-old: MPZ 509 ± 52, LPZ 693 ± 93). The serum prolactin level in the 4-week-old female rat was 19.0 ± 5.5 ng/ml (100%) and in the 7-month-old female rat 3.1 ± 1.1 ng/ml. Statistical analysis according to the Mann-Whitney U-test, $* p < 0.05$; $** p < 0.01$; $*** p < 0.002$.

vestigate. The induction of a positive feedback mechanism, e.g. controlling the synthesis of the cholinergic nicotinic receptors involved in the regulation of the tubero-infundibular DA neurons is less likely to be responsible for these effects since positive feedback mechanism is likely to lead to extreme values which was not observed.

Acknowledgements

This work has been supported by grants from the Council for Tobacco Research, New York, Svenska Tobaksmonopolet and the Swedish Medical Research Council (grant 12P-08127 and grant 12X-08275).

References

Agnati, L.K., Fuxe, K., Steinbusch, H., Zoli, M., Zini, I., Cintra, A. and Härfstrand, A. (1988) Mapping of the tuberoinfundibular dopamine neurons as studied by dopamine and tyrosine hydroxylase immunocytochemistry. In M. Sandler (Ed.), *Progress in Catecholamine Research, Part B: Central Aspects,* Alan R. Liss, New York, pp. 279 – 284.

Andersson, K. (1985) Mecamylamine pretreatment counteracts cigarette smoke induced changes in hypothalamic catecholamine neuron systems and in anterior pituitary function. *Acta Physiol. Scand.,* 125: 445 – 452.

Andersson, K., Fuxe, K., Eneroth, P. Gustafsson, J.-Å. and Agnati, L.F. (1980) Mecamylamine induced blockade of nicotine induced inhibition of gonadotrophin and TSH secretion and of nicotine induced increases of catecholamine turnover in the rat hypothalamus. *Acta Physiol. Scand. Suppl.,* 479: 27 – 29.

Andersson, K., Fuxe, K., Eneroth, P., Nyberg, F. and Roos, P. (1981) Rat prolactin and hypothalamic catecholamine nerve terminal systems. Evidence for rapid and discrete increases in dopamine and noradrenaline turnover in the hypophysectomized male rat. *Eur. J. Pharmacol.,* 76: 261 – 265.

Andersson, K., Siegel, R., Fuxe, K. and Eneroth, P. (1983) Intravenous injections of nicotine induced very rapid and discrete reductions of hypothalamic catecholamine levels associated with increases of ACTH, vasopressin and prolactin secretion. *Acta Physiol. Scand.,* 118: 35 – 40.

Andersson, K., Fuxe, K., Eneroth, P., Mascagni, F. and Agnati, L.F. (1985a) Effects of acute intermittent exposure to cigarette smoke on catecholamine levels and turnover in various types of hypothalamic DA and NA nerve terminal systems as well as on the secretion of adenohypophyseal and corticosterone. *Acta Physiol. Scand.,* 124: 277 – 285.

Andersson, K., Eneroth, P., Fuxe, K., Mascagni, F. and Agnati, L.F. (1985b) Effects of chronic exposure to cigarette smoke on amine levels and turnover in various hypothalamic catecholamine nerve terminal systems and on the secretion of pituitary hormones in the male rat. *Neuroendocrinology,* 41: 462 – 466.

Andersson, K., Fuxe, K., Eneroth, P., Härfstrand, A. and Agnati, L.F. (1988a) Involvement of D1 dopamine receptors in the nicotine-induced neuro-endocrine effects and depletion of diencephalic catecholamine stores in the male rat.

Neuroendocrinology, 48: 188 – 200.

Andersson, K., Fuxe, K., Eneroth, P, and Härfstrand A. (1988b) Effects of withdrawal from chronic exposure to cigarette smoke on hypothalamic and preoptic catecholamine nerve terminal systems and on the secretion of pituitary hormones in the male rat. *Naunyn-Schmiedebergs Arch. Pharmacol.* (in press).

Everitt, B.J., Meister, B., Hökfelt, T., Melander, T., Terenius, L., Rökaeus, Å., Theodorsson-Norheim, E., Dockray, G., Edwardson, J., Cuello, C., Elde, R., Goldstein, M., Hemmings, H., Ouimet, C., Walaas, I., Greengard, P., Vale, V., Weber, E., Wu, J.-Y. and Chang, K.-J. (1986) The hypothalamic arcuate nucleus-median eminence complex: immunohistochemistry of transmitters, peptides and DARPP-32 with special reference to coexistence in dopamine neurons. *Brain Res. Rev.,* 11: 97 – 155.

Fuxe, K., Löfström, A., Hökfelt, T., Ferland, L., Andersson, K., Agnati, L., Eneroth, P., Gustafsson, J.-Å. and Skett, P. (1978) Influence of central catecholamines on LHRH-containing pathways. In J.E. Tyson (Ed.), *Clinics in Obstetrics and Gynaecology, Vol. 45,* Saunders Company Ltd., London, pp. 251 – 269.

Fuxe, K., Andersson, K., Eneroth, P., Härfstrand, A., Nordberg, A. and Agnati, L.F. (1987) Effects of nicotine and exposure to cigarette smoke on discrete dopamine and noradrenaline nerve terminal systems of the telencephalon and diencephalon of the rat: relationship to reward mechanisms and neuroendocrine functions and distribution of nicotinic binding sites in brain. In W.R. Martin, G.R. Van Loon, E.T. Iwamoto and L. Davis (Eds.), *Tobacco, Smoking and Nicotine,* Plenum, New York, pp. 255 – 262.

Fuxe, K., Andersson, K., Eneroth, P., Härfstrand, A. and Agnati, L.F. (1988a) Neuroendocrine actions of nicotine and of exposure to cigarette smoke. Medical implications. *Psychoneuroendocrinology* (in press).

Fuxe, K., Andersson, K., Eneroth, P., Härfstrand, A., Janson, A.M., von Euler, G., Agnati, L.F., Tinner, B., Köhler, C. and Hersh, L.B. (1988b) Neuroendocrine and trophic actions of nicotine. In M.J. Rand and K. Thurau (Eds.), ISCU Press, Paris, pp. 293 – 320.

Fuxe, K., von Euler, G. and Agnati, L.F. (1988c) Reduction of (^{3}H)nicotine binding by clonidine in membrane preparations of the rat cerebral cortex. *Acta Physiol. Scand.,* 132: 265 – 266.

Fuxe, K., Agnati, L.F., Andersson, K., Cintra, A., Härfstrand, A., Zoli, M., Eneroth, P. and Goldstein, M. (1988d) D1 receptor mechanisms in the median eminence and their inhibitory regulation of LHRH release. *Neurochem. Int.* 13: 165 – 178.

Fuxe, K., Agnati, L.F., Härfstrand, A., Zoli, M., Eneroth, P., Andersson, K., Benfenati, F., Ouimet, C. and Greengard, P. (1988e) On the role of DA nerve terminals and tanycytes in the medianosomal organization of the median eminence. Foxus on the LHRH medianosome. In M. Sandler (Ed.), *Progress in Catecholamine Research, Part B: Central Aspects,* Alan R. Liss, New York, pp. 267 – 272.

Fuxe, K., Cintra, A., Agnati, L.F., Härfstrand, A. and Goldstein, M. (1988f) Studies on the relationship of tyrosine hydroxylase, dopamine and cyclic AMP-immunoreactive neuronal structures and D1 receptor antagonist binding sites in various brain regions of the male rat – mismatches indicate a role of D1 receptors in volume transmission. *Neurochem. Int.* (in press).

Härfstrand, A., Fuxe, K., Andersson, K., Agnati, L.F., Janson, A.M. and Nordberg, A. (1987) Partial di-mesencephalic hemitransections produce disappearance of ^{3}H-nicotine binding in discrete regions of the rat brain. *Acta Physiol. Scand.,* 130: 161 – 163.

Jansson, A., Andersson, K., Fuxe, K., Bjelke, B. and Eneroth, P. (1988) Effects of combined pre- and postnatal treatment with nicotine on hypothalamic catecholamine nerve terminal systems and neuroendocrine function in the four week old and adult male and female diestrous rat. *Acta Physiol. Scand.* (in press).

Meister, B., Hökfelt, T., Vale, W.W., Sawschenko, P.E., Swanson, L.W. and Goldstein, M. (1986) Coexistence of tyrosine hydroxylase and growth hormone-releasing factor in a subpopulation of tuberoinfundibular neurons of the rat. *Neuroendocrinology,* 42: 237 – 247.

Meister, B., Hökfelt, T., Steinbusch, H.W.M., Skagerberg, G., Lindvall, O., Geffard, M., Joh, T.H., Cuello, A.C. and Goldstein, M. (1988a) Do tyrosine hydroxylase-immunoreactive neurons in the ventrolateral arcuate nucleus produce dopamine or only L-dopa? *J. Chem. Neuroanat.,* 1: 59 – 64.

Meister, B., Hökfelt, T., Tsuruo, Y., Hemmings, H., Ouimet, C., Greengard, P. and Goldstein, M. (1988b) DARPP-32, a dopamine- and cyclic AMP-regulated phosphoprotein in tanycytes: distribution and relation to dopamine and LHRH neurons and other glial elements. *Neuroscience,* 27: 607 – 622.

Rao, Z.R., Yamano, M., Wanaka, A., Tatehata, T., Shiosaka, S. and Tohyama, M. (1987) Distribution of cholinergic neurons and fibers in the hypothalamus of the rat using choline acetyltransferase as a marker. *Neuroscience,* 20: 923 – 934.

Schwartz, R.D., Lehmann, J. and Kellar, K.J. (1984) Presynaptic nicotinic cholinergic receptors labeled by ^{3}H-acetylcholine on catecholamine and serotonin axons in brain. *J. Neurochem.,* 42: 1495 – 1498.

Sharp, B.M. and Beyer, H.S. (1986) Rapid desensitization of the acute stimulatory effects of nicotine on rat plasma adrenocorticotropin and prolactin. *J. Pharmacol. Exp. Ther.,* 238: 486 – 491.

Tago, H., McGeer, P.L., Bruce, G. and Hersh, L.B. (1987) Distribution of choline acetyltransferase-containing neurons of the hypothalamus. *Brain Res.,* 415: 49 – 62.

Tinner, B., Fuxe, K. Köhler, Ch., Hersh, L.B., Andersson, K., Jansson, A. and Agnati, L.F. (1988) Evidence for a subpopulation of tuberoinfundibular dopamine neurons costoring choline-acetylase and tyrosine hydroxylase immunoreactivities. *Neurosci. Lett.* in press.

A. Nordberg, K. Fuxe, B. Holmstedt and A. Sundwall (Eds.)
Progress in Brain Research, Vol. 79
© 1989 Elsevier Science Publishers B.V. (Biomedical Division)

CHAPTER 20

Regulation of brain nicotinic cholinergic recognition sites and prolactin release by nicotine

Kenneth J. Kellar[a], Bridget A. Giblin[a] and Michael D. Lumpkin[b]

Departments of [a] Pharmacology and [b] Physiology and Biophysics, Georgetown University School of Medicine, Washington D.C. 20007, U.S.A.

Nicotinic cholinergic agonists bind with high affinity to sites that have characteristics of nicotinic cholinergic receptors in mammalian brain. [^{3}H]Acetylcholine ([^{3}H]ACh), [^{3}H](−)nicotine, and [^{3}H]methylcarbamylcholine bind with an affinity of 3−10 nM and have all been used successfully to label these sites (Marks and Collins, 1982; Schwartz et al., 1982; Abood and Grassi, 1986; Boska and Quiron, 1987). The binding characteristics of [^{3}H]ACh and [^{3}H](−)nicotine indicate that these two ligands label the same site in brain (Clarke et al., 1985; Marks et al., 1986; Martino-Barrows and Kellar, 1987: Harfstrand et al., 1988) and the same is probably true for [^{3}H]methylcarbamylcholine (Boska and Quiron, 1987).

Although most binding studies have found only a single class of sites that display high affinity for nicotinic cholinergic agonists in brain (Marks et al., 1982; Schwartz et al., 1982; Lippiello and Fernandes, 1986; Abood and Grassi, 1986), studies with monoclonal antibodies (Whiting et al., 1987) and cDNA probes (Boulter et al., 1987; Wada et al., 1988) indicate there may be multiple subtypes of nicotinic cholinergic sites which are not clearly distinguishable by their affinities for presently available agonists. In addition, α-bungarotoxin, which labels muscle-type nicotinic receptors, also binds to sites in brain (Morley et al., 1977; Schmidt, 1977; Lukas and Bennett, 1979).

Although this toxin does not block nicotinic receptor-mediated transmission in many of the neuronal tissues that have been examined (Brown and Fumigalli, 1977; Patrick and Stallcup, 1977; Kouvelas et al., 1978), recent evidence suggests that a muscle-type receptor may also function in brain (de la Garza et al., 1987; Siegel and Lukas, 1988). Thus it is possible that nicotinic cholinergic neurotransmission in brain is mediated by a family of receptors which, though probably closely related genetically and structurally, still may be differentially regulated because of subtle differences in structure, subunit assembly, or other post-translational factors imposed upon them.

The nicotinic sites that bind agonists with high affinity are widely distributed in brain, and lesion studies suggest that in some rat brain regions the sites are located on dopamine, serotonin, and acetylcholine neurons (Schwartz et al., 1984; Clarke and Pert, 1985; Clarke et al., 1986). Consistent with a location on dopamine neurons or terminals, nicotine increases dopamine turnover in vivo (Lichtensteiger et al., 1982), and in vitro, nicotinic cholinergic agonists induce release of dopamine from striatal and nucleus accumbens tissues (Giorguieff-Chesselet, 1979; Sakurai, et al., 1982; Rapier et al., 1987, 1988; Rowell et al., 1987). This release has been demonstrated in synaptosomes as well as in slices, suggesting that the effects are mediated directly at dopamine axons rather than transsynaptically. Similarly, nicotinic agonist-induced release of acetylcholine has been found in mouse cerebral cortical synaptosomes (Rowell and Winkler, 1984) as well as in slices of rat cerebral cortex and hippocampus (Araujo et al., 1988).

α-Bungarotoxin does not block nicotine-induced release of dopamine from striatal synaptosomes, suggesting that the receptor mediating release is a neuronal-type rather than a muscle-type receptor (Rapier et al., 1987). On the other hand, histrionicotoxin, which blocks the ion channel of nicotinic receptors in muscle, does block release, suggesting that neuronal nicotinic receptors in brain function via an ion gating mechanism similar to that in muscle-type receptors (Rapier et al., 1987). This is consistent with electrophysiological evidence which indicates that activation of nicotinic receptors in brain increases membrane conductance to cations (Egan and North, 1986; McCormick and Prince, 1987), and it is also consistent with the concept that the assembly of the subunits of neuronal nicotinic receptors, like that of their counterparts in muscle, forms an integral membrane channel.

Regulation of nicotinic cholinergic agonist binding sites in brain

One of the interesting characteristics of nicotinic cholinergic agonist recognition sites in brain is their regulation by cholinergic drugs. Treatment of rats with cholinesterase inhibitors increases the concentration of acetylcholine in brain (Russell et al., 1981) and presumably prolongs its duration in the synapse, resulting in increased stimulation of cholinergic receptors. Following chronic treatment of rats with cholinesterase inhibitors, the density of nicotinic cholinergic recognition sites is decreased (Costa and Murphy, 1983; Schwartz and Kellar, 1983, 1985). This decrease presumably reflects down-regulation of nicotinic cholinergic receptors in response to increased exposure to acetylcholine, thus it could represent an important cellular adaptation mechanism to regulate nicotinic cholinergic transmission.

In contrast, chronic treatment of rats or mice with nicotine increases the density of these sites (Marks et al., 1983, 1985; Schwartz and Kellar, 1983, 1985; Nordberg et al., 1985). In homogenates of several grossly dissected brain areas, agonist binding sites are increased by 20 – 100%. Measurements of nicotinic cholinergic recognition sites by quantitative autoradiography confirm this and, further, indicate that chronic treatment with nicotine increases these sites to varying degrees throughout most of the rat forebrain (Fig. 1, Table I). Interestingly, increases in nicotinic cholinergic binding sites have been found also in autopsied brain samples from people who smoked cigarettes (Benwell et al., 1988), reinforcing the possibility of

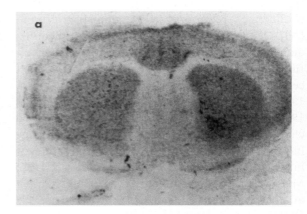

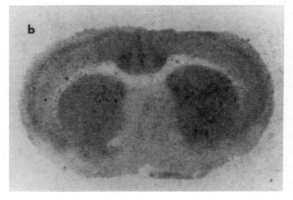

Fig. 1. Autoradiographs of nicotinic cholinergic recognition sites labelled by [3H]acetylcholine (20 nM) in brain sections (24 μm) through the striatum from (a) a control rat and (b) a rat treated with nicotine twice daily for 10 days. See Table I for further details.

a relationship between this change in binding sites and nicotine tolerance or addiction.

The increase in nicotinic cholinergic agonist recognition sites following treatment with nicotine, a presumed agonist, seems at first to be paradoxical. However, following its initial action to depolarize cells in autonomic ganglia and muscle, nicotine can produce depolarization blockade (Eccles, 1935; Thesleff, 1955); and possibly of more relevance to the increase in receptors, administra-

tion of nicotine can result in diminished receptor responses that extend beyond the period of depolarization blockade in frog ganglia (Ginsborg and Guerrero, 1964). Furthermore, in cells derived from a rat adrenal medullary tumor (PC12 cells), exposure to the nicotinic agonist carbachol results in receptor activation (as measured by ion flux) and rapid onset desensitization which is rapidly reversed when the agonist is removed. But in addition to the rapid desensitization phase, there is a

TABLE I

Quantitative autoradiographic analysis of the effect of repeated nicotine treatment on [^{3}H]acetylcholine binding to nicotinic cholinergic recognition sites in rat forebrain.

| Brain region | [^{3}H]Acetylcholine-bound (fmol/mg protein) | | % Increase |
	Control	Nicotine-treated	
Cerebral cortex			
Layer I	49 ± 7	81 ± 2	65*
Layer IV	77 ± 13	122 ± 9	58*
Layer V	58 ± 10	100 ± 7	72*
Caudate-putamen	64 ± 8	94 ± 2	47*
Substantia nigra	22 ± 5	47 ± 11	114*
Interpeduncular nucleus	369 ± 36	723 ± 79	96*
Medial habenular nucleus	344 ± 21	448 ± 54	30*
Thalamic nuclei			
Posterior	150 ± 15	217 ± 12	45*
Gelatinosus	192 ± 22	265 ± 17	38*
Ventromedial	148 ± 11	198 ± 15	34*
Laterodorsal	198 ± 22	258 ± 12	30*
Mediodorsal	161 ± 14	209 ± 16	30*
Anteriodorsal	233 ± 35	348 ± 27	49*
Ventrolateral	131 ± 15	179 ± 17	37*
Central	135 ± 15	153 ± 10	−
Lateroposterior	163 ± 30	197 ± 22	−
Anteroventral	278 ± 30	320 ± 18	−
Dorsolateral geniculate	219 ± 31	206 ± 36	−
Hypothalamic nuclei			
Anterior	25 ± 10	43 ± 1	72*
Ventromedial	18 ± 5	42 ± 3	133*
Arcuate	21 ± 7	33 ± 9	−

Rats received subcutaneous injections of nicotine bitartrate dihydrate (2 mg/kg, equivalent to 0.7 mg/kg free base) or vehicle (controls) twice daily for 10 days and were sacrificed 1 day after the last injection. Binding of [^{3}H]acetylcholine (20 nM) to nicotinic sites in frozen brain sections (24 μm), generation of autoradiograms, and quantitation of specific binding in brain areas were carried out as described by Rainbow et al. (1984). Data are the means ± S.E.M. from four rats in each group. Statistical comparisons were made with Student's t-test. * $p < 0.05$ compared to control.

212

slower onset, apparently irreversible loss of receptor responsivity (Robinson and McGee, 1985; Simasko et al., 1986). This latter process has been referred to as receptor inactivation (Simasko et al., 1986). Thus, although the initial action of nicotine is stimulatory, it is possible that its time-averaged effect is inhibitory, and therefore the predominant action of nicotine would be that of a functional antagonist. According to this hypothesis, during chronic exposure to nicotine, nicotinic cholinergic receptors would be inactivated, and this inactivation would trigger the cellular process that results in increased density of nicotinic receptors, much like denervation. The cellular process may be increased synthesis of receptors but as long as nicotine is administered, the receptors, though increased in number, would be inactivated and the cellular signals responsible for the increased number of receptors would be maintained.

It is not known why increased stimulation of nicotinic cholinergic receptors by acetylcholine during treatment with cholinesterase inhibitors does not lead to up-regulation of nicotinic cholinergic agonist binding sites by mechanisms similar to those induced by nicotine. Part of the explanation may be that the degree of nicotinic receptor stimulation during chronic inhibition of cholinesterase is less than during nicotine treatment. This could be due to several factors: (a) compensatory mechanisms evoked by excess acetylcholine, such as stimulation of muscarinic autoreceptors which inhibit release of acetylcholine; (b) a limited capacity of cholinergic axons to sustain release of transmitter; or (c) a more rapid diffusion of acetylcholine out of the synaptic space. Furthermore, and perhaps most important, it is unlikely that cholinesterases are inhibited much more than 90% even during chronic treatment with inhibitors (Russell et al., 1975, 1985; Schwartz and Kellar, 1983, 1985), and even a small amount of enzyme activity may be enough to limit the effects of acetylcholine on the receptors. Thus, cholinesterase inhibition may increase synaptic acetylcholine enough to down-regulate nicotinic cholinergic receptors, but not enough to inactivate them.

Regulation of nicotinic cholinergic receptor-mediated release of prolactin

Nicotine induces several neuroendocrine responses in rats (Fuxe et al., 1977; Andersson et al., 1983; Sharp and Beyer, 1986). Measurements of these responses do not necessarily provide direct or detailed information about molecular aspects of the receptor; but nevertheless, they can be useful for assessing an integrated response to nicotinic receptor stimulation. For example, a single injection of nicotine increases plasma prolactin levels in rats (Andersson et al., 1981, 1983; Sharp and Beyer, 1986). However, Sharp and Beyer (1986) showed that the prolactin response to a second injection of nicotine is markedly diminished for at least 6 h after the first injection. Furthermore, they found that the response to nicotine was still markedly diminished 14 h after the last of a series of chronic injections (Sharp et al., 1987).

We have measured nicotine-induced prolactin release in rats in which an indwelling jugular vein catheter allowed multiple sampling of blood over time after injection of nicotine. Relatively low doses of intravenously administered nicotine (10 – 250 μg/kg of the bitartrate dihydrate salt) in-

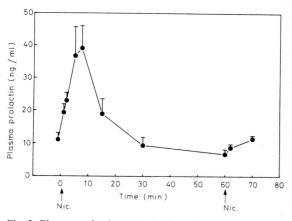

Fig. 2. Plasma prolactin concentrations after intravenous injections of nicotine. Nicotine bitartrate dihydrate (80 μg/kg) was injected at times 0 and 60 min through a jugular vein catheter inserted under anesthesia the day before the experiment. After each injection, approximately 200 μl of blood was collected at the times shown and prolactin was measured by radioimmunoassay. Values are the mean ± S.E.M. from five rats.

duce dose-dependent increases in plasma prolactin concentrations (Giblin et al., 1988). The plasma prolactin concentration peaks approximately 7 min after nicotine injection and returns to control levels within 30 min (Fig. 2). This response to nicotine is blocked by pretreatment with the

nicotinic antagonist mecamylamine (Fig. 3).

In agreement with the observations of Sharp and Beyer (1986), the prolactin response to a second injection of nicotine, administered 1 – 2 h after the first injection, is very much diminished or absent (Figs. 2 and 4). Additional studies indicate that the dose of the first injection of nicotine that reduces the prolactin response to a second injection by 50% (the ED_{50} for loss of response) is less than 25 μg/kg (Giblin et al., 1988). In contrast, prior administration of nicotine does not attenuate the prolactin response to morphine or to thyrotropin-releasing hormone (Fig. 5); and conversely, prior administration of morphine does not significantly attenuate the response to nicotine (or to thyrotropin-releasing hormone) given 1 – 2 h later (Fig. 5).

The time frame over which these measurements were made (1 – 2 h) suggested that a process more like receptor inactivation than the rapidly reversible desensitization is responsible for the relatively protracted loss of the prolactin response after a single injection of nicotine. To examine this point

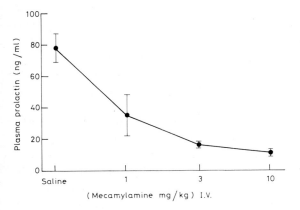

Fig. 3. Mecamylamine inhibition of nicotine-induced prolactin release. Rats were pretreated with saline or mecamylamine via a jugular vein catheter 60 min before injection of nicotine bitartrate dihydrate (75 μg/kg). Blood was collected for plasma prolactin measurements via the catheter. Values shown are 6 min after the nicotine injection and are the mean ± S.E.M. from 4 – 6 rats.

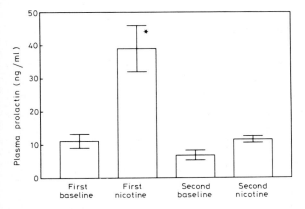

Fig. 4. Plasma prolactin concentrations following the first and second injection of nicotine. Rats received two intravenous injections of nicotine bitartrate dihydrate (80 μg/kg) 1 h apart. Blood samples were collected at 2 – 5 min intervals for 30 min after each injection. Values are the mean ± S.E.M. of the peak plasma prolactin concentrations from six rats. Baseline values were measured 2 min before each nicotine injection. * $p < 0.01$ compared to baselines and second nicotine injection.

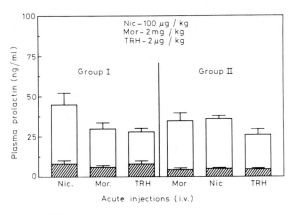

Fig. 5. Effect of nicotine and morphine on subsequent drug-induced release of prolactin. Two groups of rats received intravenous injections of nicotine bitartrate, morphine sulphate, and thyrotropin-releasing hormone (TRH) 75 min apart in the orders and doses shown. Blood samples were collected at 2 – 5 min intervals for 30 min after each injection. Plasma prolactin values are the peak concentrations after each injection and are the mean ± S.E.M. from six rats in each group. Shaded values are baseline prolactin concentrations measured 2 min before each injection.

214

further, we measured the prolactin response to nicotine (60 μg/kg, i.v.) at different time intervals after the rats had been treated for 10 days with nicotine bitartrate (2.5 mg/kg, equivalent to 0.8 mg/kg free base; subcutaneous injections twice

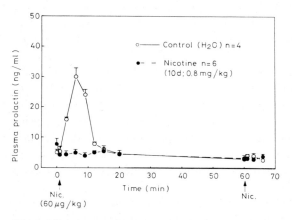

Fig. 6. Effect of chronic nicotine injections on nicotine-induced prolactin release. Rats received subcutaneous injections of water (controls) or nicotine bitartrate dihydrate (2.5 mg/kg, equivalent to 0.8 mg/kg free base) twice daily for 10 days. Six days after the last of the chronic injections, all rats received an intravenous injection of nicotine (60 μg/kg) at time 0 and again at 60 min, and blood was collected at the times shown. Values are the mean ± S.E.M. of the plasma prolactin concentrations measured by radioimmunoassay.

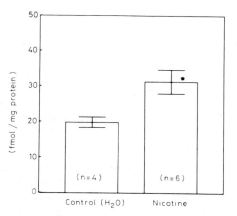

Fig. 7. [3H]Acetylcholine binding (10 mM) to nicotinic cholinergic recognition sites in the hypothalamus from rats 6 days after a 10-day treatment with nicotine or water. Two hours after the experiment shown in Fig. 6, the rats were killed and the hypothalamus was frozen until assayed for [3H]acetylcholine binding. * $p < 0.05$ compared to control.

daily). The prolactin response to the intravenous injection of nicotine was absent 2, 4, 6, and 8 days after the last of the chronic injections of nicotine. An example of this is shown in Fig. 6. Furthermore, at each of these times after the last of the chronic injections, the binding of [3H]ACh to nicotinic cholinergic recognition sites in the hypothalamus was still increased (Fig. 7). This long-term loss of responsiveness to nicotine after chronic treatment is consistent with the hypothesis that nicotine inactivates nicotinic cholinergic receptors, and that chronic inactivation of the receptors may be directly related to the nicotine-induced increase in the number of nicotinic cholinergic recognition sites.

The molecular mechanisms that underlie inactivation of nicotinic cholinergic receptors in brain are not known. But nicotinic receptors in muscle undergo phosphorylation (Miles et al., 1987; Smith et al., 1987), and the state of phosphorylation may influence the sensitivity of nicotinic receptors in both muscle (Eusebi et al., 1985) and ganglia (Downing and Role, 1987). It is thus possible that receptor phosphorylation is an important molecular mechanism involved in the regulation of nicotinic cholinergic receptors in brain as well.

Acknowledgements

Studies from our laboratory were supported by a grant from the Alzheimer's Disease and Related Disorders Association (II-86-100). Autoradiographic studies of nicotinic cholinergic recognition sites were carried out in collaboration with Drs. Thomas C. Rainbow, Rochelle D. Schwartz, and Barry B. Wolfe.

References

Abood, L.G. and Grassi, S. (1986) [3H]Methylcarbamylcholine, a new radioligand for studying brain nicotinic receptors. *Biochem. Pharmacol.*, 35: 4199–4202.

Andersson, K., Eneroth, P. and Agnati, L.F. (1981) Nicotine-induced increases of noradrenalin turnover in discrete noradrenalin nerve terminal systems of the hypothalamus and median eminence of the rat and their relationship to changes in the secretion of adenohypophyseal hormones. *Acta Physiol. Scand.*, 113: 227 – 231.

Andersson, K., Siegel, R., Fuxe, K. and Eneroth, P. (1983) Intravenous injections of nicotine induce very rapid and discrete reductions of hypothalamic catecholamine levels associated with increases in ACTH, vasopressin, and prolactin secretion. *Acta Physiol. Scand.*, 118: 35 – 40.

Araujo, D.M., Lapchak, P.A., Collier, B. and Quiron, R. (1988) Characterization of N-[^{3}H]methylcarbamylcholine binding sites and effect of N-methylcarbamylcholine on acetylcholine release in rat brain. *J. Neurochem.*, 51: 292 – 299.

Benwell, M.E.M., Balfour, D.J.K. and Anderson, J.M. (1988) Evidence that tobacco smoking increases the density of (−)[^{3}H]nicotine binding sites in human brain. *J. Neurochem.*, 50: 1243 – 1247.

Boska, P. and Quiron, R. (1987) [^{3}H]N-Methylcarbamylcholine, a new radioligand specific for nicotinic acetylcholine receptors in rat brain. *Eur. J. Pharmacol.*, 139: 323 – 333.

Boulter, J., Connolly, J., Deneris, E., Goldman, D., Heinemann, S. and Patrick, J. (1987) Functional expression of two neuronal nicotinic acetylcholine receptors from cDNA clones identifies a gene family. *Proc. Natl. Acad. Sci. U.S.A.*, 84: 7763 – 7767.

Brown, D.A. and Fumigalli, L. (1977) Dissociation of α-bungarotoxin binding and receptor block in the rat superior cervical ganglion. *Brain Res.*, 129: 165 – 168.

Clarke, P.B.S. and Pert, A. (1985) Autoradiographic evidence for nicotinic receptors on nigrostriatal and mesolimbic dopaminergic neurons. *Brain Res.*, 348: 355 – 358.

Clarke, P.B.S., Schwartz, R.D., Paul, S.M., Pert, C.B. and Pert, A. (1985) Nicotinic binding in rat brain: autoradiographic comparison of [^{3}H]acetylcholine, [^{3}H]nicotine, and [^{125}I]α-bungarotoxin. *J. Neurosci.*, 5: 1307 – 1315.

Clarke, P.B.S., Hamill, G.S., Nadi, N.S., Jacobowitz, D.M. and Pert, A. (1986) ^{3}H-Nicotine and ^{125}I-alpha-bungarotoxin labeled nicotinic receptors in the interpeduncular nucleus of rats.II. Effects of habenular deafferentation. *J. Comp. Neurol.*, 251: 407 – 413.

Costa, L.G. and Murphy, S.D. (1983) ^{3}H-Nicotine binding in rat brain: alteration after chronic acetylcholinesterase inhibition. *J. Pharmacol. Exp. Ther.*, 226: 392 – 397.

De la Garza, R., Bickford-Wilmer, P.C., Hoffer, B.J. and Freedman, R. (1987) Heterogenity of nicotine actions in the rat cerebellum: an in vivo electrophysiological study. *J. Pharmacol. Exp. Ther.*, 240: 689 – 695.

Downing, J.E.G. and Role, L.W. (1987) Activators of protein kinase C enhance acetylcholine receptor desensitization in sympathetic ganglion neurons. *Proc. Natl. Acad. Sci. U.S.A.*, 84: 7739 – 7743.

Eccles, J.C. (1935) The action potential of the superior cervical ganglion. *J. Physiol*, 85: 172 – 206.

Egan, T.M. and North, R.A. (1986) Actions of acetylcholine and nicotine on rat locus coerulus neurons in vitro. *Neuroscience*, 19: 565 – 571.

Eusebi, F., Molinaro, M. and Zani, B.M. (1985) Agents that activate protein kinase C reduce acetylcholine sensitivity in cultured myotubes. *J. Cell Biol.*, 100: 1339 – 1342.

Fuxe, K., Agnati, L., Eneroth, P., Gustafsson, J.-A, Hokfelt, T., Loftstrom, A., Skett, B. and Skett, P. (1977) The effect of nicotine on central catecholamine neurons and gonadotropin secretion. I. Studies in the male rat. *Med. Biol.*, 551: 148 – 157.

Giblin, B.A., Lumpkin, M.L. and Kellar, K.J. (1988) Repeated administration of nicotine results in long-term decrease in prolactin release by acute nicotine in rats. *Soc. Neurosci. Abstr.*, 14 (in press).

Ginsborg, B.L. and Guerrero, S. (1964) On the action of depolarizing drugs on sympathetic ganglia of the frog. *J. Physiol.*, 172: 189 – 206.

Giorguieff-Chesselet, M.F., Kemel, M.L., Wandscheer, D. and Glowinski, J. (1979) Regulation of dopamine release by presynaptic nicotinic receptors in rat striatal slices: effect of nicotine in a low concentration. *Life Sci.*, 25: 1257 – 1262.

Harfstrand, A., Adem, A., Fuxe, K., Agnati, L., Andersson, K. and Nordberg, A. (1988) Distribution of nicotinic cholinergic receptors in rat tel- and diencephalon: a quantitative receptor autoradiographical study using [^{3}H]acetylcholine, [^{125}I]α-bungarotoxin and [^{3}H]nicotine. *Acta Physiol. Scand.*, 132: 1 – 14.

Kouvelas, E.D., Dichter, M.A. and Greene, L.A. (1978) Sympathetic neurons develop receptors for α-bungarotoxin in vitro, but toxin does not block nicotinic receptors. *Brain Res.*, 154: 83 – 93.

Lichtensteiger, W., Hefti, F., Felix, D., Huwyler, T. Melamed, E. and Schlumpf, M. (1982) Stimulation of nigrostriatal dopamine neurons by nicotine. *Neuropharmacology*, 21: 963 – 968.

Lippiello, P.M. and Fernandes, K.G. (1986) The binding of l-[^{3}H]nicotine to a single class of high affinity sites in rat brain membranes. *Mol. Pharmacol.*, 29: 448 – 454.

Lukas, R.J. and Bennett, E.L. (1979) Agonist induced affinity alterations of central nervous system α-bungarotoxin receptor. *J. Neurochem.*, 33: 1151 – 1157.

Marks, M.J. and Collins, A.C. (1982) Characterization of nicotine binding in mouse brain and comparison with the binding of alpha-bungarotoxin and quinuclidinyl benzilate. *Mol. Pharmacol.*, 22: 554 – 564.

Marks, M.J., Burch, J.B. and Collins, A.C., (1983) Effects of chronic nicotine infusion on tolerance development and nicotinic receptors. *J. Pharmacol. Exp. Ther.*, 226: 817 – 825.

Marks, M.J., Stitzel, J.A., Romm, E. Wehnr, J.M. and Collins, A.C. (1986) Nicotinic binding sites in rat and mouse brain: comparison of acetylcholine nicotine and α-bungarotoxin. *Mol. Pharmacol.*, 30: 427 – 436.

Martino-Barrows, A. and Kellar, K.J. (1987) [^{3}H]Acetylcholine and [^{3}H](-)nicotine label the same recognition site in rat brain. *Mol. Pharmacol.,* 31: 169 – 174.

McCormick, D.A. and Prince, D.A. (1987) Acetylcholine causes rapid nicotinic excitation in the medial habenular nucleus of guinea pig in vitro. *J. Neurosci.,* 7: 742 – 752.

Miles, K., Anthony, D.T., Rubin, L.L., Greengard, P. and Huganir, R.L. (1987) Regulation of nicotinic acetylcholine receptor phosphorylation in rat myotubes by forskolin and cAMP. *Proc. Natl. Acad. Sci. U.S.A.,* 84: 6591 – 6595.

Morley, B.J., Lorden, J.F., Brown, G.B. and Kemp, G.E. (1977) Regional distribution of nicotinic acetylcholine receptor in rat brain. *Brain Res.,* 134: 161 – 166.

Nordberg, A., Wahlstrom, G., Arnelo, U. and Larsson, C. (1985) Effect of long-term nicotine treatment on [^{3}H]nicotine binding sites in the rat brain. *Drug Alcoh. Depend.,* 16: 9 – 17.

Patrick, J. and Stallcup, W.B. (1977) Alpha-bungarotoxin binding and cholinergic receptor function on a rat sympathetic nerve line. *J. Biol. Chem.,* 252: 8629 – 8633.

Rainbow, T.C., Schwartz, R.D., Parsons, B. and Kellar, K.J. (1984) Quantitative autoradiography of [^{3}H]acetylcholine binding sites in rat brain. *Neurosci. Lett.,* 50: 193 – 196.

Rapier, C., Wonnacott, S., Lunt, G.G. and Albuquerque, E.X. (1987) The neurotoxin histrionicotoxin interacts with the putative ion channel of the nicotinic acetylcholine receptors in the central nervous system. *FEBS Lett.,* 212: 292 – 296.

Rapier, C., Lunt, G.G. and Wonnacott, S. (1988) Stereoselective nicotine-induced release of dopamine from striatal synaptosomes: concentration dependence and repetitive stimulation. *J. Neurochem.,* 50: 1123 – 1130.

Robinson, D. and McGee, R. (1985) Agonist-induced regulation of the neuronal nicotinic acetylcholine receptor of PC12 cells. *Mol. Pharmacol.,* 27: 409 – 417.

Rowell, P.P., and Winkler, D.L. (1984) Nicotinic stimulation of [^{3}H]acetylcholine release from mouse cerebral cortical synaptosomes. *J. Neurochem.,* 43: 1593 – 1598.

Rowell, P.P., Carr, L.A. and Garner, A.C. (1987) Stimulation of [^{3}H]dopamine release by nicotine in rat nucleus accumbens. *J. Neurochem.,* 49: 1449 – 1454.

Russell, R.W., Overstreet, D.H., Cotman, C.W., Carson, V.G., Churchill, L., Dalglish, F.W. and Vasquez, B.J. (1975) Experimental tests of hypotheses about neurochemical mechanisms underlying behavioral tolerance to the anticholinesterase, DFP. *J. Pharmacol. Exp. Ther.,* 192: 73 – 85.

Russell, R.W., Carson, V.G., Booth, R.A. and Jenden, D.J. (1981) Mechanisms of tolerance to the anticholinesterase, DFP: acetylcholine levels and dynamics in rat brain. *Neuropharmacology,* 20: 1197 – 1201.

Russell, R.W., Booth, R.A., Jenden, D.J., Roch, M. and Rice, K.M. (1985) Changes in presynaptic release of acetylcholine during development of tolerance to the anticholinesterase, DFP. *J. Neurochem.,* 45: 293 – 299.

Sakurai, Y., Takano, Y., Kohjimoto, Y., Honda, K. and Kamiya, H. (1982) Enhancement of ^{3}H-dopamine release and its ^{3}H-metabolites in rat striatum by nicotinic drugs. *Brain Res.,* 242: 99 – 106.

Schmidt, J. (1977) Drug binding properties of an α-bungarotoxin binding component from rat brain. *Mol. Pharmacol.,* 13: 283 – 290.

Schwartz, R.D. and Kellar, K.J. (1983) Nicotinic cholinergic receptor binding sites in brain: regulation in vivo. *Science,* 220: 214 – 216.

Schwartz, R.D. and Kellar, K.J. (1985) In vivo regulation of [^{3}H]acetylcholine recognition sites in brain by nicotinic cholinergic drugs. *J. Neurochem.,* 45: 427 – 433.

Schwartz, R.D. McGee, R. and Kellar, K.J. (1982) Nicotinic cholinergic receptors labeled by [^{3}H]acetylcholine in rat brain. *Mol. Pharmacol.,* 22: 56 – 62.

Schwartz, R.D., Lehmann, J. and Kellar, K.J. (1984) Presynaptic nicotinic cholinergic receptors labeled by [^{3}H]acetylcholine on catecholamine and serotonin axons in brain. *J. Neurochem.,* 42: 1495 – 1498.

Sharp, B.M. and Beyer, H.S. (1986) Rapid desensitization of the acute stimulatory effects of nicotine on rat plasma adrenocorticotropin and prolactin. *J. Pharmacol. Exp. Ther.,* 238: 486 – 491.

Sharp, B.M., Beyer, H.S., Levine, A.S., Morley, J.E. and McAllen, K.M. (1987) Attenuation of plasma prolactin response to restraint stress after acute and chronic administration of nicotine to rats. *J. Pharmacol. Exp. Ther.,* 241: 438 – 442.

Siegel, H.N. and Lukas, R.J. (1988) Nicotinic agonists regulate α-bungarotoxin binding sites of TE671 human medulloblastoma cells. *J. Neurochem.,* 50: 1272 – 1278.

Simasko, S.M., Soares, J.R. and Weiland, G.A. (1986) Two components of carbamylcholine-induced loss of nicotinic acetylcholine receptor function in the neuronal cell line PC12. *Mol. Pharmacol.,* 30: 6 – 12.

Smith, M.M., Merlie, J.P. and Lawrence, J.C. (1987) Regulation of phosphorylation of nicotinic acetylcholine receptors in mouse BC3H1 myocytes. *Proc. Natl. Acad. Sci. U.S.A.,* 84: 6601 – 6605.

Thesleff, S. (1955) The mode of neuromuscular block caused by acetylcholine, nicotine, decamethonium and succinylcholine. *Acta Physiol. Scand.,* 34: 218 – 231.

Wada, K., Ballivet, M., Boulter, J. Connolly, J., Wada, E., Deneris, E.S., Swanson, L.W., Heinemann, S. and Patrick, J. (1988) Functional expression of a new pharmacological subtype of brain nicotinic acetylcholine receptor. *Science,* 240: 330 – 334.

Whiting, P., Liu, R., Morley, B.J. and Lindstrom, J.M. (1987) Structurally different neuronal nicotinic acetylcholine receptor subtypes purified and characterized using monoclonal antibodies. *J. Neurosci.,* 7: 4005 – 4012.

A. Nordberg, K. Fuxe, B. Holmstedt and A. Sundwall (Eds.)
Progress in Brain Research, Vol. 79

CHAPTER 21

Differential brain and peripheral nicotinic regulation of sympathoadrenal secretion

Glen R. Van Loon, Judith Kiritsy-Roy, Krystyna Pierzchala, Li Dong, Frances A. Bobbitt, Lesley Marson and Laura Brown

Veterans Administration Medical Center and Departments of Medicine and Pharmacology, University of Kentucky, Lexington, KY 40511, U.S.A.

Introduction

The effects of nicotine on regulation of sympathoadrenal catecholamine secretion have long been recognized (Stewart and Rogoff, 1919; Woods et al., 1956). Nicotine increases catecholamine secretion into the peripheral circulation by actions at sympathetic ganglia (Trendelenburg, 1965), sympathetic nerve terminals (Westfall and Brasted, 1972) and adrenal chromaffin cells (Westfall, 1965). In addition, nicotine crosses the blood-brain barrier readily (Schievelbein, 1984), and appears to stimulate central sympathetic outflow (Kubo, 1987; Sapru, 1987; Van Loon et al., 1987a), thus providing further mechanisms of increasing catecholamine secretion. However, these brain mechanisms for activation of peripheral sympathoadrenal secretion are poorly understood in comparison with knowledge of the peripheral mechanisms. Thus, we compared plasma catecholamine responses to systemic versus intracerebral administration of nicotine in conscious rats. To provide further support for the thesis of brain nicotinic stimulation of sympathetic outflow, thus of sympathoadrenal secretion, we also examined the plasma catecholamine responses to ICV administration of N′-methylnicotinium iodide, a quaternary analog of nicotine which does not cross the blood-brain barrier (Aceto et al., 1983).

Repeated administration of nicotine rapidly induces a state of tolerance in which pharmacologic responses to the drug are markedly reduced (Falkerborn et al., 1981; Henningfield, 1984), and this applies also with respect to stimulation of sympathoadrenal secretion (Holtz et al., 1984; Van Loon et al., 1987a). However, little is known regarding the differences in patterns and mechanisms of tolerance development to nicotine acting in brain versus periphery. This chapter describes some of our studies comparing the initial responses of plasma epinephrine and norepinephrine and the development of tolerance to intracerebroventricular (ICV) versus intra-arterial administration of nicotine in conscious rats. To determine whether alterations in sympathoadrenal sensitivity to nicotine might affect sympathoadrenal responses to other stimuli, we also compared the plasma catecholamine responses to a physiologic stress in two models in which the animals were tolerant to nicotine. We studied responses to restraint in conscious rats tolerant to either systemic or ICV nicotine and to hypoglycemia in nicotine-tolerant, pentobarbital-anesthetized dogs.

Finally, recent demonstration of endogenous opioid peptides colocalized with catecholamines in the adrenal medulla (Viveros et al., 1979) and in sympathetic nerve terminals (Schultzberg et al., 1978) has provided new impetus to study further the physiologic and pharmacologic regulation of sympathoadrenal secretion by nicotinic receptors. Thus, we describe the differences between in vivo adrenal secretory responses of not only catecholamines but also Met-enkephalin to initial and subsequent systemic administration of nicotine.

Acute sympathoadrenal catecholamine responses to systemic versus intracerebral nicotine in rats

Systemic administration of nicotine by intra-arterial injection in conscious, freely moving, chronically cannulated, adult, male, Sprague-Dawley rats produced dose-related and almost equal increases in plasma concentrations of epinephrine and norepinephrine. Plasma catecholamine concentrations became maximal within 2 min and returned to basal levels by about 10 min after drug injection. The nicotine-induced increases in plasma epinephrine and norepinephrine were almost completely inhibited by prior systemic administration of the ganglionic blocking agent, hexamethonium.

In contrast, ICV microinjection of nicotine produced a predominant plasma epinephrine response with only very small norepinephrine response. Nicotine administered ICV also produced maximal increases in plasma catecholamines within 2 min; plasma epinephrine increased in response to a dose of 30 nmol (5 μg), whereas plasma norepinephrine increased only in response to a dose of 120 nmol or greater. The ganglionic type of nicotinic receptor antagonist, hexamethonium, administered ICV also increased plasma epinephrine concentration; pretreatment with ICV hexamethonium significantly antagonized the plasma epinephrine response to ICV nicotine.

Sympathoadrenal catecholamine responses to intracerebral administration of a quaternary analog of nicotine

To support further the thesis that nicotine acts in brain to stimulate sympathetic outflow, thus to increase peripheral catecholamine secretion, we examined the plasma catecholamine responses to ICV administration of a nicotinic agonist which does not cross the blood-brain barrier. N'-methylnicotinium iodide, a quaternary analog of nicotine, administered ICV produced an increase in plasma epinephrine, comparable to that produced by equimolar nicotine (Fig. 1).

Development of tolerance of the sympathoadrenal catecholamine responses to systemic versus intracerebral nicotine in rats

The plasma catecholamine responses to nicotine administered intra-arterially repeatedly at various intervals were examined in order to establish the pattern of development of tolerance of sympathoadrenal responses to systemic nicotine. After daily intra-arterial injections of nicotine 0.1 mg/kg for one week, plasma epinephrine and norepinephrine responses to intra-arterial nicotine were similar to those seen in response to the initial intra-arterial injection of nicotine. Shortening of the interdose interval to 70 min produced small but statistically nonsignificant decreases in the plasma catecholamine responses to intra-arterial nicotine after the third dose of 0.1 mg/kg. Shortening the

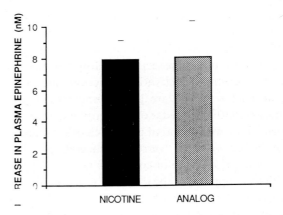

Fig. 1. Comparison of plasma epinephrine responses to ICV administration of an equimolar dose (120 nmole) of (−)-nicotine and its quaternary analog, N'-methylnicotinium iodide. Data represent the differences between the basal concentration and the peak response 2 min after drug administration.

interdose interval further to 30 min produced significant reductions in the plasma epinephrine and norepinephrine responses to intra-arterial nicotine. Increasing the dose to 0.5 mg/kg and shortening the interdose interval to 15 min produced even greater reductions in the subsequent plasma catecholamine responses to intra-arterial nicotine.

In marked contrast to this pattern of development of tolerance to systemic nicotine, sympathoadrenal stimulation by brain nicotine was much more sensitive to development of tolerance. Thirty minutes after nicotine 120 nmol ICV, the plasma epinephrine response to a second ICV injection of equal dose was decreased when compared to the response to initial nicotine administration. Even 24 h after nicotine 120 nmol ICV, the plasma epinephrine response to this dose was still reduced by 50%. Continued daily ICV injections of nicotine for one week resulted in almost complete loss of the plasma epinephrine response to ICV nicotine.

Sympathoadrenal responses to stress in nicotine-tolerant rats

To determine whether alterations in sympathoadrenal sensitivity to nicotine might affect sympathetic responses to other stimuli, we examined plasma catecholamine responses to restraint stress in rats rendered tolerant to intra-arterial or ICV nicotine. In rats who had been given nicotine 0.1 mg/kg intra-arterially either daily for one week or 3 times at 70 min intervals, and who did not show tolerance to nicotine, restraint in a wire mesh for 2 min produced the same increases in plasma epinephrine and norepinephrine as were seen in vehicle-treated rats. On the other hand, in rats given nicotine 3 times at 30 min intervals or 15 times at 15 min intervals, and who showed tachyphylaxis of their plasma catecholamine responses to nicotine, plasma epinephrine responses, and in the latter situation both plasma epinephrine and norepinephrine responses, to restraint stress were also blunted. In contrast to the situation with systemic nicotine, rats which were tolerant to ICV nicotine after daily administration for 5 days showed normal plasma epinephrine and norepinephrine responses to restraint stress.

Adrenomedullary catecholamine and Met-enkephalin responses to nicotine in anesthetized dogs: development of tolerance and cross-tolerance to hypoglycemia

In order to generalize some of these observations across species, we studied sympathoadrenal responses to nicotine in dogs. Also, to extend our findings beyond catecholamines, we were interested in comparing the catecholamine responses with nicotine-induced secretory responses of pro-enkephalin A-derived peptides which are colocalized with catecholamines in adrenal medulla and sympathetic tissues. To demonstrate adrenomedullary secretion directly, we measured epinephrine, norepinephrine and Met-enkephalin-like immunoreactivity (hereafter referred to as Met-enkephalin) in adrenal venous effluent of pentobarbital-anesthetized, adult, male, mongrel dogs. Nicotine was administered intravenously (i.v.) in two bolus doses of 40 μg/kg separated by 60 min, and an infusion of 5 μg/kg per min was administered during the interbolus interval. The initial i.v. bolus of nicotine increased adrenal epinephrine and norepinephrine secretion rates significantly when compared to either the basal secretion rates or to the responses to i.v. saline vehicle in control dogs (Fig. 2). In contrast, we observed marked blunting of the adrenal epinephrine and norepinephrine secretory responses to a second i.v. bolus of nicotine 60 min later.

To determine whether tachyphylaxis to nicotine was associated with altered adrenomedullary responsiveness to other stimuli, we examined the catecholamine responses to insulin-induced hypoglycemia during this period of tachyphylaxis to nicotine. In control dogs, i.v. insulin produced a marked fall in plasma glucose which was followed by a significant increase in adrenal secretion rate of epinephrine and norepinephrine. We ef-

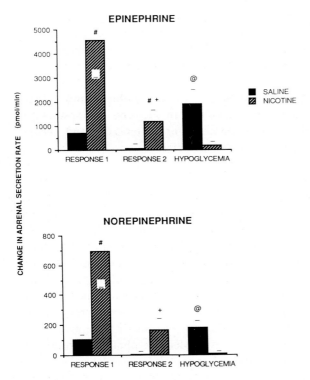

Fig. 2. Adrenal secretory responses of epinephrine and norepinephrine to intravenous bolus administration of nicotine in pentobarbital-anesthetized dogs. Tolerance to a subsequent i.v. bolus of nicotine and cross-tolerance to insulin-induced hypoglycemia. See text for further details. # represents a significant ($p < 0.05$) response from both basal secretion rate and from the response to saline vehicle. + represents a significant ($p < 0.05$) difference from the response to the initial i.v. bolus of nicotine. @ represents a significant ($p < 0.05$) overall response to hypoglycemia at 30, 45 and 60 min after insulin injection when compared to the rate before insulin injection.

fected a similar fall in plasma glucose in nicotine-treated dogs (using a slightly larger dose of insulin); however, in these dogs which showed tachyphylaxis to nicotine, hypoglycemia failed to elicit increases in plasma epinephrine and norepinephrine.

Of considerable interest, Met-enkephalin responses paralleled closely the catecholamine responses (Fig. 3). Adrenal secretion rate of Met-enkephalin increased markedly in response to the first i.v. bolus of nicotine, but did not increase significantly in response to a second i.v. bolus of nicotine 60 min later. Administration of saline vehicle in control dogs did not alter plasma concentrations of Met-enkephalin. Insulin-induced hypoglycemia produced a marked increase in adrenal secretion of Met-enkephalin in vehicle-treated dogs, but no Met-enkephalin response in dogs which showed tachyphylaxis to nicotine.

Discussion

Nicotine administered either systemically in conscious rats and anesthetized dogs or ICV in rats produces rapid increases in plasma catecholamine

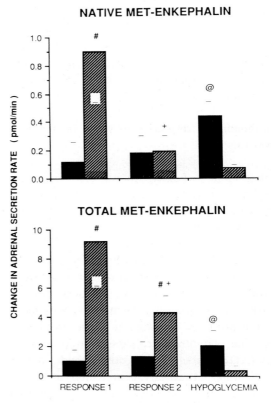

Fig. 3. Adrenal secretory responses of native and peptidase-hydrolyzable Met-enkephalin to intravenous bolus administration of nicotine in pentobarbital-anesthetized dogs. Tolerance to a subsequent i.v. bolus of nicotine and cross-tolerance to insulin-induced hypoglycemia. See text and legend to Fig. 2 for further details.

concentrations. Intra-arterial nicotine in rats increased plasma epinephrine and norepinephrine almost equally, similar to the findings of other recent reports in conscious rat (Dominiak et al., 1984) and dog (Holtz et al., 1984). These data are consistent with systemically administered nicotine acting at sympathetic ganglia and nerve terminals as well as at adrenal medulla. On the other hand, in pentobarbital-anesthetized dogs systemic nicotine produced a marked increase in plasma epinephrine but only a small increase in plasma norepinephrine (Van Loon et al., 1987a). This increase in plasma norepinephrine in response to systemic nicotine may have been small because pentobarbital anesthesia decreases the rate of norepinephrine spillover into plasma (Best et al., 1984).

In addition, however, nicotine crosses the blood-brain barrier readily and may also produce these increases in catecholamine concentrations in part by stimulating central sympathetic outflow. In support of the latter thesis, ICV administration of nicotine or its quaternary analog, N'-methylnicotinium iodide, which penetrates the blood-brain barrier only poorly (Aceto et al., 1983), increased plasma catecholamine concentrations. Since both nicotine and its analog when administered ICV produced much greater plasma increases in epinephrine than norepinephrine, it seems that they are preferentially activating central sympathetic outflow to adrenal medulla.

The effects of both systemic and ICV nicotine to increase plasma catecholamine concentrations were partially inhibited by hexamethonium, suggesting that at least some of the nicotinic receptors mediating these effects of nicotine in brain are similar to the peripheral ganglionic nicotinic receptor. Others have suggested that discriminative (Romano et al., 1981) and locomotor (Clarke and Kumar, 1983) effects of nicotine are mediated through a brain receptor with characteristics similar to the ganglionic receptor.

A single, acute ICV injection of nicotine rapidly (within 30 min − data not shown) produced a prolonged state of tolerance with respect to sym-

pathoadrenal secretion. Development of long-lasting tolerance has also been reported for other brain effects of nicotine, including nicotine-induced prostration-immobilization syndrome (Abood et al., 1979) and depression of locomotor activity (Stolerman et al., 1974). It should be noted that these nicotine-induced behavioral effects and subsequent tolerance resulted with systemic administration of nicotine, but almost certainly these effects were mediated in brain.

The pattern of tolerance development for catecholamine secretion induced by systemic nicotine is in marked contrast to that of intracerebral nicotine. Plasma epinephrine and norepinephrine responses produced by the seventh daily intra-arterial injection of nicotine were equal to those seen with the first intra-arterial injection. Tachyphylaxis of the plasma catecholamine responses to systemic nicotine was seen only when the treatment interval was decreased to 30 min. This time course for development of tolerance to systemic nicotine is similar to that reported for other effects including tremor (Holmstedt and Lundgren, 1967), respiratory stimulation and electroencephalographic and behavior arousal (Yamamoto and Domino, 1965). However, the duration of this tolerance has not been so carefully studied. From our data, it appears that tolerance for effects mediated in brain lasts considerably longer than tolerance for effects mediated in the periphery. Brain nicotinic receptors or postreceptor mechanisms appear to undergo desensitization to nicotine by a different mechanism from the peripheral mechanisms, but at present these have not been clarified. For example, at a time when we found tolerance to the catecholamine-releasing effect of a single ICV dose of nicotine, others were unable to detect changes in K_d or B_{max} for D,L-[^{3}H]nicotine binding in brain (Marks et al., 1983), whereas an increase or up-regulation in brain nicotine binding has been reported after chronic nicotine treatment (Marks et al., 1983; Schwartz and Kellar, 1983). However, changes in binding may be specific to certain brain nuclei and may be missed when only whole brain or even brain regions are examined.

222

Even within the sympathoadrenal system, tolerance to nicotine does not develop similarly for all parameters. Thus, tachyphylaxis to nicotine was not always noted for its pressor effects at a time when its catecholamine-releasing effects were significantly reduced (Van Loon et al., 1987a).

In spite of extensive studies showing release of enkephalin from adrenal chromaffin cells or from perfused adrenal glands in response to a variety of stimuli (Wilson et al., 1982; Chaminade et al., 1983), very few reports have documented similar effects in vivo (Hanbauer et al., 1982; Hexum et al., 1980). Previously, we have described the in vivo release of Met-enkephalin into plasma in rats in response to restraint stress, in dogs in response to hypoglycemia and in men in response to bicycle exercise (Van Loon et al., 1987b). In this chapter, we report the effect of i.v. nicotine to produce increases in adrenal venous secretory rates of epinephrine, norepinephrine and Met-enkephalin in pentobarbital-anesthetized dogs. A second i.v. injection of nicotine 60 min later produced much smaller increases in catecholamines and failed to increase adrenal secretion of Met-enkephalin, consistent with the rapid appearance of tachyphylaxis.

Since stress-induced sympathoadrenal catecholamine secretion involves mediation by nicotinic receptors in the periphery, it seemed reasonable to expect that tolerance of the sympathoadrenal responses to systemic nicotine might result also in inhibition of the plasma catecholamine responses to stress. We confirmed this hypothesis in two animal models. Conscious rats rendered tolerant to nicotine injected intra-arterially repeatedly at 30 min intervals demonstrated cross-tolerance of the plasma catecholamine responses to restraint stress, whereas rats given nicotine intra-arterially repeatedly at 24 h intervals showed normal plasma catecholamine responses to both nicotine and restraint stress. In addition, anesthetized dogs which showed tachyphylaxis of their adrenomedullary responses of catecholamines and Met-enkephalin to systemic nicotine also showed blunted catecholamine and Met-enkephalin responses to insulin-induced hypoglycemia. These data suggest that in the adrenal and peripheral sympathetic nerves, stress and nicotine activate sympathoadrenal secretion through common neuronal mechanisms.

In marked contrast, rats which were tolerant to repeated ICV administration of nicotine showed normal plasma catecholamine responses to restraint stress. Thus, it seems that stress and nicotine activate central sympathetic outflow through different neuronal pathways.

Similar differences in development of tolerance and cross-tolerance to nicotine acting in brain versus peripheral sites have been reported recently. Sensitization in rats of the ACTH-stimulating effect of nicotine which is thought to be mediated in brain was not associated with attenuation of the ACTH response to restraint stress, whereas tolerance to the prolactin-releasing effect of nicotine which is thought to be mediated peripherally in pituitary was associated with blunting of the prolactin response to restraint stress (Matta et al., 1987).

The relevance of these data to humans smoking tobacco and experiencing a variety of environmental stresses remains to be clarified.

Acknowledgements

These studies were supported by the University of Kentucky Tobacco and Health Research Institute and the Veterans Administration. We thank Amy Hess for typing the manuscript.

References

Abood, L.G., Lowy, K. and Booth, H. (1979) Acute and chronic effects of nicotine in rats and evidence for a non-cholinergic site of action. In N.A. Krasnegor (Ed.), *Cigarette Smoking as a Dependance Process*, NIDA Research Monograph No. 23, DHEW Pub. No. 79–800, pp. 136–149.

Aceto, M.D., Awaya, A., Martin, B.R. and May, E.L. (1983) Antinociceptive action of nicotine and its methiodide derivatives in mice and rats. *Br. J. Pharmacol.*, 79: 869–876.

Best, J.D., Taborsky, G.J., Jr., Flatness, D.E. and Halter, J.B. (1984) Effect of pentobarbital anesthesia on plasma norepinephrine kinetics in dogs. *Endocrinology,* 115: 853 – 857.

Chaminade, M., Foutz, A.S. and Rossier, J. (1983) Co-release of enkephalins and precursors with catecholamines by the perfused cat adrenal in-situ. *Life Sci.,* 33 Supp 1: 21 – 24.

Clarke, P.B.S. and Kumar, R. (1983) Characterization of the locomotor stimulant action of nicotine in tolerant rats. *Br. J. Pharmacol.,* 80: 587 – 594.

Dominiak, P., Kees, F. and Grobecker, H. (1984) Changes in peripheral and central catecholaminergic and serotoninergic neurons of rats after acute and subacute administration of nicotine. *Klin. Wochenschr.,* 62 (Suppl II): 76 – 80.

Faulkerborn, Y., Larsson, C. and Nordberg, A. (1981) Chronic nicotine exposure in rat: a behavioral and biochemical study of tolerance. *Drug Alcoh. Depend.,* 8: 51 – 60.

Hanbauer, I., Kelly, G.D., Saiani, L. and Yang H.-Y.T. (1982) [Met-5]enkephalin-like peptides of the adrenal medulla: release by nerve stimulation and functional implications. *Peptides,* 3: 469 – 473.

Henningfield, J.E. (1984) Behavioral pharmacology of cigarette smoking. *Adv. Behav. Pharmacol., 4:* 131 – 210.

Hexum, T.D., Hanbauer, I., Govoni, S., Yang, H.-Y.T. and Costa E. (1980) Secretion of enkephalin-like peptides from canine adrenal gland following splanchnic nerve stimulation. *Neuropeptides* 1: 137 – 142.

Holmstedt, B. and Lundgren, G. (1967) Arecoline, nicotine, and related compounds. Tremorigenic activity and effect upon brain acetylcholine. *Ann. N.Y. Acad. Sci.,* 142: 126 – 142.

Holtz, J., Sommer, O. and Bassenge, E. (1984) Development of specific tolerance to nicotine infusions in dogs on chronic nicotine treatment. *Klin. Wochenschr.,* 623 (Suppl II): 51 – 57.

Kubo, T. (1987) Central nicotinic regulation of arterial blood pressure. In W.R. Martin, G.R. Van Loon, E.T. Iwamoto and L. Davis (Eds.), *Tobacco Smoking and Nicotine,* Plenum Press, New York, pp. 277 – 286.

Marks, M.J., Burch, J.B. and Collins, A.C. (1983) Effects of chronic nicotine infusion on tolerance development and nicotinic receptors. *J. Pharmacol. Exp. Ther.,* 226: 817 – 825.

Matta, S.G., Beyer, H.S., McAllen, K.M. and Sharp, B.M. (1987) Nicotine elevates rat plasma ACTH by a central mechanism. *J. Pharmacol. Exp. Ther.,* 243: 217 – 226.

Romano, C., Goldstein, A. and Jewell, N.P. (1981) Characterization of the receptor mediating the discriminative stimulus. *Psychopharmacology,* 74: 310 – 315.

Sapru, H.N. (1987) Control of blood pressure by muscarinic and nicotine receptors in the ventrolateral medulla. In W.R. Martin, G.R. Van Loon, E.T. Iwamoto and L. Davis (Eds.). *Tobacco Smoking and Nicotine,* Plenum Press, New York, pp. 287 – 300.

Schievelbein, H. (1984) Nicotine resorption and fate. In D.J.K. Balfour (Ed.), *International Encyclopedia of Pharmacology and Therapeutics, Nicotine and the Smoking Habit,* Section 114, Pergamon Press, New York, pp. 1 – 15.

Schultzberg, M., Lundberg, J.M., Hokfelt, T., Terenius, L., Brandt, J., Elde, R.P. and Goldstein, M. (1978) Enkephalin-like immunoreactivity in gland cells and nerve terminals of the adrenal medulla. *Neuroscience,* 3: 1169 – 1186.

Schwartz, R.D. and Kellar, K.J. (1983) Nicotinic cholinergic receptor binding sites in the brain: regulation in vivo. *Science,* 220: 214 – 216.

Stewart, G.N. and Rogoff, J.M. (1919) The action of drugs on the output of epinephrine from the adrenals. III. Nicotine. *J. Pharmacol. Exp. Ther.,* 13: 183 – 241.

Stolerman, I.P., Bunker, P. and Jarvik, M.E. (1974) Nicotine tolerance in rats: role of dose and dose interval. *Psychopharmacology,* 34: 317 – 324.

Van Loon, G.R., Kiritsy-Roy, J.A., Brown, L.V. and Bobbitt, F.A. (1987a) Nicotinic regulation of sympathoadrenal catecholamine secretion: cross-tolerance to stress. In W.R. Martin, G.R. Van Loon, E.T. Iwamoto and L. Davies (Eds.), *Tobacco Smoking and Nicotine,* Plenum Press, New York, pp. 263 – 276.

Van Loon, G.R., Pierzchala, K., Brown, L.V. and Bobbitt, F.A. (1987b) Plasma free and cryptic Met-enkephalin responses to stress. In S. Tucek (Ed.) *Synaptic Transmitters and Receptors,* Academia, Prague, pp. 355 – 369.

Viveros, O.H., Diliberto, E.J., Hazum, E. and Chang, K.J. (1979) Opiate-like materials in the adrenal medulla: evidence for storage and secretion with catecholamines. *Mol. Pharmacol.,* 16: 1101 – 1108.

Westfall, T.C. (1965) Tobacco alkaloids and the release of catecholamines. In U.S. Von Euler (Ed.), *Tobacco Alkaloids and Related Compounds,* Pergamon, New York, pp. 179 – 203.

Westfall, T.C. and Brasted, M. (1972) The mechanism of action of nicotine on adrenergic neurons in the perfused guinea pig heart. *J. Pharmacol. Exp. Ther.,* 182: 409 – 418.

Wilson, S.P., Chang, K.J. and Viveros, O.H. (1982) Proportional secretion of opioid peptides and catecholamines from adrenal chromaffin cells in culture. *J. Neurosci.,* 2: 1150 – 1156.

Woods, L.F., Richardson, J.A., Richardson, A.K. and Bozeman, R.F. (1956) Plasma concentrations of epinephrine and arterenol following the actions of various agents on the adrenals. *J. Pharmacol. Exp. Ther.,* 116: 351 – 355.

Trendelenburg, U. (1965) Nicotine and the peripheral autonomic nervous system. In U.S. Von Euler (Ed.), *Tobacco Alkaloids and Related Compounds,* Pergamon, New York, pp. 167 – 177.

Yamamoto, K. and Domino, E.F. (1965) Nicotine-induced EEG and behavioral arousal. *Int. J. Neuropharmacol.,* 4: 359 – 373.

Behavioral Action of Nicotine
in Animal Studies

A. Nordberg, K. Fuxe, B. Holmstedt and A. Sundwall (Eds.)
Progress in Brain Research, Vol. 79
© 1989 Elsevier Science Publishers B.V. (Biomedical Division)

CHAPTER 22

Primary cholinergic and indirect dopaminergic mediation of behavioural effects of nicotine

I.P. Stolerman and C. Reavill

Department of Psychiatry, Institute of Psychiatry, De Crespigny Park, London SE5 8AF, U.K.

Introduction

The main aim of the work to be described has been to characterise the CNS mechanisms that mediate behavioural effects produced by "smoking doses" of nicotine. Primarily, the studies have focussed upon the high-affinity binding site for tritiated nicotine because this site is stereoselective and saturable; it is widely accepted that a binding site should have both of these characteristics if it is also a functional receptor. Secondarily, and more recently, the studies have begun to look beyond the nicotine recognition site, with its cholinoceptive pharmacology, to possible interactions with other neurotransmitter systems, notably dopamine and (more speculatively) to the putative adenosine system.

The behavioural basis for much of the work has been the discriminative stimulus, or cue, property of nicotine. Animals, like people, are very good at detecting the effects of psychoactive drugs in their bodies. Indeed, the ability of animals to discriminate (detect) drugs may be as good as their ability to detect conventional environmental stimuli such as sounds, smells and lights (Schuster and Balster, 1977). Many different drugs can be studied by such methods, which have become very widely used in behavioural pharmacology. In utilising these methods, we owe a great debt to the work of Rosecrans and his colleagues, who carried out many of the difficult early experiments pioneering the ap-

plication of this methodology to nicotine (overview in Rosecrans and Chance, 1977). Full technical descriptions of the present procedure may be found elsewhere (Pratt et al., 1983; Stolerman et al., 1984).

The method entails training rats to press bars to obtain food rewards; two bars are present in the test apparatus and it is the presence or absence of nicotine that tells the rat which bar to press to obtain its food. Thus, an animal can solve the task most efficiently, and thus obtain the maximum food for least effort, by identifying whether or not it has received nicotine. For a particular animal, it may be the left bar which always produces food when it is drugged and the right bar that produces food in the absence of drug. A rat exposed to such conditions, imposed in random order over 40 or more daily training sessions, starts pressing the left bar every time it receives nicotine and, in contrast, starts pressing the right bar every time it receives the saline vehicle. In any one experiment, half of the animals are trained as just described, whereas the remaining animals have the drug-bar pairings reversed, to minimise effects due to side preferences. The dose of nicotine used at the beginning of training is 0.4 mg/kg, administered subcutaneously 15 min before a 15 min training session. The dose is reduced progressively to 0.1 mg/kg as training proceeds, and all data are obtained from rats that detect this small dose of nicotine reliably.

Once rats have been trained as described above

to meet an individual criterion of 80% accurate bar-pressing, they are used in experiments where responses to different doses of nicotine, and to different agonists and antagonists, are assessed in twice-weekly test sessions. Within each experiment, tests take place in random order, with training continuing on the intervening days to maintain the baseline of discrimination performance. No food is given during tests, regardless of which bar is pressed, to ensure that only the drugs, and not any other cues, determine which bar is pressed. The reasons for using this complex procedure are multiple, of which the most important are that trained rats (1) respond to nicotine at doses well within the "smoking range" and yield clear, very reproducible dose-response curves and (2) respond to novel drugs in such a way as to indicate that the procedure has a very high degree of pharmacological specificity and because (3) the main index is one of response *choice* rather than response *rate*; this last feature minimises possible confounding effects of nonspecific stimulation or depression. In effect, the trained rats are used in a manner analogous to classical pharmacological

preparations and they provide a psychopharmacological litmus test for nicotine-like activity in the CNS. While other behavioural approaches may allow more scope for analysing the nature of nicotine's effects in psychological terms, they rarely combine the above attributes which are particularly advantageous for studies oriented to neuropharmacological mechanisms.

Nicotinic agonists

The dose-response function for nicotine itself is considered first. In the course of several years' research, dose-response curves for nicotine have been obtained at different times in eight different groups of rats. Figure 1 illustrates the similarity between the general form of these curves. The working range of doses under the conditions used is from a threshold dose of a little above 0.01 mg/kg, up to a dose of 0.1 mg/kg. The ED_{50} value, defined as the dose of drug that produces 50% of responding on the nicotine-appropriate bar, varies over an approximately 2-fold range between the different groups of rats. Figure 1 also

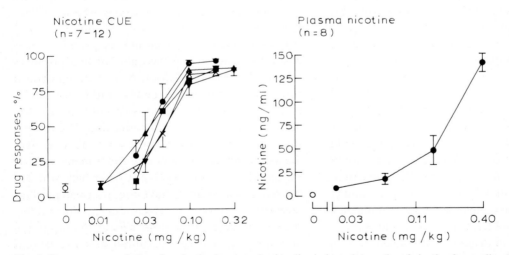

Fig. 1. Dose-response relations for nicotine in rats trained to discriminate 0.1 mg/kg of nicotine from saline (left section), and plasma concentrations of nicotine in untrained rats (right section). Discriminative stimulus (cue) effects of nicotine are shown by several curves, each representing results for a different group of 7 – 12 rats in terms of percentages of responding on the drug-appropriate bar 15 – 20 min after injection (means ± S.E.M.). Corresponding mean plasma concentrations of nicotine were determined from samples obtained 15 min after administration of nicotine (n = 6 – 7). Points above zero on abscissae show results after administration of saline.

shows the peak plasma concentrations of nicotine produced by these doses of nicotine in rats of the same sex, strain and age. The 0.1 mg/kg dose of nicotine used to maintain the discrimination is associated with a concentration of nicotine around 35 ng/ml, which is in the middle of the 5 – 70 ng/ml range for cigarette smokers who inhale (Russell et al., 1980). It may be noted that a 0.4 mg/kg dose of nicotine, widely used as a "standard" dose in many behavioural experiments, produced a plasma concentration of 145 ng/ml which was rather above the upper limit of the smoking range. At this time, plasma concentrations seem to provide the soundest basis for intra-species comparisons; when concentrations of nicotine in human brain have been measured, it may be appropriate to change the criterion in the animal studies. The data shown support the position that the nicotine cue can provide a robust behavioural assay for quantifying a behavioural effect of "smoking doses" of nicotine.

Several nicotinic agonists have been tested in rats trained to discriminate nicotine (0.1 mg/kg) from saline. Typically, these compounds have increased the percentage of drug-appropriate responding in a dose-related manner, to about the same maximum value as nicotine itself. The nicotinic agonists functioning in this way have been (+)-nicotine (the enantiomeric form of the natural material), anabasine, cytisine, (−)-nornicotine, (+)-nornicotine, isoarecolone and *N*-(3-pyridylmethyl)pyrrolidine (Stolerman et al., 1984; Garcha et al., 1986; Reavill et al., 1987; Reavill, Stolerman and Waters, unpublished data). These results confirm and extend reports from other laboratories (Chance et al., 1978; Meltzer et al., 1980; Romano et al., 1981).

It may be noted that all these nicotinic agonists potently inhibit the binding of tritiated nicotine in vitro. Furthermore, the potency of the agents as inhibitors of such binding is highly correlated with their potency in producing the nicotine cue in vivo (Reavill et al., 1988). This correlation is illustrated in Fig. 2, and the Spearman rank correlation coefficient has a value of 0.96 when the results for

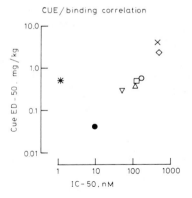

Fig. 2. Graphical representation of the correlation of IC_{50} values for displacement of (−)-[³H]nicotine in competition binding studies with ED_{50} values for producing the nicotine cue. All binding was determined on membranes prepared from homogenates of whole rat brains. Low-affinity binding could not be detected under the conditions used; high affinity binding was 80 – 90% specific as determined in presence of 10 μM unlabelled (−)-nicotine. Cytisine (*); (−)-nicotine ●; (+)-nicotine □; (−)-nornicotine ○; (+)-nornicotine △; anabasine ◇; isoarecolone ×; *N*-(3-pyridylmethyl)pyrrolidine ▽. With data for cytisine discarded, $r_s = 0.96$ ($p < 0.01$). (Modified from Reavill et al., 1988).

cytisine are omitted ($p < 0.01$). Figure 2 shows that cytisine is much less potent in the behavioural experiments than would be expected from the binding data where, of the compounds tested to date, it was the only one to be more potent than (−)-nicotine itself. The reason for omitting cytisine from the calculated correlation is evidence that it penetrates poorly into the brain (Romano et al., 1981); further data supporting this interpretation are presented below (section on Role of dopaminergic mechanisms). It must also be emphasised that cytisine is fully effective in producing the nicotine cue, and it is only potency that it lacks.

The preceding results are interpreted as evidence for mediation of the nicotine cue via the high affinity binding site for tritiated nicotine. This progress has to be weighed against the findings with lobeline, a drug that now constitutes the sole exception to the correlation between binding potency and behavioural efficacy. Lobeline has a long history of unsuccessful use as a nicotine substitute in smoking cessation and it also fails to produce nicotine-like or nicotine antagonist effects in discrimination experiments, either when administered systemically or by direct intracerebroventricular injections (Romano et al., 1981; Reavill and Stolerman, unpublished results). Lobeline is, however, a potent inhibitor of nicotine binding in vitro. The puzzle of lobeline must be solved if the high-affinity site is to be established conclusively as the main receptor involved in the nicotine cue. Certain analogues of nicotine are also inactive behaviourally, but these compounds have failed to inhibit nicotine binding (Romano et al., 1981).

The other main problem has been the frequently noted discrepancy between the nanomolar concentrations of nicotine needed to occupy the high-affinity site and the micromolar concentrations at which nicotine often seems to act. The K_D defines the concentration of nicotine needed to occupy half of the binding sites and the values we obtain, like those from several other groups, average about 6 nM (Clarke et al., 1985; Lippiello and Fernandes, 1986; Reavill et al., 1988). The ED_{50} dose of nicotine in the discrimination procedure is typically about 0.04 mg/kg; from the relationship between dose and plasma concentrations (Fig. 1), it can be calculated that the ED_{50} is associated with a peak plasma concentration of 86 nM (14 ng/ml). It seems unlikely that brain concentrations of nicotine after subcutaneous injection are much larger than plasma concentrations. Thus, tissue concentrations are some 14 times larger than the K_D. Although this is a significant discrepancy, it is smaller than has sometimes been supposed. Possible explanations have been discussed before and they will not be repeated here.

Non-nicotinic compounds

The results with nicotinic agonists must be interpreted by comparing them with non-nicotinic compounds drawn from a wide range of pharmacological classes; only if such drugs fail to produce the nicotine cue can the procedure be considered to be specific for nicotinic activity. Such drugs also fail

to potently inhibit binding of tritiated nicotine. Considerable numbers of compounds have now been tested and they have not increased drug-appropriate responding above 20 – 30% of total responses; in contrast, nicotine and its active analogues have typically produced scores around 80 – 90%.

The non-nicotinic compounds tested to date have included, but are not limited to, representatives from the following classes: muscarinic-cholinergic agonists and antagonists; 5-HT agonists and antagonists; benzodiazepine tranquillizers and inverse agonists; adrenoceptor agonists; opioids; neuroleptics; dissociative anaesthetics; convulsants (review in Stolerman, 1987; Reavill and Stolerman, unpublished data). Several nicotinic antagonists (e.g. mecamylamine) have also failed to produce nicotine-like activity. The only known exception to this outcome has been certain agents that share activity as dopaminergic agonists (section below on Role of dopaminergic mechanisms). These results generally suggest that the nicotine cue has a very high degree of pharmacological specificity for nicotine-like activity, an essential requirement if the procedure is to be useful as an assay.

Nicotinic antagonists

It has been known for many years that most, if not all, behavioural effects of nicotine are blocked by those ganglion-blocking drugs that penetrate well into the CNS. Examples include changes in locomotor activity, the nicotine cue and the rewarding effect of nicotine that maintains self-administration behaviour (Stitzer et al., 1970; Chance et al., 1978; Spealman and Goldberg, 1982; Clarke and Kumar, 1983). The drugs shown to be effective when administered systemically have been mecamylamine and pempidine; hexamethonium and chlorisondamine, which penetrate to the brain poorly, have not been found to block the nicotine cue or many of its other behavioural effects. These data suggest strongly that the nicotine cue is mediated primarily through CNS mechanisms. However, all these agents, in common with most other nicotinic antagonists, are very poor inhibitors of nicotine binding in vitro (Romano and Goldstein, 1980; Marks and Collins, 1982; Reavill et al., 1988); this suggests that they do not act at the recognition site for nicotine but should be considered to be non-competitive antagonists that act at some other site in the nicotinic receptor complex. Studies of these drugs in peripheral systems indicate that they block the ion channels linked to the nicotine recognition site (e.g. Lingle, 1983).

Behavioural evidence in support of the non-competitive hypothesis has come from experiments where the block of responses to nicotine by mecamylamine has not been reversed by increasing the dose of nicotine (Spealman et al., 1981; Spealman and Goldberg, 1982; Stolerman et al., 1983). A recent experiment of this type is shown in Fig. 3. In rats trained to discriminate nicotine (0.1 mg/kg s.c.) from saline, mecamylamine shifted the dose-response curve for nicotine downwards in a dose-related manner. A 0.75 mg/kg dose of mecamylamine was just sufficient to fully block the response to nicotine (0.1 mg/kg). Increasing

the dose of nicotine 20 times, to 2.0 mg/kg, did not significantly reverse the block. Figure 3 also shows that in the absence of mecamylamine, a 10-fold increase in the dose of nicotine is sufficient to span the full range of its dose-response curve. These findings replicate and extend those of Stolerman et al. (1983), in which an 8-fold increase in the dose of nicotine did not reverse the block in rats trained to discriminate 0.4 mg/kg of nicotine. In similar experiments on the discriminative effects of drugs from other classes, such as muscarinic-cholinergic agonists, classical competitive antagonists (e.g. atropine) have produced parallel shifts to the right in the dose-response curves for agonists (Meltzer and Rosecrans, 1981).

The reliability of mecamylamine and pempidine as antagonists of the behavioural effects of nicotine has led to suggestions that the CNS receptor mediating these effects is very similar to the peripheral, ganglionic receptor. For this proposition to be accepted, two assumptions must be justified. First, the effects of the antagonists must be specific for nicotinic agonists and second, these effects should be characteristic for ganglion-blocking drugs as a class and should not be found solely with particular drugs.

The first assumption seems to be justified. In various behavioural experiments, the ganglion-blocking drugs mecamylamine, pempidine and chlorisondamine have been found not to block the effects of the following non-nicotinic drugs: amphetamine; apomorphine; arecoline; cocaine; methysergide; midazolam; pentobarbitone; phencyclidine; quipazine; scopolamine (Clarke, 1984; Reavill and Stolerman, 1987; review in Stolerman, 1987). These data have been obtained from experiments including but not limited to those on discrimination of non-nicotinic drugs, and it appears that despite frequent assertions to the contrary, the evidence supports the view that some ganglion-blocking drugs have a very considerable degree of selectivity.

The second assumption is more questionable since most behavioural experiments have been limited to mecamylamine and pempidine, the only

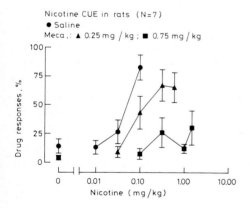

Fig. 3. Dose-response relations for nicotine in rats trained to discriminate 0.1 mg/kg of nicotine from saline (n = 7). Results show responses to nicotine 30 min after administration of the centrally active nicotine antagonist mecamylamine in doses of 0.25 mg/kg (▲) or 0.75 mg/kg (■), and in the absence of mecamylamine (●). Points above 0 on the abscissa show responses to saline only (●) and to mecamylamine in the absence of nicotine (■).

widely available ganglion-blocking drugs that penetrate well into the CNS. Quaternary agents such as hexamethonium do not penetrate well into the CNS and have been employed mainly as controls. However, in recent experiments a number of these drugs have been tested by direct intracerebroventricular microinjection (Stolerman et al., 1983; Kumar et al., 1987; Reavill and Stolerman, unpublished data). Figure 4 shows the maximum degrees of block of the nicotine cue produced by various nicotinic antagonists. It can be seen that only two drugs, the ganglion-blockers mecamylamine and chlorisondamine, produced complete blockade. The response to chlorisondamine was unusual in that it had an extremely

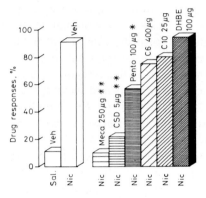

Fig. 4. Effects of different nicotinic antagonists on the nicotine cue in rats (n = 5-9). Open columns show pooled responses to saline and nicotine in the absence of any antagonist (vehicle pretreament; Veh). Hatched columns show responses to nicotine after pretreatment with 5 nicotine "antagonists" administered intracerebroventricularly in the doses shown in volumes of 1 – 4 μl (mecamylamine, meca; chlorisondamine, CSD; pentolinium, pento; hexamethonium, C6; decamethonium, C10; dihydro-β-erythroidine, DHBE). Doses of pentolinium, hexamethonium, decamethonium and dihydro-β-erythroidine were the largest that could be tested without producing convulsions, and results therefore show maximum extent of blockade produced by each drug in experiments carried out to date. Analyses of variance or t-tests were carried out to compare response to nicotine after drug pretreatment with that after vehicle pretreatment in the same rats (*, $p < 0.05$; **, $p < 0.01$). Diagram produced by combining data from Stolerman et al. (1983), Kumar et al. (1987) and Reavill and Stolerman (unpublished).

long duration of action. The ganglion-blocker pentolinium produced only partial block whereas the prototypical ganglion-blocker hexamethonium was wholly inactive. The neuromuscular blocker decamethonium and the non-selective blocker dihydro-β-erythroidine (which acts at both ganglionic and neuromuscular sites) were also inactive. All drugs were tested up to the threshold for producing myoclonic twitches; larger doses were convulsant and could not be tested.

From the preceding set of results, it appears that the ability to block the nicotine cue may not be a general attribute of ganglion-blockers. This suggests that although the CNS receptors resemble the ganglionic receptors, they may not be identical. Figure 4 is a compilation of results from several experiments carried out at different times with some variations in procedures, so the findings should be taken as a guide for further work rather than as a definitive comparison between the different drugs. It is curious that of these drugs, dihydro-β-erythroidine and decamethonium were the most potent inhibitors of nicotine binding, but the least effective as antagonists.

Role of dopaminergic mechanisms

Electrophysiological and biochemical experiments have suggested that nicotine may facilitate dopaminergic neurotransmission in the CNS. In particular, there is evidence that there are nicotinic receptors located presynaptically on terminals of both the nigral and the mesolimbic dopamine systems (review by Pert and Clarke, 1987). Experiments have been carried out to determine the functional significance of these observations with regard to behavioural effects of nicotine.

In experiments on the nicotine cue in rats, only three non-nicotinic drugs have produced substantial degrees of nicotine-appropriate responding (Rosecrans and Chance, 1977; Stolerman et al., 1984; Reavill and Stolerman, 1987). Thus, amphetamine, apomorphine and SKF 38393 have increased scores to 50 – 70%. This is a weaker response than that to direct nicotinic agonists, but it clearly differs from the total insensitivity to the drugs listed above (see section on Non-nicotinic drugs). The most obvious property that is common to amphetamine, apomorphine and SKF 38393 is the ability to act directly or indirectly as dopamine agonists. The response to SKF 38393 is the most interesting since this agent is thought to act selectively on the D-1 subtype of dopamine receptor. When used as a training drug to establish a discrimination, SKF 38393 produces a cue that can be blocked by a selective D-1 antagonist but not by a selective D-2 antagonist (Kamien et al., 1987). Thus, the partial generalization from nicotine to SKF-38393 suggests a possible role for D-1 receptors in mediating the nicotine cue.

The effects of different neuroleptics on the nicotine cue have also been compared (Reavill and Stolerman, 1987). Sch 23390 and haloperidol significantly weakened discrimination of nicotine. Sch 23390 has selectivity for D-1 receptors whereas haloperidol can act upon both D-1 and D-2 sites (Hyttel, 1978; Fujita et al., 1980). In contrast, pimozide and droperidol, two neuroleptics with selectivity for D-2 receptors, did not significantly affect discrimination of nicotine. These observations are compatible with earlier data of Rosecrans suggesting that lesions of dopamine systems affect the discriminative response to nicotine in subtle ways (Rosecrans and Chance, 1977). The studies with both the dopamine agonists and the antagonists direct attention to the D-1 subtype of receptor, but additional work with selective agonists and antagonists is needed.

Additional evidence supporting interactions of nicotine with dopamine systems have come from studies of rotational behaviour in rats with unilateral 6-hydroxydopamine lesions in the substantia nigra. Such rats provide a test system for dopaminergic activity. Drugs such as apomorphine that act as direct dopamine agonists produce rotation away from the side of the lesion. In contrast, drugs such as amphetamine that act as indirect dopamine agonists produce rotation directed towards the lesion. Figure 5 shows that nicotine can produce a modest degree of rotation towards

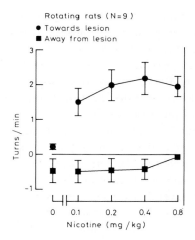

Fig. 5. Dose-response relations for rotational behaviour produced by nicotine in rats with unilateral 6-hydroxydopamine lesions of the nigrostriatal dopamine pathway. The numbers of complete rotations towards and away from the side of the lesion are shown separately. Nicotine and saline were administered subcutaneously. Each point represents the mean ± S.E.M. from eight, 1-min periods of observation from 5 to 60 min after injections.

the side of the lesion (Reavill and Stolerman, 1987). This effect was obtained with a 0.1 mg/kg "smoking dose" of nicotine but it never reached the same magnitude as the response to amphetamine. These observations support findings from other groups (Kaakkola, 1981; Lapin et al., 1987).

In order to test in which brain regions nicotine acts to produce behavioural effects, experiments have been carried out mainly with locomotor activity (Reavill and Stolerman, 1988). Rats were injected with nicotine (0.4 mg/kg s.c.) repeatedly, and placed in photocell activity cages to establish a stable baseline of response to nicotine administered systemically. Guide cannulae aimed at selected brain regions were then implanted bilaterally under brief anaesthesia. After recovery, the locomotor response to nicotine was compared after systemic and intracerebral injections of nicotine or of the nicotinic agonist cytisine. In most experiments, the guide cannulae were aimed at brain regions containing large numbers of binding sites for tritiated nicotine.

Systemic administration of nicotine has produced clear dose-related increases in locomotion from one part of the test cage to another, with the maximum response seen with a dose of 0.4 mg/kg (Clarke and Kumar, 1983). Figure 6 shows that bilateral microinjections of either nicotine (8 μg) or cytisine (3.2 μg) into the ventral tegmental area of the mesolimbic dopamine system increased locomotor activity (Reavill and Stolerman, 1988). This effect was not as marked as the activity produced by systemically administered nicotine, but it has been reproduced in several groups of rats. Similar microinjections of nicotine or cytisine into the nucleus accumbens, the substantia nigra, the dorsal hippocampus or the ventrobasal thalamus have not increased locomotor activity.

It is notable that when injections are made directly into the ventral tegmental area, cytisine is rather more potent than nicotine in eliciting locomotor activity. This observation is consistent with the potency of cytisine as an inhibitor of high-affinity binding of nicotine. It also provides further support for the view that the low potency of cytisine when it is administered by systemic injec-

tion may be attributed to its poor penetration into the CNS (see section on Nicotinic agonists). The regional specificity of the effects makes it unlikely that the drugs act primarily by diffusion to regions of the brain remote from the site of injection. The findings are consistent with recent studies from other groups that have also utilised microinjections of nicotine (Pert and Clarke, 1987). Since the ventral tegmental area contains the cell bodies from which the mesolimbic dopamine system projects, it seems possible that activation of this system contributes to the locomotor activity that nicotine produces. This hypothesis is supported by evidence that lesioning this system with the neurotoxin 6-hydroxydopamine weakens the response to nicotine (Clarke et al., 1988).

Is there a role for an adenosine system?

It has been suggested that adenosine may modulate release of several neurotransmitters. In order to evaluate in a preliminary way the possible role of adenosine mechanisms in responses to nicotine, the effects of the centrally active adenosine analogue *l*-

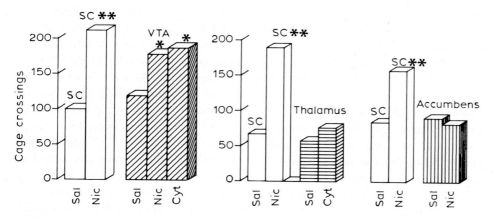

Fig. 6. Locomotor activity in rats assessed by numbers of crossings from one side to another in photocell cages. Open columns show activity after subcutaneous administration of saline or nicotine (0.4 mg/kg) in three groups of rats. Hatched columns show activity after intracerebral microinjections of nicotine (8 μg bilaterally) or of the nicotinic agonist cystine (3.2 μg bilaterally) through guide cannulae aimed at the the ventral tegmental area (VTA), ventrobasal thalamus or nucleus accumbens. Analyses of variance or *t*-tests were carried out to compare responses to microinjections of drugs with those to saline (0.5 μl bilaterally) in the same rats (*, $p < 0.05$; **, $p < 0.01$). (From Reavill and Stolerman, 1988.)

235

phenylisopropyladenosine (*l*-PIA) on the nicotine cue were determined. There was no change in the response to the 0.1 mg/kg training dose of nicotine after administration of 0.01 or 0.02 mg/kg doses of *l*-PIA. However, *l*-PIA (0.01 mg/kg) reduced the response to a submaximal, 0.032 mg/kg dose of nicotine from $62.3 \pm 16.0\%$ to $10.1 \pm 5.9\%$ ($p < 0.01$). Such an effect could be produced by *l*-PIA inhibiting the release of a transmitter that mediates the response to nicotine. However, it is a weak effect that needs further study, and it remains to be determined whether it is pharmacologically specific.

It has been known for a long time that smoking often takes place at times when beverages containing caffeine are consumed, such as after meals (Istvan and Matarazzo, 1984). It has been generally assumed that such simultaneous use of the two psychoactive agents is engendered by social factors rather than by a pharmacological interaction, although this view has been questioned recently (White, 1988). Perhaps it is not just coincidental that caffeine may act as an adenosine antagonist (Snyder et al., 1981); the possibility that it interacts in an additive or synergistic manner with nicotine would be consistent with the opposite effect of *l*-PIA described above. Definitive studies on such possible interactions have not yet been reported.

Conclusions

The behavioural effects of nicotine, as assessed in studies of the nicotine cue, appear to be mediated by a high-affinity binding site for tritiated nicotine. This binding is inhibited by cholinergic agonists, and it appears to mark the same site as the binding of acetylcholine in the presence of atropine (Clarke et al., 1985; Martino-Barrows and Kellar, 1987). When these data are taken together with knowledge of the block of the behavioural effects by certain nicotinic-cholinergic antagonists, it follows that nicotine acts upon receptors with a cholinoceptive pharmacology. There are still unresolved problems in the evidence for this conclusion, such as the findings with

lobeline and the discrepancy between K_D and ED_{50} values, as mentioned above (section on Nicotinic agonists). There is a clear need for a greater range of nicotinic compounds to be made available; the lack of such agents has inhibited the development of the area, in contrast to fields such as opioid research where the synthesis of many compounds with different profiles of action has spurred the identification of receptor subtypes.

Whether all the high-affinity sites are targets for endogenous acetylcholine is less certain due to the generally rather indirect evidence for a transmitter role of acetylcholine at these sites. Behavioural studies in which the anticholinesterase physostigmine has failed to mimic or potentiate the nicotine cue also call into doubt the presence of functional cholinergic terminals in proximity to all of the sites at which nicotine acts (Meltzer and Rosecrans, 1988). In resolving this matter, it may be necessary to distinguish between receptors that are sensitive to nicotinic-cholinergic agents (i.e. cholinoceptors) and those that may have a similar profile of pharmacological response and which are also innervated by cholinergic nerves.

The cholinoceptive site through which the nicotine cue is mediated appears to resemble the peripheral, ganglionic type of receptor but it is probably not identical with it since ganglion-blockers do not uniformly block the cue. This conclusion may have a parallel in implications from the study of CNS nicotinic receptors by means of molecular genetics, and this work is discussed elsewhere (Patrick et al., this volume). There is no evidence to date that behavioural effects of nicotine are mediated by a receptor that resembles the receptor at the neuromuscular junction more closely than that in ganglia. It is notable that in mice, ganglion-blockers have reliably been found to block convulsions produced by nicotine, whereas neuromuscular blockers have been inactive (Caulfield and Higgins, 1983).

Three lines of evidence, derived from studies of the nicotine cue, locomotor activation and rotation, all support a role for dopamine in mediating behavioural responses to nicotine. A working

236

hypothesis specifies the involvement of D-1 receptors in the mesolimbic dopamine system, but evidence in support of this conclusion is still tentative. A role for the nigrostriatal dopamine system cannot be ruled out, although biochemical and electrophysiological studies concur with the behavioural observations in suggesting a greater sensitivity of the mesolimbic system to nicotine (Imperato et al., 1986; Mereu et al., 1987). The modest degrees of substitution and antagonism in the work on the nicotine cue suggest that the role of dopamine may only be minor in the overall pattern of effects that nicotine produces. However, it may be particularly significant in view of evidence, cited before in this connection (e.g. Pert and Clarke, 1987), that several other drugs produce their addictive, rewarding effects through the same system.

There must be neurotransmitters other than dopamine involved in mediating the response to nicotine. The psychopharmacology of nicotine is clearly different from that of classical dopamine agonists despite the fact that certain points of resemblance can be identified. The evidence of the nicotine cue indicates clearly that the dopaminergic link can be no more than one component in a complex response. Whether the other components involve the adenosine system, as in the speculation above, other monoamines, the excitatory amino acids, or peptides will provide the bases for many further developments in this field.

Acknowledgements

We thank the Medical Research Council for financial support and Merck Sharpe and Dohme for donations of dihydro-β-erythroidine.

References

Caulfield, M.P. and Higgins, G.A. (1983) Mediation of nicotine-induced convulsions by central nicotinic receptors of the "C$_6$" type. Neuropharmacology, 22: 347–351.

Chance, W.T., Kallman, M.D., Rosecrans, J.A. and Spencer, R.M. (1978) A comparison of nicotine and structurally related compounds as discriminative stimuli. Br. J. Pharmacol., 63: 609–616.

Clarke, P.B.S. (1984) Chronic central nicotinic blockade after a single administration of the bisquaternary ganglion-blocking drug chlorisondamine. Br. J. Pharmacol., 83: 527–535.

Clarke, P.B.S. and Kumar, R (1983) Characterization of the locomotor stimulant action of nicotine in tolerant rats. Br. J. Pharmacol., 80: 587–594.

Clarke, P.B.S., Schwartz, R.D., Paul, S.M., Pert, C.B. and Pert, A. (1985) Nicotinic binding in rat brain: autoradiographic comparison of [^{3}H]acetylcholine, [^{3}H]nicotine and [^{125}I]-α-bungarotoxin. J. Neurosci., 5: 1307–1315.

Clarke P.B.S., Fu, D.S., Jakubovic, A. and Fibiger, H.C. (1988) Evidence that mesolimbic dopamine activation underlies the locomotor stimulant action of nicotine in rats. J. Pharmacol. Exp. Ther., 246: 701–708.

Fujita, N., Saito, K., Hirata, A., Iwatsubo, K., Noguchi, Y. and Yoshida, H. (1980) Effects of dopaminergic agonists and antagonists on [^{3}H]apomorphine binding to striatal membranes: sulpiride lack of interactions with positive cooperative [^{3}H]apomorphine binding. Brain Res., 199: 335–342.

Garcha, H.S., Goldberg, S.R., Reavill, C., Risner, M.E. and Stolerman, I.P. (1986) Behavioural effects of the optical isomers of nicotine and nornicotine, and of cotinine in rats. Br. J. Pharmacol., 88: 298P.

Hyttel, J. (1978) Effects of neuroleptics on ^{3}H-haloperidol and ^{3}H-cis(Z)-flupenthixol binding and on adenylate cyclase activity in vitro. Life Sci., 23: 551–556.

Imperato, A., Mulas, A. and Di Chiara, G. (1986) Nicotine preferentially stimulates dopamine release in the limbic system of freely moving rats. Eur. J. Pharmacol., 132: 337–338.

Istvan, J. and Matarazzo, J.D. (1984) Tobacco, alcohol, and caffeine use: a review of their interrelationships. Psychol. Bull., 95: 301–326.

Kaakkola, S. (1981) Effect of nicotinic and muscarinic drugs on amphetamine- and apomorphine-induced circling behaviour in rats. Acta Pharmacol. Toxicol., 48: 162–167.

Kamien, J.B., Goldberg, L.I. and Woolverton, W.L. (1987) Discriminative stimulus properties of D1 and D2 dopamine agonists in rats. J. Pharmacol. Exp. Ther., 242: 804–811.

Kumar, R., Reavill, C. and Stolerman, I.P. (1987) Nicotine cue in rats: effects of central administration of ganglion-blocking drugs. Br. J. Pharmacol., 90: 239–246.

Lapin, E.P., Maker, H.S., Sershen, H., Hurd, Y. and Lajtha, A. (1987) Dopamine-like action of nicotine: lack of tolerance and reverse tolerance. Brain Res., 407: 351–363.

Lingle, C. (1983) Blockade of cholinergic channels by chlorisondamine on a crustacean muscle. J. Physiol., 339: 395–417.

Lippiello, P.M. and Fernandes, K.G. (1986) The binding of L-[^{3}H]nicotine to a single class of high affinity sites in rat brain membranes. Mol. Pharmacol., 29: 448–454.

Marks, M.J. and Collins, A.C. (1982) Characterization of nicotine binding in mouse brain and comparison with binding of α-bungarotoxin and quinuclidinyl benzilate. *Mol. Pharmacol.*, 22: 554–564.

Martino-Barrows, A.M. and Kellar, K.J. (1987) [^{3}H]Acetylcholine and [^{3}H](-)-nicotine label the same recognition site in rat brain. *Mol. Pharmacol.*, 31: 169–174.

Meltzer, L.T. and Rosecrans, J.A. (1981) Discriminative stimulus properties of arecoline: a new approach for studying central muscarinic receptors. *Psychopharmacology*, 75: 383–387.

Meltzer, L.T. and Rosecrans, J.A. (1988) Nicotine and arecoline as discriminative stimuli: involvement of a noncholinergic mechanism for nicotine. *Pharmacol. Biochem. Behav.*, 29: 587–593.

Meltzer, L.T., Rosecrans, J.A., Aceto, M.D. and Harris, L.S. (1980) Discriminative stimulus properties of the optical isomers of nicotine. *Psychopharmacology*, 68: 283–286.

Mereu, G., Yoon, K-W.P., Boi, V., Gessa, G.L., Naes, L. and Westfall, T.C. (1987) Preferential stimulation of ventral tegmental area dopaminergic neurons by nicotine. *Eur. J. Pharmacol.*, 141: 395–399.

Pert, A. and Clarke, P.B.S. (1987) Nicotinic modulation of dopaminergic neurotransmission: functional implications. In W.R. Martin, G.R. Van Loon, E.T. Iwamoto and L. Davis, (Eds.), *Tobacco Smoking and Nicotine*, Plenum Press, New York, pp. 169–189.

Pratt, J.A., Stolerman, I.P., Garcha, H.S., Giardini, V. and Feyerabend, C. (1983) Discriminative stimulus properties of nicotine; further evidence for mediation at a cholinergic receptor. *Psychopharmacology*, 81: 54–60.

Reavill, C. and Stolerman, I.P. (1987) Interaction of nicotine with dopaminergic mechanisms assessed through drug discrimination and rotational behaviour in rats. *J. Psychopharmacol.*, 1: 264–273.

Reavill, C. and Stolerman, I.P. (1988) Locomotor activity in rats after intracerebral administration of nicotine. *J. Psychopharmacol. Abstr.* 2.

Reavill, C., Spivak, C.E., Stolerman, I.P. and Waters, J.A. (1987) Isoarecolone can inhibit nicotine binding and produce nicotine-like discriminative stimulus effects in rats. *Neuropharmacology*, 26: 789–792.

Reavill, C., Jenner, P., Kumar, R. and Stolerman, I.P. (1988) High-affinity binding of [^{3}H](-)-nicotine to rat brain membranes and its inhibition by analogues of nicotine. *Neuropharmacology*, 27: 235–241.

Romano, C. and Goldstein, A. (1980) Stereospecific nicotine receptors on rat brain membranes. *Science*, 210: 647–650.

Romano, C., Goldstein, A. and Jewell, N.P. (1981) Characterization of the receptor mediating the nicotine discriminative stimulus. *Psychopharmacology*, 74: 310–315.

Rosecrans, J.A. and Chance, W.T. (1977) Cholinergic and noncholinergic aspects of the discriminative stimulus properties of nicotine. In H. Lal (Ed.) *Discriminative Stimulus Properties of Drugs*, Plenum Press, New York, pp. 155–185.

Russell, M.A.H., Jarvis, M., Iyer, R. and Feyerabend, C. (1980) Relation of nicotine yield of cigarettes to blood nicotine concentrations in smokers. *Br. Med. J.*, 280: 972–976.

Schuster, C.R. and Balster, R.L. (1977) The discriminative stimulus properties of drugs. In T. Thompson and P.B. Dews (Eds.), *Advances in Behavioral Pharmacology*, Vol. 1, Academic Press, New York, pp. 85–138.

Snyder, S.H., Katims, J.J., Annau, Z., Bruns, R.F. and Daly, J.W. (1981) Adenosine receptors and behavioral actions of methylxanthines. *Proc. Natl. Acad. Sci. U.S.A.*, 78: 3260–3264.

Spealman, R.D. and Goldberg, S.R. (1982) Maintenance of schedule-controlled behavior by intravenous injections of nicotine in squirrel monkeys. *J. Pharmacol. Exp. Ther.*, 223: 402–408.

Spealman, R.D., Goldberg, S.R. and Gardner, M.L. (1981) Behavioral effects of nicotine: schedule-controlled responding by squirrel monkeys. *J. Pharmacol. Exp. Ther.*, 216: 484–491.

Stitzer, M., Morrison, J. and Domino, E.F. (1970) Effects of nicotine on fixed-interval behavior and their modification by cholinergic antagonists. *J. Pharmacol. Exp. Ther.*, 171: 166–177.

Stolerman, I.P. (1987) Psychopharmacology of nicotine: stimulus effects and receptor mechanisms. In L.L. Iversen, S.D. Iversen and S.H. Snyder (Eds.), *Handbook of Psychopharmacology, Vol. 19*, Plenum, New York, pp. 421–465.

Stolerman, I.P., Pratt, J.A., Garcha, H.S., Giardini, V. and Kumar, R. (1983) Nicotine cue in rats analysed with drugs acting on cholinergic and 5-hydroxytryptamine mechanisms. *Neuropharmacology*, 22: 1029–1037.

Stolerman, I.P., Garcha, H.S., Pratt, J.A. and Kumar, R. (1984) Role of training dose in discrimination of nicotine and related compounds by rats. *Psychopharmacology*, 84: 413–419.

White, J.M. (1988) Behavioral interactions between nicotine and caffeine. *Pharmacol. Biochem. Behav.*, 29: 63–66.

A. Nordberg, K. Fuxe, B. Holmstedt and A. Sundwall (Eds.)
Progress in Brain Research, Vol. 79
© 1989 Elsevier Science Publishers B.V. (Biomedical Division)

CHAPTER 23

Nicotine-induced tolerance and dependence in rats and mice: studies involving schedule-controlled behavior*

J.A. Rosecrans, C.A. Stimler, J.S. Hendry** and L.T. Meltzer***

Department of Pharmacology and Toxicology, Virginia Commonwealth University, School of Basic Health Sciences, MCV Station Box 613, Richmond, VA 23298, U.S.A.

Introduction

The literature is replete with evidence that nicotine can induce tolerance in a variety of behavioral tasks including spontaneous activity, Y-maze behavior and conditioned behaviors (For reviews see: Hendry and Rosecrans, 1982a; Clarke, 1987; Collins et al., 1988; Swedberg et al., 1988). In relation to observations of apparent physical dependence, however, there are relatively few reports of a measurable and reproducible withdrawal syndrome resulting from chronic nicotine administration in the rodent. Except for a couple of select examples (Balfour, 1982; Hendry and Rosecrans, 1986; Nordberg et al., 1985; Harris et al., 1987; Grunberg, 1988), an observable withdrawal syndrome resulting from the cessation of chronic nicotine treatment has been difficult to characterize.

Research conducted in these laboratories has focused on evaluating nicotine-induced tolerance and dependence using schedule-controlled behavior as a major dependent variable. The reasons for using this approach are 2-fold. First, operant behavior has been observed to be extremely sensitive to drug-induced disruption. Second, with an operant model, one has the capacity to evaluate both pharmacological (pharmacodynamic) and behavioral mechanisms of tolerance development. Contingent upon the time of drug administration in relation to the evaluation of behavior, either before (pre) or after (post) the behavioral session, one can evaluate whether tolerance was the result of an animal's ability to adapt behaviorally to the drug (pre), or whether tolerance was the result of some pharmacological mechanism (post) resulting from either an alteration in rate of drug metabolism or changes in some receptor parameter (Chen, 1968; Carlton and Wolgin, 1971). In the case of nicotine, recent research suggests that tolerance may be related to the up-regulation of nicotinic receptors suggesting a desensitization of the nicotinic receptor (Keller et al., 1987; Collins et al., 1988; Kisir et al., 1988). Additionally, it was also felt that these procedures might be useful in characterizing a withdrawal syndrome resulting from nicotine-induced physical dependence in the rat. Our thinking involved the idea that animals dependent on nicotine would be less able to bar-press due to the discomfort induced by physical withdrawal from nicotine if, indeed, it did occur. Furthermore, the possibility that animals might be behaviorally dependent because of a drug-induced state dependency can also be evaluated via the

* This research was supported by NIDA grant DA04002-2.
** *Present address:* 1271 Karen Dr. Radnor, PA 19087, U.S.A.
*** *Present address:* Parke-Davis Pharmaceutical Division, Warner-Lambert Co., 2800 Plymouth Rd., Ann Arbor, MI 48105, U.S.A.

utilization of such approaches. Such behavioral mechanisms could be sorted out by whether, either/or both treatment groups, pre- or post-nicotine administration, exhibited a disruption of behavior following the cessation of chronic nicotine injections.

The major objectives of the research conducted were as follows: (1) to characterize the development of tolerance to nicotine's disruptive effects on schedule-controlled behavior; (2) to determine whether tolerance to nicotine-induced behavioral disruption is a function of either/or both behavioral (pre-administration) or pharmacological mechanisms (post-administration); (3) to determine if tolerance development to nicotine involves an acetylcholine-sensitive cholinergic neuron; and (4) to evaluate operant procedures useful to detecting nicotine-induced dependence as measured by behavioral disruption after the cessation of chronic nicotine administration.

A characterization of nicotine-induced tolerance in rats and mice

Experimental approaches and methods

Male adult rats (Sprague-Dawley) and mice (ICR) were used throughout these studies. Experimental subjects were housed singly and maintained on a 12 h light-dark cycle (0600 – 1800 h). Animals were food deprived to 85% of their growing body weights and water was available ad libitum. Experimental subjects were initially shaped to bar-press for food in standard operant chambers (Hendry and Rosecrans, 1982; Meltzer and Rosecrans, 1982) in which behavioral control was maintained by either Colbourne solid state programming equipment or by Commodore 64 computers.

Once experimental subjects learned to bar-press, schedule control was implemented during 30 min daily behavioral sessions. Rats were trained to bar-press for sweetened milk using a 15 s variable interval (VI-15) schedule of reinforcement while mice were trained to bar-press for the same reinforcement using a fixed ratio 25 (FR-25) schedule. Drug studies were initiated once behavior was stabilized.

The initital objective of this research was to determine a dose-response relationship between nicotine and the disruption of behavior. These studies yielded approximate ED_{80} disruptive subcutaneous (s.c.) doses of 0.8 mg/kg (free base) of nicotine in the rat and 1.2 mg/kg (free base) of nicotine in the mouse. These experiments were conducted during 30 min behavioral sessions in which rats were placed in their respective operant chamber immediately after the s.c. administration of a dose of nicotine. Dose-response studies were conducted randomly amongst all animals using a counter-balanced design; nicotine doses were administered every 3rd session to prevent the development of premature tolerance and to prevent a cumulative dose effect.

Once dose-response behavioral disruption relationships were established, rats ($n = 10$) and mice ($n = 20$) were placed in two treatment groups in which baseline response rates were equivalent. Each treatment group of each species was designated as either a pre- or post-session nicotine treatment group and chronic nicotine treatment initiated. In this regimen the pre-session nicotine group received 0.8 (rats) or 1.2 (mice) mg/kg s.c. immediately prior to being exposed to a daily 30 min behavioral session. The post-session nicotine groups received nicotine within 10 – 30 min after being exposed to each daily 30 min behavioral session. Tests of tolerance in both treatment groups were conducted when the response rates in the nicotine pre-session treatment group had returned to baseline rates of behavior; behavioral rates observed prior to the administration of the chronic nicotine regimen. Dose-response behavioral disruption relationships were redetermined in a manner similar to that in the phase prior to chronic nicotine dosing, that is, both pre- and post-session nicotine groups were administered nicotine immediately prior to a behavioral session. The only difference was that chronic dosing was maintained at 0.8 or 1.2 mg/kg respectively for rats and mice between the administration of challenge doses of nicotine. Dose-response curves generated were compared statistically using the appropriate ANOVA.

Results and discussion

Initial studies conducted in the mouse indicated that nicotine disrupted behavior in a dose-related manner in both treatment groups. Doses between 0.8 and 1.6 mg/kg disrupted behavior almost completely; 1.2 mg/kg was the dose used in the chronic evaluation of nicotine-induced tolerance. Tolerance to nicotine's disruptive (1.2 mg/kg s.c.) effects when administered immediately prior to each behavioral session, occurred within 30 daily sessions in the mouse when compared to baseline rates of behavior (Fig. 1). Post-session nicotine administration had no effect on response rates. Redetermination of the nicotine-induced behavioral disruption dose-response curve indicated that tolerance to nicotine had occurred in both treatment groups (Fig. 2). That is, the dose-response disruption curve was shifted to the right in both the

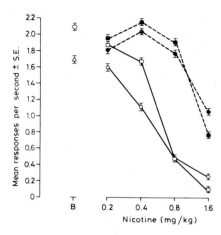

Fig. 2. The effects of single injections of nicotine on response rates per sec administered to mice in pre-group (○, ●) and post-group (□, ■) prior to (open symbols) and during (closed symbols) chronic treatment with 1.2 mg/kg nicotine. The symbols above B represent baseline control responding during four vehicle sessions, each of which preceded a prechronic test of nicotine in each treatment group. Pre-group: ○, initial determination; ●, redetermination; post-group: □, initial determination; ■, redetermination. (Reproduced with permission from Hendry and Rosecrans, 1982b.)

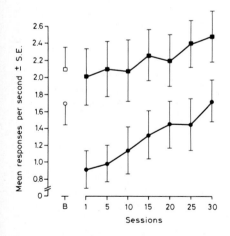

Fig. 1. The effects of the chronic administration of 1.2 mg/kg nicotine on responses per sec in mice for pre-session nicotine (●) and post-session nicotine treatment groups. The symbols above B represent baseline responding during four vehicle sessions obtained during the initial nicotine dose-effect determinations for each treatment group. Pre-group, ○, ●; post-group, □, ■. (Reproduced with permission from Hendry and Rosecrans, 1982b.)

pre- and post-session nicotine treatment groups when administered before the behavioral session. The approximate acute ED_{50} disruption dose of 0.6 mg/kg was increased to about 1.6 mg/kg. These data clearly suggest that tolerance to nicotine involves both behavioral as well as pharmacological mechanisms. Extinction of tolerance was apparent in both treatment groups over a 3-week period (testing once/week).

A similar pattern of tolerance development to nicotine's disruptive effects was also evident in the rat. Tolerance to the pre-session nicotine disruption of behavior was evident over a 36-day period. As was evident in the mouse study, post-session s.c. nicotine injections had no effect on behavior 24 h later. Challenge doses of nicotine administered pre-session to both groups after 36 daily nicotine doses yielded similar results to that observed with the mouse (Figs. 3 and 4). The results clearly indicate that pre- and post-session nicotine treatment groups were tolerant to 0.8

242

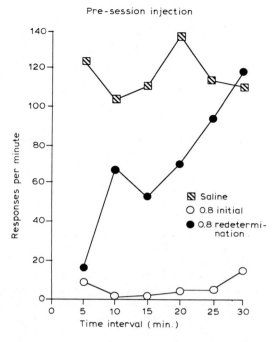

Fig. 3. A comparison of the effects of challenge doses of nicotine (0.8 mg/kg s.c.) administered prior to (○) and during (●) the chronic administration of 0.8 mg/kg (s.c.) of nicotine administered pre-session daily for 36 days. Rats were trained on a VI-15 s schedule of reinforcement. Saline response rates were determined 24 h prior to the nicotine challenge: $n = 5$.

mg/kg of nicotine when compared to its initial determination. Response rates were almost totally suppressed over 30 min in the naive rat, while in nicotine-treated rat, this disruption profile was significantly altered when the same dose of nicotine was re-administered just prior to the behavioral session; under these conditions disruption was significantly attenuated.

Both studies, in mouse and rat, suggest that nicotine-induced tolerance is the result of an alteration in the pharmacodynamics of nicotine's effects since tolerance occurred regardless of when it was administered, before or after the behavioral session. While there is some evidence that an alteration in the metabolism of nicotine can occur in chronic smokers, there is little evidence to suggest that this may occur in the rat (Swedberg et al., 1988). An explanation of the mechanism of the

observed nicotine-induced tolerance is difficult, but may be related to an up-regulation of central nicotinic receptors induced by chronic nicotine administration as suggested by others (Collins et al., 1988; Keller et al., 1988; Kisir et al., 1988). This is a reasonable explanation as an up-regulation suggests that the nicotinic receptor may be in a desensitized state which correlates with our findings, that is, a reduced nicotine-induced behavioral disruptive effect during its chronic administration.

Preliminary studies involving the mechanism of nicotine-induced pharmacological tolerance

Experimental approaches and methods

The approaches used in this study are similar to that described above except that only chronic post-

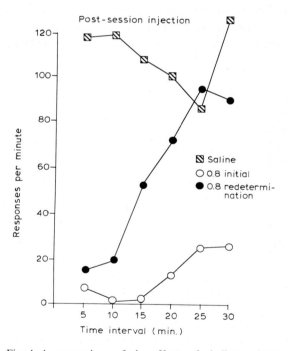

Fig. 4. A comparison of the effects of challenge doses of nicotine (0.8 mg/kg s.c.) administered prior to (○) and during (●) the chronic administration of 0.8 mg/kg (s.c.) of nicotine administered post-session daily for 36 days. Rats were trained on a VI-15 s schedule of reinforcement. Saline response rates were determined 24 h prior to the nicotine challenge; $n = 5$.

session nicotine regimens were employed. In these studies, rats were trained to bar-press for food pellets using a VI-20 s schedule of reinforcement. After rats had stabilized behaviorally, they were separated randomly into various treatment groups ($n = 8$). A challenge dose of 0.8 mg/kg was administered just prior to being placed in the operant chamber and response rates monitored for 20 min. After this first challenge session, chronic dosing was begun in each drug treatment group. Rats were maintained behaviorally 4 – 5 days/week and doses of drugs administered 1 h after behavioral exposure; drug dosing continued 7 days/week.

The drugs administered chronically post-session to behavioral evaluation were as follows: saline, 1 ml/kg; nicotine, 0.8 mg/kg; mecamylamine, 1 mg/kg; physostigmine, 0.5 mg/kg; nicotine + mecamylamine; and physostigmine + mecamylamine. All drugs were co-administered for 14 days s.c. Each treatment group was challenged pre-session with 0.8 mg/kg nicotine administered 14 days after the initiation of the administration of a specific drug regimen, and 14 days following the cessation of a specific drug treatment combination.

Results and discussion

The ability of repeated nicotine doses to induce tolerance to its disruptive effects following 14 daily injections is clearly evident in Fig. 5. Rates of responding were suppressed to 15 – 30% of baseline levels over a 5 – 20-min period following an acute disruptive dose of nicotine (0.8 mg/kg s.c.). Fourteen days of nicotine dosing, however, significantly attenuated the disruptive effects of nicotine. Interestingly, as observed in the chronic saline treatment group, single nicotine doses also appeared to attenuate its disruptive effects when administered at least 14 days later (Fig. 5). However, extinction to the development of tolerance was not evident in any specific experimental group. Thus, as observed in our initial studies (Figs. 1 – 4), post-session nicotine administration will induce tolerance to its disruptive effects within 14 days suggesting an operative pharmacological mechanism of tolerance.

To begin to sort out potential mechanisms involved in the development of pharmacological tolerance two experiments were conducted. In the

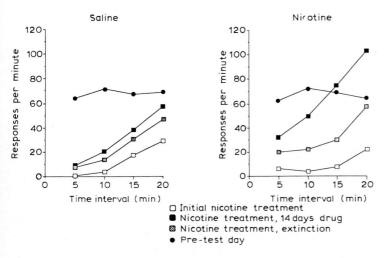

Fig. 5. A comparison of the effects of challenge doses of nicotine (0.8 mg/kg s.c.) on the response rates of rats administered 14 daily doses of nicotine (0.8 mg/kg s.c.) or saline (1 ml/kg s.c.). Challenges were conducted either, prior to daily dosing, after 14 doses of drug, or 14 days after the cessation of saline or nicotine administration. Subjects in each group ($n = 8$) were trained to bar-press for food pellets using a VI-20 s schedule of reinforcement and all drugs were injected post to any behavioral exposure.

first we attempted to attenuate the development of tolerance by the post-session co-administration of mecamylamine with nicotine. Preliminary studies showed that doses as low as 0.5 mg/kg of mecamylamine will block nicotine's acute behavioral disruptive effects. The results of these experiments indicated that mecamylamine was unable to antagonize nicotine-induced tolerance (Table I). The results of this interaction indicated that 14 days of nicotine dosing will reduce its initial behavioral disruptive effects by at least 50% while mecamylamine was unable to significantly attenuate the development of this tolerance. Mecamylamine alone, on the other hand, also appeared to induce cross-tolerance to nicotine. This effect was statistically significant at a $p > 0.029$ even though nicotine-induced cross tolerance was half of that observed when nicotine was administered alone.

The second experiment conducted involved the post-session injection of physostigmine alone, and in combination with mecamylamine. This experiment was designed to determine whether an increase of central acetylcholine (ACh) levels (via the inhibition of acetylcholinesterase by physostigmine) chronically would after nicotine-induced tolerance. The results indicated that physostigmine did induce cross-tolerance to nicotine's behaviorally disruptive effects but that mecamylamine was unable to significantly attenuate physostigmine's effects (Table I). If anything, there appeared to be an additive effect between these drugs.

These studies indicate that physostigmine, but not mecamylamine, was able to alter nicotine-induced tolerance in the rat. Additional studies in the mouse have provided similar findings indicating that physostigmine can induce cross-tolerance to nicotine which was reduced by mecamylamine co-administration (Rosecrans, 1988). Thus, these studies provide information that acetylcholine may play an important role in the development of nicotine tolerance. At this point it is difficult to draw any significant conclusions except that a nicotinic receptor sensitive to acetylcholine may be involved in the development of nicotine-induced tolerance.

These findings are in contrast to that obtained by Meltzer and Rosecrans (1988) who were unable to mimic the nicotine cue with physostigmine. But as suggested by several investigators, nicotine appears to affect several different sets of cholinergic receptors which may have different ligands. Thus, these studies are consistent with research which

TABLE I

Cross-tolerance to nicotine in rats administered specific cholinergic drugs*

Treatment group	Challenge 1	Challenge 2	Change in response	Level of significance
Saline (1 ml/kg)	70.0	55.0	−15.0	n.s.
Nicotine 1 (0.8 mg/kg)	60.6	15.6	−45.0	$p > 0.002$
Nicotine 2 (0.8 mg/kg)	80.1	27.5	−52.6	$p > 0.034$
Mecamylamine (1.0 mg/kg)	85.9	61.8	−24.1	$p > 0.016$
Physostigmine (0.50 mg/kg)	65.2	38.6	−26.8	$p > 0.013$
Mecamylamine + nicotine	84.9	39.5	−45.4	$p > 0.029$
Mecamylamine + physostigmine	94.8	58.8	−36.0	n.s.

* Data is presented as a comparison of the % disruption of behavior between the first nicotine challenge (0.8 mg/kg) (before dosing with a specific drug regimen) with the second nicotine challenge (after 14 days of repeated drug dosing).
The change in response is a measure of the degree of attenuation of the nicotine-induced behavioral disruption between challenges induced by repeated drug dosing. This factor is a measure of the level of nicotine-induced tolerance and drug-induced cross-tolerance to nicotine as produced by a specific drug regimen.
"p" values compared degree of nicotine-induced disruption between challenges using a within subject comparison statistic.

suggests that nicotine may have effects on one or more receptors which are selectively sensitive to acetylcholine. At this point, it would be plausible to suggest that nicotine-induced tolerance may be correlated to the desensitization of nicotinic receptors as evidenced by the up-regulation of these binding sites (Kellar et al., 1987). Furthermore, it is also suggested that physostigmine-induced cross tolerance to nicotine may be correlated to the down-regulation of nicotinic binding sites as observed after the chronic administration of the irreversible acetylcholinesterase inhibitor DFP (Kellar et al., 1987). Thus, in this latter situation, nicotine tolerance may be related to a reduction in the number of available nicotinic receptors.

Approaches to evaluating nicotine-induced physical dependence

Experimental approaches and methods

The procedures used to evaluate nicotine physical dependence (behavioral dependence) involved a continuation of experiments designed to evaluate tolerance. These experiments evaluated behavioral responding after the cessation of the chronic nicotine regimen once tolerance was achieved. In some studies extinction of tolerance was evaluated by administering a challenge dose of nicotine at weekly intervals.

Results and discussion

A major question to be answered, which directly involves tobacco and health, concerns the issue of nicotine-induced physical dependence and whether an animal will exhibit a withdrawal syndrome like that observed with morphine following the cessation of chronic nicotine administration (Kallman et al., 1979). The first evidence of a behavioral withdrawal syndrome following the cessation of nicotine occurred in mouse tolerance studies (Hendry and Rosecrans, 1982). In these studies, response rates declined by about 50% over the 3-week period following the cessation of daily nicotine dosing, suggesting a nicotine withdrawal syndrome. Even though there are several missing studies, such as a concurrent control group, and the completion of a study to determine if response rates would return to normal over time, this study did provide some provocative conclusions. An additional study was conducted using a separate group of mice dosed in a similar manner and sacrificed at 1, 7 and 14 days after the cessation of chronic nicotine administration (Rosecrans et al., 1985). The results of this study indicated that chronic nicotine administration induced a significant biphasic alteration of brain hypothalamic β-endorphin levels over this period of evaluation. Many hypotheses were generated, but the most important conclusion, at least to us, was that the changes in β-endorphin appeared to correlate with the behavioral disruption induced by the cessation of chronic nicotine dosing.

Similar studies have been attempted in rats as an extension of the above tolerance studies. In contrast to what was anticipated, few changes in response rates were observed following the cessation of nicotine dosing. More recently, a similar investigation was conducted in which rats were administered nicotine post-session over 3 weeks with no evidence of an apparent behavioral withdrawal syndrome. In this latter study rats were administered s.c. nicotine twice daily; 1 week 0.8 mg/kg per dose and 2 weeks 1.6 mg/kg per dose. Again, there were few alterations in response/reinforcement rates over a 3 week withdrawal period.

In addition to these above studies, a withdrawal study was conducted as part of an experiment designed to determine if neonatal dopamine depletion could alter the development of tolerance to either barbital, (+)-amphetamine or nicotine, when compared to morphine. Morphine withdrawal was quite evident as animals stopped bar-pressing for food (VI-15 s) within 24 h after the last dose of morphine; 120 mg/kg s.c., of 30 days (Kallman et al., 1979a, 1979b). Even though dopamine depletion was observed to alter the responses of (+)-amphetamine and nicotine, but not that of barbital, there was no evidence of a drug-induced withdrawal syndrome with any drug studied regardless of dopamine level (Fig. 6). Thus, while we may be discouraged in our attempts evaluate nicotine withdrawal, it may be that the rat may not be an appropriate model in which to evaluate physical dependence, at least via scheduled-controlled behavioral approaches.

Other approaches to evaluating nicotine-induced withdrawal in the rat have been attempted and two studies deserve special attention. In the first, Mor-

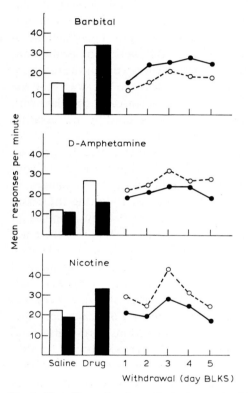

Fig. 6. The effect of saline substitution and withdrawal of chronic drug treatment. For comparison, pre-chronic saline rates and rates on the last block of chronic treatment are presented. The open bars represent vehicle control groups whereas the closed bars represent rats depleted of dopamine neonatally: neonate rats 14 days of age received 150 μg/kg 6-OH dopamine intracisternally 30 min after a 30 mg/kg (i.p.) dose of desmethylimipramine (DMI), control received vehicle + DMI. Withdrawal data is plotted as 2 day means; dotted lines represent the 6-OH dopamine group (DA-depleted). Doses of drugs administered chronically (36 daily doses beginning at 120 days of age) were: barbital 100 mg/kg, (+)-amphetamine 0.9 mg/kg, and nicotine 0.8 mg/kg. All doses were given i.p.

rison (1974) trained rats to avoid shock under the influence of repeated nicotine dosing and observed a disruption in performance after the cessation of the nicotine regimen. Interestingly, the greatest reduction in responding occurred 3 days after substituting saline for nicotine. The second study involved the administration of a chronic nicotine regimen to rats trained to discriminate metrazol from saline (Harris et al., 1986). Rats were withdrawn from the nicotine regimen and placed in an operant chamber to determine drug-lever preference; metrazol vs. saline. In this situation rats selected the metrazol-correct lever 50% of the time 3 days into withdrawal suggesting that these rats experienced subjective effects similar to that induced by metrazol. The authors concluded that these nicotine-treated rats had experienced a level of anxiety following the cessation of nicotine injections. Thus, in addition to the mouse withdrawal paradigm described (Hendry and Rosecrans, 1982), we have at least two examples of a nicotine syndrome which might be described as an indication that these subjects were physically dependent (or behaviorally dependent) upon nicotine. The lack of a vigorous model of nicotine-induced withdrawal again, reminds us of the subtleness of nicotine dependence processes. Thus, we need to pursue this problem further, and most importantly, need to replicate and further characterize the above three dependence models before we can begin to understand the nature of the nicotine dependence process in subhuman species.

Summary and conclusions

Tolerance to nicotine's disruptive effects on operant responding develops rapidly over a 14–36 day repeated dosing period in both rats and mice. This occurred regardless of whether nicotine was administered pre- or post- to each behavioral exposure. Thus, tolerance development appeared to depend on both behavioral as well as pharmacological mechanisms.

It is suggested that the pharmacological mechanism(s) involved in the development of tolerance may be related to an up-regulation of brain area nicotinic receptors. As observed with receptor binding studies, mecamylamine did not appear to attenuate the development of pharmacological tolerance to nicotine (does not attenuate nicotinic receptor up-regulation) even though this cholinergic antagonist will antagonize nicotine's acute behavioral disruptive effects completely. However, the fact that mecamylamine may induce some cross-tolerance to nicotine does complicate our interpretation of these data.

The development of nicotine tolerance, in part, appears to depend upon an interaction at some acetylcholine-sensitive nicotinic receptor as evidenced by the ability of physostigmine to induce cross-tolerance to nicotine in both the rat and mouse. These data support the view that nicotine may be inducing its effects via at least two separate nicotinic receptors, one of which may be acetylcholine sensitive. Furthermore, binding data suggest that physostigmine's effects were related to a reduction of available central nicotinic receptor sites.

In contrast to what humans experience, the rat does not appear as sensitive to nicotine-induced physical dependence, at least when operant behavior is utilized as the dependent variable used to measure withdrawal signs. Other approaches such as drug discrimination and conditioned avoidance paradigms may provide a better alternative to the evaluation of nicotine-induced dependence. Research utilizing schedule-controlled behavior in the mouse, on the other hand, has provided us with an additional model of a nicotine-induced withdrawal syndrome which may be of value in evaluating mechanisms of nicotine dependence. However, as with all of these findings, much work is needed to confirm and further characterize each model in so far as they may provide us with a reliable and specific measure of nicotine dependence.

Acknowledgement

The authors would like to dedicate this article to the memory of Dr. Paul S. Larson, Professor Emeritus and former Chairman of the Department of Pharmacology of Virginia Commonwealth University, who died in early 1988. Dr. Larson was active in the area of Nicotine, especially as a writer and promoter of nicotine research. His monographs are well known and he had a special influence on the research of many of us, especially the first author of this paper.

References

Balfour, D.J.K. The pharmacology of nicotine dependence: a working hypothesis. *Pharmacol. Ther.* 15: 239 – 250.

Carlton, P.L. and Wolgin, D.L. (1971) Contingent tolerance to anorexigenic effects of Amphetamine. *Physiol. Behav.*, 7: 221 – 223.

Chen, C.S. (1968) A study of the alcohol-tolerance effect and an introduction of a new behavioral technique. *Psychopharmacology*, 12: 433 – 440.

Clarke, P.B.S. (1987) Nicotine and smoking: a perspective from animals studies. *Psychopharmacology*, 92: 135 – 143.

Collins A.C., Miner, L.C. and Marks, M.J. (1988) Genetic influences on acute responses to nicotine and nicotine tolerance in the mouse. *Pharmacol. Biochem. Behav.*, 30: 269 – 278.

Grunberg, N.E., Popp, K.A. and Winders, S.E. (1988) Effects of nicotine on body weight in rats with access to "junk" foods. *Psychopharmacology*, 94: 526 – 539.

Harris, C.M., Emmett-Oglesby, Robinson, N.G. and Lal, H. (1986) Withdrawal from chronic nicotine substitutes partially for the interoceptive stimulus produced by pentylentrazol. *Psychopharmacology*, 90: 85 – 89.

Hendry, J.S. and Rosecrans, J.A. (1982a) Effects of nicotine on conditioned and unconditioned behaviors. *Pharmacol. Ther.*, 17: 431 – 454.

Hendry, J.S. and Rosecrans, J.A. (1982b) The development of pharmacological tolerance to the effect of nicotine on schedule-controlled responding in mice. *Psychopharmacology*, 77: 339 – 343.

Kallman, M.J., Spencer, R.M., White, A.T., Chance, W.T. and Rosecrans, J.A. (1979a) Morphine-induced behavioral disruption in rats chronically depleted of brain dopamine. *Res. Comm. Chem. Pathol. Pharmacol.*, 24: 115 – 125.

Kallman, M.J., White, A.C. and Rosecrans, J.A. (1979b) Chronic and acute effects of barbital and amphetamine in dopamine in dopamine-depleted rats. *Pharmacologist*, 19: 228.

Kellar, K.J., Schwartz, R.D. and Martino, A.M. (1987) Nicotinic cholinergic binding recognition sites in brain. In W.R. Martin, G.R. van Loon, E.T. Iwamoto and L. Davis (Eds.) *Tobacco Smoking and Nicotine: A Neurobiological Approach*, Plenum Press, New York, pp. 467 – 480.

Kisir, C., Hakan, R.L. and Kellar, K.J. (1987) Chronic nicotine and locomotor activity: influences of exposure and test dose. *Psychopharmacology*, 92: 25 – 29.

Nordberg, A., Wahlstrom, G., Arnelo, U. and Larsson, L. (1985) Effect of long-term nicotine treatment on [3-H]-nicotine binding sites in the brain. *Drug Alcoh. Depend.*, 16: 9 – 17.

Meltzer, L.T. and Rosecrans, J.A. (1982) Tolerance to the disruptive effects of arecoline on schedule-controlled behavior *Psychopharmacology*, 77: 85 – 93.

Meltzer, L.T. and Rosecrans, J.A. (1988) Nicotine and Arecoline as discriminative stimuli: involvement of a non-cholinergic mechanism for nicotine. *Pharmacol. Biochem.*

Behav., 29: 587 – 593.

Morrison, C.F. (1974) Effects of nicotine and its withdrawal on the performance of rats on signalled and unsignalled avoidance schedules. *Psychopharmacology*, 38: 25 – 35.

Rosecrans, J.A. (1988) In vivo approaches to studying cholinergic receptors. In M. Rand and K. Thrau (Eds.), *The Pharmacology of Nicotine*, ICSU Symposium Series, IRL Press, Washington, D.C., pp. 207 – 226.

Rosecrans, J.A., Hendry, J.S. and Hong, J.S. (1985) Biphasic effects of chronic nicotine treatment on hypothalamic immunoreactive beta-endorphin in the mouse. *Pharmacol. Biochem. Behav.,* 23: 141 – 143.

SECTION VIII

Trophic Mechanisms and Nicotine in Brain

A. Nordberg, K. Fuxe, B. Holmstedt and A. Sundwall (Eds.)
Progress in Brain Research, Vol. 79

CHAPTER 24

Muscle-derived trophic factors influencing cholinergic neurons in vitro and in vivo

Stanley H. Appel, James L. McManaman, Ron Oppenheim, Lanny Haverkamp and Kenneth Vaca

Department of Neurology, Program in Neuroscience, Baylor College of Medicine, Houston, Texas, U.S.A.

Introduction

Neurons depend on continual contact with their target tissue for survival and maturation during embryonic and early postnatal development (Hollyday and Hamburger, 1976). This information transfer is due in part to the action of trophic factors released by neural target tissues and exerting a retrograde effect on the innervating cells (Hamburger, 1975). The best studied examples of such retrograde trophic effects are those mediated by nerve growth factor (NGF) (Levi-Montalcini and Angeletti, 1963; Hamburger and Yip, 1984). Nerve growth factor is synthesized in the innervated target tissue of sympathetic and dorsal root ganglia neurons, as well as in the hippocampal targets of medial septal cholinergic neurons (Bradshaw, 1978; Heumann, et al., 1984; Shelta and Reichardt, 1984; Whittemore, et al., 1986). In immature animals, NGF antiserum produces destruction of sympathetic and dorsal root ganglia neurons. During the period of naturally occurring cell death administration of exogenous NGF reduces the loss of neurons in both sympathetic and sensory ganglia. Finally, NGF can prevent the deleterious consequences of axotomy in developing sympathetic and sensory neurons as well as medial septal cholinergic neurons.

Motor neurons also depend on their target mus-

cle for survival and maturation. Evidence for the existence of neurotrophic factors affecting motor neurons comes from studies demonstrating that conditioned media or soluble extracts of skeletal muscle enhance the survival, stimulate the growth of neurites, and accelerate the development of cholinergic properties in cultured spinal cord neurons (Giller et al., 1977; Dribin and Bennett, 1980; Hardeman et al., 1981; Smith and Appel, 1983). These neurotrophic effects of skeletal muscle extracts are regulated by the age and innervation state of the muscle tissue, and are tissue-specific (Smith et al., 1985). NGF cannot reproduce these effects on ventral spinal cord neurons, and antibodies to NGF do not block any of the neurotrophic effects of skeletal muscle extracts (Smith and Appel, 1983). Prior reports from our laboratory demonstrated the presence in muscle extracts of several distinct species with differing effects on motor neuron cultures. An acidic glycoprotein of 35 000 daltons was found to induce process outgrowth in both motoneurons and non-motoneurons of the ventral spinal cord, but had no effect either on acetylcholine synthesis or choline acetyltransferase (CAT) activity (Smith et al., 1986). In this chapter we review the effects of two different factors derived from muscle which enhance cholinergic properties of neurons in vitro.

CAT development factor

In the absence of muscle extract the development of CAT in tissue culture follows a biphasic time course in which there is an initial decrease in CAT activity followed by a period of steadily increasing activity. The addition of muscle extract to the cultures prevents the initial decline and stimulates the subsequent development of CAT activity (McManaman et al., 1988b). A factor (cholinergic development factor, CDF) in the muscle extract which produces such effects has been purified and partially characterized (McManaman et al., 1988a).

Muscle tissue obtained from the limbs of 14-day-old Sprague-Dawley rats was used as the source for purification since activity in the crude extract was found to peak at 2 weeks postnatal age. The tissue was homogenized in phosphate buffered saline containing a number of protease inhibitors to minimize proteolytic digestion (Table I). The supernatant of the crude extract was adjusted to pH 5 with 1 M acetic acid.Extraction of the acid precipitate at pH 9.2 results in the recovery of all of the precipitated activity and a tenth of the precipitated protein, leading to a 5.6-fold increase in specific activity. The CDF activity was then subjected to gel permeation chromatography on Sephadex G-100, followed by hydroxylapatite chromatography at pH 9.0. All of the activity in the 20 mM Na_2HPO_4 eluent from hydroxylapatite bound to DEAE, and a single peak of activity was eluted between 0.15 and 0.2 M NaCl. This activity was adsorbed to hydroxylapatite at pH 7.0 and eluted with 10 mM ethanolamine, 30 mM Na_2HPO_4, pH 9.2. Final purification was achieved by preparative SDS-PAGE under non-reducing conditions. However, in the presence of reducing conditions, the gel position of the CDF activity was not altered and no additional activity migrating at a lower molecular weight was noted. Most of the biological activity migrated with an apparent molecular mass of 20 – 22 kDa. Electrophoresis of the SDS purified material followed by silver stain, or autoradiography of radioiodine labelled purified CDF, yielded a molecular mass of 20 ± 0.1 kDa.

The peak activity fractions from preparative SDS-PAGE migrated as a single component at pH 4.8 by isoelectric focusing (Fig. 1). In agreement with its acidic pI, CDF is enriched in aspartic and glumatic acid residues (Table II). The purified CDF is sensitive to heating and to proteases. Incubating the purified factor at either 25° or 56°C for 20 min destroys 66 and 90% of activity respectively, while boiling for 5 min destroys all activity. Activity is also destroyed with either trypsin or proteinase K. The CDF activity bound neither to lectin nor heparin affinity columns (McManaman et al., 1988a).

TABLE I

Motor neuron neurotrophic factor purification. Summary of the purification of CDF from crude skeletal muscle extract

Purification step	Units	Protein (mg)	Specific activity (U/mg)	Purification	Activity recovery (%)
100 KS	$2.4 \pm 1.9 \times 10^5$	4800	50	× 1	100
pH 5-P	$1.5 \pm 1.1 \times 10^5$	525	286	× 5	62
PH5-S	$1.2 \pm 0.9 \times 10^5$	3249	37	× 0.5	50
G-100B	$0.7 \pm 0.5 \times 10^5$	50	1400	× 20	29
HAP-9	$0.3 \pm 0.2 \times 10^5$	17	1764	× 20	12
DEAE	$0.3 \pm 0.2 \times 10^5$	2.8	10 714	× 214	12
HAP-7	$1.1 \pm 0.8 \times 10^4$	0.47	23 404	× 468	5
SDS-PAGE	$0.11 \pm 472 \times 10^4$	0.004	275 000	× 5500	0.5

Our CDF does not resemble the neurotrophic factor purified from heart cell conditioned medium (Fukada, 1985). The heart CDF is isolated as a 45 kDa glycoprotein which can be converted to an active 22 kDa polypeptide by exhaustive deglycosylation. However, both the 45 kDa and the 22 kDa forms of heart cell factor exhibit charge microheterogeneity with pI values ranging from 5.7 to 7.5, while CDF demonstrates a single species at pI = 4.8.

CDF does not appear to be a member of the heparin-binding growth factors (Burgess et al., 1986; Walicke et al., 1986; Unsicker et al., 1987) such as acidic and basic FGF since it is not retained on heparin-sepharose and has no mitogenic activity in cultured 3T3 cells (McManaman et al., submitted). Basic FGF enhances CAT activity in cultured ventral horn cells as does CDF, but at saturating concentrations of either constituent

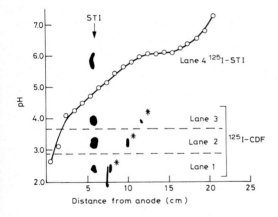

Fig. 1. Isoelectric focusing of [125I]CDF purified by preparative SDS-PAGE. The peak activity fractions obtained from preparative SDS-PAGE of 125I-labelled HAP-7 were analyzed by isoelectric focusing in 7% acrylamide gels using pH 3-7 ampholines. Samples were loaded at three different positions on the gel (lanes, 1, 2, and 3) indicated by the asterisks. [125I]Soybean trypsin inhibitor ([125I]STI) and control unlabelled soybean trypsin inhibitor (STI) were loaded in separate lanes, and the gel was then focused at 1500 V, 30 W for 30 min. The focusing position of the bands is plotted as a function of pH (0) and distance migrated. The focusing position of unlabelled soybean trypsin inhibitor is indicated by the arrow. For convenience of presentation the positive images from the autoradiograms are shown; however, all bands visible on the original autoradiograms are also visible on these positive images.

TABLE II

Amino acid composition of purified rat skeletal muscle CDF determined after 24-h hydrolysis in 6 N HCl at 150°C

Amino acid	Residues/mole Mean ± range
Aspartic acid	19 ± 1.0
Threonine	8 ± 0
Serine	14 ± 0.5
Glutamic acid	28 ± 1.5
Proline	8 ± 0.5
Glycine	26 ± 1.0
Alanine	13 ± 0
Valine	11 ± 0
Methionine	N.D.
Isoleucine	7 ± 0
Leucine	14 ± 0.5
Tyrosine	6 ± 0.5
Phenylalanine	8 ± 0.5
Histidine	10 ± 0.5
Tryptophan	N.D.
Lysine	12 ± 0.5
Arginine	8 ± 0.5
Cysteine	N.D.

The values are the means of separate determinations of two SDS-PAGE slices. The number of residues per mole of each amino acid is based on a molecular weight of 20 000. N.D. = not determined.

there is a total additivity of CAT activity with the other moiety.

Skeletal muscle CDF appears to differ from known neurotrophic growth factors both by its action on ventral spinal cord cultures and its physical and chemical properties. Further studies will be required to determine if CDF is identical to members of any one of the other known classes of growth factors.

Our prior studies demonstrated that motoneurons could be retrogradely labelled with wheat germ agglutinin-lucifer yellow conjugate, with persistence of the label in culture (Smith et al., 1986). This population of cells could also be visualized by CAT immunohistochemistry. Thus the purification from muscle of a factor which enhances CAT activity in ventral cord in vitro argues for its physiologic role in the development or maintenance of motor neurons.

Basic FGF isolated from human skeletal muscle

While CDF was being purified from rat skeletal muscle, attempts were made to purify neurotrophic activity from autopsied human skeletal muscle. Chick ciliary neurons were used to assay the enhancement of acetylcholine synthesis, CAT activity, and neurite elongation. Following several purification steps it became apparent that the predominant trophic agent of the adult human muscle had properties which resembled basic fibroblast growth factor (bFGF) (Vaca et al., 1988). Basic FGF enhanced cholinergic activity (ACh synthesis and CAT activity) and neurite elongation in cultured chick ciliary neurons. Furthermore, the cholinergic enhancing activity of human skeletal muscle bound to heparin-sepharose columns and was mitogenic on 3T3 cells.

In vivo effects of muscle-derived neurotrophic factors

Although neurotrophic factors may exhibit a range of specific behaviors in vitro such as neuron survival or enhancement of cholinergic activity, for complete validation such factors must be shown to be capable of influencing neuronal survival in vivo. An in vivo effect could eliminate a trivial explanation of the in vitro phenomena, i.e. that purified extracts may provide essential components missing or perturbed in the tissue culture environment, or that they act in a manner normally inoperative in vivo.

The natural death of motoneurons during development of the chick has many attractive features as an in vivo model to assay trophic factors. In the chick embryo 50% or more of the somatic motoneurons innervating skeletal muscle at limb as well as nonlimb regions degenerate between embryonic day (E) 5.5 and E12 (Hamburger and Oppenheim, 1982). Hindlimb bud removals at day E2.5 enhance the motoneuron loss, and addi-

tion of an extranumerary limb decreases motoneuron loss. Furthermore, agents such as d-tubocurare which inhibit neuromuscular activity also diminish motoneuron loss. These results can best be explained by a competition among motoneurons for a target-derived factor that is in limited supply.

Treatment of chick embryos in ovo with crude and partially purified extracts from embryonic hindlimbs (E8 to E9) during the normal cell death period (E5 to E9) rescues a significant number of motoneurons from degeneration (Oppenheim et al., 1988). Before cell death, approximately 23 000 to 24 000 motoneurons innervate a single hindlimb. By day E8 to E9, approximately 14 000 motoneurons innervate the single hindlimb. Daily application of crude muscle extract yielded 18 000 motoneurons on day E8 to E9. Ammonium sulfate fractionation of the crude extract indicated that the bulk of the 25 – 75% $AmSO_4$ fraction contained the bulk of the recovered activity and had the highest specific activity (Fig. 2, Table III). By gel filtration chromatography virtually all of the survival activity could be recovered in a fraction composed of agents less than 30 000 daltons. Further-

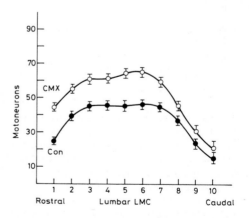

Fig. 2. The number (mean ± S.E.M.) of surviving motoneurons in the lumbar lateral motor column (LMC) along the rostral-caudal axis on E9. The eight lumbar segments were divided into 10 equal lengths. Embryos were treated daily with partially purified CMX (250 μl of 25 to 75% $AmSO_4$ fraction) beginning on E6. $p < 0.01$ between CMX and control for all rostral-caudal points except 9 and 10 (t tests).

TABLE III

The number (mean ± S.E.M.) of lumbar motoneurons in control (saline), crude muscle extract (CMX), and three $AmSO_4$ fraction groups on E8 to E9 (experiment A), and on E10 after treatment with 2.5, 25, or 250 μl of the 25 to 75% $AmSO_4$ CMX fraction (experiment B)

Treatment	n	Motoneurons
Experiment A (AmSO₄ fractions)		
Saline (E8 to E9)	9	13 600 ± 237
CMX	8	16 120 ± 226[a]
AmSO₄		
0 – 25%	6	14 827 ± 353[b]
25 – 75%	6	17 364 ± 185[c]
75 – 100%	5	11 070 ± 168[b]
Experiment B (dose response)		
Saline (E10)	5	13 387 ± 221
CMX		
2.5 μl	6	13 017 ± 255
25.0 μl	7	16 108 ± 309[d]
250.0 μl	8	17 650 ± 462[e, f]

[a] $p < 0.002$ compared to saline.
[b] $p < 0.05$ compared to saline.
[c] $p < 0.02$ compared to saline.
[d] $p < 0.05$ compared to CMX.
[e] $p < 0.005$ compared to saline.
[f] $p < 0.05$ compared to 25 μl of CMX (t-test with Bonferroni correction).
Embryos were treated daily beginning on either E5 (experiment A) or E6 (experiment B) (Oppenheim et al., 1988.)

more, the activity was heat labile and trypsin sensitive.

The partially purified hindlimb extract was relatively specific for motoneurons. Neurons in the dorsal root ganglia, ciliary ganglia, and sympathetic preganglionic column were unaffected by the hindlimb extract. By contrast, NGF could alter the survival of some of these neuronal populations but had no effect on motoneuron survival. Furthermore, the factor is developmentally regulated since muscle preparations from E16 embryos were ineffective in promoting motoneuron survival or in reducing the number of degenerating motoneu-

rons. Similarly lung, liver, and kidney extracts were ineffective in promoting motoneuron survival.

The identity of the putative neurotrophic factor remains to be determined. It is tempting to suggest that CDF purified from rat muscle may possess some of the characteristics of the chick motoneuron survival factor despite the fact that the 14th postnatal day rat is beyond the period of natural lumbar motoneuron cell death. Future experiments will examine CDF as well as other possible candidates. Regardless of its class, the motoneuron survival factor could provide new important information about motoneuron biology and its potential perturbation in disease. What is most important for the present discussion is that the demonstration of skeletal muscle factors influencing the development and cholinergic differentiation of motor neurons in vitro and in vivo may provide a mechanism for retrograde information transfer from muscle to motoneuron.

Acknowledgements

We are grateful to grants from the Muscular Dystrophy Association, the National Institutes of Health and the M. H. Jack Wagner Memorial Fund for ALS Research.

References

Bradshaw, R.A. (1978) Nerve growth factor. Ann. Rev. Biochem., 47: 191–216.
Dribin, L.B. and Barrett, J.N. (1980) Conditioned medium enhances neuritic outgrowth from rat spinal cord explants. Dev. Biol., 74: 184–195.
Fukada, K. (1985) Purification and partial characterization of a cholinergic neuronal differentiation factor. Proc. Natl. Acad. Sci. U.S.A., 82: 8795–8799.
Giller, E.L., Jr., Neale, J.H., Bullock, P.N., Schrier, B.K. and Nelson, P.G. (1977) Choline acetyltransferase activity of spinal cord cell cultures increased by co-culture with muscle and by muscle-conditioned medium. J. Cell. Biol., 74: 16–29.
Hamburger, V. (1975) Cell death in the development of the lateral motor column of the chick embryo. J. Comp. Neurol., 160: 535–546.

Hamburger, V. and Oppenheim, R.W. (1982) Naturally occurring neuronal death in vertebrates. *Neurosci. Comment.*, 1: 39–55.

Hamburger, V. and Yip, J.W. (1984) Reduction of experimentally induced neuronal death in spinal ganglia of the chick embryo by nerve growth factor. *J. Neurosci.*, 4: 767–774.

Henderson, C.E., Huchet, M. and Changeux, J.P. (1981) Neurite outgrowth from embryonic chicken spinal neurons is promoted by media conditioned by muscle cells. *Proc. Natl. Acad. Sci. U.S.A.*, 78: 2625–2629.

Hermann, R., Korshing, S., Scott, J. and Thoenen, H. (1984) Relationship between levels of nerve growth factor (NGF) and its messenger RNA in sympathetic ganglia and peripheral target tissues. *EMBO J.*, 3: 3183–3189.

Hollyday, M. and Hamburger, V. (1976) Reduction of the naturally occurring motor neuron loss by enlargement of the periphery. *J. Comp. Neurol.*, 170: 311–320.

Levi-Montalcini, R. and Angeletti, P.U. (1963) Essential role of nerve growth factor in the survival and maintenance of dissociated sensory and sympathetic nerve cells in vitro. *Dev. Biol.*, 7: 653–659.

McManaman, J.L., Crawford, F.G., Stewart, S.S. and Appel, S.H. (1988a) Purification of a skeletal muscle polypeptide which stimulates choline acetyltransferase activity in cultured spinal cord neurons. *J. Biol. Chem.*, 263: 5890–5897.

McManaman, J.L., Haverkamp, L.J. and Appel, S.H. (1988b) Developmental discord among markers for cholinergic differentiation: in vitro time courses for early expression and responses to skeletal muscle extract. *Dev. Biol.*, 125: 311–320.

McManaman, J.L., Crawford, F., Clark, R. and Richker, J.

(1988c) Multiple neurotrophic factors from skeletal muscle: demonstration of effects of bFGF and comparisons with the 22-K dalton CAT development factor. Identification basic FGF as a cholinergic growth factor from human muscle. *J. Neurosci. Res.* (in press).

Oppenheim, R.W., Haverkamp, L.J., Prevette, D., McManaman, J.L. and Appel, S.H. (1988) Reduction of naturally occurring motoneuron death in vivo by a target-derived neurotrophic factor. *Science*, 240: 919–922.

Shelton, D.L. and Reichardt, L.F. (1984) Expression of the β nerve growth factor gene correlates with the density of sympathetic innervation in effector organs. *Proc. Natl. Acad. Sci. U.S.A.*, 81: 7951–7955.

Smith, R.G. and Appel, S.H. (1983) Extracts of skeletal muscle increase neurite outgrowth and cholinergic activity of fetal rat spinal motor neurons. *Science*, 219: 1079–1081.

Smith, R.G., McManaman, J. and Appel, S.H. (1985) Trophic effects of skeletal muscle extracts on ventral spinal cord neurons in vitro: separation of a protein with morphologic activity from proteins with cholinergic activity. *J. Cell. Biol.*, 101: 1608–1621.

Smith, R.G., Vaca, K., McManaman, J. and Appel, S.H. (1986) Selective effects of skeletal muscle extract fractions on motoneuron development. *J. Neurosci.*, 6: 439–447.

Vaca, K., Stewart, S.S. and Appel, S.H. (1989). *J. Neurochem.* (submitted).

Whittemore, S.R., Ebendal, T., Larkfors, L. et al. (1986) Developmental and regional expression of β-nerve growth factor messenger RNA and protein in the rat central nervous system. *Proc. Natl. Acad. Sci. U.S.A.*, 83: 817–821.

A. Nordberg, K. Fuxe, B. Holmstedt and A. Sundwall (Eds.)
Progress in Brain Research, Vol. 79
© 1989 Elsevier Science Publishers B.V. (Biomedical Division)

CHAPTER 25

Protective effects of chronic nicotine treatment on lesioned nigrostriatal dopamine neurons in the male rat

A.M. Janson[a], K. Fuxe[a], L.F. Agnati[b], A. Jansson[a], B. Bjelke[a], E. Sundström[a], K. Andersson[b], A. Härfstrand[a], M. Goldstein[c] and C. Owman[d]

[a] Department of Histology and Neurobiology, Karolinska Institutet, P.O. Box 60400, S-10401 Stockholm, Sweden, [b] Department of Human Physiology, University of Modena, Modena, Italy, [c] Department of Psychiatry, New York University Medical Center, New York, U.S.A. and [d] Department of Medical Cell Research, University of Lund, Lund, Sweden

Introduction

In view of the negative association between smoking and Parkinson's disease independent of other associated factors (Baron, 1986), we have evaluated the possible antiparkinsonian activity of nicotine by analyzing whether chronic nicotine treatment can protect the nigrostriatal neurons from undergoing a retrograde and anterograde degeneration following partial hemitransection (Janson et al., 1988a, 1986) and from the neurotoxic actions of 1-methyl-4-phenyl-1,2,3,6-tetrahydropyridine (MPTP) (Janson et al., 1988b). The latter study was performed in a sensitive strain of black mouse (C57-B16) (Hallman et al., 1985). Catecholamine fluorescence histochemistry, tyrosine hydroxylase immunocytochemistry in combination with image analysis and tests on striatal dopamine (DA) function were employed in these studies (Fuxe et al., 1988; Janson et al., 1988a, b).

Chronic nicotine treatment and partial hemitransection of the nigrostriatal DA neurons in the male rat

The partial hemitransection was made at the mesodiencephalic junction. Chronic nicotine treatment was performed by means of Alzet minipumps loaded with nicotine hydrogene (+)-tartrate and im-

planted subcutaneously. The dose of nicotine was 0.125 mg/kg per h producing a serum nicotine level of 50.0 ± 5.1 ng/ml (Fig. 1). As seen in Figs. 2 and 3, the partial hemitransection resulted in a substantial reduction in [^{3}H]nicotine binding in the medial and especially lateral part of the nucleus caudatus putamen (rostral to the lesion) and in the ventral tegmental area and the substantia nigra zona compacta (caudal to the lesion). This analysis

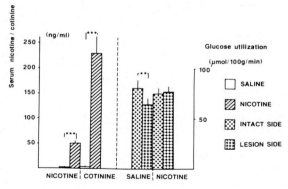

Fig. 1. Concentrations of nicotine and cotinine in serum (left part of figure) and glucose utilization in caudate nucleus (right part of figure) of partially hemitransected rats. The animals were given saline or nicotine during 2 weeks (see text). Means ± S.E.M. are shown. Two-tailed Mann-Whitney U-test was used in the serum analysis ($n = 6-7$) and two-tailed Student's paired t-test in the glucose utilization experiment ($n = 5$). Significance levels: *** $p < 0.002$ (saline vs. nicotine); ** $0.001 < p < 0.01$ (intact vs. lesioned side).

258

was performed by means of quantitative receptor autoradiography (Härfstrand et al., 1987, 1988). Reduction in body temperature and body weight induced by the partial hemitransection was similar in the saline- and nicotine-treated groups (Janson et al., 1988a).

As seen in Figs. 4 and 5, the lesion induced disappearance of tyrosine hydroxylase (TH) immunoreactive cell body and dendritic profiles in the substantia nigra was significantly counteracted in the nicotine-treated animals, especially with regard to the dendritic profiles. The lesion-induced

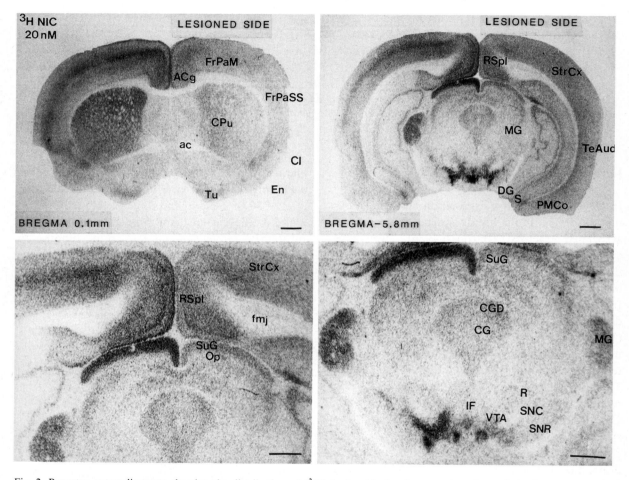

Fig. 2. Receptor autoradiograms showing the distribution of [³H]nicotine binding in coronal sections of the forebrain (Bregma +0.1 mm) and the midbrain (Bregma −5.8 mm) in hemitransected rats receiving no treatment. A marked disappearance of nicotine binding is seen both rostrally and caudally to the lesion. The marked reduction is especially clear within the cerebral cortex, in the nucleus caudatus putamen, in the superficial layer of the superior collicle, in the medial geniculate body and in the substantia nigra and lateral part of the ventral tegmental area. However, the effects in the latter two areas are less pronounced in the animal seen in the upper right figure. Abbreviations: ACg = anterior cingulate cortex; FrPaM = frontoparietal cortex, motor area; FrPaSS = frontoparietal cortex, somatosensory area; CPu = caudate putamen; ac = anterior commissure; Tu = olfactory tubercle; Cl = claustrum; En = endopirifrom nucleus; RSpl = retrosplenial cortex; StrCx = striate cortex; TeAud = temporal cortex, auditory area; PMCo = posteromedial cortical amygdaloid nucleus; MG = medial geniculate cortex; DG = dentate gyrus; S = subiculum; SuG = superficial grey layer of superior colliculus; Op = optic nerve layer of superior colliculus; fmj = forceps major corpus callosum; CGD = dorsal part of central grey; CG = central grey; IF = interfascicular nucleus; VTS = ventral tegmental area; R = red nucleus; SNC = substantia nigra, zona compacta; SNR = substantia nigra, zona reticulata.

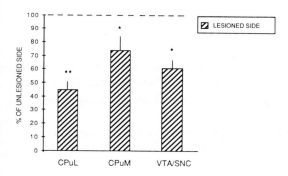

Fig. 3. The effects of partial hemitransection on [³H]nicotine binding in caudate putamen and substantia nigra. The values are expressed in percent of the corresponding area on the intact side. Means ± S.E.M. are shown, $n = 3$. Abbreviations: CPuL = caudate putamen, lateral part; CPuM = caudate putamen, medial part. For other abbreviations, see Fig. 2. Specific binding (fmol/mg protein) on the intact side (100%) was: CPuL 140 ± 4, CPuM 142 ± 9, VTA/SNC 186 ± 28. Two-tailed paired Student's-test. * $p < 0.05$; ** $p < 0.01$.

disappearance of TH immunoreactive nerve terminal profiles in medial and intermediate parts of the neostriatum was also in part counteracted by the chronic nicotine treatment (Janson et al., 1988a). Similar observations were made within the substantia nigra after Nissl staining. These morphometric results are compatible with the hypothesis that chronic nicotine treatment can in part protect the nigrostriatal DA neurons from undergoing retrograde and anterograde degeneration upon partial hemitransection.

The functional experiments demonstrated that chronic nicotine treatment resulted in an enhancement of the apomorphine-induced ipsilateral rotational behavior, an action which was positively correlated with the serum nicotine levels. Furthermore, chronic nicotine treatment was found to eliminate the asymmetry in striatal glucose utiliza-

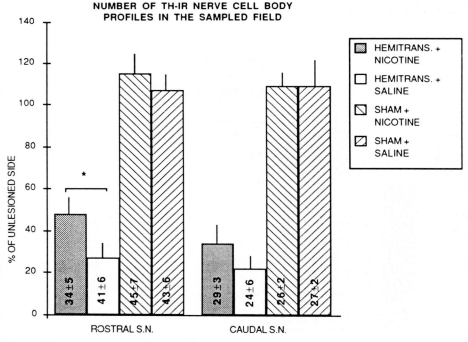

Fig. 4. Effects of chronic nicotine treatment on the lesion-induced decrease of the number of TH-IR nerve cell body profiles in the sampled field of the medial substantia nigra at the rostral and caudal levels. All values are expressed as a percentage of the respective intact side. Means ± S.E.M. are shown, $n = 4$ or 5 rats. The absolute number of profiles on the respective intact side are found within each bar. The mean profile area for the intact side was at the rostral level 148 ± 6 μm² and at the caudal level 137 ± 6 μm². For the lesioned sides the corresponding values were 118 ± 10 μm² rostrally and 100 ± 11 μm² caudally. Two-tailed Mann-Whitney U-test. * $p < 0.05$.

tion seen following the partial hemitransection (Fig. 1) (Owman et al., 1989). It seems possible that the reduction in glucose utilization in part reflects an increased survival of striatal DA nerve terminals containing not only DA but also comodulators and unknown trophic factors, which may improve neuronal survival and thus the associated glucose metabolism.

All the above morphometric and functional studies can be explained by a hypothesis postulating that the protective effect of nicotine on the lesioned nigrostriatal DA systems is due to a desensitization of excitatory nicotinic cholinoceptors located on the nigral DA nerve cells leading to reduction of firing rate and energy demands (Janson et al., 1988a). Such an action may importantly

contribute to a putative antiparkinsonian effect of chronic nicotine treatment in man.

Chronic nicotine treatment and DA utilization in surviving forebrain DA nerve terminal systems after partial di-mesencephalic hemitransection

If the above hypothesis is correct chronic nicotine treatment as reported above should markedly and preferentially reduce DA utilization in surviving forebrain DA nerve terminal systems after di-mesencephalic hemitransection. Such a study has now been performed (Fuxe et al., 1989) using catecholamine fluorescence histochemistry in combination with the tyrosine hydroxylase inhibition method (Andén et al., 1969). DA stores in various

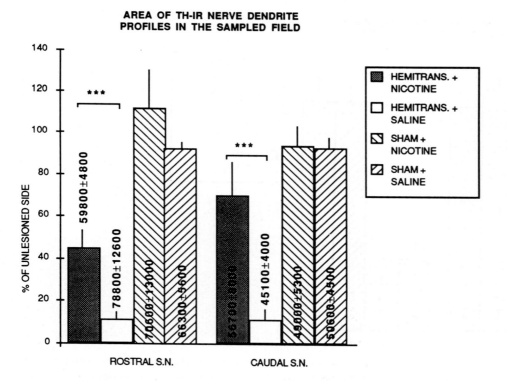

Fig. 5. Effects of chronic nicotine treatment on the lesion-induced decrease of the area of TH-IR nerve dendrite profiles in the sampled field of the medial substantia nigra at the rostral and caudal levels. All values are expressed as a percentage of the respective intact side. Means ± S.E.M. are shown, $n = 4-5$ rats. The absolute field area (expressed in μm^2) for profiles on the respective intact side are found within each bar. Two-tailed Mann-Whitney U-test. *** $p < 0.002$.

DA nerve terminal systems of the forebrain were determined by quantitative histofluorimetry and as a tyrosine hydroxylase inhibitor α-methyl-tyrosine-methylester (α-MT) was used.

As seen in Fig. 6, the results show that a partial hemitransection leads to a marked reduction in DA fluorescence on the lesioned side within the nucleus caudatus putamen, the anterior nucleus accumbens and the lateral posterior tuberculum olfactorium, while the DA nerve terminals of the dotted type in the medial and posterior nucleus accumbens and tuberculum olfactorium appear to be unaffected by the lesion. On the non-operated side, a compensatory increase of DA stores was observed in the medial part of the nucleus caudatus putamen and in the anterior nucleus accumbens compared with the sham-operated group mean value. Studies on

DA utilization demonstrated an increased DA utilization in the surviving DA nerve terminal systems on the hemitransected side in comparison with the operated side of the sham-operated animals as revealed by an increased α-MT induced depletion on the DA stores of the hemitransected side (Fuxe et al., 1989). Furthermore, on the hemitransected side, the interesting observation was made that following chronic nicotine treatment a marked reduction of DA utilization was seen in the various DA nerve terminal systems of the forebrain compared with the DA utilization found on the operated side of sham-operated animals treated chronically with nicotine (Fig. 7). As seen in Fig. 7, it was found that also on the non-operated side of hemitransected rats treated chronically with nicotine, reductions of DA utiliza-

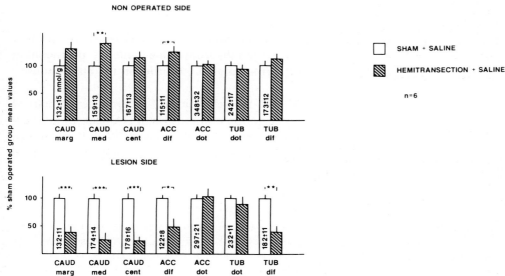

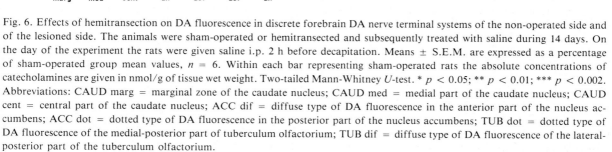

Fig. 6. Effects of hemitransection on DA fluorescence in discrete forebrain DA nerve terminal systems of the non-operated side and of the lesioned side. The animals were sham-operated or hemitransected and subsequently treated with saline during 14 days. On the day of the experiment the rats were given saline i.p. 2 h before decapitation. Means ± S.E.M. are expressed as a percentage of sham-operated group mean values, $n = 6$. Within each bar representing sham-operated rats the absolute concentrations of catecholamines are given in nmol/g of tissue wet weight. Two-tailed Mann-Whitney U-test. * $p < 0.05$; ** $p < 0.01$; *** $p < 0.002$. Abbreviations: CAUD marg = marginal zone of the caudate nucleus; CAUD med = medial part of the caudate nucleus; CAUD cent = central part of the caudate nucleus; ACC dif = diffuse type of DA fluorescence in the anterior part of the nucleus accumbens; ACC dot = dotted type of DA fluorescence in the posterior part of the nucleus accumbens; TUB dot = dotted type of DA fluorescence of the medial-posterior part of tuberculum olfactorium; TUB dif = diffuse type of DA fluorescence of the lateral-posterior part of the tuberculum olfactorium.

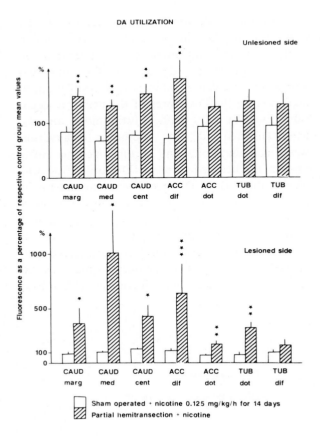

DA UTILIZATION

Unlesioned side

Lesioned side

☐ Sham operated + nicotine 0.125 mg/kg/h for 14 days
▨ Partial hemitransection + nicotine

Fig. 7. Effects of nicotine on α-MT-induced DA fluorescence disappearance in discrete forebrain DA nerve terminal systems of the intact side and lesioned side. The animals were sham-operated or hemitransected and subsequently treated with nicotine (0.125 mg/kg per h) during 14 days α-MT (250 mg/kg i.p.) was given 2 h before decapitation of the rats. In order to correct the effects of nicotine for the effects of the operation, the sham-operated + nicotine + α-MT treated group (given as % of the sham-operated + nicotine-treated group) is expressed as a percentage of the sham-operated + saline + α-MT treated group mean values (given as % of the sham-operated + saline group). By analogy with this the hemitransected + nicotine + α-MT treated group (given as % of the hemitransected + nicotine group) is expressed as a percentage of the hemitransected + saline + α-MT treated group mean values (given as % of the hemitransected + saline group). Means ± S.E.M., $n = 6$ in all groups except for CAUD marg, CAUD med, CAUD cent on the lesioned side in hemitransected + nicotine + α-MT-treated group where $n = 3$ and in ACC dif where $n = 4$. * $p < 0.05$, ** $p < 0.01$, *** $p < 0.002$. Abbreviations, see Fig. 6. The absolute concentrations of catecholamines representing 100% expressed in nmol/g of tissue wet weight were as follows: non-operated side in the

tion were observed although not as marked as on the hemitransected side in the striatal and anterior nucleus accumbens DA nerve terminal systems in comparison with the non-lesioned side of sham-operated rats treated chronically with nicotine.

From the present results it seems clear that chronic nicotine treatment can markedly reduce the DA utilization in surviving DA nerve terminal systems of the nucleus caudatus putamen, nucleus accumbens and tuberculum olfactorium. Thus, these results support the hypothesis that the protective action of chronic nicotine treatment on ascending DA systems after mechanical lesions may be produced via a desensitization of excitatory nicotinic cholinoceptors located on nigral and ventromedial tegmental DA nerve cells and/or on DA forebrain terminals. These effects lead to reduced firing rates and reduced local release of DA in the forebrain and thus to reduced energy demands (Fuxe et al., 1989).

Chronic nicotine treatment and 1-methyl-4-phenyl, 1,2,3,6-tetrahydropyridine-induced degeneration of nigrostriatal DA neurons in the black mouse

MPTP is well known to produce a marked degeneration of nigrostriatal DA neurons in a sensitive strain of black mouse (C57-Bl/6) (Hallman et al., 1985). In the present study, a dose of 50

group sham-operated + nicotine treatment, CAUD marg = 147 ± 9; CAUD med = 193 ± 15; CAUD cent = 181 ± 16; ACC dif = 140 ± 9; ACC dot = 335 ± 33; TUB dot = 244 ± 22; TUB dif = 183 ± 21; non-operated side in the group hemitransection + nicotine treatment, CAUD marg = 147 ± 14; CAUD med = 197 ± 21; CAUD cent = 180 ± 20; ACC dif = 114 ± 10; ACC dot = 388 ± 32; TUB dot = 236 ± 10; TUB dif = 186 ± 8; Lesioned side in the group sham-operated + nicotine treatment, CAUD marg = 155 ± 10; CAUD med = 180 ± 14; CAUD cent = 176 ± 19; ACC dif = 130 ± 6; ACC dot = 330 ± 36; TUB dot = 235 ± 20; TUB dif = 198 ± 15; lesioned side in the group hemitransection + nicotine treatment, CAUD marg = 32 ± 17; CAUD med = 49 ± 40; CAUD cent = 39 ± 32; ACC dif = 33 ± 27; ACC dot = 211 ± 21; TUB dot = 117 ± 26; TUB dif = 51 ± 26.

mg/kg body weight was given subcutaneously. On the day before, Alzet minipumps, model 2002 were implanted subcutaneously. The minipumps contained nicotine hydrogene (+)-tartrate in amounts calculated to produce a dose of 0.125 mg/kg per h for 2 weeks. As reported above, this dose will result in serum nicotine levels similar to those found in smokers (40 – 50 ng/ml). Immediately following the MPTP injections, the mice received four injections of nicotine with 30 min intervals in order to obtain high nicotine concentrations in the brain in relation to the MPTP injection.

The results are shown in Fig. 8. MPTP produced a substantial disappearance of TH immunoreac-

tivity (IR) within the striatum as seen especially from the total IR (T-IR) value which is the product of the mean grey value (GV) and the immunoreactive area in the sampled field (FA = field area). Chronic nicotine treatment was found to significantly counteract the MPTP-induced disappearance of TH IR within the striatum. Biochemical analysis showed that the chronic nicotine treatment also significantly counteracted the MPTP-induced disappearance of the nigral DA stores.

These results indicate that chronic nicotine treatment may lead to an increase in the survival of nigrostriatal DA neurons after an MPTP-induced lesion of these neurons. In line with the above hypothesis, it is suggested that the protective action against the MPTP-induced toxicity may be desensitization of excitatory nicotinic cholinoceptors located on the nigral cell bodies and terminals (Janson et al., 1988b) which may help the DA nerve cells to maintain their ion homeostasis since the energy demands may be reduced due to reduced firing rates. These results strongly underline the

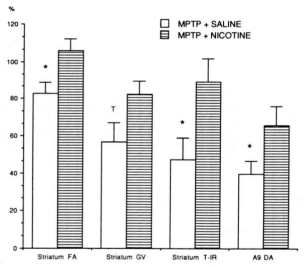

Fig. 8. Effects of 14 days chronic nicotine treatment (0.125 mg/kg per h) on the MPTP-induced disappearance of TH IR in the striatum and of DA stores in the substantia nigra of the black mouse. Means ± S.E.M. are shown, $n = 4 - 6$ for the striatal immunohistochemical parameters (FA, GV and T-IR – see text) and $n = 11 - 12$ for the analysis of DA levels in the substantia nigra (A9). All values are expressed as a percentage of the respective control groups not receiving MPTP. The absolute values for the groups treated only with saline or nicotine were (not significantly different from each other): striatum FA saline: 2.21 ± 0.70 mm^2, nicotine: 2.20 ± 0.58 mm^2; striatum GV saline: 26.8 ± 1.8, nicotine: 28.7 ± 1.8; striatum T-IR saline: 61 ± 6, nicotine: 63 ± 5; A9 DA saline: 1109 ± 117 ng · g^{-1} of tissue wet weight, nicotine: 948 ± 86 ng · g^{-1} of tissue wet weight. Two-tailed Mann-Whitney U-test. * $p <$ 0.05; $T = 0.05 < p < 0.10$.

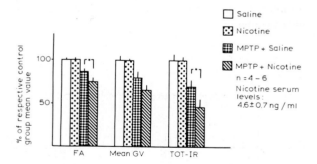

Fig. 9. Effects of chronic nicotine treatment on TH-IR in the neo-striatum of MPTP lesioned animals expressed in % of the respective saline-treated group. Morphometric and microdensitometric analysis of TH-IR terminal profiles in the striatum of C57 black mice. FA = mean GV and TOT-IR are given (see text). Means ± S.E.M., $n = 4 - 6$. Nicotine serum levels 4.6 ± 7 ng/ml. Two-tailed Mann-Whitney U-test. * $p < 0.05$. The absolute values in the 100% groups were as follows: FA = 0.248 ± 0.005 mm^2 (saline), 0.251 ± 0.008 mm^2 (nicotine); mean GV = 24.8 ± 1.3 (saline), 24.6 ± 0.5 (nicotine); TOT-IR = 61.6 ± 4.1 (saline), 62.0 ± 2.8 (nicotine).

view that chronic nicotine treatment may exert antiparkinsonian actions in man.

It should be noticed (Fig. 9) that when a 10 times lower dose is used, leading to serum nicotine levels in the order of 4–5 ng/ml evidence has been obtained that chronic nicotine treatment produced an increase in the disapperance of TH IR within the striatum as seen from the measurements of the FA and of the total IR. These findings may be explained on the basis that with this low dose of nicotine no desensitization of the excitatory nicotinic cholinoceptors is obtained. Instead excitation may be induced. The analysis of the TH immunoreactive cell body profiles in the substantia nigra of these animals shows no effects of the nicotine treatment on number of cell body profiles, field area, mean grey value nor on total IR (Fig. 10). However, the reduction of the mean area of the TH cell body profiles seen with MPTP alone was counteracted by chronic nicotine treatment in the low doses indicated above (Fig. 10). The dendritic profiles were not significantly affected by this low dose treatment with nicotine as seen from the number of profiles, the FA and the total IR measurements.

Summary

The present results demonstrate that chronic nicotine treatment can in part protect against mechanically-induced and neurotoxin-induced degeneration of nigrostriatal DA neurons. These results indicate that in sufficient doses chronic treatment with nicotine may be considered in the pharmacological treatment of Parkinson's disease. It remains to be demonstrated whether these protective actions can be extended to include also other injured neurons such as the cholinergic neurons, known to be severely affected in Alzheimer's disease.

Acknowledgements

This work has been supported by a Grant (1762) from the Council for Tobacco Research, New York, U.S.A. and by a grant from the Swedish Tobacco Monopoly. We are grateful for the excellent technical assistance of Ulla Altamimi, Beth Andbjer, Kerstin Lundberg and Maria Nilsson, and for the excellent secretarial assistance of Monia Särd.

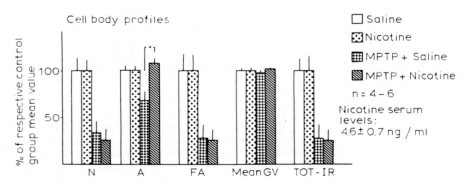

Fig. 10. Effects of chronic nicotine treatment on nigral TH IR cell body profiles in MPTP lesioned animals expressed in % of the respective saline-treated group. Morphometric and microdensitometric analyses of TH-IR cell body profiles in the substantia nigra of C57 black mice. N = number of cell body profiles; A = mean area of of cell body profiles. FA = mean GV and TOT-IR (see text). Nicotine serum levels 4.6 ± 0.7 ng/ml. Two-tailed Mann-Whitney U-test. * $p < 0.05$. The absolute values in the 100% groups were as follows: N = 104 ± 15 (saline), 99 ± 12 (nicotine); A = 87.1 ± 5.8 μm^2 (saline), 76.0 ± 3.6 μm^2 (nicotine); FA = 10900 ± 2000 μm^2 (saline), 8900 ± 1500 μm^2 (nicotine); mean GV = 105.5 ± 1.9 (saline), 106.9 ± 2.9 (nicotine); TOT-IR = 1157000 ± 215000 (saline), 945000 ± 154000 (nicotine).

References

Andén, N.E., Corrodi, H. and Fuxe, K. (1969) Turnover studies using synthesis inhibition. In B. Hooper (Ed.), *Metabolism in Amines in the Brain*, MacMillan & Co, London, pp. 38 – 47.

Baron, J.A. (1986) Cigarette smoking and Parkinson's disease. *Neurology*, 36: 1490 – 1496.

Fuxe, K., Andersson, K., Eneroth, P., Härfstrand, A., Janson, A.M., von Euler, G., Agnati, L.F., Tinner, B., Köhler, C. and Hersh, L.B. (1988) Neuroendocrine and trophic actions of nicotine. In M.J. Rand and K. Thurau (Eds.), *The Pharmacology of Nicotine*, IRL Press, Oxford, pp. 293 – 320.

Fuxe, K., Janson, A.M., Jansson, A., Andersson, K. and Agnati, L.F. (1989) Chronic nicotine treatment reduces dopamine utilization in surviving forebrain dopamine nerve terminal systems after partial di-mesencephalic hemitransections. *Naunyn-Schmiedeberg's Arch. Pharmacol.* (submitted).

Hallman, H., Lange, L., Olson, L., Strömberg, I. and Jonsson, G. (1985) Neurochemical and histochemical characterization of neurotoxic effects of 1-methyl-4-phenyl-1,2,3,6-tetrahydropyridine on brain catecholamine neurons in the mouse. *J. Neurochem.*, 44: 117 – 120.

Härfstrand, A., Fuxe, K., Andersson, K., Agnati, L.F., Janson, A.M. and Nordberg, A. (1987) Partial di-mesencephalic hemitransections produce disappearance of ^{3}H-nicotine binding in discrete regions of rat brain. *Acta Physiol. Scand.*, 130: 161 – 163.

Härfstrand, A., Adem, A., Fuxe, K., Agnati, L.F., Andersson, K. and Nordberg, A. (1988) Distribution of nicotinic cholinergic receptors in the rat tel- and diencephalon: a quantitative receptor autoradiographical study using (^{3}H)-acetylcholine, (α-^{125}I)-bungarotoxin and (^{3}H)-nicotine. *Acta Physiol. Scand.*, 132: 1 – 14.

Janson, A.M., Fuxe, K., Kitayama, I., Härfstrand, A. and Agnati, L.F. (1986) Morphometric studies on the protective action of nicotine on the substantia nigra dopamine nerve cells after partial hemitransection in the male rat. *Neurosci. Lett*, Suppl. 26: S.88.

Janson, A.M., Fuxe, K., Agnati, L.F., Kitayama, I., Härfstrand, A., Andersson, K. and Goldstein, M. (1988a) Chronic nicotine treatment counteracts the disappearance of tyrosine-hydroxylase-immunoreactive nerve cell bodies, dendrites and terminals in the mesostriatal dopamine system of the male rat after partial hemitransection. *Brain Res.*, 455: 332 – 345.

Janson, A.M., Fuxe, K., Sundström, E., Agnati, L.F. and Goldstein, M. (1988b) Chronic nicotine treatment partly protects against the 1-methyl-4-phenyl-1,2,3,6-tetrahydropyridine-induced degeneration of nigro-striatal dopamine neurons in the black mouse. *Acta Physiol. Scand.*, 132: 589 – 591.

Owman, C., Fuxe, K., Janson, A.M. and Kåhrström, J. (1989) Chronic nicotine treatment eliminates asymmetry in striatal glucose utilization following unilateral transection of the meso-striatal pathways in rats. *Neurosci. Lett.* (in press).

A. Nordberg, K. Fuxe, B. Holmstedt and A. Sundwall (Eds.)
Progress in Brain Research, Vol. 79
© 1989 Elsevier Science Publishers B.V. (Biomedical Division)

CHAPTER 26

Studies of protective actions of nicotine on neuronal and vascular functions in the brain of rats: comparison between sympathetic noradrenergic and mesostriatal dopaminergic fiber systems, and the effect of a dopamine agonist

Ch. Owman[a], K. Fuxe[b], A.M. Janson[b] and J. Kåhrström[a]

[a] Department of Medical Cell Research, Section of Neurobiology, University of Lund, Lund, and [b] Department of Histology and Neurobiology, Karolinska Institutet, Stockholm, Sweden

Introduction

There are several epidemiological studies indicating an inverse relationship between Parkinson's disease and smoking, irrespective of the involvement of other factors (Nefzger et al., 1968; Kessler and Diamond, 1971; Kessler, 1972; Baumann et al., 1980; Godwin-Austen et al., 1982; Baron, 1986). This is supported in experiments on animals demonstrating that administration of nicotine or exposure to cigarette smoke enhances the utilization of dopamine in the neostriatum, and especially in the anterior parts of the nucleus accumbens and of tuberculum olfactorium (Andersson et al., 1981a,b).

With the use of a rat model in which partial hemitransection of the ascending dopamine system was performed at the meso-diencephalic level, it has been shown in a recent report (Janson et al., 1988) that chronic infusion of nicotine counteracts histochemical, biochemical and functional

evidence of retrograde and anterograde degeneration in the meso-striatal dopamine neurons. Similar effects have been obtained by treatment with the ganglioside fraction, GM1 (Agnati et al., 1983). The action of GM1 was associated with a substantial reduction also in the difference between striatal glucose utilization and blood flow in the operated compared to the unoperated side (Agnati et al., 1985) as determined by computer-assisted quantitative autoradiography (cf. Sokoloff, 1981a; Owman and Diemer, 1985). Indeed, preliminary results have indicated that chronic nicotine treatment will counteract the side-to-side asymmetry in the metabolic rate of glucose in the caudate-putamen of the partially hemitransected rats (Owman et al., 1988).

There may be several explanations for this remarkable effect of nicotine on the energy metabolism in the striatum. It could be due to a general neuronal activation by nicotine mediated by the nicotine-type of cholinergic receptors, locally or via neuronal circuits. It could be directly related to the suggested protective action of nicotine, not necessarily linked to activation of nicotinic receptors; If so, the enhancement in

Address for correspondence: Professor Christer Owman, PhD, MD, Department of Medical Cell Research, Biskopsgatan 5, S-223 62 Lund, Sweden glucose utilization might be part of the beneficial effect itself, or it might reflect an improved functional activity of the dopamine neurons as a result of amelioration of dopaminergic neurotransmission. Another question is whether this action is restricted to the lesioned dopaminergic system, or if it in a more general manner engages catecholaminergic systems.

These possibilities were elucidated in a series of experiments on rats in which hemitransection of the dopaminergic meso-striatal pathway was carried out and the effect of nicotine treatment on regional blood flow and metabolism was compared with the action of a dopamine agonist with prejunctional action, or surgical interference with the cranial sympathetic nervous system was performed in order to elucidate the influence of chronic nicotine treatment on noradrenergic neurons.

Experimental procedures

Experiments were performed on male Sprague-Dawley rats, both 4 weeks old juvenile animals (50 g body weight) and adult animals (215 g).

Hemitransected animals.   Partial unilateral hemitransection of the ascending meso-striatal dopamine pathway was performed as previously described in detail by Agnati et al. (1983). During chloral hydrate (30 mg of a 7% solution per 100 g b.w.) anesthesia, a 4 mm wide knife was inserted in the coronal plane of the right side − 1 mm lateral to the midline and 1 mm posterior to the bregma − and lowered at the angle of 70° to the horizontal plane, in order to reach the ventral border at level A3200 of König and Klippel (1974). The level is located immediately in front of the substantia nigra (Fig. 1). This large lesion produces an axotomy of both ascending and descending pathways, preferentially of the lateral component of the meso-striatal dopamine system. Immediately following the lesion, one group of animals received 4 i.p. injections of nicotine hydrogen(+) tartrate (0.5 mg/kg) every 30 min in order to achieve prompt high plasma and brain levels of

nicotine. For the prolonged continued treatment, Alzet Model 2002 Minipumps were implanted s.c. in the neck during the same surgical session. The pumps contained 42 mg nicotine base, calculated to produce a dose of 0.125 mg/kg per h during 2 weeks. This would result in plasma nicotine levels similar to those measured in smokers (Wilkins et al., 1982). Control animals received the same volume of saline throughout. One control group also included non-operated, saline-treated animals.

Another group of animals received two daily injections for 14 days of the dopamine agonist EMD

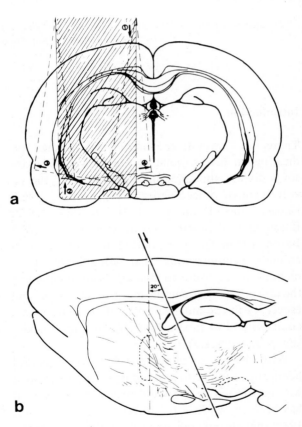

Fig. 1. Illustration of the partial hemitransection. After linear craniotomy a 4 mm wide knife was inserted perpendicular to the sagittal plane (a) with a 20° angle to the frontal plane (b). The medial edge was located 0.5 mm caudal to the bregma and 0.5 mm from the midline. After the knife had been submerged into the brain it was lifted slightly and then moved laterally and medially as indicated in (a).

269

23 448 (6 mg/kg i.p., the first injection given immediately following the transection). Because of the animals' bad condition in the course of the second week, the dose had to be halved during the last 3 days. Control animals received the same volume of solvent (10% 1,2-propandiol in distilled water).

In the analysis of regional cerebral blood flow and glucose metabolism, a double-isotope autoradiographic method was used (Owman and Diemer, 1985). After the 2 weeks of drug treatment the animals were anesthetized with 2% fluothane, and PVC catheters (PP 50 dimension) were inserted into both femoral arteries and one femoral vein, after which the animal was placed in a restraining cage. During the experiment, which started 1 h after recovery from anesthesia, PaO_2, $PaCO_2$, pH, and rectal temperature were monitored. The non-anesthetized rats were injected with a bolus of 600 μCi of [2-^{3}H]deoxyglucose (15 mCi · mmol^{-1}, Amersham, TRK 383). During the experimental period, arterial blood samples were taken for the determination of the 2-deoxyglucose plasma integral and plasma glucose concentration, according to the schedule of Sokoloff et al. (1977). After the 45 min plasma sample, one arterial catheter was connected to a constant-velocity withdrawal pump (Gjedde et al., 1980). At the same time, 60 μCi of [^{14}C]iodoantipyrine (53 mCi · mmol^{-1}, New England Nuclear, Boston, Mass., U.S.A., Nec-712) was injected as an i.v. bolus (cf. Sakurada et al., 1978). Twenty seconds later the animal was decapitated and the brain rapidly dissected out and frozen in isopentane chilled with liquid nitrogen. Blood was collected from the neck for determination of plasma nicotine and cotinine concentrations using capillary column gas chromatography with nitrogen-sensitive detection (Curvall et al., 1982). Scintillation counting of ^{3}H and ^{14}C, and measurement of plasma glucose concentration, were performed as described in detail by Gjedde et al. (1980).

The frozen brains were cut in 20 μm frontal sections on a cryostat. Two sets were produced: the sections from the first were immediately immersed in 2,2-dimethoxypropane for 2 × 1 min, dried and then placed on LKB Ultrofilm for 16 days together with methylmethacrylate standards, which had been calibrated against dimethoxypropane-treated brain tissue. The sections from the other set were exposed to a Kodak X-ray film together with appropriate standards for 14 days. The procedures have been described in more detail by Owman and Diemer (1985).

The computer-assisted autoradiographical analysis was performed on a Leitz TAS Plus image analyzer by means of a double-isotope program, which allowed the exact alignment of sections and regions of interest when determination of optical density was made.

Sympathetically denervated animals. Juvenile rats were subjected to sympathetic denervations under ethyl ether anesthesia according to the following protocol (Fig. 2): (1) unilateral excision of the entire superior cervical ganglion (6 animals); (2) unilateral removal of the caudal one-third of the ganglion (6 animals); (3) transection of the postganglionic trunk 5 mm cranial to the ganglion (6 animals). The other side was either left intact (3 of the animals from each group), or the preganglionic trunk was excised from the caudal pole of the ganglion and 10 mm below (3 of the animals from each group).

Nicotine was administered as above, but the infusion time was extended to 4 weeks. Because of the limited duration of the pump it had to be exchanged after 2 weeks; the dose in the second pump was adjusted to an approximated 50% increase in body weight to compensate for the growth of the juvenile animals. The same number of age-matched, untreated animals served as controls.

The animals were decapitated under ethyl ether anesthesia after 4 weeks' nicotine treatment. The Alzet pumps were found to have emptied adequately in both steps. Pial vessels at the base of the brain and the iris from both sides were prepared as whole-mounts, dried, and subjected to formalde-

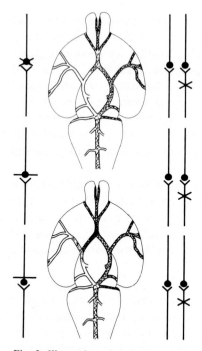

Fig. 2. Illustration of the operation carried out on the cervico-cranial sympathetics. On the right side of the neck (left in the illustration) the superior cervical sympathetic ganglion was excised either completely or partly, or the postganglionic nerve trunk was divided above the ganglion. On the contralateral (control) side, illustrated to the right in the figure, the sympathetic ganglion was either left intact or was decentralized by removal of the preganglionic trunk. The insert of the basal aspect of the brain shows schematically the almost complete perivascular sympathetic denervation resulting from ipsilateral ganglionectomy or postganglionic axotomy (above), and the tendency to regenerative overgrowth of fibers from the intact side 4 weeks after ganglionectomy (below).

hyde treatment for visualization of noradrenergic nerves according to the Falck-Hillarp method (Björklund et al., 1972). The superior cervical sympathetic ganglia were fixed in ice-cold modified Stefanini's solution consisting of 4% paraformaldehyde and 0.2% picric acid. Cryostat sections were prepared for immunohistochemical demonstration of tyrosine hydroxylase and neuropeptide Y (NPY).

Results and comments

It can be seen from Table I that there were no statistically significant differences between the groups in the various physiological parameters measured in the 45-min course of the flow/metabolism experiment. Table I also shows the high plasma levels of nicotine and its primary metabolite, cotinine, following nicotine treatment.

There was considerable individual variation in the absolute values for the glucose utilization in the different animals, which is illustrated in Fig. 3 for animals receiving nicotine and saline, respectively. It is still evident, though, that in all control animals the glucose utilization was relatively lower on the lesioned side and that this asymmetry was eliminated following nicotine administration; in some animals the glucose use even tended to be higher on the operated than on the intact side. From the mean values in Table II it can be seen that hemitransection significantly reduced striatal

TABLE I

Values for the physiological parameters measured frequently in the course of the 45-min metabolism/flow experiments, and levels of plasma nicotine and cotinine in animals receiving nicotine or saline

Parameter measured	Saline ($n = 6$)	Nicotine ($n = 8$)	EMD 23 448 ($n = 3$)
Mean arterial blood pressure (mmHg)	117.0 ± 17.4	111.5 ± 4.9	121.7 ± 2.9
Arterial pO_2 (mmHg)	105.7 ± 3.1	101.5 ± 4.2	112.0 ± 8.7
Arterial pCO_2 (mmHg)	39.0 ± 3.5	36.9 ± 1.9	39.5 ± 1.7
Arterial pH	7.445 ± 0.046	7.460 ± 0.047	7.487 ± 0.005
Plasma nicotine (ng/ml)	1.1 ± 0.3	40.8 ± 4.3	–
Plasma cotinine (ng/ml)	1.3 ± 0.4	330 ± 38.1	–

Significance level for saline vs. nicotine in the determinations of plasma nicotine and cotinine (two-tailed Mann-Whitney U-test): $p < 0.001$. Values are given as means ± S.D., n = number of animals.

glucose metabolism by 20.2% compared with the unoperated side, and that the side-to-side difference was eliminated in the nicotine-treated animals, which even had 1.4% higher glucose metabolism on the operated side. Saline-treated operated animals had almost identical values as the unoperated animals as measured on the left side (corresponding to the intact side of the operated animals). There was no significant difference in the rate of glucose use on this side between the two groups of hemitransected animals.

So far, only a small number of animals (three in each group) have been studied with regard to simultaneous regional fluctuations in both glucose

utilization and blood flow. The preliminary data shown in Fig. 4 confirmed the asymmetry in glucose metabolism for the caudate-putamen in the control animals, and how this is blunted by nicotine treatment. This was associated with corresponding side-to-side differences and improvement, respectively, in local blood flow. In substantia nigra, on the other hand, the metabolism was considerably higher on the operated side. This pattern was not seen in the flow measurements. Both metabolism and flow were much lower on the lesioned side of the ventrolateral thalamus, and it did not appear to be affected by nicotine, in contrast to the situation for the striatum. In the hippocampus and nucleus accumbens the glucose utilization was largely unaffected by the hemitransection, and nicotine treatment did not seem to differ from controls. The blood flow in these regions was somewhat reduced and did not entirely match the

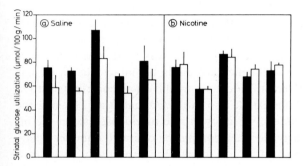

Fig. 3. Pairs of bars showing striatal glucose utilization on the intact (filled bars) and transected (open bars) side following 2 weeks' infusion of (a) saline or (b) nicotine. Mean + S.D.

TABLE II

Glucose utilization (μmol/100 g per min) in caudate-putamen of unoperated and partially hemitransected rats given saline or nicotine during 2 weeks

Experimental group	n	Intact side	Lesioned side
Unoperated + saline	5	80.0 ± 7.5[a]	–
Hemitransected + saline	5	80.2 ± 7.9	64.0 ± 5.8[b]
Hemitransected + nicotine	5	75.4 ± 5.2	76.8 ± 4.7[c]

[a] In unoperated animals the value is from the left side, corresponding to the intact side of the operated animals.
[b] Paired t-test: $0.001 < p < 0.01$ (intact vs. lesioned side).
[c] Non-significant.
Values are means ± S.E.M., n = number of animals.

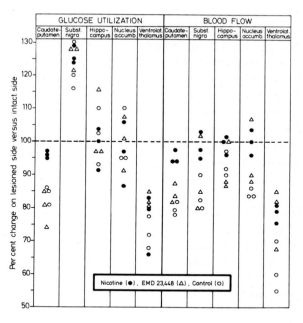

Fig. 4. Preliminary data showing percent changes in regional glucose utilization or blood flow on the transected side related to the intact side, with the 100% level (= no change) indicated, following 2 weeks' infusion of nicotine or EMD 23 448 (three animals in each group). The control group includes two animals receiving saline infusion and one animal receiving matching injections of the solvent for EMD 23 448.

metabolic figures. The material is not large enough to allow for any interpretation of this possible, slight mis-match between local metabolism and blood flow. If anything, nicotine tended to improve the local flow response. The results from the experiments with the dopamine agonist EMD 23 448 resembled those of the saline-solvent-treated controls in the above-mentioned regions.

Removal of the cervical sympathetics on one side will result in a complete and almost strictly unilateral denervation of the pial vessels on the same side (Fig. 2). When this denervation is performed in young animals it can be seen from Fig. 2 that there is a considerable tendency for compensatory overgrowth of adrenergic fibers from the contralateral side, a process that is inhibited by decentralization of the intact ganglion (see further Kåhrström et al., 1986). Well-known retrograde changes are seen in the ganglion following postganglionic axotomy. These features, together with partial lesion of the superior cervical ganglion, were used to elucidate any trophic and/or protective effects of nicotine on a catecholaminergic system different from the meso-striatal dopaminergic. Nicotine treatment during 4 weeks had, however, no effect on the process of reinnervation seen with the Falck-Hillarp histofluorescence method in the pial vessels after unilateral superior cervical sympathetic ganglionectomy. Nor was there any alteration in the remaining perivascular nerve plexus following partial ganglionectomy. Changes in the tyrosine hydroxylase and NPY immunoreactivities in the ganglia following axotomy, decentralization, or partial ganglionectomy were not overtly affected by the nicotine administration.

Discussion

The results show that partial hemitransection of the meso-striatal dopamine pathway of rats leads to a significantly lower glucose utilization in the striatum of the operated compared to the intact side, and that continuous nicotine infusion for 2 weeks eliminates this side-to-side asymmetry. The almost identical values for the striatum of the unoperated animals compared to the intact side of hemitransected animals (both groups treated with saline) show that the lower values reflect a true, net reduction in metabolism. This would conform with observations on parkinsonian patients in which central glucose utilization appears to be reduced by some 20% (Kuhl et al., 1984), and with the reported reduction in the caudate nucleus of the MPTP primate model of Parkinson's disease (Schwartzman and Alexander, 1985).

The present observations agree with earlier similar studies in which the trophic effect of ganglioside GM1 was investigated (Agnati et al., 1985). In those studies it was found that also GM1, like nicotine, is able to counteract the imbalance in glucose metabolism between the striata following partial hemitransection of the dopamine pathways. In the case of GM1 this was partly due to a lowering of the metabolism on the intact side. A similar effect of nicotine might have contributed to a slight extent also in the present study, though the tendency to a lower rate of glucose use on the intact side when comparing operated animals receiving saline or nicotine, respectively, was not statistically significant.

The reduction in striatal glucose metabolism was associated with a lowering also of the local blood flow, in agreement with the tight coupling between flow and metabolism that exists in the brain (cf. Sokoloff, 1981b; Owman and Diemer, 1985). The flow reduction, too, was counteracted by nicotine.

Of areas functionally associated with the meso-striatal dopamine system, hippocampus and nucleus accumbens showed no clearcut change in flow or metabolism in any of the groups studied. The ventro-lateral thalamus, on the other hand, had markedly lower values of both parameters on the side of axotomy. In contrast to the situation for caudate-putamen, nicotine treatment had no effect on this imbalance. This would indicate, firstly, that the axotomy as such may affect also certain surrounding regions through functional connections by fibers that are not directly included in the lesion and, secondly, it supports the notion that the action of nicotine is probably restricted to certain neuron systems, such as the dopamine system.

The glucose utilization was markedly elevated in the substantia nigra on the operated compared to the unoperated side. This may be related to an increased protein synthesis in the cell bodies of this region as a retrograde response to the lesion during a regenerative phase. It was notable that this increase in metabolism was not associated with an enhanced local flow, which would support an unusual neuronal reaction different from an increased neurotransmission activity. In agreement with this EMD 23 448 – which reduces the firing rate in nigral dopamine cells (Chiodo and Bunney, 1983) – did not counteract the increased glucose use. It cannot be excluded that a failing flow response is a direct consequence of an impairment of the vascular bed of this region, which is situated near the site of the lesion.

There is much evidence to suggest that nicotine exerts a protective effect on the lesioned mesostriatal dopamine system (Janson et al., 1988) in line with the antiparkinson effect of tobacco smoking mentioned in the Introduction. Functional observations and results from studies on dopamine utilization (Fuxe et al., 1989) suggest that the protection is related to a desensitization of the excitatory nicotinic cholinoceptors located on the nigral dopamine nerve terminals and cell bodies leading to a reduced dopaminergic neurotransmission activity and energy requirement. The recorded improvement of the glucose utilization of the striatum on the lesioned side would thus reflect the high functional activity and/or increased number of surviving and regenerating dopamine fibers within this region. This is in line with observations of a marked increase in central glucose metabolism following L-dopa injections in MPTP-treated monkeys (Porrino et al., 1987). In accordance with this, the dopamine agonist EMD 23 448 – which is an indolyl-3-butylamine acting primarily on the presynaptic autoreceptors (Seyfried et al., 1982; Goldstein et al., 1987) – had no effect on the glucose utilization following the partial hemitransection. This does not exclude the possibility that also EMD 23 448 – through its ability to lower the firing rate of nigral dopamine neurons (Chiodo and Bunney, 1983) – exerts a neuroprotective action, but that the component of dopaminergic activity (corresponding to the effect seen after the 2 weeks of nicotine treatment) is outbalanced by the autoreceptor inhibition at the striatal level. When comparing the action of EMD 23 448 with that of nicotine it should also be considered that chronic nicotine treatment can desensitize nicotinic cholinoceptors located also on the striatal nerve cells, which may lead to increased survival of such cells undergoing retrograde changes in response to the hemitransection.

The effect of acute nicotine administration on the metabolic rate of glucose in various brain regions is well correlated with the distribution of specific binding sites for radiolabelled nicotine (London et al., 1985; Grünwald et al., 1987). This may not necessarily be true for the protective or trophic action of nicotine. Further interpretation may have to await results from any blocking effects of mecamylamine, which is a ganglionic blocking agent that can cross the blood-brain barrier and is able to block the vast majority of the effects of nicotine in brain (cf. Andersson, 1985; London et al., 1985). Also studies on other neuronal systems suggest that, indeed, the nicotine-type of cholinergic receptors are involved in trophic effects (cf. Bear and Singer, 1986; Lipton et al., 1988), besides their role in intercellular communication.

The failure of nicotine to affect survival and growth in the cranial sympathetic nerves would further support the view that the effect is not related to a more general, non-specific action in terms of neuronal protection and repair but rather linked to a receptor-mediated protective effect, particularly involving the dopaminergic system. The neurons may contain not only dopamine but also dopamine co-modulators and unknown trophic factors which might, in addition, help to reduce the retrograde degeneration of nerve cells in the striatum. This can be expected to lead to improved metabolic activity of the nerve cells in this region, and thus account for part of the observed normalization in the rate of glucose utilization.

Summary

Neuroprotective and possible trophic actions of nicotine were studied in two types of experimental models: (1) one in which the meso-striatal dopamine system was subjected to partial hemitransection, and regional glucose utilization (using 2-[^{3}H]deoxyglucose) and blood flow (using [^{14}C]iodoantipyrine) were measured by computer-assisted quantitative autoradiography based on a double-isotope technique; and (2) another where the sympathetic cranial nervous system supplying the brain vasculature was subjected to decentralization, axotomy, and partial or complete ganglionectomy, and the neuronal survival and fiber regeneration were elucidated by fluorescence histochemistry of noradrenaline, tyrosine hydroxylase, and neuropeptide Y.

Continuous nicotine infusion for 4 weeks failed to significantly affect the neuronal response to the surgical interference of the sympathetic noradrenergic system. The same nicotine treatment for 2 weeks significantly improved glucose utilization and blood flow in caudate-putamen on the side in which the meso-striatal dopamine system had been transected, thus eliminating the 16% side-to-side asymmetry in the metabolism caused by the axotomy. The dopamine agonist, EMD 23 448, was without significant effect on this asymmetry. The hemitransection produced marked reduction in metabolism and flow also in the ventro-lateral thalamus. In substantia nigra, glucose utilization was markedly elevated — perhaps as a consequence of a regenerative increase in protein synthesis — opposite to a considerable reduction in nigral blood flow. Little or no effect of the hemitransection was seen in hippocampus or nucleus accumbens. In neither of these four regions did nicotine (or EMD 23 448) have any overt influence on glucose metabolism or blood flow. It is concluded that nicotine, mainly through its protective action on the meso-striatal dopaminergic system, is able to improve striatal glucose utilization and associated blood flow, probably reflecting a tendency to amelioration of neurotransmission function of surviving terminals belonging to the nigro-striatal dopamine neurons.

Acknowledgements

We are greatly indebted to Dr. Nils H. Diemer, Department of Neuropathology, University of Copenhagen, Denmark for providing access to the Leitz TAS Plus equipment in the quantitative autoradiographic measurements. Beth Andbjer, Ulla-Britt Andersson, Anders Nylén, Sverker Nystedt and Per Svensson have kindly assisted in the experiments and measurements, and Fredrik Kuylenstierna (Pharmacia-Leo) in the determinations of nicotine and cotinine. The work has been supported by grant No. 1762 from the Council for Tobacco Research, New York, by a grant from the Swedish Tobacco Company; and by the Swedish Medical Research Council (grant No. 14X-5680/732).

References

Agnati, L.F., Fuxe, K., Calza, L., Benfenati, F., Cavicchioli, L., Toffano, G. and Goldstein, M. (1983) Gangliosides increase the survival of lesioned nigral dopamine neurons and favour the recovery of dopaminergic synaptic function in striatum of rats by collateral sprouting. *Acta Physiol. Scand.*, 199: 347 – 363.

Agnati, L.F., Fuxe, K., Benfenati, F., Zoli, M., Owman, Ch., Diemer, N.H., Kåhrström, J., Toffano, G. and Cimino, M. (1985) Effects of ganglioside GM1 treatment on striatal glucose metabolism, blood flow, and protein phosphorylation of the rat. *Acta Physiol. Scand.*, 125: 43 – 53.

Andersson, K. (1985) Mecamylamine pretreatment counteracts cigarette smoke-induced changes in hypothalamic catecholamine neuron systems and in anterior pituitary function. *Acta Physiol. Scand.*, 125: 445 – 452.

Andersson, K., Fuxe, K. and Agnati, L.F. (1981a) Effects of single injections of nicotine on the ascending dopamine pathways in the rat. *Acta Physiol. Scand.*, 112: 345 – 347.

Andersson, K., Fuxe, K., Agnati, L.F. and Eneroth, P. (1981b) Effects of acute central and peripheral administration of nicotine on ascending dopamine pathways in the male rat brain. Evidence of nicotine induced increases of dopamine turnover in various telencephalic dopamine nerve terminal systems. *Med. Biol.*, 59: 170 – 176.

Baron, J.A. (1986) Cigarette smoking and Parkinson's disease. *Neurology*, 36: 1490 – 1496.

Baumann, R.J., Jameson, H.D., McKean, H.E., Haack, D.G. and Weisberg, L.M. (1980) Cigarette smoking and Parkinson's disease. 1. A comparison of cases with matched neighbors. *Neurology*, 30: 839 – 843.

Bear, M.F. and Singer, W. (1986) Modulation of visual cortical plasticity by acetylcholine and noradrenaline. *Nature (Lond.)*, 320: 172 – 176.

Björklund, A.H., Falck, B. and Owman, Ch. (1972) Fluorescence microscopic and microspectrofluorometric techniques for the cellular localization and characterization of biogenic amines. In J.E. Rall and I.J. Kopin (Eds.), *Methods in Investigative and Diagnostic Endocrinology. The Thyroid and Biogenic Amines*, Elsevier North-Holland, Amsterdam, pp. 318 – 368.

Chiodo, L.A. and Bunney, B.S. (1983) Electrophysiological studies on EMD 23,448, an indolyl-3-butylamine, in the rat: a putative selective dopamine autoreceptor agonist. *Neuropharmacology*, 22: 1087 – 1093.

Curvall, M., Kazemi-Vala, E. and Enzell, C.R. (1982) Simultaneous determination of nicotine and cotinine in plasma using capillary column gas chromatography with nitrogen-sensitive detection. *J. Cromatogr.*, 232: 283 – 293.

Fuxe, K., Janson, A.M., Jansson, A., Andersson, K. and Agnati, L.F. (1989) Chronic nicotine treatment reduces dopamine utilization in surviving forebrain dopamine nerve terminal systems after partial di-mesencephalic hemitransections. *Naunyn-Schmiedeberg's Arch. Pharmacol.* (submitted).

Gjedde, A., Hansen, A.J. and Siemkowicz, E. (1980) Rapid simultaneous determination of regional blood flow and blood-brain glucose transfer in brain of rat. *Acta Physiol. Scand.*, 118: 231 – 330.

Godwin-Austen, R.B., Lee, P.N., Marmot, M.G. and Stern, G.M. (1982) Smoking and Parkinson's disease. *J. Neurol. Neurosurg. Psychiatr.*, 45: 577 – 581.

Goldstein, M., Fuxe, K., Meller, E., Seyfried, C.A., Agnati, L. and Mascagni, F.M. (1987) The characterization of the dopaminergic profile of EMD 23,448, an indolyl-3-butylamine: selective actions on presynaptic and supersensitive postsynaptic DA receptor populations. *J. Neural. Trans.*, 70: 193 – 215.

Grünwald, F., Schrock, H. and Kuschinsky, W. (1987) The effect of an acute nicotine infusion on the local cerebral glucose utilization of the awake rat. *Brain Res.*, 400: 232 – 238.

Janson, A.M., Fuxe, K., Agnati, L.F., Kitayama, I., Härfstrand, A., Andersson, K. and Goldstein, M. (1988) Chronic nicotine treatment counteracts the disappearance of tyrosine-hydroxylase-immunoreactive nerve cell bodies, dendrites and terminals in the mesostriatal dopamine system of the male rat after partial hemitransection. *Brain Res.*, 455: 332 – 345.

Kessler, I.I. (1972) Epidemiologic studies in Parkinson's disease. 3. A community based survey. *Am. J. Epidemiol.*, 96: 242 – 254.

Kessler, I.I. and Diamond, K.L. (1971) Epidemiologic studies of Parkinson's disease. 1. Smoking and Parkinson's disease: a survey and explanatory hypothesis. *Am. J. Epidemiol.*, 94: 16 – 25.

König, F.F. and Klippel, R.A. (1974) *The Rat Brain. A Stereotaxic Atlas of the Forebrain and Lower Parts of the Brain Stem*, Robert E. Krieger, New York.

Kuhl, D.E., Metter, E.J., Riege, W.H. and Markham, C.H. (1984) Patterns of cerebral glucose utilization in Parkinson's disease and Huntington's disease. *Am. J. Neurol.*, 15: Suppl. S119 – 125.

Lipton, S.A., Frosch, M.P., Phillips, M.D., Tauck, D.L. and Aizenman, E. (1988) Nicotinic antagonists enhance process outgrowth by rat retinal ganglion cells in culture. *Science*, 239: 1293 – 1296.

London, E.D., Connolly, R.J., Szikszay, M. and Wamsley, J.K. (1985) Distribution of cerebral metabolic effects of nicotine in the rat. *Eur. J. Pharmacol.*, 110: 391 – 392.

Nefzger, M.D., Quaddfasel, F.A. and Karl, V.C. (1968) A retrospective study of smoking and Parkinson's disease. *Am. J. Epidemiol.*, 88: 149 – 158.

Owman, Ch. and Diemer, N.H. (1985) Studies of local blood flow and glucose utilization in brain by computer assisted autoradiography. In K. Fuxe, L.F. Agnati and T. Hökfelt (Eds.), *Quantitative Neuroanatomy in Transmitter Research*, MacMillan, England, pp. 71 – 87.

Owman, Ch., Fuxe, K., Janson, A.M. and Kåhrström, J. (1988) Enhanced striatal glucose metabolism by chronic nicotine treatment following meso-striatal lesions in rats. In M. Rand and K. Thurau (Eds.), *The Pharmacology of Nicotine*, IRL Press, Oxford and Washinton DC., pp. 266 – 267.

Porrino, L.J., Burns, R.S., Crane, A.M., Palombo, E., Kopin, I.J. and Sokoloff, L (1987) Local cerebral metabolic effects of L-dopa therapy in 1-methyl-4-phenyl-1,2,3,6-tetrahydro-pyridine-induced parkinsonism in monkeys. *Proc. Natl. Acad. Sci. U.S.A.*, 84: 5995 – 5999.

Sakurada, O., Kennedy, C., Jehle, J., Brown, J.D., Carbin, G.L. and Sokoloff, L. (1978) Measurement of local cerebral blood flow with iodo-(^{14}C)antipyrine. *Am. J. Physiol.*, 234: H59-H66.

Schwartzman, R.J. and Alexander, G.M. (1985) Changes in the local cerebral metabolic rate for glucose in the 1-methyl-4-phenyl-1,2,3,6-tetrahydropyridine (MPTP) primate model of Parkinson's disease. *Brain Res.*, 358: 137 – 143.

Seyfried, C.A., Fuxe, K., Wolf, H.-P and Agnati, L.F. (1982) Demonstration of a new type of dopamine receptor agonist: an indolyl-3-butylamine. Actions at intact versus supersensitive dopamine receptors in the rat forebrain. *Acta Physiol. Scand.*, 116: 465 – 468.

Sokoloff, L. (1981a) Localization of functional activity in the central nervous system by measurement of glucose utilization with radioactive deoxyglucose. *J. Cereb. Blood Flow Metab.*, 1: 7–36.

Sokoloff, L. (1981b) Relationships among local functional activity, energy metabolism, and blood flow in the central nervous system. *Fed. Proc.*, 40: 2311–2316.

Sokoloff, L., Reivich, M., Kennedy, C., Des Rosiers, M.H., Patlak, C., Pettigrew, K.D., Sakurada, O. and Shinohara, M. (1977) The (^{14}C)deoxyglucose method for the measurement of local cerebral glucose utilization: theory, procedure, and normal values in the conscious and anaesthetized albino rat. *J. Neurochem.*, 28: 897–916.

Wilkins, J.N., Carlsson, H.E., Van Vunakis, H., Hill, M.A., Gritz, E. and Jarvik, M.E. (1982) Nicotine from cigarette smoking increases circulating levels of cortisol, growth hormone and prolactin in male chronic smokers. *Psychopharmacology*, 78: 305–308.

Human Pharmacology and Nicotine Dependence

A. Nordberg, K. Fuxe, B. Holmstedt and A. Sundwall (Eds.)
Progress in Brain Research, Vol. 79

CHAPTER 27

Nicotine dependence and tolerance in man: pharmacokinetic and pharmacodynamic investigations

Neal L. Benowitz, Herve Porchet and Peyton Jacob, III

The Clinical Pharmacology Unit of the Medical Service, San Francisco General Hospital Medical Center, and The Department of Medicine and Langley Porter Psychiatric Institute, University of California, San Francisco, CA 94143, U.S.A.

Introduction

Although vigorously debated over the years, there is now compelling evidence that tobacco use results in addiction to nicotine. Criteria for drug dependence, based on concepts developed by the World Health Organization and other scientific and medical organizations, have been summarized in the recent U.S. Surgeon General's Report (US. Department of Health and Human Services, 1987) (Table I). Tobacco use meets these criteria and nicotine is clearly the dependence-producing component.

Pharmacokinetic and pharmacodynamic characteristics of a drug are important determinants of dependence liability, the temporal patterns of drug use, and the level of drug use. Ultimately, an understanding of pharmacokinetics and pharmacodynamics may be useful in developing effective treatment strategies. Several pharmacokinetic and pharmacodynamic characteristics appear to be necessary or optimal for a drug to produce dependence: the drug must be effectively absorbed into the blood stream, the drug must rapidly enter into the brain, and the drug must be psychoactive and that psychoactivity related to levels of the drug in the brain. These characteristics allow for the drug abuser to manipulate the dose of his or her drug to optimize mood and psychological functioning and are most likely to result in the behavior described as criteria for drug dependence. This chapter will

examine the pharmacokinetic and pharmacodynamic characteristics of nicotine and how these characteristics might determine cigarette smoking behavior and tobacco addiction.

Absorption of nicotine

Nicotine is a tertiary amine composed of a pyridine and a pyrrolidine ring. Nicotine is a weak base with a pKa of 8.0. At physiological pH, about 31% of

TABLE I

Criteria for drug dependence (from U.S. Department of Health and Human Services, 1987)

Primary criteria
  Highly controlled or compulsive use
  Psychoactive effects
  Drug-reinforced behavior
Additional criteria
  Addictive behavior often involves:
    stereotypic patterns of use
    use despite harmful effects
    relapse following abstinence
    recurrent drug cravings
  Dependence-producing drugs often produce:
    tolerance
    physical dependence
    pleasant (euphoriant) effects

nicotine is non-ionized such that it readily crosses cell membranes. The pH of tobacco smoke is important in determining the absorption of nicotine from different sites within the body. The pH of smoke from flue-cured tobaccos found in most cigarettes is acidic. At this pH, the nicotine is primarily ionized. In its ionized state, such as in acidic environments, nicotine does not rapidly cross membranes. As a consequence, there is little buccal absorption of nicotine from cigarette smoke, even when it is held in the mouth. The pH of smoke from aircured tobaccos, such as in pipes, cigars and in a few European cigarettes, is alkaline and nicotine is primarily un-ionized. Smoke from these products is well absorbed through the mouth.

When tobacco smoke reaches the small airways and alveoli of the lung the nicotine is rapidly absorbed independent of pH of the smoke. Blood concentrations of nicotine rise quickly during and peak at the completion of cigarette smoking (Fig. 1). The rapid absorption of nicotine from cigarette smoke through the lungs, presumably because of the huge surface area of the alveoli and small air-

ways and dissolution of nicotine into fluid of pH in the human physiologic range, facilitates transfer across cell membranes.

Chewing tobacco, snuff, and nicotine gum are buffered to alkaline pH to facilitate absorption of nicotine through mucous membranes. Concentrations of nicotine in the blood rise gradually with the use of smokeless tobacco and tend to plateau at about 30 min with levels persisting and declining only slowly over 2 h or more (Fig. 1). Based on studies comparing concentrations of nicotine in the blood following tobacco use with concentrations achieved after intravenous administration of nicotine, the average dose of nicotine delivered to the circulation is estimated at 1 mg after smoking a cigarette and 3 – 5 mg following a dose of oral snuff or chewing tobacco, the latter held in the mouth for 30 min (Benowitz et al., 1988). In a study of 22 cigarette smokers, smoking an average of 36 cigarettes per day (range 20 to 62), we found an average daily intake of 37 mg, with a range of 10 to 79 mg nicotine (Benowitz and Jacob, 1984). This daily dose in some cases exceeds the single dose estimate of 60 mg which has been reported to be fatal.

Nicotine blood levels during tobacco use

Blood or plasma concentrations of nicotine sampled in the afternoon in smokers generally range from 10 to 50 ng/ml ($0.6 - 3 \times 10^{-7}$ M). The increment in blood nicotine concentration after smoking a single cigarette ranges from 5 to 30 ng/ml, depending on how the cigarette is smoked. Peak blood concentrations of nicotine are similar, although the rate of rise of nicotine is slower for cigar smokers and users of snuff and chewing tobacco, compared to cigarette smokers.

Nicotine is rapidly and extensively metabolized, primarily in the liver, but also to a small extent in the lung. Renal excretion depends on urinary pH and urine flow, and accounts for 2 – 35% of total elimination. The total clearance of nicotine averages 1300 ml/min; renal clearance averages 100 – 200 ml/min. (Benowitz et al., 1982a). The

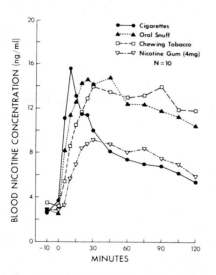

Fig. 1. Blood concentrations during and after cigarette smoking for 9 min (1 – 3 cigarettes), oral snuff (2.5 g), chewing tobacco (average 7.9 g), and nicotine gum (two 2 mg pieces). Average values for 10 subjects (± S.E.M.). Smokeless tobacco and nicotine gum exposure was 30 min.

half-life of nicotine averages 2 h, although there is considerable individual variability (range 1 to 4 h). The relatively long half-life of nicotine in the face of rapid intrinsic metabolism is due to extensive tissue distribution (average volume of distribution 180 l).

Smoking is commonly considered to be a process of intermittent dosing of nicotine, which is in turn rapidly eliminated from the body. There is considerable peak to trough oscillation levels from cigarette to cigarette. However, consistent with a half-life of 2 h, nicotine accumulates over 6 – 8 h of regular smoking and nicotine levels persist overnight, even as the smoker sleeps (Benowitz et al., 1982b). Thus, smoking results not in intermittent exposure but in exposure to nicotine that lasts 24 h of each day.

In summary, nicotine is well absorbed from all tobacco products and substantial concentrations of nicotine are achieved in the blood. These blood concentrations tend to be similar in people who use different forms of tobacco. With regular use, there is accumulation of nicotine in the body throughout the day with persistence of significant concentrations overnight.

Entry of nicotine into the brain

Smoking is a unique form of systemic drug administration in that entry into the circulation is through the pulmonary rather than the portal or systemic venous circulations. The lag time between smoking and entry of nicotine into the brain is shorter than after intravenous injection. Nicotine rapidly moves from blood into the brain. Nicotine is rapidly and extensively taken up from the blood in its first pass through the brains of rats (Oldendorf, 1974). By autoradiographic techniques, it has been shown that high levels of nicotine are present in the brain 5 min after i.v. injections in mice and that most nicotine is cleared from the brains by 30 min (Hansson and Schmiterlow, 1962). Intravenously injected [14C]nicotine is immediately taken up in the brains of mice, reaching a maximum concentration within 1 min after injection (Stalhand-

ski and Slanina, 1970). Similar findings based on positron emission tomography of the brain were seen after injections of [11C]nicotine in monkeys (Maziere et al., 1976).

Brain concentrations decline quickly as nicotine is distributed to other body tissues. Using partition coefficients (the ratio of concentration in tissue compared to blood at steady state) derived from experiments in rabbits, along with organ weights and blood flows taken from people, one can use perfusion models to simulate the concentrations of nicotine in various organs after smoking a cigarette (Fig. 2). Concentrations in the lungs, arterial blood and the brain are seen to increase sharply following exposure, then decline over 20 – 30 min as nicotine is redistributed to other body tissues, particularly skeletal muscle. Thereafter, concentrations of nicotine decline more slowly as nicotine is distributed out of storage sites and eliminated.

The rapid absorption of nicotine through the lungs into the blood and into the brain provides the possibility for the smoker to precisely control levels of nicotine in the brain and, hence, to

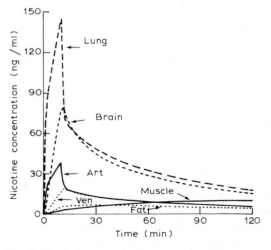

Fig. 2. Perfusion model simulation of the distribution of nicotine in various tissues following infusion of 1.5 mg of nicotine into the lung compartment over 10 min. The dose, route and dosing regimen were intended to mimic smoking a cigarette. Typical human organ weights and blood flows and partition coefficients from rabbits were used for the simulation.

modulate pharmacologic effects. To better understand this process, one must examine the relationship between levels of nicotine and effects in the brain.

Pharmacodynamics of nicotine in the brain

That nicotine binds to specific receptors in the brain and affects cerebral metabolism in the areas of specific binding has been reviewed elsewhere in this symposium proceedings. The issue to be examined here is the relationship between doses of nicotine and effects and levels of nicotine and effects over time.

Nicotine is known to have a complex dose-response relationship. In experimental preparations, nicotine in low doses causes ganglionic stimulation and in high doses causes ganglionic blockade following brief stimulation. This biphasic response pattern is observed in the intact organism as well, although the mechanisms are far more complex. With regard to central nervous system effects, Ashton has provided evidence that nicotine exerts biphasic actions in humans (Ashton et al., 1980). Examining contingent negative variation (CNV, a component of the evoked EEG potential), nicotine in low doses increased while high doses decreased the amplitude of the CNV. It was proposed that low doses produced arousal or increased attention and vigilance, while higher doses produced relaxation and reduced stress. Presumably, the smoker, by titrating the level of nicotine in the brain, can achieve the desired mental state.

The time course of nicotine absorption strongly influences the actions of nicotine, although it is not clear whether this is a result of a more rapid rate of rise or higher absolute levels of nicotine in the brain. We and others have found that heart rate acceleration, a reflection of the level of sympathetic neural discharge and a sensitive physiologic response, tends to correlate with the magnitude of subjective effect (Rosenberg et al., 1980; West and Russell, 1987). Heart rate acceleration appears to be mediated by the central nervous system either through actions on chemoreceptor afferent pathways or via direct effects on the brain stem (Comroe, 1960). With repeated intravenous injections of nicotine, dosed to simulate cigarette smoking, heart rate acceleration and subjective effects were greatest with the first series of injections and then declined over time with successive injections (Rosenberg et al., 1980).

To examine the importance of rate of dosing of nicotine and resultant actions, and the relationship between effects and brain levels of nicotine, we compared the consequences of rapid and slow loading infusions of nicotine (Porchet et al., 1988b). The study was performed in seven volunteer smokers who had abstained from smok-

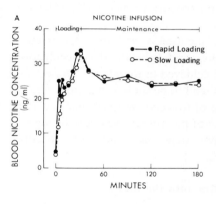

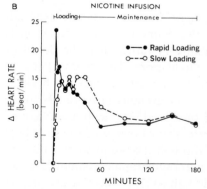

Fig. 3. Average blood nicotine concentration (A) and increase in heart rate (B) in seven subjects. On the rapid loading day, subjects received 17.5 μg/kg per min nicotine for 1.5 min, then 1.75 μg/kg per min for 30 min; on the slow loading infusion day, 2.5 μg/kg per min for 30 min. Maintenance infusion on both days was 0.35 μg/kg per min from 30 – 180 min. (Reprinted with permission from Porchet et al., 1988b.)

ing overnight. Venous blood concentrations were not markedly different with the two infusion rates except for the first 5 min, where the increase in blood concentration was greater with rapid infusion (Fig. 3). Peak concentrations were seen at 30 min and were similar for both infusions. However, the heart rate increased much more quickly and to a greater level, and was associated with transient dizziness and euphoria after rapid infusion compared with slow infusion. During the slower infusion, heart rate increased slowly, peaking at 30 min, and subjective effects were minimal. Using kinetic data derived from subjects of this study and studies in rabbits and with the use of a perfusion model, we estimated levels of nicotine in the human brain during the rapid and slow infusions. Predicted brain concentrations rose much more rapidly and to a higher level after rapid infusion, with a concentration time course similar to that as seen for heart rate acceleration (Fig. 4). Brain concentrations versus heart rate acceleration tended to change in parallel. These data suggest that effects of nicotine track brain concentrations over time and illustrate the concept that the concentration in the brain is augmented and psychoactive effects are greater when the drug is dosed more rapidly. This may explain why smoking a cigarette may be

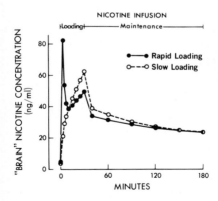

Fig. 4. Predicted mean "brain" concentrations of nicotine over time (based on a physiological kinetic model, fit to the rabbit data and scaled to humans). Infusion rates are as described in Fig. 3. Note that "Brain concentrations" peak at the same time as heart rate (see Fig. 3). (Reprinted with permission from Porchet et al., 1988b.)

preferred over the use of smokeless tobacco (where absorption is slow) and why nicotine gum is not as satisfying as smoking a cigarette.

Tolerance to nicotine

Another pharmacologic issue of importance in drug dependence is the development of tolerance. Tolerance is defined as when, after repeated doses, a given dose of a drug produces less effect. In some cases, increasing the dose of a drug in the presence of tolerance may achieve the same effect which was obtained after the first dose. Pharmacodynamic tolerance can be further defined as when a particular concentration of a drug at a receptor site (in the intact organism approximated by the concentration in the blood) produces less of an effect than it did after a prior exposure.

Studies in animals demonstrate rapid development of tolerance to many effects of nicotine, although tolerance may not be complete (Marks et al., 1985). Smokers know that tolerance develops to some effects of smoking. Smoking the first cigarette as a teenager is commonly associated with dizziness, nausea and/or vomiting, effects to which the smoker rapidly becomes tolerant. Smoking a cigarette after 24 h of abstinence increases heart rate, whereas smoking an identical cigarette during the course of the day fails to increase heart rate (West and Russell, 1987). Pharmacodynamic tolerance to heart rate acceleration and subjective effects is seen within the course of a 30 min intravenous infusion of nicotine (Benowitz et al., 1982a). Tolerance may develop to toxic effects, such as nausea, vomiting and pallor, even during the 8 h course of an accidental nicotine poisoning, despite persistence of nicotine in the blood in extremely high concentrations (200–300 ng/ml) (Benowitz et al., 1987).

These and other findings indicate that tolerance is gained rapidly, within minutes (tachyphylaxis). Tolerance is lost, at least to some degree, after overnight abstinence. To characterize the time course of tolerance development and regression, we conducted a study in which subjects received paired intravenous infusions of nicotine, separated by different time intervals (Porchet et al., 1988a). Despite higher blood concentrations, heart rate acceleration and subjective effects were much less when a second infusion was given at 60 or 120 min after the first infusion (Fig. 5). At 240 min, the response was fully restored. The pharmacodynamics were modeled using a pharmacokinetic-pharmacodynamic model (Fig. 6). This modeling indicates a half-life of development and regression of

tolerance of 35 min (Table II). Such modeling suggests that at three half-lives or one and one-half hours after a cigarette, nearly full sensitivity should have been regained. The interval at which cigarette smokers smoke cigarettes may be determined to a considerable degree by the kinetics of regression of tolerance. The model also predicts that, at a steady state concentration of 30 ng/ml, the response to another dose of nicotine will be attenuated to 20% of the initial response. Tolerance

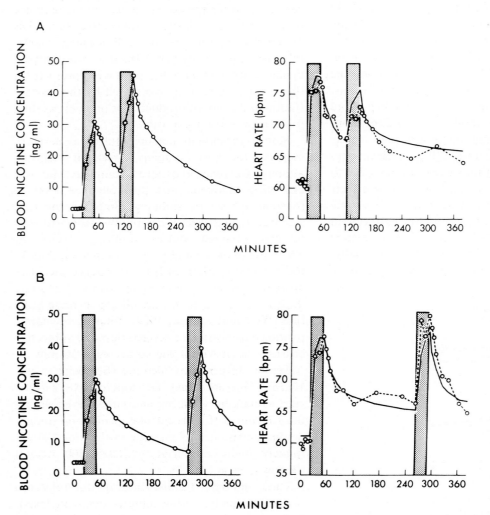

Fig. 5. Mean blood concentrations of nicotine (A) and the corresponding mean heart rate (B) in eight subjects after two 30 min intravenous infusions of 2.5 µg/kg per min of nicotine separated by 60 min (A) or 210 min (B). The shadowed area indicates the periods during which nicotine was infused. The solid line shows the fit of the model to the effect data. (Reprinted with permission from Porchet et al., 1988a.)

is not complete. Some persistent effect of nicotine may be expected despite a substantial degree of tolerance.

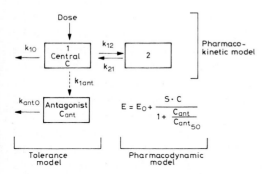

Fig. 6. Diagrammatic representation of the pharmacokinetic and pharmacodynamic model for nicotine tolerance. The k_{ij} are the intercompartmental and elimination rate constants, C is the concentration of the agonist, C_{ant} is the concentration of the hypothetical antagonist, S is the slope of the linear relationship between effect and concentration, E is the effect, and E_0 is the baseline effect (Reprinted with permission from Porchet et al., 1988.)

TABLE II

Pharmacodynamic model of tolerance: parameter estimates

$$E = E_0 + \frac{SC}{(1 + C_{ant}/C_{ant_{50}})}$$

$S = 1.31$ bpm/ng per ml (12.8, C.V.%)
$C_{ant_{50}} = 7.72$ ng/ml (18.8)
$E_0 = 61.2$ bpm (0.5)
$K_{ant_0} = 0.0195 \cdot$ min^{-1} (13.8)
$t\text{-}1/2_{tolerance} = 35$ min

E	=	effect
E_0	=	baseline effect
S	=	slope of linear relationship between effect and concentration
C	=	concentration of nicotine
C_{ant}	=	concentration of the hypothetical antagonist
$C_{ant_{50}}$	=	concentration of the hypothetical antagonist that results in loss of 50% of the maximal effect due to tolerance.
K_{ant_0}	=	first order rate of appearance or disappearance of tolerance

* C.V% = coefficient of variation.

Pharmacokinetics and pharmacodynamics of nicotine dependence

The reinforcing properties of a drug are expected to be strongest when a psychoactive effect, usually a pleasant one, follows in close temporal proximity to the self-administration of a drug. In pharmacokinetic or pharmacodynamic terms, after dosing the drug should enter the bloodstream rapidly and move rapidly from the bloodstream into the brain. The appearance of the drug in the brain should be associated with the desired psychoactive effects. If the effect of a drug is delayed after appearance of the drug in the brain or if tolerance to the drug effect is already present, the drug is less likely to be reinforcing.

That the user can easily control the dose of the drug and modulate the resultant psychoactivity would be expected to further strengthen the reinforcing nature of the drug. As discussed previously, nicotine obtained by cigarette smoking demonstrates these characteristics, as do other drugs of abuse which are inhaled, such as crack, cocaine or marijuana, and drugs which are intravenously injected, such as heroin and cocaine.

Tolerance is most likely to develop to the effects of a drug when its receptors are continuously exposed to the drug. Frequent and sustained dosing and/or a long half-life of a drug would favor development of tolerance (Busto and Sellers, 1986). Barbiturates and caffeine are good examples of drugs with long half-lives and high degrees of tolerance. The half-life of nicotine in the brain following a single dose exposure is short, probably about 10 min in humans, due to redistribution out of the brain to other body tissues. Some degree of tolerance does develop rapidly (tachyphylaxis), even after brief exposures. Such tolerance also regresses relatively quickly (half-life, 35 min) such that a smoker may learn that smoking a cigarette at particular intervals is more reinforcing than smoking more frequently.

With repetitive dosing, levels of nicotine build up in the body (and in the brain) in accordance with the terminal half-life of 2 h. Thus, there is substantial accumulation of nicotine in the brain and an increasing general level of tolerance throughout the day. Presumably, however, smoking individual cigarettes still results in peaks of nicotine in the brain, which may overcome the underlying level of tolerance and produce some of the desired effects.

Withdrawal symptoms are most likely to occur when there has been a sustained effect in the brain and the drug is then rapidly removed (Busto and Sellers, 1986). Thus, rapid exit from specific brain regions would favor more severe withdrawal symptoms. Nicotine, as it is dosed by cigarette smokers, meets both of these characteristics. Due to repetitive dosing, the brain is exposed to nicotine for prolonged periods of time such that neuroadaptation occurs during the smoking day. When exposure is terminated, nicotine rapidly exits the brain and withdrawal symptoms are experienced. It is likely, therefore, that relief of withdrawal symptoms plays an increasingly important role in determining smoking behavior as the day progresses.

TABLE III

Pharmacokinetic properties of psychoactive drugs associated with dependence (adapted from Busto and Sellers, 1986)

Reinforcement	Tolerance	Withdrawal syndrome
Rapid absorption into blood stream	Cumulative exposure (dose frequency, duration)	Rapid exit from specific brain regions (redistribution, elimination)
Rapid entry into brain (lipid solubility, molecular size, transport processes)	Long half-life	Short half-life
Psychoactivity related to brain levels	Low clearance	High clearance

Nicotine and the daily smoking cycle

The pharmacokinetic and pharmacodynamic considerations discussed thus far help us understand the development of nicotine dependence, human cigarette smoking behavior as well as adverse effects of cigarette smoking (Table III). The daily smoking cycle can be conceived as follows. The first cigarette of the day produces substantial pharmacologic effects, primarily arousal, but at the same time tolerance begins to develop. A second cigarette may be smoked at a later time, at which the smoker has learned there is some regression of tolerance. With subsequent cigarettes, there is accumulation of nicotine in the body, resulting in a greater level of tolerance and withdrawal symptoms become more pronounced between cigarettes. Transiently high brain levels of nicotine following smoking individual cigarettes may partially overcome tolerance. But the primary (euphoric) effects of individual cigarettes tend to lessen throughout the day. Overnight abstinence allows considerable resensitization to actions of nicotine. Because of the dose response and tolerance characteristics, habitual smokers need to smoke at least 15 cigarettes and consume 20–40 mg nicotine per day to achieve the desired effects of cigarette smoking and minimize withdrawal discomfort throughout the day.

Acknowledgements

Supported in part by US Public Health Service grants DA02277, CA32389 and DA01696. These studies were carried out in part in the General Clinical Research Center (RR-00083) with support of the Division of Research Resources, National Institutes of Health.

References

Ashton, H., Marsh, V.R., Millman, J.E., Rawlins, M.D., Telford, R. and Thompson, J.W (1980) Biphasic dose-related responses of the CNV (contingent negative variation) to i.v. nicotine in man. *Br. J. Clin. Pharmacol.*, 10: 579–589.

Benowitz, N.L. and Jacob, P. III. (1984) Daily intake of nicotine during cigarette smoking. *Clin. Pharmacol. Ther.,* 35: 499 – 504.

Benowitz, N.L., Jacob, P. III, Jones, R.T. and Rosenberg, J. (1982a) Interindividual variability in the metabolism and cardiovascular effects of nicotine in man. *J. Pharmacol. Exp. Ther.,* 221: 368 – 372.

Benowitz, N.L., Kuyt, F. and Jacob, P. III. (1982b) Circadian blood nicotine concentrations during cigarette smoking. *Clin. Pharmacol. Ther.,* 32: 758 – 764.

Benowitz, N.L., Lake, T., Keller, K.H. and Lee, B.L. (1987) Prolonged absorption with development of tolerance to toxic effects following cutaneous exposure to nicotine. *Clin. Pharmacol. Ther.,* 42: 199 – 120.

Benowitz, N.L., Porchet, H., Sheiner, L. and Jacob, P. III (1988) Nicotine absorption and cardiovascular effects with smokeless tobacco use: comparison to cigarettes and nicotine gum. *Clin. Pharmacol. Ther.,* 44: 23 – 28.

Busto, U. and Sellers, E.M. (1986) Pharmacokinetic determinants for drug abuse and dependence. A conceptual perspective. *Clin. Pharmacokin.,* 11: 144 – 153.

Comroe, J.H. (1960) The pharmacological actions of nicotine. *Ann. N.Y. Acad. Sci.,* 90: 48 – 51.

Marks, M.J., Stitzel, J.A. and Collins, A.C. (1985) Time course study of the effects of chronic nicotine infusion on drug response and brain receptors. *J. Pharmacol. Exp. Ther.,* 235: 619 – 628.

Maziere, M., Comar, D., Marazano, C. and Berger, G. (1976) Nicotine-[11]C: synthesis and distribution kinetics in animals. Eur. J. Nucl. Med., 1: 255 – 258.

Oldendorf, W.H. (1974) Lipid solubility and drug penetration in the blood-brain barrier. *Proc. Soc. Exp. Biol Med.,* 147: 813 – 816.

Hansson, E. and Schmiterlow, C.G. (1962) Physiological disposition and fate of ^{14}C-labelled nicotine in mice and rats. *J. Pharmacol. Exp. Ther.,* 137: 91 – 102.

Porchet, H.C., Benowitz, N.L. and Sheiner, L.B. (1988a) Pharmacodynamic model of tolerance: application to nicotine. *J. Pharmacol. Exp. Ther.,* 244: 231 – 236.

Porchet, H.C., Benowitz, N.L., Sheiner, L.B. and Copeland, J.R. (1988b) Apparent tolerance to the acute effect of nicotine results in part from distribution kinetics. *J. Clin. Invest.,* 80: 1466 – 1471.

Rosenberg, J., Benowitz, N.L., Jacob, P. III and Wilson, K.M. (1980) Disposition kinetics and effects of intravenous nicotine. *Clin. Pharmacol. Ther.,* 28: 516 – 522.

Stalhandski, T. and Slanina, P. (1970) Effect of nicotine treatment on the metabolism of nicotine in the mouse liver in vitro. *Acta Pharmacol. Toxicol.,* 28: 75 – 80.

U.S. Department of Health and Human Services. (1987) The Health Consequences of Smoking: Nicotine Addiction. A Report of the Surgeon General. U.S. Dept. of Health and Human Services, Public Health Service, Centers for Disease Control Office on Smoking and Health. DHHS (CDC) Publication No. 88 – 8406.

West, R.J. and Russell, M.A.H. (1987) Cardiovascular and subjective effects of smoking before and after 24 h of abstinence from cigarettes. *Psychopharmacology,* 92: 118 – 121.

A. Nordberg, K. Fuxe, B. Holmstedt and A. Sundwall (Eds.)
Progress in Brain Research, Vol. 79
© 1989 Elsevier Science Publishers B.V. (Biomedical Division)

CHAPTER 28

Subjective and behavioural effects of nicotine in humans: some sources of individual variation

M.A.H. Russell

Addiction Research Unit, Institute of Psychiatry, University of London, U.K.

Introduction

It is not my purpose to attempt to review all the well documented pharmacological effects of nicotine or its effects on the mood and performance of human smokers (see U.S. Surgeon General's 1988 Report). Over recent years, the work of those of us concerned with smoking in humans has tended to focus on *whether* people smoke mainly for nicotine. Now that this has been established, it is appropriate to begin to address the more difficult question of *why* they smoke for nicotine. Which of its many effects do they seek or find reinforcing? Which effects mediate satisfaction and the relief of craving? How are these affected by blood nicotine levels and chronic exposure, by tolerance and pharmacokinetic factors, by learning and conditioning and by personality and social factors? To what extent are individual differences in the response to nicotine innate or acquired? How much do differences in subjective and pharmacological responses to nicotine simply reflect past smoking behaviour, or do they determine who smokes and who becomes highly dependent? I have no answers to these questions and can only identify some of the major gaps and problems in this area. It is clear, however, that understanding of the subjective and behavioral effects of nicotine in humans will only be achieved within the context of these multiple sources of individual variation, among which prior exposure to nicotine and multi-

ple experiences of its effects are likely to be the most crucial.

Nicotine as the controlling factor in smoking

It has been widely assumed that nicotine, as the major psychoactive substance in tobacco, is the main factor underlying its widespread use. Until recently the evidence has been largely circumstantial, but nicotine's role as the primary and controlling factor in smoking has now been firmly established by evidence from three main areas of behavioural research. These involve its role as a primary reinforcer, its controlling influence in the self-regulation of smoke intake and its effects on the cigarette withdrawal syndrome.

One serious gap in the evidence was an inability to demonstrate that nicotine can, like other drugs of abuse, act as a primary reinforcer. This matter has now been settled by Goldberg and his colleagues (Goldberg et al. 1981; Spealman and Goldberg, 1982; Risner and Goldberg, 1983). In a series of studies in several species they have shown unequivocally that animals will press a lever to self-administer nicotine. Why the early studies failed can be explained by use of excessive doses and accumulation of nicotine to aversive levels. Goldberg and his colleagues got around this problem by use of schedules which included time-out periods and the requirement of multiple responses to elicit each injection.

With the advent of reliable methods for measuring nicotine in blood, it has been shown that most smokers absorb sufficient nicotine to obtain pharmacological effects and tend to regulate the way they smoke to maintain their individually characteristic blood nicotine levels which are kept fairly constant from one day to the next. It is now clear that blood, and presumably brain, levels of nicotine exercise a controlling influence on smoking behaviour and smoke inhalation independent of the effects of sensory factors and those of other smoke components. These issues were reviewed at a meeting in Lexington 3 years ago (Russell, 1987) and in Australia last year (Russell, 1988b), so will not be repeated in this chapter. There are, however, major areas which are not yet understood. While it is clear, for example, that smokers down-regulate their nicotine intake with great precision to avoid the toxic aversive effects of exceeding their usual blood nicotine levels, up-regulation to maintain usual smoking levels is far less precise. Although pharmacologists have identified a host of nicotine effects that are potentially rewarding and reinforcing, indeed an abundance of such effects has been on offer for many years (Murphree and Weyer, 1967), it is still unknown which of these effects are of motivational importance to smokers and whether or not satisfaction and relief of withdrawal are mediated by the same or different processes.

Perhaps the most important contribution to establishing the pharmacological role of nicotine in the motivation of human smoking has come from the development, in Sweden, of the nicotine polacrilex gum (Ferno et al., 1973). This product and the scientific studies it has generated have made clinicians, psychologists and the public at large more fully aware of the nature of smoking as a form of drug dependence. Its effects, compared with placebo, on the alleviation or reversal of cigarette withdrawal effects and as an aid to smoking cessation have provided new evidence of the role of nicotine in smoking, and have stimulated a search for more effective forms of nicotine replacement for the treatment of dependent smokers (Pomerleau and Pomerleau, 1988).

In view of the progress made in these three areas of behavioural research, together with the surge of progress on the neuropharmacology of nicotine and nicotinic receptors, the central role of nicotine as the major factor responsible for tobacco use is now clearly established. Through its action on the nicotinic receptors of acetylcholine it has widespread direct effects throughout the nervous system as well as numerous indirect effects caused by the release of many other neurotransmitters and hormones. It can act as a primary reinforcer, it induces tolerance, and physical as well as psychological changes occur on withdrawal. Chronic exposure to nicotine induces an increase in the number of nicotinic cholinergic receptors in many parts of the brain, indicating the presence of structural as well as functional changes in the brain and nervous system of smokers (Benwell et al., 1988). Nicotine therefore displays all the classical hallmarks of an addictive drug. The strength of this evidence has led the recent U.S. Surgeon General's Report (1988), significantly entitled "Nicotine Addiction," to state quite categorically that cigarettes and other forms of tobacco are addicting, that nicotine is the drug in tobacco that *causes* addiction, and that the underlying pharmacological and behavioural processes are similar to those that determine addiction to drugs such as heroin and cocaine.

Natural history of smoking behaviour

In those who take up smoking, the behaviour has a fairly typical and almost stereotyped natural history. The onset of smoking tends to occur during the teenage years, the proportion who start smoking after the age of 20 being relatively small (McKennell and Thomas, 1967). A striking feature of the recruitment process is that it is largely settled after the first few cigarettes. Of those who smoke more than one cigarette some 82% go on to become regular smokers as adults, and of those who smoke at least four cigarettes 94% escalate to regular smoking (McKennell and Thomas, 1967). Having bevome a regular smoker, the behaviour is then highly refractory and the majority continue smoking for the next 40 years or more, despite wanting to stop within the first 10 years and, as the years go by and the health risks assume closer personal relevance, making numerous abortive attempts to do so. Of those who smoke regularly only about 35% stop permanently before the age of 60 (Jarvis and Jackson, 1988). This outline of the typical smoking career testifies to the addictiveness of the behaviour.

Recruitment to smoking

The uptake of smoking is mediated by psychosocial factors which are fairly well known. Awareness of the health risks is not a strong deterrent at this stage. They are presumably too remote in time to compete with the more immediate psychosocial motives and pressures to smoke. The sensory and pharmacological effects of nicotine are initially aversive, but sensory adaptation to the local irritancy of nicotine and other smoke components appears to develop rapidly and skill may be acquired to regulate smoke intake to an acceptable level. Tolerance to aversive effects such as nausea may also develop rapidly, but it is not known whether the absence of nausea after the first few cigarettes is attributable to rapid development of chronic tolerance, to quickly learnt self-titration of nicotine intake or to selective dropping out. In some novices the predominant subjective nicotine effect is dizziness or light headedness, while in others it is nausea. There is some evidence that the former group are more likely to smoke again than are those who experience nausea or no drug effect (Hirschman et al., 1984).

Chronic tolerance to the inhibitory effect of nicotine on the locomotor activity of rats has been shown to develop rapidly, within 3 to 4 days, after which it is replaced by a stimulant effect (Morrison and Stephenson, 1972; Stolerman et al., 1973; Clarke and Kumar, 1983). Anecdotal evidence suggests that smokers develop chronic tolerance to the nausea-inducing effect of nicotine, but no systematic studies have been done. Indeed, the only study to systematically compare abstinent chronic smokers with non-smokers found no difference in their sensitivity to the effects of intravenous nicotine on heart rate, blood pressure and EEG (Murphree, 1979). There is, therefore, no evidence as yet that the rapid development of chronic tolerance to initially aversive effects of nicotine plays a part in facilitating the uptake or maintenance of smoking.

Whatever mechanisms may be involved, the aversiveness of the first one or two cigarettes declines rapidly on repetition. If the psychosocial pressure to smoke is sufficient for the act to be repeated despite its initial aversiveness, it is understandable that smoking will continue once the aversiveness subsides and the effects shift from producing nausea and dizziness to become calming and pleasurable. There are no published data on blood nicotine levels immediately after a cigarette in novice smokers, but estimates based on expired-air carbon monoxide and saliva cotinine levels suggest that the nicotine intake per cigarette of children in the early stages of smoking 2 – 3 cigarettes per week is similar to that of more regular daily smokers (McNeill et al., 1986, 1987).

Maintenance of smoking

Expressed motives for smoking

To find out why people smoke it is reasonable to take some account of the reasons they give and the situations in which they are most likely to smoke. Some years ago we searched the literature for self-reported data of this kind and borrowing freely from previously published work and making use of our own clinical experience with smokers, we developed a 34-item questionnaire covering a wide variety of reasons and motives for smoking (Russell et al., 1974). The responses of 175 normal smokers were factor analysed and the seven main smoking motives are shown in Table I, together with their correlations with age and cigarette consumption. The "sedative" smoking factor was the least robust. It can be seen that the "stimulation", "sedative", "addictive" and "automatic" factors all correlated most highly with cigarette consumption. They also correlated with each other to form a higher order "pharmacological-addictive" dimension, and differentiated the main sample of normal smokers from a criterion sample of 103 addicted heavy smokers attending withdrawal clinics (Fig. 1). These results are therefore broadly compatible with the known pharmacological effects of nicotine.

Three-stage model of dependence

The fact that two-thirds of smokers do not stop before the age of 60, despite awareness of the health risks and despite wanting to stop and trying to stop, underlines the extent to which dependence is a key feature of the maintenance of smoking. The development of dependence can be viewed as proceeding through three main overlapping stages. After a *first stage* of smoking for psychosocial motives, most smokers progress fairly rapidly to a *second stage* of smoking for the pleasurable, calming and mildly stimulant effects of nicotine and its interaction with the sensorimotor components of the habit (positive reinforcement). Some smokers may then progress to a *third stage* where few of the cigarettes smoked are positively pleasurable, where smoking continues (when permitted) in most situations, and where the behaviour is largely controlled by the need to stave off or avoid withdrawal effects

TABLE I

Correlation of smoking motivation factors with age and daily cigarette consumption ($n = 175$)

Motive to smoke	Age	Consumption
Pyschosocial	−0.23*	0.16
Sensorimotor	0.09	0.08
Indulgent	−0.08	0.27*
Stimulation	0.16	0.50**
Sedative	−0.11	0.45**
Addictive	0.11	0.63**
Automatic	−0.03	0.56**

* $p < 0.01$, ** $p < 0.001$. Abstracted from Russell et al. (1974) J. R. Statist. Soc. A, 137: 313–333, 1974.

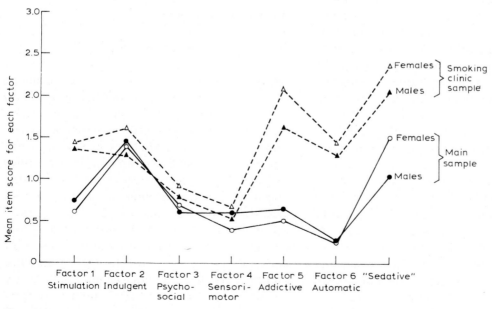

Fig. 1. Mean factor scores by sex for main sample of normal smokers ($n = 175$) and criterion sample of addicted heavy smokers attending a smokers clinic ($n = 103$). (From Russell et al., 1974.)

(negative reinforcement). The study described above gives some support to this model.

Primary effects and acquired withdrawal effects

Although it is unlikely that many remain life-long psychosocial smokers, the proportion of smokers falling predominantly into stages two or three, or between the two, is unknown. Indeed, it is by no means clear to what extent the well-known effects of nicotine on mood and performance represent primary positive effects or whether they reflect a reversal of the negative effects of withdrawal. The problem is illustrated by comparison of Tables II and III. The main subjective and behavioral effects

TABLE II

Commonly reported behavioral and subjective effects of smoking (from Pomerleau, 1986)

Improved concentration/ability to tune out irrelevant stimuli
Improved memory (recall)
Improved psychomotor performance
Increased alertness/arousal
Decreased anxiety/tension
Pleasure/facilitation of pleasure
Decreased body weight (reduced preference for sweet tastes)

TABLE III

Main features of the tobacco withdrawal syndrome

Subjective effects	Physical and objective effects
Craving for tobacco	EEG changes
Increased tension	Drop in heart rate and blood pressure
Restlessness	Increased peripheral circulation
Irritability	Drop in urinary adrenaline,
Aggressiveness	noradrenaline and cortisol
Depression	Sleep disturbances
Difficulty with	Increased weight
concentration	Decreased performance on:
Increased hunger	vigilance tasks
	simulated driving
	selective attention tasks
	memory tasks

of nicotine reported in the literature and summarised by Pomerleau (1986) are shown in Table II, while the main features of the tobacco withdrawal syndrome are listed in Table III. Many of these effects of cigarette withdrawal are nicotine-related in that they are reversed or alleviated by nicotine more effectively than by placebos or nicotine-free cigarettes. It can be seen that the withdrawal effects (Table III) are largely a mirror image of the positive effects or benefits of smoking (Table II). Since most of the data on the beneficial effects of nicotine on human mood and performance are derived from studies in smokers after short-term deprivation, it is not known whether they are primary effects or attributable to the relief of acquired withdrawal effects. Conversely, it is unclear how much the effects of withdrawal are due to simple lack of the primary effects of nicotine, to acquired factors based on learning and conditioning, or to pharmacological "rebound" phenomena based on the neuronal adaptive changes underlying chronic tolerance.

Relatively few studies are available to provide clues to these matters. The limited evidence suggests that some of the effects of nicotine are primary while others appear to be based on the relief of acquired withdrawal effects. Various approaches have been used. One is to compare the responses or performance of deprived and non-deprived smokers with those of non-smokers. Where this has been done the performance of deprived smokers is inferior to that of non-smokers. When allowed to smoke or given nicotine their performance is improved to reach, but does not surpass, the performance of non-smokers (Heimstra et al., 1967). Similar findings have been demonstrated in the ability of non-smokers and smokers to tolerate electric shocks and their degree of irritation and annoyance to interruptions during viewing a television drama (Schachter, 1978). Deprived smokers also have a slower EEG alpha rhythm than do non-smokers indicating a lower resting level of arousal (Knott and Venables, 1977). Such studies show that the major differences between smokers and non-smokers emerge when the smokers are deprived. Whether this is due to acquired withdrawal effects or to innate differences is an interesting question.

A second approach, also seldom done, is to study the effects of nicotine in non-smokers (or possibly long-term ex-smokers). Non-smokers find it difficult to inhale cigarettes or to chew nicotine gum. Crushed nicotine tablets held in the mouth to allow buccal absorption appear to enhance the performance of non-smokers on prolonged symbol recognition tasks (Wesnes and Warburton, 1983). The performance of non-smokers on a simple motor task (rate of finger tapping) was increased after taking 2 mg nicotine via a nasal solution (West and Jarvis, 1986). This effect was blocked by mecamylamine. With repeated dosage hourly for 6 h there was no evidence of acute tolerance. These two studies of the effects of nicotine in non-smokers indicate that these effects at least are primary and cannot be attributed to the relief of withdrawal.

A third approach used to distinguish between primary effects and those due to reversal of acquired withdrawal effects is to look for evidence of a "rebound" element. Rebound phenomena following acute withdrawal are believed to depend on various adaptive mechanisms including those that underlie the development of chronic tolerance. In general, therefore, "rebound" withdrawal phenomena are more likely to occur in relation to those effects of nicotine that are subject to chronic tolerance. Although there is no proven evidence of chronic tolerance to nicotine in humans, it has been systematically investigated in only one study as has been mentioned earlier (Murphree, 1979). So there are no leads as to which withdrawal effects might be expected to show a rebound element. To demonstrate a rebound element it is necessary to show that a withdrawal effect, evident in the early stages of abstinence, is followed by a return towards baseline as abstinence is prolonged. Such studies are difficult in free-ranging smokers who find it difficult to abstain for longer than 24 h and failure to do so confounds the results. It is also selective in that those with more severe with-

drawal effects are less likely to comply. Only one study has been reported (West et al., 1984) in which 17 smokers abstained for 10 days. Urinary adrenaline concentrations dropped significantly in the first 3 days after stopping smoking. This was followed by a significant rise after 10 days, back to pre-abstinence levels, indicating that this withdrawal effect is mediated by rebound phenomena. It does not follow from this, however, that the effect has motivational significance. Heart rate, finger-skin temperature and urinary cortisol also showed changes at 3 days but there was no significant tendency to return to baseline after 10 days.

Of all the withdrawal effects listed in Table III, it is probable that the urge or craving to smoke is the most important for the smoker. The anatomical site and neurophysiological mechanisms that mediate pleasure, reward, reinforcement etc. are not well understood. Currently prevailing views might place this at the nucleus accumbens and its dopaminergic input from the ventral tegmental area (Wise and Bozarth, 1984) both of which are stimulated directly by smoking doses of nicotine presumably through its action on pre-synaptic nicotinic cholinergic receptors (Imperator et al., 1986). It is possible that the effects in nicotine at such sites play some part in the pleasure and reinforcement derived from smoking and that craving is partly mediated by rebound effects following its withdrawal. Although harder for the outside observer to detect and measure than are the florid and dramatically obvious rebound effects in the peripheral autonomic nervous system following opiate withdrawal, similar effects within the mesolimbic dopamine system might be subjectively more important but manifest only in terms of the severity of craving. To complicate the issue, the pharmacological mechanisms are confounded with the effects of learning and conditioning, as is the case with other addictive drugs.

In summary, the subjective and behavioural effects of nicotine are determined by many factors. Despite many similarities there are wide individual differences, some of which are innate, others acquired. A major factor is the smoking history and the pattern of prior exposure and experience, which through interactions between pharmacological and learning mechanisms, determines the stage and level of dependence reached by the subject. Probably through lack of systematic study, chronic tolerance to nicotine has not been demonstrated in humans (acute tolerance is a different matter and will be discussed later). Only one study so far has looked for and found evidence of physiological "rebound" effects after withdrawal. It is still not clear how much the effects of nicotine on human mood and performance are primary effects or determined by acquired factors involved in the relief of withdrawal. The answer seems to be both, some are primary, others based on withdrawal relief. As far as the status of smoking as an addiction is concerned, the issue is largely one of semantics. Outmoded notions that compulsive drug use is not an addiction in the absence of chronic tolerance and pharmacological rebound after withdrawal fail to recognise that dependence is essentially a learnt phenomenon. It is the strength of the reinforcement that matters, not its nature. Powerful psychosocial rewards or strongly reinforcing primary pharmacological effects may be more reinforcing and addictive than the relief of pharmacological rebound phenomena that are only mildly aversive.

Pharmacokinetic factors and acute tolerance

Besides individual differences due to the stage and degree of dependence and other factors, the effects of nicotine are greatly influenced by pharmacokinetic factors and the development of acute tolerance. It is well known that some of the effects of nicotine are subject to acute tolerance (tachyphylaxis) whereas others are less sensitive. Acute tolerance develops to the increase in heart rate and the acute subjective feeling of dizziness or light-headedness caused by nicotine, but not to its effect on finger skin temperature (a measure of peripheral vasoconstriction) or rate of finger tapping (Benowitz et al., 1982; West and Jarvis, 1986; West and Russell, 1987).

It has been suggested that it is the effects at postsynaptic nicotinic cholinergic receptors that are most susceptible to acute tolerance and that those at presynaptic receptors are less sensitive (Russell, 1988a). Acute tolerance results from prolonged neuronal depolarisation or tachyphylaxis which persists for at least as long as the receptors are occupied by nicotine, full sensitivity being restored after nicotine depletion is complete. On the other hand, those effects of nicotine mediated by its action at presynaptic receptors on nerve terminals, causing the release of neurotransmitters, are unlikely to cause prolonged neuronal depolarisation and tachyphylaxis. However, other forms of desensitisation can no doubt occur and the effect at presynaptic sites may also be attenuated by depletion of the neurotransmitter. So little is known of the mechanisms of desensitisation processes that to suggest that acute nicotine tolerance of the tachyphylactic type is confined to its actions at postsynaptic receptors is at present somewhat speculative.

The efficiency of the modern cigarette as a vehicle for getting nicotine to the brain in the form of a series of high-concentration boli after each inhaled puff is now well understood. Blood nicotine profiles derived from the concentrations in mixed venous blood provide an incomplete picture but much can be inferred from them. It is well known that the nicotine intake from a cigarette depends on the intensity of puffing and the depth of inhalation. Rapid absorption of nicotine through the lungs together with its rapid distribution to the brain and other tissues give inhaling smokers the capacity to obtain a series of sharp blood nicotine peaks coinciding with each cigarette. There are, however, two factors which limit the frequency with which a cigarette can be inhaled deeply. One is pharmacodynamic, the other pharmacokinetic, both subject to marked individual variation. Even chronic smokers have limited tolerance to nicotine and avoid exceeding their usual peak blood nicotine levels. This suggests that their usual peak levels are set by their individual toxic aversive barriers and that they regulate their smoking to get as

296

much nicotine as they can without exceeding this barrier. Peak blood nicotine levels just after a cigarette under steady-state conditions in the afternoon of a normal smoking day average between 30 and 40 ng/ml, and even among samples of heavy smokers levels over 60 ng/ml are relatively rare.

The decline of nicotine in blood is biexponential. The half-life of the initial distributional phase ranges from 7 to 10 min (mean 9 min) and the terminal or elimination phase averages 2 h but ranges from 1 to 4 h (Benowitz and Jacob, 1984; Feyerabend et al., 1985). In view of the 2-h elimination half-life, significant cumulative build-up of nicotine levels occurs even at a modest frequency of smoking (Fig. 2). To maintain a profile of such prominent blood nicotine peaks at high frequencies of smoking (around 2 – 3 per hour) would be pharmacokinetically impossible without reaching aversive blood nicotine levels. In consequence, very heavy smokers tend to take in less nicotine from each cigarette and to have less prominent blood nicotine peaks (Fig. 3). These constraints imposed by the elimination half-life are compounded by those of the initial distributional phase half-life. For example, at a smoking rate of three cigarettes per hour, the interval between cigarettes (allowing for the time taken to smoke them) is less than twice the initial half-life resulting in further

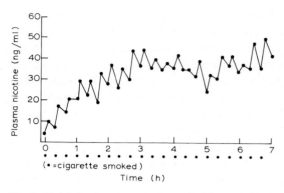

Fig. 3. Blood nicotine concentrations of a heavy smoker, smoking three cigarettes an hour. (From Russell and Feyerabend, 1978.)

reduction of the prominence of the blood nicotine peaks. Distribution of nicotine from one cigarette will not have been completed by the start of the next one.

The two profiles in Figs. 2 and 3 are clearly very different and might by expected to have different pharmacological effects. It is unlikely that smokers of the type shown in Fig. 3 would get much pharmacoligical impact from individual cigarettes. This type of profile may characterise those at stage 3 of the dependence typology, described earlier, who get little positive pleasure and smoke mainly to stave off withdrawal effects. Those at stage 2 who report positive effects from smoking may have blood nicotine profiles more similar to that in Fig. 2. The two profiles shown are extreme cases. On average, the blood nicotine boost per cigarette during steady-state in the few smokers we have tested is about 10 ng/ml which reflects the average "smoking dose" of 1 mg nicotine per cigarette of systemically available nicotine. Very few data are available on full profiles of different smokers. It is clear from comparing the two profiles contrasted here that a single measure of steady-state peak or trough nicotine (or even cotinine) level would not detect their marked differences. This may in part account for the low correlation ($r = 0.2$) between cigarette consumption and steady-state peak blood nicotine levels found in smokers who smoke 15 or more cigarettes per day (Russell et al., 1980).

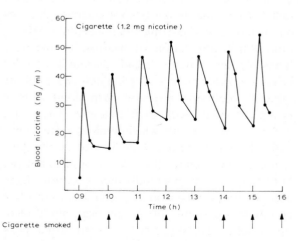

Fig. 2. Blood nicotine concentrations of an inhaling smoker, smoking one cigarette an hour. (From Russell et al., 1976.)

It is possible that smokers who get prominent blood nicotine peaks may partly overcome some of the effect of acute tolerance despite repeated smoking. Bearing in mind the puff-by-puff intermittent dosage from inhaled smoking and that the blood nicotine concentrations in post-inhalation boli must be many times higher than those in mixed venous blood, the contrast between the bolus levels and the steady-state trough levels may be sufficiently great to enable nicotine to reach sufficient numbers of unoccupied receptors to produce an appreciable effect. It is unlikely, however, that smokers with insignificant blood nicotine peaks experience any of those effects of nicotine that are subject to acute tolerance.

One final point concerns the interaction between the rate of absorption and acute tolerance in determining the pharmacological effects of nicotine. Effects that are subject to acute tolerance will not become manifest if the rate of absorption and rise in blood nicotine concentration is so slow that acute tolerance develops at the same pace.

In summary, the pharmacokinetic profile of most smokers is such that for most of the day and much of the night they have high levels of acute tolerance to nicotine. In other words, there is chronic partial blockade of the agonist effects of nicotine at postsynaptic cholinergic receptors. This may explain why chronic nicotine exposure gives rise to upregulation rather than downregulation of nicotinic receptors. Such changes might be expected to be most evident in heavy smokers with high trough blood nicotine levels just before and relatively small peaks after each cigarette.

Subjective effects, acute tolerance and severity of withdrawal

One of the puzzling features of nicotine and tobacco use is to reconcile its high addictiveness with the relative low intensity, compared with other addictive drugs, of its acute subjective effects and the relatively small changes it produces in mood, performance and level of arousal. It causes no striking euphoriant effects and any acute subjective effects

it has appear to be susceptible to acute tolerance.

Using the Addiction Research Center's established questionnaires for testing drugs for their abuse potential, Henningfield et al. (1985) found dose-related increases in subjective "liking" scores and similar changes on the "euphoria" scale in response to intravenous doses of 0.75, 1.5 and 3.0 mg nicotine administered rapidly over 10 s. It must be said that these are generous doses and that all of the eight subjects had histories of drug abuse. Indeed, six of them identified the nicotine as cocaine, and one as amphetamine. This contrasts with the well known ability of rodents to discriminate nicotine from other drugs with great precision.

In a subsequent study, using nicotine gum, Nemeth-Coslett et al. (1987) successfully manipulated the systemically available dose by getting subjects to chew 2, 4 or 8 mg nicotine gum in a standard way for 20 min. Despite respective mean blood nicotine concentrations of about 27, 31 and 34 ng/ml after chewing compared with 23 ng/ml in the placebo condition (data abstracted from figure), the nicotine gum produced dose-related decreases, rather than increases, in liking and acceptability. Heart rate and blood pressure were unaffected by the gum, but there was a dose-related decrease in the desire to smoke and an inhibitory effect on subsequent smoking behaviour. The authors attribute the "attenuation" of the nicotine effects from the gum compared with intravenous dosage to the slower rate of administration and absorption. This is no doubt a crucial factor, but another explanation is the confounding effect of acute tolerance. Prior to gum dosage subjects had refrained from smoking for only 50 min compared with 9 h in the intravenous study described above. Furthermore, the average blood nicotine concentration carried over from prior smoking was about 23 ng/ml at the end of the placebo gum condition, which is sufficient to ensure a high degree of acute tolerance and might partly explain the lack of subjective "liking" and any heart rate response.

We have recently started to examine the extent

to which acute tolerance might affect subjective and other responses to smoking during the course of a normal smoking day (West and Russell, 1987). As expected, on smoking a cigarette after a period of abstinence of 24 h there was an increase in heart rate, averaging 14 beats per minute, a drop in finger skin temperature, and just over half the subjects experienced clearcut subjective effects from the cigarette (Table IV). In contrast, a cigarette smoked during the course of a normal smoking day caused no change in heart rate and no subjective effect indicating acute tolerance to these effects of nicotine. Finger skin temperature still dropped significantly, confirming that this effect is less sensitive to acute tolerance. Subjective effects were assessed by asking the subjects whether they had "felt anything from the cigarette". All 11 who experienced effects from the post-abstinence cigarette reported feeling "dizzy" or "light headed". Of these, four also felt nauseous, two felt palpitations, one had a feeling of tremor, one a buzzing in the ears and one also reported feeling sweaty. The strength of the subjective response

TABLE IV

Effects of a cigarette smoked during a normal smoking day and after 24 h abstinence ($n = 21$) (abstracted from West and Russell, 1987)

	Normal smoking day			Post-abstinence		
	Pre-	Post-	Change	Pre-	Post-	Change
Heart rate (mean b.p.m.)	79.4	80.2	+0.8	64.6	78.2	+13.6**
Finger skin temperature (mean °C)	31.4	30.0	−1.4*	31.6	29.2	−2.4**
Number of subjects who felt subjective effects			0			11
Expired-air CO (p.p.m.)	37.0	41.0	+4.0**	6.0	12.0	+6.0**

Significant change * $p < 0.05$, ** $p < 0.01$.

was coded as follows: no effect = 0, dizzy or light headed = 1, any additional effect = 2. There was a positive correlation between the strength of the subjective effect and the rise in heart rate which remained when rise in CO was taken into account (partial correlation = 0.54, $p < 0.02$). There was also a correlation of 0.28 with drop in skin temperature, but this was not statistically significant. Blood nicotine concentrations were not measured in this study, but none of the responses to the post-abstinence cigarette were related to the CO boost it produced (highest correlation $r = -0.16$ with change in finger skin temperature).

It seems clear from this study that due to the development of acute tolerance these smokers experienced little positive impact from individual cigarettes smoked under steady-state conditions. In the nicotine-depleted state, however, after 24 h abstinence, the striking feature was the marked individual variation in response to a cigarette. Some subjects felt no effect while others felt nauseous and three even asked if they could put the cigarette out before it was completed. These differences did not appear to be attributable to differences in nicotine intake from the post-abstinence cigarette, which suggests that the subjects varied considerably in their sensitivity to nicotine. Although further studies will need to be directed specifically at this question, we have some data suggesting that it is the more dependent smokers who experience more severe withdrawal symptoms who are also more sensitive to their first cigarette following 24 h abstinence (Table V).

Data from two different studies are shown in Table V. One of them (West and Russell, 1985) was focused on pre-abstinence measures of smoke intake and scores on a smoking motivation questionnaire (SMQ, based on the factors shown in Table I) as predictors of the severity of cigarette withdrawal. The other study (West and Russell, 1988) focused on the relation of the severity of withdrawal to the strength of the response to the first post-abstinence cigarette. Severity of withdrawal was assessed by questionnaire ratings of craving and other subjective effects which form

TABLE V

Correlations of severity of withdrawal effects during 24 h abstinence from smoking with pre-abstinence variables and responses to the first post-abstinence cigarette (abstracted from West and Russell, 1985, 1988)

Variable	1985 Study (n = 29)	1988 Study (n = 21)
Pre-abstinence variables		
Daily cigarette consumption	− 0.09	
Plasma nicotine	0.57*	
Expired-air carbon monoxide	0.23	
Smoking motivation scores		
indulgent	0.33	
stimulation	0.13	
sedative	0.15	
automatic	0.01	
addictive	0.51*	0.61*
Post-abstinence cigarette		
Subjective effects		0.56*
Heart rate increase		0.41

* $p < 0.05$. Correlations with effects of post-abstinence cigarette are partial, the estimated nicotine boost from the cigarette having been partialled out.

the core of the withdrawal syndrome and which changed significantly during abstinence (e.g. depression, irritability, restlessness, hunger, difficulty concentrating). It can be seen from Table V that blood nicotine level measured prior to abstaining was the best predictor of severity of withdrawal and that the "addictive smoking" subscale of the smoking motivation questionnaire also predicted the severity of subsequent withdrawal. In the right-hand column of Table V, the positive correlations between the severity of withdrawal and the strength of response to the first post-abstinence cigarette indicate that the more dependent smokers who experience more severe withdrawal effects appear to be more sensitive to the effects of nicotine when the confounding effects of acute tolerance have been largely eliminated by 24 h prior abstinence. The increased, rather than decreased, sensitivity to nicotine in the absence of acute tolerance further indicates that these more depen-

dent smokers have not developed chronic tolerance to the acute subjective effects of nicotine or to its effect on heart rate.

These results may at first sight appear inconsistent with those of Hughes and Hatsukami (1987) who found a negative correlation between heart rate boost per unit of nicotine intake from a cigarette smoked prior to abstinence and the severity of subsequent withdrawal. Their results were, of course, confounded by acute tolerance. It is consistent with the view expressed earlier that the more addicted type of heavy smokers (stage 3) with high blood nicotine troughs and relatively small peaks would, during most of their smoking day, have a high degree of acute tolerance to the effects of nicotine at postsynaptic sites. This virtually chronic blockade may lead to upregulation of nicotinic receptors at these sites, which would in turn account for the greater sensitivity of the more dependent smokers to some of the effects of nicotine when in a nicotine-depleted state. On the other hand, individual differences in the responses of naive smokers to their first experiences with nicotine, as alluded to earlier, could be mediated by inherent differences in densities of nicotinic receptors, to differences in the capacity to upregulate, or to differences resulting from prior exposure in utero or to passive smoking. Such differences could in turn determine who becomes dependent on smoking. It is thus unknown at present how much the differences between regular smokers in their degree of dependence, severity of withdrawal and sensitivity to nicotine just after withdrawal are the consequence or the cause of their smoking behaviour.

The animal studies of Collins, Marks and collaborators are highly relevant to these questions. They have found in studies of mice of different strains that the initial sensitivity to nicotine is related to the number of nicotinic receptors. They have shown further that those strains that are initially more sensitive develop tolerance more rapidly and that the development of tolerance and its regression are reflected by upregulation and downregulation respectively of the density of nicotinic receptors in the brain (e.g. Marks et al., 1986; Collins, this volume).

What about those effects of nicotine that are not subject to acute tolerance of the tachyphylactic type? It was suggested earlier that these effects may be mediated by nicotine's action at presynaptic sites on nerve terminals. Since the more addicted heavier smokers with less prominent blood nicotine peaks are mostly in a state of acute tolerance to nicotine's effects at postsynaptic sites it is among the effects at presynaptic sites that their main source of reinforcement is likely to be found. It is not yet known whether postsynaptic, as opposed to presynaptic, nicotinic receptors are more responsive to upregulation, or indeed whether nicotine induces downregulation at the postsynaptic sites of neurotransmitters released by its action at presynaptic sites.

Summary and conclusions

Despite its addictiveness, the subjective effects of nicotine in "smoking doses" are of low intensity compared with those of other addictive drugs. Although mildly pleasurable to many regular users, it causes no striking euphoriant effects and its effects on mood, performance and the level of arousal are relatively small. This chapter does not attempt to list or review the numerous effects of nicotine, but focuses instead on some of the multiple sources of individual variation. The subjective and behavioral effects of nicotine in humans differ markedly, not only between individuals but also within individuals, according to the stage of their smoking career, their level of dependence and the time since their last few doses. Some of the influences and mechanisms discussed include innate and acquired factors, pharmacokinetic factors, acute and chronic tolerance, learning and conditioning.

It is not clear to what extent the effects of nicotine are primary, or how much they reflect reversal or relief of acquired withdrawal effects. Only one study has found a "rebound" element in the effects of withdrawal and although chronic exposure to nicotine induces an increase in the number of nicotinic receptors, chronic tolerance to nicotine has not been demonstrated in humans. Acute tolerance (tachyphylaxis) develops rapidly to many of the effects of nicotine and is completely reversible after nicotine depletion. Other effects of nicotine are less sensitive to acute tolerance. It is suggested that it is the effects of nicotine at postsynaptic receptors that are most susceptible to acute tolerance and that those mediated by its action at presynaptic receptors are less sensitive to it.

Due to accumulation of nicotine and other pharmacokinetic factors, for most of the day and much of the night, regular smokers have high levels of acute tolerance to nicotine. In other words, there is a chronic partial blockade of its agonist effects at postsynaptic receptors. This explains why nicotinic receptors are upregulated rather than downregulated and why heavy smokers experience no subjective effects from a cigarette smoked during the course of a normal smoking day. When the effects of acute tolerance are unmasked after abstinence for 24 h, it is the more addicted heavy smokers who experienced more severe withdrawal effects who also have stronger subjective and heart rate effects following the first post-abstinence cigarette. Their greater sensitivity to nicotine after abstinence may reflect their higher density of unoccupied nicotinic receptors. On the other hand, those who have higher innate sensitivity may be more likely to take up smoking and to become more dependent if they do.

Acknowledgements

I thank the Medical Research Council and the Imperial Cancer Research Fund for financial support, and Wilhemima Boyle for secretarial help.

References

Benowitz, N.L. and Jacob, P. (1984) Daily intake of nicotine during cigarette smoking. *Clin. Pharmacol. Ther.*, 35: 499 – 504.

Benowitz, N.L., Jacob, P., Jones, R.T. and Rosenberg, J. (1982) Inter-individual variability in metabolism and cardiovascular effects of nicotine in man. *J. Pharmacol. Exp.*

Ther., 221: 368 – 372.

Benwell, M.E.M., Balfour, D.J.K. and Anderson, J.M. (1988) Evidence that tobacco smoking increases the density of nicotine binding sites in human brain. *J. Neurochem.*, 50: 1243 – 1247.

Clarke, P.B.S. and Kumar, R. (1983) The effects of nicotine on locomotor activity in non-tolerant and tolerant rats. *Br. J. Pharmacol.*, 78: 329 – 337.

Ferno, O., Lichneckert, S.J.A. and Lundgren, C.E.G. (1973) A substitute for tobacco smoking. *Psychopharmacologia*, 31: 201 – 204.

Feyerabend, C., Ings, R.M.J. and Russell, M.A.H. (1985) Nicotine pharmacokinetics and its application to intake from smoking. *Br. J. Clin. Pharmacol.*, 19: 239 – 247.

Goldberg, S.R., Spealman, R.D. and Goldberg, D.M. (1981) Persistent behavior maintained by intravenous self-administration of nicotine. *Science*, 214: 573 – 575.

Heimstra, N.W., Bancroft, N.R. and Dekock, A.R. (1967) Effects of smoking upon sustained performance in a simulated driving task. *Ann. N.Y. Acad. Sci.*, 142: 295 – 307.

Henningfield, J.E., Moyasato, K. and Jasinski, D.R. (1985) Abuse liability and pharmacodynamic characteristics of intravenous nicotine and inhaled nicotine. *J. Pharmacol. Exp. Ther.*, 234: 1 – 12.

Hirschman, R.S., Leventhal, H. and Glynn, K. (1984) The development of smoking behavior: conceptualisation and supportive cross-sectional survey data. *J. Appl. Soc. Psychol.*, 14: 184 – 206.

Hughes, J.R. and Hastukami, D.K. (1986) Signs and symptoms of tobacco withdrawal. *Arch. Gen. Psychiatry*, 43: 289 – 294.

Imperato, A., Mulas, A. and De Chiara, G. (1986) Nicotine preferentially stimulates dopamine release in the limbic system of freely moving rats. *Eur. J. Pharmacol.*, 132: 337 – 338.

Jarvis, M.J. and Jackson, P.H. (1988) Cigar and pipe smoking in Britain: implications for smoking prevalence and cessation. *Br. J. Addict.*, 83: 323 – 330.

Knott, V.J. and Venables, P.H. (1977) EEG alpha correlates of non-smokers, smoking and smoking deprivation. *Psychophysiology*, 14: 150 – 156.

Marks, M.J., Romm, E., Gaffney, D.K. and Collins, A.C. (1986a) Nicotine-induced tolerance and receptors changes in four mouse strains. *J. Pharmacol. Exp. Ther.*, 237: 809 – 819.

Marks, M.J., Stitzel, J.A. and Collins, A.C. (1986b) Dose-response analysis of nicotine tolerance and receptor changes in two inbred mouse strains. *J. Pharmacol. Exp. Ther.*, 239: 358 – 364.

McKennell, A.C. and Thomas, R.K. (1967) *Adults and Adolescents, Smoking Habits and Attitudes,* Government Social Survey, HMSO, London.

McNeill, A.D., West, R., Jarvis, M.J. and Russell, M.A.H. (1986) Do older children take in more smoke from their cigarettes? Evidence from carbon monoxide levels. *J. Behav. Med.*, 9: 559 – 565.

McNeill, A.D., Jarvis, M.J., West, R.J., Russell, M.A.H. and Bryant, A. (1987) Saliva cotinine as an indicator of cigarette smoking in adolescents. *Br. J. Addict.*, 82: 1355 – 1360.

Morrison, C.F. and Stephenson, J.A. (1972) The occurrence of tolerance to a central depressant effect of nicotine. *Br. J. Pharmacol.*, 46:141 – 156.

Murphree, H.B. and Weyer (Eds.) (1967) The effects of nicotine and smoking on the central nervous system. *Ann. N.Y. Acad. Sci. Vol 142, Art. 1, N.Y. Acad. Sci.*

Murphree, H.D. (1979) EEG effects in humans of nicotine, tobacco smoking, withdrawal from smoking and possible surrogates. In A. Remond and C. Izard, (Eds.), *Electrophysiological Effects of Nicotine,* Elsevier, Amsterdam, pp. 227 – 243.

Nemeth-Coslett, R., Henningfield, J.E., O'Keefe, J. and Griffiths, R.R. (1987) Nicotine gum; dose-related effects on cigarette smoking and subjective ratings. *Psychopharmacology*, 92: 424 – 430.

Pomerleau, O.F. (1986) Nicotine as a psychoactive drug: anxiety and pain reduction. *Psychopharmacol. Bull.*, 22: 865 – 869.

Pomerleau, O.F. and Pomerleau, C.S. (Eds.) (1988) *Nicotine Replacement: A Critical Evaluation,* Alan R. Liss, Inc., New York.

Report of the U.S. Surgeon General (1988) *Nicotine Addiction*, U.S. Department of Health and Human Services.

Risner, M.E. and Goldberg, S.R. (1983) A comparison of nicotine and cocaine self-administration in the dog: fixed ratio and progressive-ratio schedules of intravenous drug infusion. *J. Pharmacol. Exp. Ther.*, 224: 319 – 326.

Russell, M.A.H., (1987) Nicotine intake and its regulation by smokers. In: W.R. Martin, G.R. Van Loon, E.T. Iwamoto and L. Davis (Eds.), *Tobacco Smoking and Nicotine: A Neurobiological Approach*, Plenum Press, New York, pp. 25 – 50.

Russell, M.A.H. (1988a) Nicotine replacement: the role of blood nicotine levels, their rate of change, and nicotine tolerance. In: O.F. Pomerleau and C.S. Pomerleau, (Eds.), *Nicotine Replacement: A Critical Evaluation,* Alan R. Liss, Inc., New York, pp. 63 – 94.

Russell, M.A.H. (1988b) Nicotine intake by smokers; are rates of absorption or steady-state levels more important? In: M. Rand and K. Thurau, (Eds.), *The Pharmacology of Nicotine,* IRL Press, Oxford, pp. 375 – 402.

Russell, M.A.H. and Feyerabend, C. (1978) Cigarette smoking: a dependence on high-nicotine boli. *Drug Metab. Rev.*, 8: 29 – 57.

Russell, M.A.H., Peto, J. and Patel, U.A. (1974) The classification of smoking by factorial structure of motives. *J. R. Statis. Soc. A.*, 137: 313 – 333.

Russell, M.A.H., Feyerabend, C. and Cole, P.V. (1976) Plasma nicotine levels after smoking and chewing nicotine gum. *Br.*

302

Med. J., 1: 1043 – 1046.

Russell, M.A.H., Jarvis, M.J., Iyer, R. and Feyerabend, C. (1980) Relation of nicotine yield of cigarettes to blood nicotine concentrations in smokers. *Br. Med. J.,* 280: 972 – 975.

Schachter, S. (1978) Pharmacological and psychological determinants of smoking. In: R.E. Thornton (Ed.), *Smoking Behaviour: Physiological and Psychological Influences,* Churchill Livingstone, London, pp. 208 – 228.

Spealman, R.D. and Goldberg, S.R. (1982) Maintenance of schedule-controlled behavior by intravenous injection of nicotine in squirrel monkeys. *J. Pharmacol. Exp. Ther.,* 223: 402 – 408.

Stolerman, I.P., Fink, R. and Jarvik, M.E. (1973) Acute and chronic tolerance to nicotine measured by activity in rats. *Psychopharmacologia,* 30: 329 – 342.

Wesnes, K. and Warburton, D.M. (1983) Smoking, nicotine and human performance. *Pharmacol. Ther.,* 21: 189 – 208.

West, R.J. and Jarvis, M.J. (1986) Effects of nicotine on finger tapping rate in non-smokers. *Pharmacol. Biochem. Behav.,* 25: 727 – 731.

West, R.J. and Russell, M.A.H. (1985) Preabstinence smoke intake and smoking motivation as predictors of severity of cigarette withdrawal symptoms. *Psychopharmacology,* 87: 334 – 336.

West, R.J. and Russell, M.A.H. (1987) Cardiovascular and subjective effects of smoking before and after 24h abstinence from cigarettes. *Psychopharmacologia,* 92: 118 – 121.

West, R.J. and Russell, M.A.H. (1988) Loss of acute tolerance and severity of cigarette withdrawal. *Psychopharmacology,* 94: 563 – 565.

West, R.J., Russell, M.A.H., Jarvis, M.J., Pizzie, R. and Kadam, B. (1984) Urinary adrenaline concentrations during 10 days of smoking abstinence. *Psychopharmacology,* 84: 141 – 142.

Wise, R.A. and Bozarth, M.A. (1984) Brain reward circuitry: four circuit elements "wired" in apparent series. *Brain Res. Bull.,* 12: 203 – 208.

A. Nordberg, K. Fuxe, B. Holmstedt and A. Sundwall (Eds.)
Progress in Brain Research, Vol. 79
© 1989 Elsevier Science Publishers B.V. (Biomedical Division)

CHAPTER 29

Behavioral and physiologic aspects of nicotine dependence: the role of nicotine dose*

Jack E. Henningfield and Phillip P. Woodson

Addiction Research Center, National Institute on Drug Abuse, P.O. Box 5180, Baltimore, MD 21224, U.S.A.

Introduction

Studies of the behavioral and physiologic actions of nicotine have been conducted over nearly a century, during which time the relationship between the dose administered and the response produced have been extensively studied (Langley, 1905; Johnston, 1942; U.S. Department of Health and Human Services, 1988). These studies have revealed that the biobehavioral mechanisms by which nicotine controls behavior are similar to those of other dependence-producing drugs. For both animals and humans, central nervous system effects of nicotine can be discriminated, can either reinforce or punish behavior, and can elicit behavioral and physiologic responses. Moreover, following chronic exposure to nicotine, physical dependence develops such that acute abstinence is accompanied by a cascade of neurochemical effects that can also control behavior (i.e. nicotine withdrawal syndrome). Together, these effects of nicotine administration, and of deprivation following chronic exposure, lend it to readily controlling the behavior of those who administer it, including the behavior of self-administration itself. The implications of these observations for both the understanding and treatment of tobacco dependence have been extensively reviewed in the Report of the Sur-

geon General on the Health Consequences of Smoking: Nicotine Addiction (U.S. DHHS, 1988).

Implicit to all of the foregoing observations, however, is the concept of drug dose. That is, the behavioral and physiologic actions of nicotine that lead to controlled behavior and to physical dependence are determined by the dose of nicotine. The presence or absence of a given nicotine-associated response, the magnitude of a given response, and even the qualitative nature of the response, may all be determined by the dose of nicotine (Table I). In this chapter, the term dose refers to the amount of nicotine that was systemically absorbed; not indicated is the nicotine which may have been present in the tobacco or polacrilex (gum) vehicle but not extracted, or that which was extracted but then exhaled or swallowed.

These relationships between the dose of nicotine administered and the resulting response of the person or animal are fundamental to understanding and treating tobacco dependence. Research from which such observations have been derived has been exhaustively reviewed in the compendia by Larson and his colleagues (Larson et al., 1961; Larson and Silvette, 1968, 1971, 1975) and in a recent Report of the Surgeon General (U.S. DHHS, 1988). This chapter will review studies which illustrate each of the dose-related effects of nicotine summarized in Table I, and then will propose implications of these observations for pharmacologic replacement approaches used to treat tobacco dependence.

---

* The present discussion is based on an earlier published review by Henningfield and Woodson (1989).

TABLE I

Summary of dose-related actions of nicotine

Dose is a determinant of response magnitude
Responses are differentially affected by changes in nicotine dose
The vehicle of nicotine delivery can affect the nature and magnitude of the nicotine-associated response
The magnitude of a response to a given dose of nicotine is inversely related to the degree of tolerance resulting from prior exposure
Dose is a determinant of the nature of the response

Dose is a determinant of response magnitude

The magnitude of complex responses to nicotine is related to nicotine dose. For example, the level of physical dependence, as indicated by the severity of nicotine abstinence-associated withdrawal signs and symptoms, is directly related to the level of nicotine intake (U.S. DHHS, 1988). Analogously, the difficulty in quitting smoking is also directly, although complexly, related to the level of nicotine intake (U.S. DHHS, 1988). Experimental dose-response manipulations have also been extensively conducted using somewhat more simple response measures. Illustrations of some of this work will be provided in the present section.

Most responses to nicotine that have been systematically studied are directly, although sometimes complexly, related to the magnitude of the nicotine dose level. That is, they are not "all-or-one" or "threshold" responses (cf. Discussion of dose-response relationships in Gilman et al., 1985). Most biological responses also show a ceiling effect, and, as will be discussed further on, some responses change in qualitative ways as the dose continues to increase. Early studies by Langley, Dixon, and others, near the turn of the 20th century showed that a wide range of physiological responses in a variety of species are directly related to the magnitude of the nicotine dose level (Langley and Dixon, 1889; Langley, 1914). In 1942, Johnston showed that the magnitude of subjective responses to nicotine was also related to the dose

level of nicotine which was administered (Johnston, 1942).

The unit dose level of nicotine obtained when smoking cigarettes is a determinant of rate of cigarette smoking and of the amount of nicotine obtained. A thorough review of such studies by Gritz (1980) revealed that in general, when nicotine dose level is increased, cigarette smoking tends to decrease. The relationships between nicotine dose of cigarettes and amount of smoking generally fit this pattern although the relationships tended to be variable across studies (Henningfield, 1984). Such findings lead to diverse interpretations. For example, Schachter (1977) concluded that decreases in cigarette consumption by approximately 25% when yield ratings of the cigarettes increased by 433% was evidence of titration to maintain relatively stable nicotine intakes. Other theoreticians (e.g. Russell, 1979) suggested that rather than titration of reinforcing levels of nicotine, what was being observed was diminished responding in response to high and aversive dose levels of nicotine. An interpretation of these and other data by Kozlowski and his colleagues (Kozlowski and Herman, 1984; Benowitz et al., 1986) provided a reconciliation: they suggested that the data were sufficient to indicate that people did, in fact, adjust their nicotine dose intake in response to changes in delivery but that the relations were not precise and were best described by what they termed the "boundary model". The boundary model is a descriptive summary of data in which it is hypothesized that self-administration of high dose levels may be acutely aversive leading to diminished intake, whereas low dose levels may lead to aversive symptoms of nicotine withdrawal leading to increased intake. It was also observed (Henningfield, 1984) that (1) across studies varied and sometimes insensitive dependent variables were used to assess changes in nicotine dose, and (2) attempted manipulations of nicotine dose may have often been thwarted by unmeasured changes in smoking behavior by subjects.

In one study, cigarettes with a range of nicotine delivery yield ratings, and intravenous injections at varying levels of nicotine were given to human volunteers (Henningifeld et al., 1985). A variety of measures used to assess the dependence potential (abuse liability) of drugs was used to characterize nicotine. Figure 1 (Henningfield et al., 1985) shows a variety of changes in response measures as a function of nicotine dose levels. As shown in the figure, nicotine given by either route produced dose-related changes in several self-reported and observer-reported responses. In addition, several

physiologic measures were also changed as a function of the nicotine dose level.

Responses are differently affected by changes in nicotine dose

A second point illustrated in Fig. 1 is that the values of the changes in response as a function of changes in dose vary across response measures. One possible explanation for this across-measure

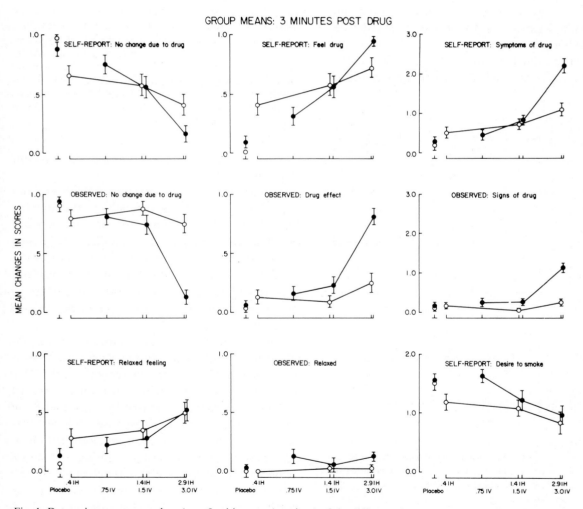

Fig. 1. Data points are mean values (*n* = 8 subjects × 4 sessions) of the difference between the the predrug values and those taken approximately 3 min following drug administration. ○, Inhaled nicotine from research cigarettes (doses refer to nicotine yield ratings of the cigarettes); ●, intravenous nicotine. (From Henningfield et al., 1985.)

306

variation is that the biologic responses are differentially sensitive to nicotine and to nicotine dose manipulations.

Figures 2 and 3 provide additional illustrations of this phenomenon. Volunteer cigarette smokers were permitted to smoke without restriction (except to use a cigarette holder that was used to collect data) during 90 min sessions in which they had access to a television and reading materials (Nemeth-Coslett et al., 1987). Thirty minutes before each session, the subjects were given two pieces of polacrilex gum to chew. The gum contained either 0, 2, 4 or 8 mg nicotine. Analysis of chewed gum and blood samples confirmed that the standardized chewing procedure (see Nemeth-Coslett et al., 1987) had resulted in orderly increases in nicotine dose level administered. Post-session plasma nico-

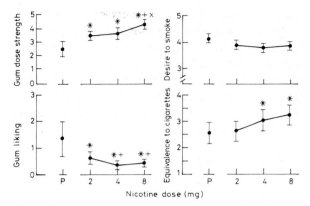

Fig. 3. Data points are mean values (n = 8 subjects × 4 sessions) of subject ratings of variables shown on the y axis following 90 min test sessions; brackets indicate 1 S.E.M. The x axis shows the nicotine gum dose; P = placebo gum. Asterisks indicate that the value was significantly different from that obtained at placebo ($p < 0.05$); plus signs indicate that value was significantly different from that obtained at 2 mg; x indicates that the value was significantly different from that obtained at 4 mg. (From Nemeth-Coslett et al., 1987.)

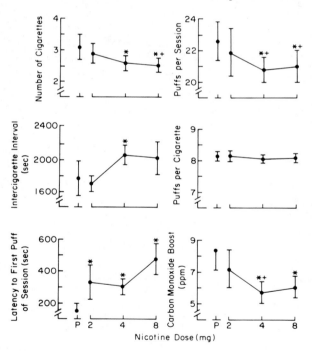

Fig. 2. Data points are mean values (n = 8 subjects × 4 sessions) of measures of cigarette smoking collected during 90 min test sessions; brackets indicate 1 S.E.M. The x axis shows the nicotine gum dose; P = placebo gum. Asterisks indicate that the value was significantly different from that obtained at placebo ($p < 0.05$); plus signs indicate that value was significantly different from that obtained at 2 mg. (From Nemeth-Coslett et al., 1987.)

tine levels were also increased as a direct function of pretreatment dose.

Figure 2 shows that cigarette smoking was decreased as a function of increases in the nicotine dose, however, the sensitivity of the response varied across measures. For example, only latency to the first puff of the session was affected at the 2 mg dose level; significant decreases in puffing, cigarettes smoked, and carbon monoxide (CO) intake, required 4; puffs per cigarette did not change under any condition.

Figure 3 shows self-report subjective data obtained by questionnaire. In this study, the subjective response of rating the "strength" of the dose was quite sensitive and was significantly increased by 2 mg and further increased by the 4 and 8 mg doses. Another subjective response, "liking" was similarly responsive but as an inverse function of dose level. In contrast, however, ratings of "desire to smoke" were not significantly changed by any of the doses although there was a weak inverse relationship that approached significance at the 8 mg dose level.

Of the many studies of the effects of variations

in nicotine dosing of cigarettes, or of nicotine pretreatments (see reviews, Gritz, 1980; Henningfield, 1984; U.S. DHHS 1988), none have utilized such a broad range of behavioral measures, and many have relied only upon measures that this study showed to be of low sensitivity. For example, the most commonly used measures are cigarettes smoked which showed less that a one-half cigarette per session decrease even at the 8 mg dose, and "craving" related indices such as desire to smoke which was not significantly decreased in this study.

The vehicle of nicotine delivery can affect the nature and magnitude of the nicotine-associated response

Inextricably linked to the drug itself is the route and vehicle of drug delivery. Route and vehicle are factors which can influence the resulting response by (1) affecting the kinetics (pattern of drug absorption and elimination), (2) determining the amount of administered drug which is systemically absorbed (via entry into the blood), and (3) providing ancillary stimulation due to the presence of other chemicals. In addition, response measures can be altered by previous histories, or lack of histories, of exposure to that route and social attributions of the route. These factors are not mutually exclusive and may often be co-operative.

Nicotine preloading via the routes of cigarette smoke inhalation, nicotine polacrilex chewing, intravenous injection, and swallowing of nicotine-containing capsules have all been shown capable of decreasing the subsequent behavior of cigarette smoking. However, of these, the route of smoke inhalation (Benowitz, 1986; Svensson, 1987) appears most efficacious and potent, whereas the gastrointestinal route (e.g. Jarvik et al., 1970) is certainly the least potent and possibly less efficacious. The findings with swallowed nicotine capsules may be explained by both the poor absorption of nicotine given via this route as well as the relative absence of sensory stimulation which may be important in the reduction of smoking and satiation of urges to smoke (U.S. DHHS, 1988).

With regard to measures of desire to smoke, Fig. 1 shows that at the highest nicotine dose, cigarette smoke inhalation and intravenous delivery both produced reductions. However, cigarette smoke delivery also reduced desire at low dose levels; in fact, the reduction in desire to smoke was similar in magnitude at the 0.4 mg dose delivered via tobacco smoke as it was at the 1.5 mg dose delivered intravenously and was nearly as effective as the 3.0 mg dose delivered intravenously. Presumably, these differences are related to the more efficient delivery of nicotine via the intrapulmonary route of smoke inhalation as well as the additional contribution of the array of sensory stimuli provided by cigarette smoke inhalation.

The magnitude of a response to a given dose of nicotine is inversely related to the degree of tolerance resulting from prior exposure

The reduction in responsiveness occurring when nicotine doses are repeatedly given has been systematically studied for nearly a century. Pioneers in the systematic study of tolerance to

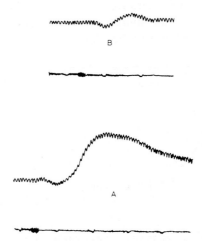

Fig. 4. The data presented are continuous recordings of the blood pressure of decerebrate rabbits. The effects of a nicotine injection on an animal which had been pretreated with nicotine every other day for one month prior to the test are shown in (B); (A) data from an untreated (nontolerant) rabbit. (From Dixon and Lee, 1912.)

drugs in general, and to nicotine in particular, were Langley and Dixon. Langley reported diminished responsiveness to repeated doses of nicotine in a variety of species and response measures (Langley, 1905, 1907–1908; Langley and Dixon, 1989). Figure 4 shows data from a study by Dixon and Lee (1912) on cardiovascular responses to nicotine administration in a study using rabbits as subjects. With respect to the electroencephalographic activating effects of nicotine, rapid tolerance occurs with at least 30 min being needed between nicotine injections to get the effect (Domino, 1983).

Multiple mechanisms account for the development of tolerance to nicotine and have been reviewed (see U.S. DHHS, 1988). There have also been many studies of tolerance to the effects of nicotine in human subjects (see reviews, U.S. DHHS, 1988). Jones et al. (1978) reported that both nicotine-induced increases in heart rate and subjective effects decreased when doses were repeated at hourly intervals. Similarly, as shown in Fig. 5, Henningfield showed that nicotine injections, repeated at 10 min intervals resulted in pronounced tolerance to the "positive" ("liking") effects of nicotine in a human subject.

Studies of nicotine tolerance suggest that the basic phenomena are not unlike tolerance-related phenomena observed when other dependence pro-

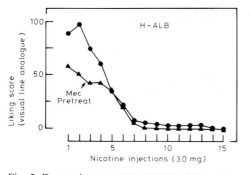

Fig. 5. Data points are scores on visual line analogue scales of positive ("liked") effects by a human volunteer following 3.0 mg intravenous injections of nicotine. These injections were given at 10-min intervals. The function indicated as "Mec Pretreat" shows data collected on a different test day in which the test session occurred 60 min following an oral dose of 10 mg mecamylamine. (From Henningfield, 1984.)

ducing drugs are given in other species and using a range of response measures (see reviews, Martin, 1977; Abood, 1984; U.S. DHHS, 1988). A few generalities might be drawn from such observations: (1) the degree of tolerance development varies across response measure, (2) increasing dose levels of the drug are often partially but not always completely effective at restoring responsiveness, (3) tolerance development is often partial, such that some level of responsiveness generally remains, (4) development of tolerance to a drug implies that diminished responsiveness will be observed when alternate formulations of the drug are given. Implications of these observations for replacement therapy of tobacco dependence are discussed further on in this chapter.

Dose is a determinant of the nature of the response

Nicotine may produce stimulating actions when given at low dose levels, whereas depressant-like actions may be produced when the nicotine dose levels are high and/or when exposure is prolonged. Observations of dual stimulant/depressant actions are not unique to nicotine but rather are typical of drugs categorized either as sedatives or stimulants (e.g. Thompson and Schuster, 1968; Gilman et al., 1985) and are consistent with early differences of opinion as to whether nicotine should be categorized primarily as stimulant-like (Lewin, 1964) or sedative-like (Amstrong-Jones, 1927). Some examples of how the nicotine dose, in combination with the response measure itself, as well as the timing of the measures, may determine whether nicotine effects are termed "stimulant-like" or "depressant-like" are provided below.

Responses mediated by the central nervous system may appear to be either stimulant- or depressant-like. For example, Ashton et al. (1980) showed that low intravenous doses of nicotine produced stimulant-like effects on the electroencephalographically-measured contingent negative variation (CNV) response in human subjects; higher intravenous doses of nicotine produced depressant-like effects on the CNV response. Woodson et al.

(1986) showed how smoking can produce simultaneous actions of diminished skin conductance amplitudes (depressant-like action) yet also induce heart rate acceleration, peripheral vasoconstriction and increased pulse velocity (stimulant-like activity); in turn, these effects of smoking were accompanied by a suppression of noise-induced tachycardia as well as a partial inhibition of noise-induced vasoconstriction. Such studies indicate that stimulation or depression depends not only on dose (CNV) but also on the response system measured (cardiovascular vs. electrodermal).

An example of how reliance upon a single response or use of a single dose level might be misleading in an effort to categorize nicotine as either predominantly a stimulant or as a depressant is the ganglionically-mediated actions of nicotine on skeletal muscle activity. Domino and Von Baumgarten (1969) found that the patellar reflex of human subjects was decreased in a dose-dependent manner. These effects were presumably mediated, however, by nicotine's initial action of stimulating Renshaw cells in the spinal cord which lead to inhibition of responsiveness of certain skeletal muscles (U.S. DHHS, 1988). To further complicate the interpretation (1) tolerance occurs, such that repeated or prolonged exposure to nicotine results in restoration of the responsiveness of the patellar reflex, and (2) different muscle groups are affected (e.g. muscle tonus may increase in trapezious muscles; Fagerström and Götestam, 1977).

Behavioral effects of nicotine can also be either increased or decreased depending upon the specific response measured and the dose given. One study compared the effects of nicotine and d-amphetamine on the behavior of squirrel monkeys which were working under different contingencies (reinforcement schedules) (Spealman et al., 1981). In one experiment the monkeys pressed a lever which led to food delivery, whereas in another, they pressed a lever which prevented the delivery of impending electric shocks. Different doses of either d-amphetamine or nicotine were given prior to the test sessions. In general, the effects of nicotine and

amphetamine were similar, and it mattered little whether the animals were working to avoid electric shock or to obtain food. Major determinants of the resulting behavior were (1) the dose of the drug and (2) the specific schedule of reinforcement [(e.g. whether the response-shock or response-food relationship was determined by the number of responses emitted (fixed-ratio schedule) or the temporal pattern of responding (fixed-interval schedule)]. As shown in Fig. 6, increasing doses of both nicotine and d-amphetamine first increased then decreased response rates in the fixed-interval schedule, and both drugs produced dose related

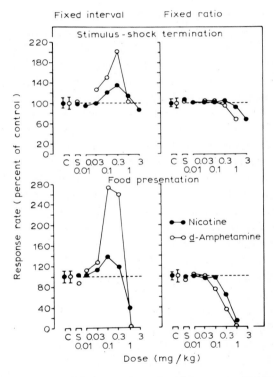

Fig. 6. Data points are mean values from one or two test sessions from each of three or four squirrel monkeys. Each value is the overall rate of lever pressing which occurred during either the fixed-interval (left-side panels) or fixed-ratio (right-side panels) schedules of electric shock avoidance (upper panels) or food reinforcement (lower panels). The dashed lines indicate the mean values obtained when no injection was given ("C") or when saline ("S") injections were given. Injections of saline, nicotine or d-amphetamine were given intramuscularly immediately before the sessions. (From Spealman et al., 1981.)

decreases in responding on the fixed-ratio schedule.

Understanding the determinants of drug categorization, such as dose level and response measure may help avoid misunderstanding of theoretical formulations designed to integrate and unify nicotine's manifold effects on physiology, mood, and behavior. Most notably, Schachter and others (Nesbitt, 1973; Schachter, 1973; Gilbert, 1979; Woodson, 1984) have developed theories to reconcile nicotine's predominantly excitatory effects on the autonomic nervous system (i.e. sympathomimetic effects resulting in end-organ arousal such as tachycardia, increased blood pressure, increased vasoconstriction, increased release of circulating catecholamines) and brain (i.e. electroencephalographic alpha blocking coupled with induction of hippocampal theta) with its predominantly tranquilizing effects on mood and behavior, especially when the organism is under stress. For example, treatment strategies for cigarette smoking ranging from the administration of stimulants to sedatives have been proposed, in part, based upon a misunderstanding of these theories (cf. Jarvik and Henningfield, 1988). While such drugs may ultimately be found to be of utility in the treatment of some cigarette smokers, indiscriminate use may actually increase the tendency to smoke (Jarvik and Henningfield, 1988).

As indicated by the preceding discussion, nicotine is neither a simple stimulant nor a simple depressant. Its categorization, rather, depends upon the measures of interest. For example, by tests of dependence potential, nicotine is an "addicting drug" (U.S. DHHS, 1988), and when peripheral mechanisms of action are considered, nicotine is categorized a ganglionic agonist (Gilman et al., 1985).

Conclusions

As the preceding review shows, the dose of nicotine is a fundamental determinant of nicotine-related behavioral and physiologic responses, including those which comprise nicotine dependence

or addiction. Analogous data have been obtained in studies of dependence to opioids, sedatives, stimulants, and other drugs (U.S. DHHS, 1988). Besides enhancing the understanding of drug dependencies, studies of various dose-response relations may also be useful in helping to understand earlier seemingly paradoxical conclusions regarding the pharmacology of nicotine, e.g. whether nicotine exposure led to physical dependence or whether nicotine was categorized as a stimulant with some paradoxical relaxant actions or a relaxant with some paradoxical stimulant actions. We may now conclude that, rather than a paradoxical drug, the pharmacology of nicotine is complex but orderly, with many effects being determined by factors such as the dose administered.

References

Abood, L.G. (1984) Mechanisms of tolerance and dependence: an overview. In C.W. Sharp (Ed.), *Mechanisms of Tolerance and Dependence,* NIDA Research Monograph 54, DHHS Publication No. (ADM) 84-1330, U.S. Government Printing Office, Washington, D.C., pp. 4–11.

Amstrong-Jones, R. (1927) Tobacco, its use and abuse: from the nervous and mental aspect. *Practitioner (Lond.),* 118: 6–19.

Ashton, H., Marsh, V.R., Millman, J.E., Rawlins, M.D., Telford, R. and Thompson, J.W. (1980) Biphasic dose-related responses of the CNV (contingent negative variation) to i.v. nicotine in man. *Br. J. Clin. Pharmacol.,* 10: 579–589.

Benowitz, N.L. (1986) The human pharmacology of nicotine. In H. Kappell (Ed.), *Research Advances in Alcohol and Drug Problems, Vol. 9,* Plenum Press, New York, pp. 1–52.

Benowitz, N.L., Jacob, P., III, Kozlowski, L.T. and Yu, L. (1986) Influence of smoking fewer cigarettes on exposure to tar, nicotine, and carbon monoxide. *N. Engl. J. Med.,* 315: 1310–1313.

Dixon, W.E. and Lee, W.E. (1912) Tolerance to nicotine. *Q. J. Exp. Physiol. (Lond.),* 5: 373–383.

Domino, E.F. (1983) Neuropsychopharmacology of nicotine and tobacco smoking. *Psychopharmacol. Bull.,* 19: 398–401.

Domino, E.F. and Von Baumgarten, A.M. (1969) Tobacco cigarette smoking and patellar reflex depression. *Clin. Pharmacol. Ther.,* 10: 72–79.

Fagerström, K.O. and Götestam, K.G. (1977) Increase of muscle tonus after tobacco smoking. *Addict. Behav.,* 2: 203–206.

Gilbert, D.G. (1979) Paradoxical tranquilizing and emotion-reducing effects of nicotine. *Psychol. Bull.,* 86: 643 – 661.

Gilman, A.G., Goodman, L.S., Rall, T.W. and Murad, F. (Eds.) (1985) *Goodman and Gilman's The Pharmacological Basis of Therapeutics,* 7th edn., Macmillan, New York, 1839 pp.

Gritz, E.R. (1980) Smoking behavior and tobacco abuse. In N.K. Mello (Ed.), *Advances in Substance Abuse: Behavioral and Biological Research, Vol. 1,* JAI Press, Greenwich, Connecticut, pp. 91 – 158.

Henningfield, J.E. (1984) Behavioral pharmacology of cigarette smoking. In T. Thompson, P.B. Dews and J.E. Barrett (Eds.), *Advances in Behavioral Pharmacology, Vol. 4,* Academic Press, New York, pp. 131 – 210.

Henningfield, J.E. and Woodson, P.P. (1989) Dose-related actione of nicotine on behavior and physiology: review and implication for replacement therapy for nicotine dependence. *J. Subst. Abuse,* 1: 301 – 317.

Henningfield, J.E., Miyasato, K. and Jasinski, D.R. (1985) Abuse liability and pharmacodynamic characteristics of intravenous and inhaled nicotine. *J. Pharmacol. Exp. Ther.,* 234: 1 – 12.

Jarvik, M.E., Glick, S.D. and Nakamura, R.K. (1970) Inhibition of cigarette smoking by orally administered nicotine. *Clin. Pharmacol. Ther.,* 11: 574 – 576.

Jarvik, M.E. and Henningfield, J.E. (1988) Pharmacological treatment of tobacco dependence. *Pharmacol. Biochem. Behav.,* 30: 279 – 294.

Johnston, L.M. (1942) Tobacco smoking and nicotine. *Lancet,* 2: 742.

Jones, R.T., Farrell, T.R., III and Herning, R.I. (1978) Tobacco smoking and nicotine tolerance. In N.A. Krasnegor (Ed.), *Self-Administration of Abused Substances: Methods for Study,* NIDA Research Monograph 20, DHEW Publication No. (ADM) 78-727, U.S. Government Printing Office, Washington, D.C., pp. 202 – 208.

Kozlowski, L.T. and Herman, C.P. (1984) The interaction of psychosocial and biological determinants of tobacco use: More on the boundary model. *J. Appl. Soc. Psychol.,* 14: 244 – 256.

Langley, J.N. (1905) On the reaction of cells and of nerve-endings to certain poisons, chiefly as regards the reaction of striated muscle to nicotine and to curari. *J. Physiol. (Lond.),* 33: 374 – 413.

Langley, J.N. (1907 – 1908) On the contraction of muscle, chiefly in relation to the presence of "receptive" substances. Part. I. *J. Physiol. (Lond.),* 36: 347 – 384.

Langley, J.N. (1914) The antagonism of curari and nicotine in skeletal muscle. *J. Physiol. (Lond.),* 48: 73 – 108.

Langley, J.H. and Dixon, W.L. (1889) On the local paralysis of the peripheral ganglia and on the connexion of different classes of nerve fibres with them. *Proc. R. Soc. Lond. (Biol.),* 46: 423 – 431.

Larson, P.S., Haag, H.B. and Silvette, H. (1961) *Tobacco: Experimental and Clinical Studies. A Comprehensive Account of the World Literature.* Williams & Wilkins, Baltimore.

Larson, P.S. and Silvette, H. (1968) *Tobacco: Experimental and Clinical Studies. A Comprehensive Account of the World Literature, Supplement I,* Williams & Wilkins, Baltimore, 803 pp.

Larson, P.S. and Silvette, H. (1971) *Tobacco: Experimental and Clinical Studies. A Comprehensive Account of the World Literature, Supplement II,* Williams & Wilkins, Baltimore, 563 pp.

Larson, P.S. and Silvette, H. (1975) *Tobacco: Experimental and Clinical Studies. A Comprehensive Account of the World Literature, Supplement III,* Williams & Wilkins, Baltimore.

Lewin, L. (1964) *Phantastica: Narcotic and Stimulating Drugs, Their Use and Abuse* (Translated by P.H.A. Wirth from the Second German edition, 1927), E.P. Dutton, New York, 335 pp.

Martin, W.R. (1977) Assessment of depressant abuse potentiality. In T. Thompson and K.R. Unna (Eds.), *Predicting Dependence Liability of Stimulant and Depressant Drugs,* University Park Press, Baltimore, pp. 9 – 15.

Nesbitt, P.D. (1973) Smoking, physiological arousal, and emotional response. *J. Pers. Soc. Psychol.,* 25: 137 – 144.

Nemeth-Coslett, R., Henningfield, J.E., O'Keeffe, M.K. and Griffiths, R.R. (1987) Nicotine gum: dose-related effects on cigarette smoking and subjective ratings. *Psychopharmacology (Berlin),* 92: 424 – 430.

Pomerleau, O.F. and Pomerleau, C.S. (1984) Neuroregulators and the reinforcement of smoking: towards a biobehavioral explanation. *Neurosci. Biobehav. Rev.,* 8: 503 – 513.

Russell, M.A.H. (1979) Tobacco dependence: Is nicotine rewarding or aversive? In N.A. Krasnegor (Ed.), *Cigarette Smoking as a Dependence Process,* NIDA Research Monograph 23, DHEW Publication No. (ADM) 79-800, U.S. Government Printing Office, Washington, D.C., pp. 100 – 122.

Schachter, S. (1973) Nesbitt's paradox. In W.L. Dunn, Jr. (Ed.), *Smoking Behavior: Motives and Incentives,* V.H. Winston, Washington, D.C., pp. 147 – 155.

Schachter, S. (1977) Nicotine regulation in heavy and light smokers. *J. Exp. Psychol. (Gen.),* 106: 5 – 12.

Spealman, R.D., Goldberg, S.R. and Gardner, M.L. (1981) Behavioral effects of nicotine: schedule-controlled responding by squirrel monkeys. *J. Pharmacol. Exp. Ther.,* 216: 484 – 491.

Svensson, C.K. (1987) Clinical pharmacokinetics of nicotine. *Clin Pharmacokinet,* 12: 30 – 40.

Thompson, T. and Schuster, C.R. (1968) *Behavioral Pharmacology,* Prentice-Hall, Englewood Cliffs, New Jersey, 297 pp.

U.S. Department of Health and Human Services (1988) *The Health Consequences of Smoking: Nicotine Addiction, A Report of the Surgeon General,* DHHS (CDC) 88-8406, U.S. Government Printing Office, Washington, D.C., 618 pp.

Woodson, P.P. (1984) *The Neuropharmacological Double Action of Nicotine: Mediation of Cigarette Smoking's Energizing yet Stabilizing Effects on Psychophysiological Function,* Swiss Federal Institute of Technology Dissertation No. 7560, ADAG Administration & Druck AG, Zürich.

Woodson, P.P., Buzzi, R., Nil, R. and Bättig, K. (1986) Effects of smoking on vegetative reactivity to noise in women. *Psychophysiology,* 23: 272 – 282.

A. Nordberg, K. Fuxe, B. Holmstedt and A. Sundwall (Eds.)
Progress in Brain Research, Vol. 79
© 1989 Elsevier Science Publishers B.V. (Biomedical Division)

CHAPTER 30

Attempts to visualize nicotinic receptors in the brain of monkey and man by positron emission tomography

H. Nybäck[a], A. Nordberg[c], B. Långström[d], C. Halldin[b], P. Hartvig[e], A. Åhlin[a], C-G. Swahn[a] and G. Sedvall[a]

[a] Department of Psychiatry and Psychology, [b] Karolinska Pharmacy, Karolinska Hospital, Stockholm and [c] Department of Pharmacology, [d] Department of Organic Chemistry and [e] Hospital Pharmacy, Uppsala, Sweden

Introduction

The widespread use of nicotine in various forms of tobacco has attracted scientists from different disciplines to study actions of this addictive compound in the central nervous system. A rapid accumulation of radiolabelled nicotine in mouse and rat brain following its i.v. administration was shown already in 1962 (Appelgren et al., 1962). Specific receptor binding sites for nicotine in association with brain cholinergic transmission has since then been demonstrated both in rodents (Nordberg and Larsson, 1980; Romano and Goldstein, 1980; Schwartz et al., 1982; Larsson and Nordberg, 1985) and humans (Flynn and Mash, 1986; Nordberg and Winblad, 1986; Whitehouse et al, 1986; Adem and Nordberg, 1988). The topographic distribution and the biochemical characteristics of these receptors has been investigated by autoradiography in animals (Clarke et al., 1984; London et al., 1985; Härfstrand et al., 1988).

With the increased research activities using new techniques a complex picture of the nicotinic receptor system and its regulation has emerged. The cholinergic nicotinic receptors appear, perhaps to a greater extent than other receptors, to be characterized by heterogeneity, diversity and plasticity which may be brought about through different combinations of receptor subunits (Changeux et al., 1988; Patrick et al., 1988).

Following repeated administration of nicotine to rodents an increase of the number of high affinity nicotinic receptors has been found (Marks et al., 1983; Schwartz and Kellar, 1983; Larsson et al., 1986). This increase may be brought about by an interconversion of low to high affinity binding sites (Romanelli et al., 1988). Also in humans nicotine appears to induce its own receptor as an increased density of [^{3}H]nicotine binding sites has been found in postmortal brain from smokers as compared to non-smokers (Benwell et al., 1988).

By use of positron emission tomography (PET) it has now become possible to visualize receptor binding in the brain in vivo. By labelling specific ligands with positron emitting radionuclides such as ^{11}C or ^{18}F, images of different neuroreceptor populations can be obtained from living animals or human subjects (Comar et al., 1979; Wagner et al., 1983). Thus the technique contains a great potential for further research in psychopharmacology and neuropathology (Sedvall et al., 1986). Recently Långström et al. (1982) succeeded to synthesize racemic [^{11}C]nicotine with high specific activity and further on also the (+)-(R-) and the (−)-(L-)-enantiomers have been labelled with ^{11}C. The first attempt to visualize central nicotine receptors in vivo by use of PET was done in monkeys (Nordberg et al., 1989). In this chapter we report on the extension of these experiments to human subjects.

Experimental procedures

Nine female Rhesus monkeys (6 – 10 kg) were anesthetized with 100 mg of ketamine (Ketalar®, Parke-Davis) and 50 – 100 mg of diazepam (Diazemuls®, Kabi-Vitrum). The animal was positioned in the positron emission tomograph (PC 384-3B, Scanditronix, Sweden) so that the lowest of four horizontal transsections included the cerebellum and the basal parts of the temporal lobes. $(+)-[^{11}C]$ and $(-)-[^{11}C]$nicotine, the naturally occurring enantiomer, were synthesized (10 – 100 Ci/mmol) from the $(+)-$ and $(-)-$forms of nornicotine (Jacob, 1982) by means of $[^{11}C]$methyl iodide (Långström et al., 1982). The $[^{11}C]$nicotine (16 – 150 MBq) was given i.v. as a bolus and radioactivity was recorded for periods of 12, 40 and 600 s up to 60 min in brain, blood and extra-cranial tissue. To prevent the binding of $[^{11}C]$nicotine to peripheral nicotinic receptors two monkeys received trimetaphan (Arfonad®) as an i.v. infusion (0.1 mg/ml per min) for 35 min following the tracer injection.

Two healthy male volunteers of 32 years, a smoker and a non-smoker, were the first humans to be studied. The non-smoker had never used tobacco in any form, whereas the smoker consumed 20 – 25 cigarettes per day, each containing approximately 1.0 mg of $(-)$-nicotine, since about 15 years. Both were normal on physical examination and had normal blood and urine tests. They were drug-free and they received no medication prior to the PET examination. The smoker was allowed to smoke one cigarette early in the morning and another at 1.5 h before the PET examinations which were performed at 10.30 or 13.30.

A sterile saline solution containing 200 – 220 MBq (about 5 μg) of $(+)-$ or $(-)-[^{11}C]$nicotine was injected (170 – 180 Ci/mmol in 2 – 6 ml) in the right antecubital vein and the radioactivity in the brain was measured for periods of 1 to 6 min by a PC-384 PET camera (Scanditronix) during 54 min (Litton et al., 1984). Radioactivity was also measured in the arterial blood from the left brachial artery. The PET instrument gives images from seven brain slices from the base of the skull to the bregma. The subjects were also examined by computed X-ray tomography at the same brain levels for the identification of anatomical regions of interest. By use of a display cursor cortical and subcortical areas were delineated for determination of the regional sum activity and sequential activity.

Following the measurement of radioactivity in plasma aliquots were put on Dowex 50 ion exchange columns for the separation of unchanged $[^{11}C]$nicotine from the major metabolite $[^{11}C]$cotinine.

Results

The uptake of $(-)-[^{11}C]$nicotine in brain, eye and extra-cranial tissue of the monkeys is illustrated in Fig. 1A. The brain activity reached a peak within 1 – 2 min and then rapidly declined for 25 – 30 min followed by a slower decline. The activity in eye and extra-cranial tissue showed no peak and remained low and constant throughout the examination. The brain peak was higher following $(+)-[^{11}C]$nicotine but the decline of activity during 60 min was slower for $(-)-[^{11}C]$nicotine. However, the regional distribution of radioactivity in the brain was not different between the two enantiomers. Also the blood levels of activity were the same following $(+)-$ and $(-)-$nicotine.

Pretreatment of the monkeys with nicotine (10 μg/kg) decreased the total uptake of ^{11}C-radioactivity to brain by 30%. Following blockade of peripheral receptors by trimetaphan the uptake of $(+)-[^{11}C]$nicotine decreased and the regional distribution was lost while the uptake and the regional distribution of the $(-)$-enantiomer was retained (Fig. 1B). The difference in uptake between the $(+)-$ and the $(-)$-enantiomers may indicate a high affinity binding of $(-)-[^{11}C]$nicotine.

In the human subjects there were, as expected, no subjective or objective evidence of a pharmacological effect of the tracer upon its injection. The ^{11}C-radioactivity in arterial plasma reached a peak at about 2 min and then declined to a steady-state level which was maintained from 4 to 5 min through the examination (Fig. 2). There were no differences between the (+)- and (−)-forms within

the individuals but the steady-state level of both enantiomers was slightly higher in the plasma of the smoker.

The radioactivity in human brain peaked within 4 − 6 min and then declined slowly throughout the measuring periods up to 54 min (Fig. 3). The regional distribution of radioactive (+)- and (−)-nicotine were similar with high accumulations in both cortical and subcortical regions (Fig. 4). The (−)-form reached a higher activity level than the (+)-form, particularly in the smoker (Figs. 4 and

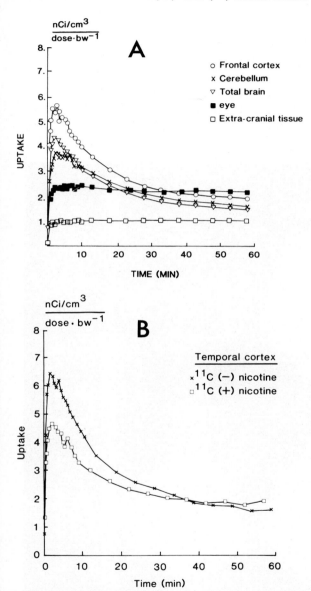

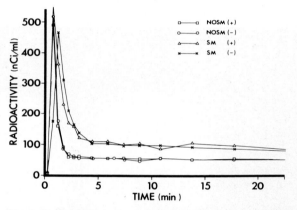

Fig. 2. Radioactivity in arterial plasma (nCi/ml) during the first 25 min following the i.v. injection of (+)-[^{11}C] and (−)-[^{11}C]nicotine to a non-smoker (upper two curves) and a smoker (lower two curves).

Fig. 1. (A) Uptake of radioactivity in brain and extra-cranial tissues of a monkey during 60 min following the intravenous injection of (−)-[^{11}C]nicotine. (B) The uptake and disappearance of (+)- and (−)nicotine in temporal cortex of a monkey during peripheral nicotine receptor blockade by trimetaphan (0.1 mg/kg per min).

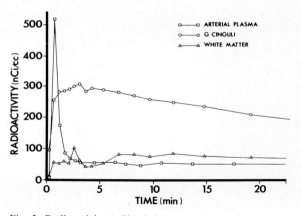

Fig. 3. Radioactivity (nCi/ml) in arterial plasma (□) as compared to white matter (○) and cingulate cortex (△) in the brain of a non-smoker following the i.v. injection of (−)-[^{11}C]nicotine.

5). The regions with the highest activity in the smoker were the insular, the cingulate and the frontal cortex. In the non-smoker the highest activities were found in the cingulate, parietal and frontal cortices. The lowest activities were in the pons, the cerebellum, the occipital cortex and the white matter of centrum semiovale. In the white matter the activity was close to the plasma level throughout the examination.

Separation of [^{11}C]nicotine from its major metabolite [^{11}C]cotinine showed that an increasing fraction, up to about 25% of the ^{11}C-activity

following the (+)-form, appeared as metabolite during the 54 min of the study. Following the (−)-[^{11}C]nicotine only 15% of the ^{11}C-activity was in the metabolite fraction at the end of the examination.

Discussion

The present results from PET studies on monkeys and men support the evidence from previous pharmacological and neurophysiological studies that specific nicotinic binding sites are present in the

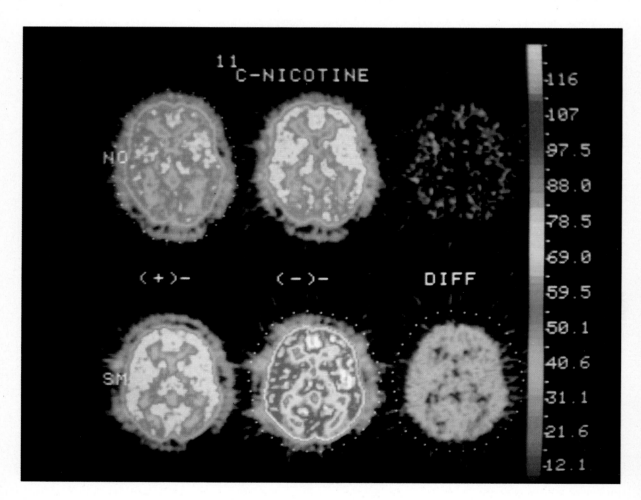

Fig. 4. PET sections through the basal ganglia of the brain of a non-smoker (upper three images) and a smoker (lower three images). The scale indicates the collected ^{11}C-activity (nCi/cc) during 54 min following the i.v. injection of (+)-[^{11}C]nicotine (left) and (−)-[^{11}C] nicotine (middle). The images received by subtracting the (+)-images from the (−)-images are also seen (right).

brain. The time course of blood levels of [^{11}C]nicotine and its rapid uptake into the brain were similar in the monkeys and the human subjects. The regional distribution of [^{11}C]nicotine within the human brain shows great similarities with the results obtained by in vitro mapping of receptor distributions in animal and human brains

(Clarke et al., 1984; London et al., 1984; Shimohama et al., 1986; Adem and Nordberg, 1988; Härfstrand et al., 1988). Thus the frontal and the cingulate cortex and the thalamus showed higher activities than the occipital cortex, the cerebellum and the pons.

When comparing the (+)- and the (−)-

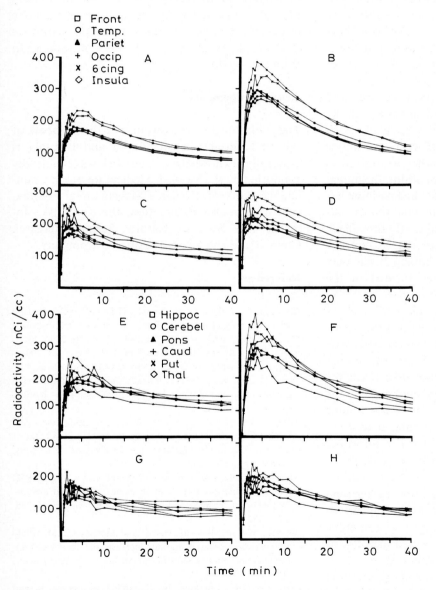

Fig. 5. Radioactivity (nCi/cc) in six cortical (A – D) and six subcortical (E – H) brain regions following the i.v. injection of (+)-[^{11}C]nicotine (A, C, E, G) and (−)-[^{11}C]nicotine (B, D, F, H) to a smoker (A,B,E,F) and a non-smoker (C, D, G, H).

enantiomers of [^{11}C]nicotine both accumulated rapidly in the brain but the unnatural form, (+)-nicotine, showed in the monkeys as well as in the men a lower peak and a slower decline during the 50 min of study. This may indicate a different binding prophile of the (+)-form which according to binding studies also has affinity for nicotinic receptors (Aceto et al., 1979; Abood et al., 1985). The more rapid decline in brain activity following (−)-[^{11}C]nicotine (Fig. 5) may be due to a short drug-receptor interaction time (Larsson and Nordberg, 1985) and a rapid clearance of the tracer through the blood flow.

Nicotine is rapidly metabolized in the body and so is probably the ^{11}C-labelled nicotine used in the present study. We found evidence for a 15 − 25% conversion of [^{11}C]nicotine in blood to the major metabolite [^{11}C]cotinine, and other neutral compounds, during the PET examination. In animal experiments small amounts of labelled cotinine has been found within the brain following the i.v. administration of labelled nicotine (Stålhandske, 1970). Thus a minor fraction of the ^{11}C-activity measured in the monkey and human brain may come from labelled metabolites formed in the brain or in the liver.

The reason for the higher plasma level of [^{11}C]nicotine in the smoker may be the presence of nicotine in pheripheral tissues due to the daily self-administration from tobacco. From the higher plasma level an increased amount may reach the brain as compared to the non-smoker. On the other hand, also in the brain of the smoker some unlabelled nicotine may be accumulated to compete for receptor sites with the [^{11}C]nicotine. Thus, if the presence in the smoker of nicotine in blood and tissues would influence the accumulation of [^{11}C]nicotine in brain, a lower rather than a higher uptake would be expected. The higher uptake found in the brain of the smoker is compatible with the recent report of increased density of nicotinic binding sites in postmortal brain of smokers as compared to non-smokers (Benwell et al., 1988). However, as we have so far analyzed the PET results from only two subjects we have to abs-

tain from further interpretations until a larger material of smokers and controls has been collected. Nevertheless these preliminary results are encouraging and indicate that PET with [^{11}C]nicotine as tracer may be used for the study of human brain nicotinic cholinergic receptors in vivo. Thus the technique is promising for further understanding of receptor mechanism in tobacco dependence as well as for research on degenerative diseases affecting central cholinergic pathways (Perry et al., 1978; Coyle et al., 1983; Nordberg and Winblad, 1986).

Acknowledgements

The skillful technical assistance of the members of the PET groups in Stockholm and Uppsala is gratefully acknowledged. The work was financially supported by the Swedish Medical Research Council, the Bank of Sweden's Tercentennial Foundation, Osterman's Foundation, the Karolinska Institute, the Swedish Natural Science Research Council and the Swedish Tobacco Company.

References

Abood, L.G., Grassi, S. and Noggle, H.D. (1985) Comparison of the binding of optically pure (−)- and (+)- (^{3}H) nicotine to rat brain membranes. Neurochem. Res., 10: 259 − 267.

Aceto, M.D., Martin, B.R., Uwaydah, I.M., May, E.L., Harris, L.S., Izazola-Conde, C., Dewey, W.L., Bradshaw, T.J. and Vincek, W.C. (1979) Optically pure (+)-nicotine from (+)-nicotine and biological comparisons with (−)-nicotine. J. Med. Chem., 22: 174 − 177.

Adem, A. and Nordberg, A. (1988) Nicotinic cholinergic receptor heterogeneity in mammalian brain. In M.J. Rand and K. Thurau (Eds.), The Pharmacology of Nicotine, IRL Press, Oxford, pp. 227 − 248.

Appelgren, L.E., Hansson, E. and Schmitterlöw, C.G. (1962) The accumulation and metabolism of ^{14}C-labelled nicotine in the brain of mice and cats. Acta Physiol. Scand., 56: 249 − 257.

Benwell, M.E.M., Balfour, D.J.K. and Anderson, J.M. (1988) Evidence that tobacco smoking increases the density of (−)-(^{3}H)nicotine binding sites in human brain. J. Neurochem., 50: 1243 − 1247.

Changeux, J.P., Fontaine, B., Klarsfeld, A., Laufer, R. and Chartaud, J. (1989) Molecular biology of acetylcholine receptors: "Long-term" evolution during motor end-plate morphogenesis. In A. Nordberg, K. Fuxe, B. Holmstedt and A. Sundwall (Eds.), *Nicotinic Receptors in the CNS — Their Role in Synaptic Transmission, Progress in Brain Research,* Elsevier, Amsterdam (this volume).

Clarke, P.S., Pert, C.B. and Pert, A. (1984) Autoradiographic distribution of nicotine receptors in rat brain. *Brain Res.,* 323: 390 – 395.

Comar, D., Mazier, M., Godot, J.M., Berger, G., Soussaline, F., Menini, C., Arfel, G. and Naquet, R. (1979) Visualization of ^{11}C-flunitrazepam displacement in the brain of the live baboon. *Nature,* 280: 329 – 331.

Coyle, J.T., Price, D.L. and DeLong, M.R. (1983) Alzheimer's disease: a disorder of cortical cholinergic innervation. *Science,* 219: 1184 – 1190.

Flynn, D.D. and Mash, D.C. (1986) Characterization of L-(^{3}H)nicotine binding in human cerebral cortex: Comparison between Alzheimer's disease and the normal. *J. Neurochem.,* 47: 1948 – 1954.

Härfstrand, A., Adem, A., Fuxe, K., Agnati, L., Andersson, K. and Nordberg, A. (1988) Distribution of nicotinic cholinergic receptors in the rat tel- and diencephalon: a quantitative receptor autoradiographic study using (^{3}H)-acetylcholine, (alpha-^{125}I)bungarotoxin and (^{3}H)nicotine. *Acta Physiol. Scand.,* 132: 1 – 14.

Jacob, P. (1982) Resolution of (+)-5-bromonornicotine. Synthesis of (R)- and (S)-nornicotine of high enentiomeric purity. *Org. Chem.,* 47: 4165 – 4167.

Långström, B., Antoni, G., Halldin, C., Svärd, H. and Bergson, G. (1982) The synthesis of some ^{11}C-labelled alkaloids. *Chemica Scripta,* 20: 46 – 48.

Larsson, C. and Nordberg, A. (1985) Comparative analysis of nicotine-like receptor ligand interactions in rodent brain homogenate. *J. Neurochem.,* 45: 24 – 31.

Larsson, C., Nilsson, L., Halen, A. and Nordberg, A. (1986) Subchronic treatment of rats with nicotine: effects on tolerance and on ^{3}H-acetylcholine and ^{3}H-nicotine binding in the brain. *Drug Alcohol Depend.,* 17: 37 – 45.

Litton, J., Bergström, M., Eriksson, L., Bohm, C., Blomqvist, G. and Kesselberg, M. (1984) Performance study of the PC-384 positron camera system for emission tomography of the brain. *J. Comput. Assist. Tomogr.,* 8: 74 – 87.

London, E.D., Waller, S.B. and Wamsley, J.K. (1985) Autoradiographic localization of (^{3}H)-nicotine binding sites in the rat brain. *Neurosci. Lett.,* 53: 179 – 184.

Marks, M.J., Burch, J.B. and Collins, A.C. (1983) Effects of chronic nicotine infusion on tolerance development and nicotinic receptors. *J. Pharmacol. Exp. Ther.,* 226: 817 – 825.

Nordberg, A. and Larsson, C. (1980) Studies of muscarinic and nicotinic binding sites in brain. *Acta Physiol. Scand. Suppl.,* 479: 19 – 23.

Nordberg, A. and Winblad, B. (1986) Reduced number of (^{3}H)nicotine and (^{3}H)acetylcholine binding sites in the frontal cortex of Alzheimer brains. *Neurosci. Lett.,* 72: 115 – 119.

Nordberg, A., Hartvig, P., Lundqvist, H., Antoni, G., Ulin, J. and Långström, B. (1989) Uptake and regional distribution of (+)-(R)- and (−)-(S)-N-(methyl-^{11}C)-nicotine in the brain of Rhesus monkey — an attempt to study nicotinic receptors in vivo. *J. Neural Trans.,* 1: 195 – 205.

Patrick, J., Boulter, J., Connolly, J., Deneris, E., Wada, K., Wada, E., Swanson, L. and Heinemann, S. (1989) Structure and function of neural nicotinic acetylcholine receptors deduced from cDNA clones. In A. Nordberg, K. Fuxe, B. Holmstedt and A. Sundwall (Eds.), *Nicotinic Receptors in the CNS — Their Role in Synaptic Transmission, Progress in Brain Research,* Elsevier, Amsterdam (this volume).

Perry, E.K., Tomlinson, B.E., Blessed, G. Bergmann, K., Gibson, P.H. and Perry, R.H. (1978) Correlation of cholinergic abnormalities with senile plaques and mental test scores in senile dementia. *Br. Med. J.,* 2: 1457 – 1459.

Romanelli, L., Öhman, B., Adem, A. and Nordberg, A. (1988) Subchronic treatment of rats with nicotine: Interconversion of nicotinic receptor subtypes in brain. *Eur. J. Pharmacol.,* 148: 289 – 291.

Romano, C. and Goldstein, A. (1980) Stereospecific nicotine receptors on rat brain membranes. *Science,* 210: 647 – 650.

Schwartz, R.D. and Kellar, K.J. (1983) Nicotinic cholinergic receptor binding sites in the brain: regulation in vivo. *Science,* 220: 214 – 216.

Schwartz, R.D., McGee, R. and Kellar, K.J. (1982) Nicotinic cholinergic receptors labelled by (^{3}H)acetylcholine in rat brain. *Mol. Pharmacol.,* 22: 56 – 62.

Shimohama, S., Taniguchi, T., Fujiwara, M. and Kameyama, M. (1986) Changes in nicotinic and muscarinic cholinergic receptors in Alzheimer-type dementia. *J. Neurochem.,* 46: 288 – 293.

Sedvall, G., Farde, L., Persson, A. and Wiesel, F.A. (1986) Imaging of neurotransmitter receptors in the living human brain. *Arch. Gen. Psychiatr.,* 43: 995 – 1005.

Stålhandske, T. (1970) Effects of increased liver metabolism of nicotine on its uptake, elimination and toxicity in mice. *Acta Pshysiol. Scand.,* 80: 222 – 234.

Wagner, H.N., Burns, H.D., Dannals, R.F., Wong, D.F., Långström, B., Duelfer, T., Frost, J.J., Raert, H.T., Links, J.M., Rosenbloom, S.B., Lukas, S.E., Kramer, A.V. and Kuhar, M.J. (1983) Imaging dopamine receptors in the human brain by positron tomography. *Science,* 221: 1264 – 1266.

Whitehouse, P.J., Martino, A.M., Antuono, P.G., Lowenstein, P.R., Coyle, J.T., Price, D.L. and Kellar, K.J. (1986) Nicotinic acetylcholine binding sites in Alzheimer's disease. *Brain Res.,* 371: 146 – 151.

A. Nordberg, K. Fuxe, B. Holmstedt and A. Sundwall (Eds.)
Progress in Brain Research, Vol. 79

CHAPTER 31

Treatment of nicotine dependence

Karl Olov Fagerström

Pharmacia LEO Therapeutics AB, Box 941, S-251 09 Helsingborg, Sweden

Introduction

Addictions to drugs are not only dependent on the substance taken but also on the behavioural factors. This is probably never more true than with nicotine where the drug has to be extracted from tobacco requiring a substantial amount of behaviour.

Usually, the nicotine intake, most commonly by cigarette smoking, starts in social conditions where it is reinforced by, for example, peer pressure. At the very beginning the tobacco smoking is aversive but soon habituation to the local and tolerance to the systemic effects develops. At the beginning also social environmental reinforcement is thought to be common and important, but soon the smoking behaviour becomes sufficiently reinforced in its own to be maintained without social reinforcement. Smoking will be generalized to many situations according to operant learning principles. The drug user learns more and more about the effects of the drug and how s/he can use it to the best.

The smoker gradually finds out which needs the nicotine can satisfy. The neuroadaptation proceeds, and gradually a physical dependence develops. Drug intake comes more under control of classical conditioning like principles, manifesting itself as automatic smoking in the presence of smoking-associated cues. It is important to realize that these cues can be both external and internal.

An optimal treatment therefore has to address both the behavioural and pharmacological factors of nicotine dependence. In this report the purpose, however, will be to describe how the pharmacological part of the dependence can be pharmacologically treated.

Before arriving at the current state of the art, a catalogue of drugs, divided into prescription and non-prescription drugs that have been used, will be given.

Over-the-counter drugs

There are good reasons to believe that the list of over-the-counter drugs is incomplete. Probably many more drugs have tried the market but soon disappeared. Already in the beginning of the century, lobeline (a nicotine analogue) was used. Also in the first smoking cessation clinic, started in Stockholm in 1955 by Börje Ejrup (1963), injections of lobeline hydrochloride formed the major treatment. There are many lobeline-containing products sold over the counter, but the scientific evidence that they really work better than placebo is still not at hand.

In Eastern Europe, anabasine and cytisine, alkaloids with similarities to nicotine are used. While these drugs are pharmacologically active (Sloan et al., 1987) and may relatively well mimic the peripheral effects of nicotine, they probably do not mimic the central nervous effects of nicotine as well. Lobeline and anabasine may have different pharmacological actions than cytisine. The former binds to lower affinity sites in the brain, whereas cytisine binds to high affinity sites.

Other over-the-counter products are those containing silver acetate that in contact with sulphides in the tobacco smoke alters the taste perception of the smoke, pH modifying agents like sodium bicarbonate for prolonging the circulation time of nicotine, several vitamins and fibres that swell in the stomach giving some fulness sensations.

Of these products lobeline is the most researched drug. The efficacy of these drugs is not proven with a possible exception of lobeline which may have a minute effect. It should however be mentioned that several of the drugs, which also is applicable to the prescription drugs, have simply not been exposed to well controlled research.

Prescription drugs

It is interesting that some of the earliest attempts to treat nicotine dependence involved stimulants like amphetamine (Ross, 1967) and tranquillizers (Schwartz and Dubinsky, 1968). Despite these drugs' opposite actions, it should come as no surprise that clinicians have tried to treat both the major effects of nicotine: stimulation and sedation.

In more recent days, clonidine, an α_2-noradrenergic agonist, has received some attention since it has been found to effectively reduce withdrawal symptoms among opiate and alcohol addicts (Jasinsky et al., 1985; Björkqvist, 1975). Glassman et al. (1984) were the first to try clonidine on tobacco withdrawal symptoms They found clonidine to be significantly better in alleviating withdrawal symptoms than alprazolam and placebo in a 3 day crossover study. Alprazolam also reduced several withdrawal symptoms better than placebo.

In a second study by Glassman et al. (1988), using a daily dose around 200 μg, they found that clonidine also produced better long-term quit rates than placebo. After 6 month 27% were classified as successes compared with only 5% in the placebo group. In a French study by Grimaldi et al. (1987), with a daily dose of 100 μg, the long-term effect of clonidine was not significantly different from the placebo. Tønnesen (personal communication) treated 47 smokers with clonidine in an open study. At first he used a daily dose of 100 μg, but

because of mostly severe sedative side effects, he reduced the dose to 50 μg daily. He concluded that the frequent side effects clearly outweighed the possible positive effects. More research is needed before clonidine can be identified as an effective treatment for tobacco dependence. If proven effective, its use has to be weighed against the potential hazardous side effects.

The sedating effect of clonidine may be disadvantageous in smoking cessation, since some of the more common withdrawal symptoms are sleepiness and loss of energy. Since these symptoms might be more pronounced with clonidine, it has been suggested to combine with a stimulating component; phentermine resin (Cooper, 1988). Nicotine replacement may be another valuable method to counteract these effects.

From a theoretical point of view, a centrally acting nicotine blocker would be very appealing. Mecamylamine is such a blocker that has been used widely in animal research and also found to effectively block the effects of nicotine, e.g. the reinforcing consequences (Goldberg et al., 1983). In humans, the experience is much more limited. In laboratory, research mecamylamine seems to influence, for example, smoking intake (Stolerman et al., 1973). Whether it would help in quitting smoking is a different question since it would not help to cope with withdrawal. In one clinical trial, Tennant and Tarver (1984) used mecamylamine in two different dosage schedules on heavy smokers. Unfortunately, there was no treatment control employed, and the impression that it reduced the craving has therefore to be taken cautiously. However, the peripheral ganglionic blocking by mecamylamine makes it anyway difficult to deal with in terms of side effects. A more longacting and central nervous system selective antagonist would be more suitable.

Also β-blockers have been tried although the rationale has not been clear. Dow and Fee (1984) in a controlled study obtained no effect of a cardioselective agent with no intrinsic sympathomimetic activity and a non-selective agent with intrinsic sympathomimetic activity.

In the treatment of tobacco dependence, palliative treatments for various symptoms also occur. That can be, for example, giving hypnotics to insomniacs and antidepressants to depressed patients. Anticholinergics have been used to curb hunger sensations and dry out the mouth to alter the taste of tobacco. Corticotrophin also has a relatively long history in smoking cessation. It has usually been injected and claimed to be of some value with, e.g. difficulties to concentrate. More recently, Bourne (1985), in a pilot study obtained positive findings.

As with the over-the-counter products none of the prescription drugs have a record of proven efficacy today. Clonidine presently seems to be the most promising candidate for such a status.

Because of the growing awareness of the need to stop smoking, it can be hypothesized that several of today's available drugs and others under development that have some connection with the mechanisms determining dependence to tobacco or alleviating withdrawal symptoms, will be tested. As an example of that development, serotonine reuptake inhibitors originally developed as antidepressants, are now also being studied in smoking cessation (e.g. Scrip, 1986; Edwards et al., 1988).

Nicotine replacement

The only pharmacological treatment so far for tobacco dependence is nicotine replacement with nicotine chewing gum (Nicorette®). It has been approved for marketing in about 50 countries and is extensively researched with at least 65 trials in over 20 countries.

Rationale

The idea with nicotine replacement treatment is to enable the tobacco-dependent subjects to break the total dependence (behavioural and pharmacological) in two steps. The first step is the extinction of the behavioural dependence, while nicotine is still taken from an alternative source so that the smoker does not need to be troubled by physical withdrawal. In the second step the nicotine also is gradually faded out.

Thus, treatment with nicotine gum recognizes that cigarette smoking results from two interrelated problems; behavioural and pharmacological dependence.

Efficacy

In rough terms, nicotine chewing gum almost doubles the long-term success rates (Fagerström, 1988). The results obtained with nicotine replacement vary considerably and some of the important variables are dose used, the smoker's degree of nicotine dependence and education received on how to appropriately use nicotine chewing gum properly. The best results are usually obtained by experienced therapists in smoking cessation clinics. The reason may be specific experience with smoking cessation in general, including behavioural methods but also an ability to present the idea of nicotine replacement so as to give patients correct expectations and instructions for use.

Too little research has been devoted to exploring process variables in nicotine gum treatment. It has not been determined, e.g. whether nicotine gum should be taken ad libitum or on a fixed schedule. More basic questions have to do with the contribution of the nicotine level. It is important that nicotine is delivered in amounts that are high enough to be effective. It can, however, be questioned whether optimum doses of nicotine have been employed in most of the research so far.

The vast majority of the studies have been performed with the weaker strength of the gum containing 2 mg nicotine. This dose has been found to replace approximately one third of the nicotine levels that the smokers were used to when smoking (Fagerström, 1988). The optimum level of replacement is not known, but it will most likely be higher than one third of the nicotine level that is assessed in the venous blod. Firstly, because the pharmacokinetics of the gum have a much less aggressive peak effect and secondly, because concentrations reaching the brain when smoking are higher as a consequence of the direct access to blood going directly from the lung via the heart up to the brain arteries. With gum use it is unlikely that there would be any difference in concentration gradients between arterial and venous blood.

The average replacement when the higher 4 mg strength is used is two thirds of the nicotine level when smoking. Therefore, much of the later research has been concentrated on the 4 mg strength of the gum. In laboratory research, the 4 mg preparation has reliably been found to better alleviate withdrawal symptoms than 2 mg. This also includes decreases in the smoking behaviour, and the 4 mg is rated as more equivalent to cigarettes but interestingly not increases liking scores compared with 2 mg, which suggests low abuse liability (Henningfield and Jasinsky, 1988).

In clinical research, it is not clear whether 4 mg is generally superior to the 2 mg strength. It seems to be an interaction with degree of nicotine dependence, so that lower-nicotine-dependent subjects do just as well on 2 mg. As dependence increases, the need for the 4 mg strength becomes obvious and for the really highly dependent subjects an even stronger preparation of the gum could be indicated. In one of the recent studies employing 4 mg, low/medium-dependent subjects were randomly assigned to each of placebo or 2 mg and high-nicotine-dependent patients to 2 or 4 mg. Long-term results showed that 38% of 2 mg and 23% of placebo users were able to stop and maintain abstinence in the low/medium-dependent subjects. In the high-dependent subjects the rates were 44% for 4 mg and 12% for 2 mg (Tønnesen et al., 1988).

The future with nicotine replacement

The need for giving nicotine in other formulations than chewing gum has been increasingly recognized, some of these needs being, for example, alternatives to individuals who cannot or do not want to chew and weaning off from nicotine replacement.

The nasal route has been tried by Russell's group, where it has been found that nicotine in the form of a gel better mimicked the pharmacokinetic profile of a cigarette than gum does (Russell et al., 1983) and can be of potential in smoking cessation (Jarvis et al., 1987).

The same research group has also looked into inhaled nicotine vapor. With normal puffing behaviour only a very small increase in nicotine levels were obtained, but when puffing as rapid as possible for 20 min a substantial level of 17 ng/ml was reached (Russell et al., 1987).

Lastly, transdermal nicotine delivery has been advocated. Rose et al., (1985) were the first to show that transdermally administered nicotine entered the circulation and had an effect on smoking behaviour and craving. In a pilot trial with a prototype nicotine patch, Russell (1988) found that very high nicotine levels could be reached (60 ng/ml). The rise time was, however, very slow and skin irritation was another problem.

These nicotine delivery systems vary greatly in terms of behaviour involvement, where the patch requires the least behaviour and nicotine vapor inhaler the most. In pharmacokinetic profiles the nasal and transdermal applications are the extremes. Clearly, a system of drugs is building up giving great opportunities to satisfy different needs.

Future challenges and developments

In the trials where nicotine dependence has been recorded it has usually been done with paper-and-pencil tests, most frequently the Fagerström Tolerance Questionnaire (Fagerström, 1978). One of the future challenges for any pharmacological treatment will be to determine with better accuracy the smoker's nicotine dependence. It varies considerably from smoker to smoker, thus giving fire to the dispute of tobacco or nicotine as being a "real" drug or not. The use pattern does not differ from several other drugs like alcohol and cocaine where most users cannot be termed physically addicted. Some tobacco users are satisfied with nicotine levels of less than 10 ng/ml, while others need over 70 ng/ml (Russell, 1986). Another sign of the large variation in dependence is an observation made by Nemeth-Coslett and Henningfield (1985) where they found an extreme tolerance to nicotine in one subject who could not distinguish a 16 mg dose (4 × 4 mg) gum from placebo. Pharmacological treatment has to be applied with sense, and for this purpose better instruments are needed to determine the strength of the nicotine dependence.

It may be that smokers suffer from different classes of withdrawal reactions. While some may experience a loss in cognitive functions being the major obstacle towards abstinence, others may be troubled by autonomic visceral malfunctions and others may relapse because of difficulties to control emotions. If different medications will have different effect profiles, it could also be something that the therapist should have in mind when s/he is prescribing an optimal treatment.

In the future development of drugs, the mode of administration also has to be considered, not only from the aspects of pharmacokinetics − which is important − but also from the possibility of giving a combined pharmacological and behavioural replacement. The more effectively we can treat tobacco dependence initially, the more work we have to address to subjects who will have difficulties to come off our effective medicines. Systems of medication, varying the degree of behavioural involvement, and the pharmacokinetic aggressiveness are wanted for an optimum transference from tobacco use in the first place and from the medicine in the second place.

At no time in history has so much attention been directed at helping tobacco-dependent persons to free themselves from tobacco. Today we have one effective treatment − nicotine gum − that most likely already within a couple of years can be followed by further improvements.

References

Björkqvist, S.E. (1975) Chlonidine in alcohol withdrawal. *Acta Psychiatr. Scand.,* 52: 256 – 263.

Bourne, S. (1985) Treatment of cigarette smoking with short-term high dosage corticotrophine therapy: preliminary communication. *J. R. Soc. Med.,* 78: 649 – 650.

Cooper, Y. (1988) Means and method for aiding individuals to stop smoking. In M. Aoki, S. Hisamichi and S. Tominaga (Eds.), *Smoking and Health,* Elsevier, Amsterdam, pp. 889 – 890.

Dow, R.J. and Fee, W.M. (1984) Use of beta-blocking agents with group therapy in smoking withdrawal clinic. *J. R. Soc. Med.,* 77: 648 – 651.

Edwards, N.B., Simmons, C.R., Rosenthal, T.L., Hoon, P.V. and Downs, J.N. (1988). Dopexin in the treatment of nicotine withdrawal. *Psychosomatics,* 29: 203 – 206.

Ejrup, B. (1963) Breaking the cigarette habit. *Cancer J. Clin.,* 13: 183 – 186.

Fagerström, K.O. (1978) Measuring degree of physical dependence to tobacco smoking with reference to individualization of treatment. *Addict. Behav.,* 3: 235 – 241.

Fagerström, K.O. (1988) Efficacy of nicotine chewing gum. In O.F. Pomerleau and C.S. Pomerleau (Eds.), *Nicotine Replacement: A Critical Evaluation,* Alan R. Liss, New York, pp. 109 – 128.

Glassman, A.H., Jackson, W.K. Walsh, B.T. and Roose, S.P. (1984) Cigarette craving, smoking withdrawal, and chlonidine. *Science,* 226: 864 – 866.

Glassman, A.H., Stetner, F., Walsh, B.T., Raizman, P.S., Fleiss, J.L., Cooper, T.B. and Covey, L.S. (1988) Heavy smokers, smoking cessation, and chlonidine: results of a doubleblind, randomized trial. *J. Am. Med. Assoc.,* 259: 2863 – 2866.

Goldberg, S.R., Spealman, R.D., Risner, M.E. and Henningfield, J.E. (1983) Control of behavior by intravenous nicotine injections in laboratory animals. *Pharmacol. Biochem. Behav.,* 19: 1011 – 1020.

Grimaldi, B., Demaria, C., Loufranie, E., Bang, F., Greslin, P. and Lagrue, G. (1987) Results of a controlled study of chlonidine versus placebo for cessation of smoking. *Sem. Hop.,* 43: 3369 – 3370.

Henningfield, J.E. and Jasinsky, D.R. (1988) Pharmacologic basis for nicotine replacement. In O.F. Pomerleau and C.S. Pomerleau (Eds.), *Nicotine Replacement: A Critical Evaluation,* Alan R. Liss, New York, pp. 35 – 61.

Jarvis, M.J., Hajek, P., Russell, M.A.H., West, R.J. and Feyerabend, C. (1987) Nasal nicotine solution as an aid to cigarette withdrawal: a pilot clinical trial. *Br. J. Addict.,* 82: 983 – 988.

Jasinsky, D.R., Johnson, R.E. and Kocher, T.R. (1985) Chlonidine in morphine withdrawal: differential effects on signs and symptoms. *Arch. Gen. Psychiatry,* 42: 1063 – 1066.

Nemeth-Coslett, R. and Henningfield, J.E. (1985) Rational basis for chemo-therapy of tobacco dependence. In J. Grabowsky and S.M. Hall (Eds.), *Pharmacological Adjuncts in Smoking Cessation,* NIDA, Research Monograph No. 53, Washington D.C., pp. 15 – 26.

Rose, J.E., Herzkovic, J.E., Trilling, Y. and Jarvik, M.E. (1985) Transdermal nicotine reduces cigarette craving and nicotine preference. *Clin. Pharmacol. Ther.,* 38: 450 – 456.

Ross, C.A. (1967). Withdrawal Research Clinics. In: S.V. Zagona (Ed.), *Studies and Issues in Smoking Behavior,* University of Arizona Press, Tucson, pp. 111 – 113.

Russell, M.A.H. (1986) Conceptual framework for nicotine substitution. In J.K. Ockene (Ed.), *The Pharmacologic Treatment of Tobacco Dependence: Proceedings of the World Congress, November 4-5, 1985,* Institute for the Study of Smoking Behavioral and Policy, Cambridge, Massachusetts, pp. 90 – 107.

Russell, M.A.H. (1988) Nicotine replacement: the role of blood nicotine levels, their rate of change and nicotine tolerance. In O.F. Pomerleau and C.S. Pomerleau (Eds.), *Nicotine Replacement: A Critical Evaluation,* Alan R. Liss, New York.

Russell, M.A.H., Jarvis, M.J., Feyerabend, C. and Fernö, O (1983) Nasal nicotine solution: a potential aid to giving up smoking. *Br. Med. J.,* 286: 683 – 684.

Russell, M.A.H., Jarvis, M.J., Sutherland, G. and Feyerabend, C. (1987). Nicotine replacement in smoking cessation. *J. Am. Med. Assoc.,* 257: 3282 – 3285.

Schwartz, J.L. and Dubitzky, M. (1968) One-year follow-up results of a smoking cessation program. *Can. J. Public Health,* 59: 161 – 165.

Scrip (1986) CNS disease, 1085, 11.

Sloan, J.M., Martin, W.R., Bostwick, M., Hook, R., and Wala, E. (1987) The comparative binding characteristics of nicotine ligands and their pharmacology. *Pharmacol. Biochem. Behav.,* 26.

Stolerman, I.P., Goldfarb, T.L., Fink, R. and Jarvik, M.E. (1973) Influencing cigarette smoking with nicotine antagonists. *Psychopharmacologia,* 28: 247 – 259.

Tennant, F.S. and Tarver, A.L. (1984) Withdrawal from nicotine dependence using mecamylamine. *Drug Abuse Monogr. Series 55,* Washington D.C., pp. 291 – 297.

Tønnesen, P., Fryd, V., Hansen, M., Helsted, J., Gunnersen, A-B., Forchammer, H. and Stockner, M. (1988) Effect of nicotine chewing gum in combination with group counseling on the cessation of smoking. *N. Engl. J. Med.,* 318: 15 – 18.

SECTION X

Nicotine and Degenerative Brain Disorders

A. Nordberg, K. Fuxe, B. Holmstedt and A. Sundwall (Eds.)
Progress in Brain Research, Vol. 79
© 1989 Elsevier Science Publishers B.V. (Biomedical Division)

CHAPTER 32

Parkinson's disease — etiology and smoking

Sten-Magnus Aquilonius

Department of Neurology, University Hospital, S-751 85 Uppsala, Sweden

In his classical "Essay on the Shaking Palsy" James Parkinson (1817) refers to an early description of the symptomatology by Galen and the disorder was also known in ancient Indian medicine (Manyam, 1988). Accordingly, when causative factors are considered it is to remember that modern technology, diet or life style are not etiological prerequisites. Furthermore, it should be added that clinical manifestations under the headline Parkinson's disease might rather represent a syndrome, including forms with non-dopaminergic degeneration and symptoms of dementia and autonomic dysfunction, than a well-defined entity.

Etiology

Although the cause of Parkinson's disease is still unknown, a lot of new information has been obtained in the last few years. A starting point was the report by Langston and colleagues (1983) on parkinsonism in drug addicts due to minor amounts of 1-methyl-4-phenyl-1,2,3,6-tetrahydropyridine (MPTP), thereby reiterating and extending pioneer work on this subject by Davis and coworkers (1979). Today the mechanisms of action of MPTP have largely been worked out (see review by Kopin, 1987 and thesis by Sundström, 1988). Briefly the lipophilic MPTP which readily enters the brain is extraneuronally converted to the active toxin 1-methyl-4-phenylpyridine (MPP$^+$), a reaction catalyzed by MAO-B. MPP$^+$, quarternary and highly polar, is extremely slowly eliminated

from the brain where it is taken up into dopaminergic neurons by the dopamine re-uptake mechanism. The preferential specificity of MPTP toxicity for dopaminergic neurons is probably due to the high affinity of MPP$^+$ for the dopamine carrier. The toxicity of MPTP can be reduced by drugs blocking MAO-B such as *l*-selegiline and by compounds binding to the dopamine re-uptake carrier such as nomifensine.

The ultimate intraneuronal toxic mechanism might be multifactorial involving depletion of mitochondrial ATP, inhibition of tyrosine hydroxylation, production of reactive oxygen and binding to neuromelanin (reviews by Kinemuchi, et al., 1987; Nagatsu, 1987; Riederer and Youdim, 1987). In animal experiments, MPTP toxicity has been reduced by antioxidants and by metal chelators. If binding to neuromelanin and gradual intraneuronal release of MPP$^+$ plays a toxic role, this mechanism might be one explanation of the high MPTP vulnerability of certain species, notably primates having extensive melaninization of mesencephalic nuclei.

It was recently demonstrated by Calne and coworkers (1985) by means of positron emission tomography (PET) that MPTP exposed but asymptomatic individuals accumulate radioactivity in the striatum following [^{18}F]fluorodopa administration to an extent higher than parkinsonian patients but lower than controls. This indicates a preclinical situation which may develop into a symptomatic stage when a certain degree of age-related neuronal loss has taken place.

The interest in neurotoxic factors in the cause of Parkinson's disease has also been challenged by the relative lack of support for theories of infectious or heritable origin. The pandemic of encephalitis letargica in the decade 1916–1926 characterized by a high number of postencephalitic parkinsonism had a natural impact on the search for a viral cause for Parkinson's disease. However, presently there is no convincing evidence in favour of an association between a known viral disorder and Parkinson's disease.

In studies performed in the U.S.A. (Duvoisin, 1986) and in the United Kingdom (Marsden, 1986) in pairs of monozygotic and dizygotic twins, one of whom had Parkinson's disease, 3.7% were concordant among identical twins and 3.3% among non-identical pairs. These results should rule out Parkinson's disease as a heritable disorder. However, the high occurrence of Parkinson's disease in siblings in this and in other investigations might favour the existence of a familiar susceptibility for some etiological factors.

If exposure to a neurotoxic factor or factors plays an important role in the etiology, certain risk groups should be expected to be delimited. Except for the MPTP exposed individuals, earlier mentioned, this is not the case. However, the appearance of symptoms in middle or late life is compatible with a theory of neurodegeneration induced by exposure to a toxin, eventually in combination with age-related neuronal loss and increased sensitivity. Further, uneven geographic distribution of environmental factors should result in place-related prevalence rates. Parkinson's disease exists world-wide and in western countries the prevalence is in the order of $150/10^5$ inhabitants. Recently, regional differences in prevalence have been reported from Canada, the people's republic of China, the U.S.A. and Sweden. Barbeau and colleagues (1986) demonstrated an uneven distribution of Parkinson's disease in Quebec and proposed an association to the use of pesticides in the high prevalence area. In the Canadian province Saskatchewan, Rajput and collaborators (1988a) obtained evidence in favour of childhood exposure

to a rural environment as a risk factor for developing Parkinson's disease later in life. There are indications from mortality data that Parkinsons's disease in the U.S.A. might be more common in northern than in southern states (Lux and Kutzke, 1987). Using a door-to-door technique, a prevalence study in urban populations in China (Li et al., 1985) revealed figures well below those obtained in Europe and the United States. In a recent case control study in China (Tanner et al., 1988) it was concluded that the Parkinson's disease cases were significantly more likely to have had exposure to chemical, pharmaceutical, herbicide or pesticide industries before the onset of the disease. Using utilization of L-dopa containing drugs as a tracer for estimation of the prevalence of Parkinson's disease in Sweden, one county with unexpectedly high utilization was identified (Aquilonius and Hartvig, 1986). Unique to this county are three large steel alloy factories and a high deposition of heavy metals.

Taken together, the afore-mentioned neuroepidemiological studies represent a new and important approach in research into the cause of Parkinson's disease. Such studies should direct investigations of post-mortem concentrations of neurotoxins in brains of individuals from areas with high and low prevalence of the disease. Indeed, a recent investigation from Canada reports a trend that individuals living in rural areas have accelerated loss of nigral neurons as compared to those from urban communities (Rajput et al., 1988b). Some laboratories have started a search for exogenous and endogenous MPTP-like compounds in the brain (Collins et al., 1986) and the concentration of tetrahydroisoquinoline has been reported to be much higher in parkinsonian than in normal brain (Niwa et al., 1987).

Smoking

The observation that Parkinson's disease is less common in smokers than in non-smokers was first made in 1966 by Hammond and by Kahn in studies based on death certificate information and this trend was confirmed in some retrospective and case control studies (Nefzger et al., 1968; Kessler, 1972; Bauman et al., 1980; Martilla and Rinne, 1980; Goodwin-Austen et al., 1982). When the subject was reviewed by Baron (1986) it was stated that although many of the studies had limitations, the negative association between cigarette smoking and Parkinson's disease seemed established and a disease-protective effect of constituents within smoke was discussed. In view of the afore-mentioned theories of neurotoxic etiological mechanisms in Parkinson's disease, it is attractive to assume that regular administration of nicotine or other compounds within tobacco-smoke would block and counteract a neurotoxin. Several modes of action have been discussed:

(1) Nicotine derivatives such as 4-phenylpyridine present in smoke might act as a competitive inhibitor of exogenous or endogenous neurotoxins.
(2) Carbon monoxide in smoke might act to neutralise nigral lesions by free radicals.
(3) Direct effects of nicotine might stimulate turnover of dopamine and thereby ameliorate early symptoms of Parkinson's disease.

However, based on the information presently available there are reasons to severely question epidemiological evidence in favour of smoke being protective against Parkinson's disease. In some studies based on comparison of parkinsonian patients to matched neighbours and claiming an inverse relationship between smoking and Parkinson's disease, the differences between the groups is not impressive. In the investigation by Baumann and colleagues (1980), 63% of parkinsonian patients and 47% of the controls were never smokers. This should be compared to the figures obtained in recent studies by Rajput and co-workers (1984, 1987) demonstrating no significant difference between the groups in which the corresponding values were 63 and 55%, respectively. Further, in a population study in 70-year-old people in Gothenburg, no support for the hypothesis could be obtained as 40% of the parkinsonian individuals were smokers or former smokers as compared to 43% among controls (Granérus and Mellström, 1987). If constituents in smoke act protectively against Parkinson's disease, a dose-dependent effect should be expected. However, in the two studies performed addressing such an effect (Haack et al., 1981; Golbe et al., 1986) no correlations were found between age at onset or severity of disease and extent of nicotine exposure. In a recent study, an excessive number of smokers were found among demented parkinsonian patients (Korczyn and Salganik, 1988). Thus, based on clinical epidemiological data, smoking seems not to be of benefit in preventing or delaying Parkinson's disease. The trend to an inverse relationship between smoking and Parkinson's disease which is obvious in some investigations might be related to premorbid constitutional factors (Golbe et al., 1986). It is interesting to note that the nicotine-protecting hypothesis based on the aforementioned epidemiological studies inspired experiments in experimental animals. In some models with mechanical and neurotoxic lesions of meso-striatal dopamine neurons, nicotine treatment seems to increase neuronal survival (Jansson et al., this volume).

Although nicotine (2 mg i.v.) in the experimental situation has been claimed to reduce parkinsonian tremor (Marshall and Schnieden, 1966) amelioration following smoking is not information obtained in clinical work. In a controlled study, Nicorette® 4 mg was recently demonstrated not to induce or significantly effect different forms of tremor including parkinsonian (Koller and Herbster, 1988).

References

Aquilonius, S.-M. and Hartvig, P. (1986) A Swedish county with unexpectedly high utilization of anti-Parkinsonian drugs. *Acta Neurol. Scand.,* 74: 379 – 382.

Barbeau, A., Roy, M., Cloutier, T., Plasse, L. and Paris, S. (1986) Environmental and genetic factors in the etiology of Parkinson's disease. In M.D. Yahr and K.J. Bergmann (Eds.), *Advances in Neurology, Vol. 45,* Raven Press, New

332

York, pp. 299 – 306.

Baron, J.A. (1986) Cigarette smoking and Parkinson's disease. *Neurology,* 36: 1490 – 1496.

Baumann, R.J., Jameson, H.D., McKean, H.E., Haak, D.G. and Weisberg, L.M. (1980) Cigarette smoking and Parkinson disease. 1. A comparison of cases with matched neighbors. *Neurology,* 30: 839 – 843.

Calne, D.B., Langston, J.W., Martin, W.R.W., Stoessl, A.J., Ruth, T.J., Adam, M.J., Pate, B.D. and Schulzer, M. (1985) Positron emission tomography after MPTP: observations relating to the cause of Parkinson's disease. *Nature (Lond),* 317: 246 – 248.

Collins, M.A., Neafsey, E., Cheng, B.Y., Hurley-Gius, K., Ung-Chhun, N.A., Pronger, D.A., Christensen, M.A. and Hurley-Gius, D. (1986) Endogenous analogs of N-methyl-4-phenyl-1,2,3,6-tetrahydropyridine: Indoleamine-derived tetrahydro-beta-carbolines as potential causative factors in Parkinson's disease. In M.D. Yahr and K.J. Bergmann (Eds.), *Advances in Neurology, Vol. 45,* Raven Press, New York, pp. 179 – 182.

Davis, C.G., Williams, A.C., Markey, S.P., Ebert, M.H., Caine, E.D., Reichert, C.M. and Kopin, I.J. (1979) Chronic parkinsonism secondary to intravenous injection of meperidine analogues. *Psychiatr. Res.,* 1: 249 – 254.

Duvoisin, R.C. (1986) Genetics of Parkinson's disease. In M.D. Yahr and K.J. Bergmann (Eds.), *Advances in Neurology, Vol. 45,* Raven Press, New York, pp. 307 – 312.

Godwin-Austen, R.B., Lee, P.N., Marmot, M.G. and Stern, G.M. (1982) Smoking and Parkinson's disease. *J. Neurol. Neurosurg. Psychiatr.,* 45: 577 – 581.

Golbe, L.I., Cody, R.A. and Duvoisin, R.C. (1986) Smoking and Parkinson's disease. Search for a dose-response relationship. *Arch. Neurol.,* 43: 774 – 778.

Granérus, A.-K. and Mellström, D. (1987) Smoking and Parkinson's disease (in Swedish). *Läkartidningen,* 84: 2025 – 2026.

Haack, D.G., Baumann, R.J., McKean, H.E., Jameson, H.D., and Turbek, J.A.(1981) Nicotine exposure and Parkinson disease. *Am. J. Epidemiol.,* 114: 191 – 200.

Hammond, E.C. (1966) Smoking in relation to the death rates of one million men and women. In W. Haenszel (Ed.), *Epidemiological Approaches to the Study of Cancer and Other Chronic Diseases,* Department of Health, Education and Welfare, National Cancer Institute, Monograph 19, U.S.

Kahn, H.A. (1966) The Dorn study of smoking and mortality among U.S. veterans: report on eight and one-half years of observation. In W. Haenszel (Ed.), *Epidemiological Approaches to the Study of Cancer and Other Chronic Diseases,* Department of Health, Education and Welfare, National Cancer Institute, Monograph 19, U.S.

Kessler, I.I. (1972) Epidemiologic studies of Parkinson's disease. III. A community-based survey. *Am. J. Epidemiol.,* 96: 242 – 254.

Kinemuchi, H., Fowler, C.J. and Tipton, K.F. (1987) The neurotoxicity of 1-methyl-4-phenyl-1,2,3,6-tetrahydropyridine (MPTP) and its relevance to Parkinson's disease. *Neurochem. Int.,* 11: 359 – 373.

Koller, W. and Herbster, G. (1988) Nicotine and tremor. *Neurology,* 38 (Suppl. 1): 313.

Kopin, I.J. (1987) MPTP: an industrial chemical and contaminant of illicit narcotics stimulates a new era in research on Parkinson's disease. *Environ. Health Perspect.,* 75: 45 – 51.

Korczyn, A.D. and Salganik, I. (1988) Risk factors for dementia in Parkinson's disease. *Neurology,* 38 (Suppl. 1): 321.

Langston, J.W., Ballard, P., Tetrud, J.W. and Irwin, I. (1983) Chronic parkinsonism in humans due to a product of meperidine-analog synthesis. *Science,* 219: 979 – 980.

Li, S.-C., Schoenberg, B.S., Wang, C.-C., Cheng, X.-M., Rui, D.-L., Bolis, C.L. and Schoenberg, D.G. (1985) A prevalence survey of Parkinson's disease and other movement disorders in the People's Republic of China. *Arch. Neurol.,* 42: 655 – 657.

Lux, W.E. and Kurtzke, J.F. (1987) Is Parkinson's disease acquired? Evidence from a geographic comparison with multiple sclerosis. *Neurology,* 37: 467 – 471.

Manyam, B.V. (1988) Parkinson's disease and levodopa in "Ayurveda", ancient indian medical treatise. *Neurology,* 38 (Suppl. 1): 385.

Marsden, C.D. (1986) Homén lecture. The cause of Parkinson's disease. In E. Kivalo and U.K. Rinne (Eds.), *Uusia Nakokohtia Neurologiassa,* pp. 9 – 17.

Marshall, J. and Schnieden, H. (1966) Effect of adrenaline, noradenaline, atropine and nicotine on some types of human tremor. *J. Neurol. Neurosurg. Psychiatr.,* 29: 214 – 218.

Marttila, R.J. and Rinne, U.K. (1980) Smoking and Parkinson's disease. *Acta Neurol. Scand.,* 62: 322 – 325.

Nefzger, M.D., Quadfasel, F.A. and Karl, V.C. (1968) A retrospective study of smoking in Parkinson's disease. *Am. J. Epidemiol.,* 88: 149 – 158.

Niwa, T., Takeda, N., Kaneda, N., Hashizume, Y and Nagatsu, T. (1987) Presence of tetrahydroisoquinoline and 2-methyl-tetrahydroquinoine in parkinsonian and normal human brains. *Biochem. Biophys. Res. Commun.,* 144: 1084 – 1089.

Parkinson, J. (1817) *An Essay on the Shaking Palsy,* Whittingham and Rowland, London, pp. 1 – 66.

Rajput, A.H. (1984) Epidemiology of Parkinson's disease. *Can. J. Neurol. Sci.,* 11: 156 – 159.

Rajput, A.H., Offord, K.P., Beard, C.M. and Kurland, L.T. (1987) A case-control study of smoking habits, dementia, and other illnesses in idiopathic Parkinson's disease. *Neurology,* 37: 226 – 232.

Rajput, A.H., Bennett, V.L. and Uitti, R.J. (1988a) Early rural exposure as a risk to Parkinson's disease onset in later age, *9th International Symposium on Parkinson's Disease,* Jerusalem, Book of Abstracts, p. 11.

Rajput, A.H., Thiessen, B., Munoz, D., Laverty, D. and Desai, H. (1988b) Consideration of age and environments and number of substantia nigra cells, *9th International Symposium on Parkinson's disease,* Jerusalem, Book of Abstracts, p. 11.

Sundström, E. (1988) The parkinsonism-inducing neurotoxin MPTP (1-methyl-4-phenyl-1,2,3,6-tetrahydropyridine). Effects and mechanisms of action in the mouse central nervous system, Thesis, Stockholm.

Tanner, C.M., Chen, B., Wang, W.-Z., Gilley, D.W., Peng, M.-L., Liu, Z.-L., Liang, X.-L., Kao, L.C. and Goetz, C.G. (1988) Environmental factors and Parkinson's disease (PD): a case-control study in China. *Neurology,* 38 (Suppl. 1): 349.

A. Nordberg, K. Fuxe, B. Holmstedt and A. Sundwall (Eds.)
Progress in Brain Research, Vol. 79
© 1989 Elsevier Science Publishers B.V. (Biomedical Division)

CHAPTER 33

The cholinergic receptor system of the human brain: neurochemical and pharmacological aspects in aging and Alzheimer

Ezio Giacobini[a], Patrizia DeSarno[a], Brent Clark[b] and Michael McIlhany[c]

Departments of [a] Pharmacology, [b] Pathology and [c] Surgery, Southern Illinois University School of Medicine, P.O. Box 19230, Springfield, IL 62794-9230, U.S.A.

Cholinergic receptor changes in CNS degenerative disorders leading to dementia

The cholinergic system of the brain has been implicated in memory function (Drachman and Leavitt, 1974; Drachman, 1978, 1983; Drachman et al., 1983). Evidence for this role is anatomical, physiological and pharmacological (Davies, 1979; Rossor et al., 1982; Bird et al., 1983; Coyle et al., 1983; Drachman, 1983; Parent et al., 1984; Danielsson et al., 1988). In addition, three major degenerative disorders of the nervous system which lead to various types of severe dementia with memory impairment such as Alzheimer disease (AD) in all cases (Coyle et al., 1983), Parkinson disease in 10 – 15% of cases (Fahn, 1986), and Korsakoff syndrome in all cases (Warrington and McCarthy, 1986) also show involvement of the cholinergic system. Besides AD, which demonstrates a severe cholinergic cell degeneration, other CNS disorders leading to dementia such as the Gerstmann-Straussler syndrome, Parkinson dementia of Guam and the Dementia Pugilistica also affect this neurotransmitter system.

In all these diseases, pathological and biochemical lesions in the cholinergic neuron are localized mainly to the presynaptic region. Alterations include biochemical and morphological changes in the structures storing, synthesizing and releasing acetylcholine (ACh) and possibly also in cholinergic receptors regulating ACh release. In a comparison between cholinergic terminals in the CNS of normally aged and AD patients, it is apparent that muscarinic and nicotinic receptors are not equally affected.

Changes in muscarinic receptors

Among the 15 studies of frontal cortex samples from autopsies and biopsies (Table I) we found in the literature, spanning a 5-year period (1983 – 1988) and using mainly 1-quinuclidinyl-phenyl-4-benzilate ($[^3H]QNB$) as a muscarinic ligand, 8 showed no difference, 4 reported decreases from 14 to 30% and 2 reported increases from 20 to 39%. It seems therefore clear, that using this ligand alone, only a modest change can be seen in receptor binding in this region. Since muscarinic receptors are localized both pre- and postsynaptically and QNB does not differentiate between the two sites, it is difficult to establish a precise localization of the lesion. It is also difficult to estimate the effect of down- or up-regulation phenomena due to presynaptic degeneration and denervation reactions present in degenerative disorders such as AD.

Vanderheyden et al. (1987) found that if the proportion of M_1/M_2 muscarinic receptor subtypes is defined by their affinity for pirenzepine using [^{3}H]QNB as a non-subtype antagonist in competition binding experiments, the human neocortex contains 67% M_1 and 33% M_2 sites. In comparison with control values, the total amount of [^{3}H]QNB binding sites found in this study is lower in the frontal cortex of three patients who died at 39–42 years of early onset AD. The proportion of M_2 sites was also slightly diminished in these patients. This finding confirms the results of Mash et al. (1985) who also found some reduction of the proportion of M_2/M_1 sites in the neocortex of AD patients. Activation of muscarinic autoreceptors decreases the release of ACh in the rodent cerebral cortex (Szerb, 1979; Raiteri et al., 1984) and modulates its spontaneous release in basal forebrain slices (Suzuki et al., 1988). Probably both M_1 and M_2 receptors are involved in mechanisms of ACh release regulation. If these receptors have a similar function in human neocortex they would be important for the regulation of cholinergic neurotransmission in the surviving or sprouting terminals in AD. In addition, any pharmacological intervention directed to correct

cholinergic deficits should take into consideration the particular conditions of receptor sensitivity present in the diseased brain. Cholinesterase inhibition in particular, by increasing ACh levels, could elicit long-lasting inhibition of ACh release.

Changes in nicotinic receptors

A survey including percent changes in nicotinic receptors in the frontal cortex of AD patients reveals a striking difference from aging normal controls (Table II). Nicotinic receptor studies are mainly from autopsies using three different ligands (nicotine, α-bungarotoxin and ACh). Out of eight

TABLE I

Percent changes in muscarinic receptors of frontal cortex of Alzheimer patients as compared with controls

	Ligand	Author/year
No difference	QNB	Lang and Henke, 1983
6/11 Decrease	QNB	Wood et al., 1983
−14%	QNB	Mash et al., 1985
No difference	QNB	Rinne et al., 1985
+20%	QNB	Nordberg and Winblad, 1986
No difference	QNB	Shimohama et al., 1986
No difference	QNB	Jenni-Eiermann et al., 1984
No difference	QNB	London and Waller, 1986
−30%	QNB	Reinikainen et al., 1987
No difference	QNB	Waller et al., 1986
Decrease	QNB	Vanderheyden et al., 1987
No difference	QNB	Kellar et al., 1987
+23%	QNB	Danielsson et al., 1988
No difference (autopsy)	QNB	De Sarno et al., 1988
+39% (Biopsy)	QNB	De Sarno et al., 1988

TABLE II

Percent changes in nicotinic receptors of frontal cortex of Alzheimer patients as compared with controls

	Ligand	Author/year
No difference	α-Bungarotoxin	Davies and Feisullin, 1981
−65	ACh	Nordberg and Winblad, 1986
−53	Nicotine	Nordberg and Winblad, 1986
No difference	Nicotine	Shimohama et al., 1986
−62	ACh	Whitehouse et al., 1986
−56	Nicotine	Whitehouse et al., 1986
−56	ACh	Kellar et al., 1987
−44 (Autopsy)	Nicotine	De Sarno et al., 1988
−55 (Biopsy)	Nicotine	De Sarno et al., 1988

TABLE III

Cholinergic terminals in CNS: changes in receptor binding

	Physiological aging	Pathological aging: Alzheimer disease
Muscarinic receptors [^{3}H]QNB, [^{3}H]ACh [^{3}H]pirenzepine [^{3}H]oxotremorine	Reduced (M_1 or M_2?)	Slightly reduced (M_2) Normal or increased (M_1)
Nicotinic receptors [^{3}H]ACh [^{3}H]nicotine	Not reduced or increased	Significant loss

autopsy studies (1981 – 1988), including the one from our laboratory (DeSarno et al., 1988; Fig. 2), six show decreases from 44 to 65% and two show no difference. Our study of cholinergic receptor binding confirms and extends the studies found in the literature. Using [³H]QNB as ligand, we found no differences in specific binding in autopsy material but a significant 39% increase in the biopsies (Fig. 1 and Table I). With regard to nicotinic receptors using [³H]nicotine as ligand, we found a 44% decrease in the autopsies and a 55% decrease in the biopsies (Fig. 2 and Table II). The differences between biopsy and autopsy material suggest that different degrees of preservation of the tissue might explain differences between the two preparations. The results can be interpreted as indicative of selective presynaptic changes. These decreases correlate to enzymatic [cholineacetyltransferase (ChAt) and acetylcholinesterase

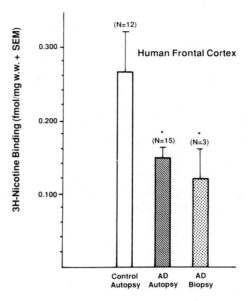

Fig. 2. Comparison of [³H]nicotine binding (fmol/mg w.w. ± S.E.M.) in samples of human frontal cortex from normal controls and Alzheimer patients (biopsies and autopsies). *Significantly different from controls, $p \leq 0.05$.

(AChE) activity] changes (Davies, 1979; Rossor et al., 1982; Bird et al., 1983; and this study, Table V) supporting a presynaptic localization of nicotinic receptor changes. The changes seen in cholinergic terminals in CNS during physiological and pathological aging are summarized in Table III.

Bioptic vs. autoptic analysis of human cortical tissue: a comparison

In order to study changes in nicotinic cholinergic binding in AD, we utilized human biopsies of patients undergoing intracerebroventricular (i.c.v.) treatment with physostigmine (Phy) (Giacobini et al., 1988). These biopsies consist of a 1 cm³ prism of right frontal cortex (premotor cortex, area 6) removed during the insertion of an intraventricular catheter for infusion of the drug. The general features of our frontal lobe biopsies are represented in Fig. 3. The access to sensitive methods of biochemical analysis makes it possible

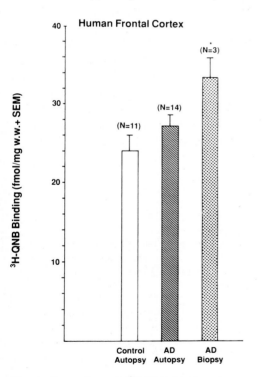

Fig. 1. Comparison of [³H]QNB binding (fmol/mg w.w. ± S.E.M.) in samples of human frontal cortex from normal controls and Alzheimer patients (biopsies and autopsies). *Significantly different from controls, $p \leq 0.05$.

to use μg samples of brain tissue (Giacobini, 1987). In our study, several types of enzyme activities, receptor binding and studies of ACh release were performed in the same bioptic sample (Fig. 3). The major advantages of using bioptic instead of autoptic cortical tissues are listed in Table IV. Better preservation of function and metabolic activity of the tissue and homogeneous time of analysis makes biopsy a more reliable and more precise tool for research and diagnosis. Another important concern has been the instability of enzymes, transmitters and receptors in tissue following death in the interval before freezing the tissue. For com-

parison, data obtained from three patients aged 76–82 suffering of different degrees of severity of dementia are reported in Table V. A good general agreement is seen among clinical diagnosis, pathology (density of neuritic plaques) and biochemistry (AChE and ChAt activity and nicotinic binding) in relation to controls. Particularly striking is the decrease of nicotinic receptors which correlates to similar findings in a larger group of autopsy samples examined in our laboratory with the same methods (Table II and Fig. 2). Other studies (Neary and Snowden, 1988) have shown a significant correlation between reduction in ChAt activity in cortex and senile plaque frequency, thereby linking changes in the subcortical projection system of the nucleus basalis with cortical pathology. On the other hand, the same authors reported that biochemical markers of the serotonergic and noradrenergic nerve endings were reduced but did not correlate with clinical findings.

One should note that our biopsy data do not demonstrate a one-to-one correlation between cholinergic neurochemistry (enzyme activities and receptor binding) and number of neuritic plaques, duration of illness or clinical dementia rating. For a statistical analysis of these observations, a larger number of bioptic samples is necessary.

TABLE IV

Major advantages of bioptic vs. autoptic human cortical tissue

(1) Preservation of functional activity of nerve cells due to:
   (a) Avoidance of ischemic pre-agonal effects
   (b) Avoidance of damage to cellular membranes due to freezing and thawing
   (c) Immediate oxygenation of the tissue which preserves metabolism
(2) Instability of enzymes, transmitters and receptors in tissues following death in the interval before freezing of the tissue
(3) Minimal variation in post-mortem times of experiment

TABLE V

Frontal biopsies of Alzheimer patients: comparison of biochemistry, pathology and dementia rating

Patient no.	Age (yrs)	Duration of illness (yrs)	Neuritic plaques (mm²)	AChE activity (μmol/g prot/h)	ChAt activity (nmol/100 mg prot/min)	[³H]nicotine specific binding (fmol/mg prot)	[³H]QNB specific binding (fmol/mg prot)	Clinical dementia rating
1	76	10	15–20	326	8.2	3.0	577	II
2	82	3	8	241	11	0.9	545	II–III
3	79	7	10–15	378	9.2	2.5	750	III
Normal controls	75 ± 3	–	–	443 ± 25	12.5	4.7 ± 0.9	430 ± 33	–
n	12	–	–	12[a]	10[b]	12[a]	11[a]	–

[a] DeSarno et al., 1988, autopsy.
[b] DeKosky et al., 1988, autopsy.

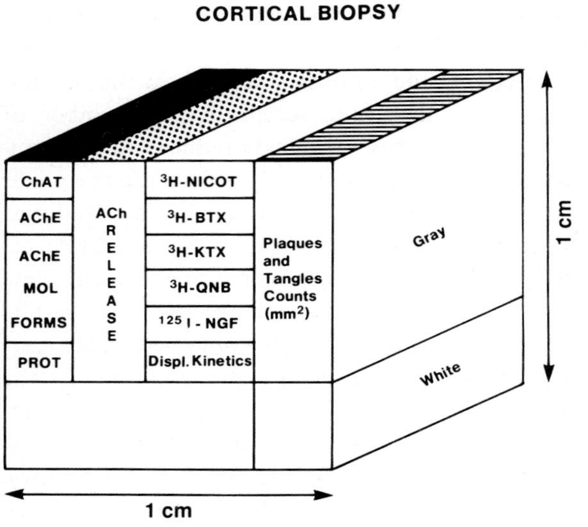

CORTICAL BIOPSY

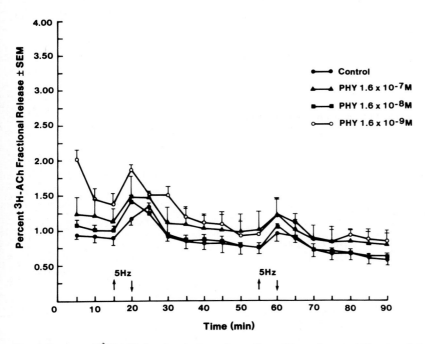

Fig. 3. Schematic diagram of a prefrontal lobe biopsy (area 6) 1 × 1 cm) indicating the relative volume of the various sections used for neurochemical and histopathological analysis and the biochemical tests performed.

Pharmacological testing of drugs in human biopsies

As conditions of the cholinergic innervation are widely different among patients, depending on the severity, the pathological subgroups and the progression of the disease (Rossor et al., 1982; Bird et al., 1983), it is important to determine the effect of the drug in vitro before experimenting the drug in vivo. A second major advantage of the biopsy is that it provides a unique possibility of testing in vitro the drug to be administered to the patient on a cortical sample of the same individual.

Activation of presynaptic nicotinic autoreceptors increases the release of ACh in the rodent cerebral cortex (Rowell and Winkler, 1984). Nilsson et al. (1987) found that Phy at high concentrations ($10^{-5} - 10^{-4}$ M) stimulated release of ACh in post-mortem cortical slices of AD patients. To examine differences in drug effects related to differences between biopsy and autopsy tissue we

Fig. 4. Percent of [^{3}H]ACh fractional release from slices of frontal cortical biopsies of three Alzheimer patients. Effect of different concentrations of physostigmine ($1.6 \times 10^{-7} - 10^{-9}$ M) administered in vitro.

studied the effect of three concentrations (1.6 × 10^{-7}, 10^{-8}, 10^{-9} M) of Phy which were considered to be close to therapeutic concentrations reached in the brain (Hallak and Giacobini, 1987). The effect of Phy was studied on the electrically stimulated release of ACh from 350 μm thick slices from frontal cortex biopsies of AD patients. A typical experiment is shown in Fig. 4.

The cortical biopsy tissue was found to respond maximally at a frequency of 5 Hz for 5 min. Slices from autopsy samples did not respond to electrical stimulation but only to increased K^+ concentrations in the medium. Physostigmine did not influence the electrically evoked release of ACh (Fig. 4). However, an increase in baseline release was seen at the lowest dose (1.6 × 10^{-9} M) of Phy. These results are particularly important as they suggest that "therapeutic" concentrations of Phy produced by i.c.v. administration may not negatively influence ACh release in the patient's cortex. In order to establish optimal effects and "best drug" concentrations, we are now regularly testing our i.c.v. patients with this "in vitro" technique.

Importance of biopsy studies and conclusions

Both biopsy and autopsy studies show that in contrast with muscarinic receptors, the number of nicotinic receptors is strongly reduced in the frontal cortex of AD patients. This decrease correlates closely to pathological and clinical findings. Biochemical findings such as AChE and ChAt activity and nicotinic receptor binding also correlate with pathology and clinical findings. Our results suggest that biochemical data obtained from a cortical biopsy may correlate more closely to the clinical picture than histological results based purely on counts of neuritic plaques and tangles. Biochemical data could provide useful diagnostic information at stages of the disease earlier than those examined in our study. Thus, we propose that an accurate "biochemical diagnosis" of AD could be established using enzymatic and receptor analysis in bioptic tissues of earlier cases of AD. Studies in the literature strongly support our data on presynaptic cholinergic dysfunction obtained by analysis of cortical biopsies of AD patients. Bowen (1983) and Sims et al. (1983) found

TABLE VI

Biochemical changes in cholinergic markers which correlate with a positive diagnosis of Alzheimer

	Neural		Extraneural		
	Cortical biopsy or autopsy	CSF	Lymphocytes	Erythrocytes	Plasma
Nicotinic receptor binding	Decrease		Decrease		
Muscarinic receptor binding	No difference		Decrease		
ChAt activity	Decrease	Not measured			
AChE activity	Decrease	Decrease		No difference	BuChE, no difference
Release of ACh	Decrease				
ACh synthesis	Decrease				
H.A. Ch uptake	Decrease	Increase Ch levels			
Altered Ch metabolism	Decrease	Increase Ch levels		Increase Ch levels	Ch, no difference

decreases of ACh synthesis, choline uptake and ChAt activity in biopsy samples of temporal and frontal cortex. Gauthier et al. (1986) also found a good correlation between clinical diagnosis and ChAt activity of fronto-temporal biopsies. Finally, Martin et al. (1987) reported a 100% overlapping of clinical and neuropsychological data with pathological findings in 11 AD patients with a left frontal biopsy. Many AD patients die in cachectic state (Hardy et al., 1985), therefore, there is a possibility that decreases in enzymatic activities or receptor bindings may represent an artefact. In addition, a prolonged agonal state may cause biochemical changes in the brain (Hardy et al., 1985). Acidotic states after a long terminal phase may cause profound changes in biochemical measurements on autopsy tissues (Hardy et al., 1985). The use of fresh (3 – 5 min from dissection in the surgery room) and oxygenated cortical tissue makes this less likely. Thus, cortical biopsy with combined histological and biochemical analysis could provide an accurate pre-mortem diagnosis of AD.

Cholinergic markers may not be useful for day-to-day clinical applications, however, they correlate with a positive diagnosis of AD. Such markers are found in (a) cortical bioptic or autoptic tissue; (b) CSF and (c) extraneuronal elements (lymphocytes) (Table VI). Biopsy data, in particular, represent important research findings for understanding and evaluating the progress of the disease. Biochemical data could prove useful as biochemical markers of disease progression. With regard to data variability and diagnostic significance, no presently available non-invasive measures of disease progression (e.g. CT head scan, PET scan, MRI, EEG, CSF or blood study) are less variable or more diagnostically reliable. Even histological examination of the brain (e.g. plaques, tangles and cell counts) results in highly variable data. From the pathophysiological standpoint the present data on cholinergic receptor changes raise important questions regarding the function of cholinergic transmission and its relation to dementia in AD patients.

Acknowledgements

The authors wish to thank Elizabeth Williams for her technical assistance and Diana Smith for the preparation and editing of the manuscript. Martha Downen-Keomala performed the ChAt activity measurements. Supported in part by the National Institutes of Aging AG05416 and by a grant from R.J. Reynolds Tobacco Company.

References

Bird, T.D., Stranahan, S., Sumi, M. and Raskind, M. (1983) Alzheimer's disease: choline acetyltransferase activity in brain tissue from clinical and pathological subgroups. Ann. Neurol, 14: 284 – 293.

Bowen, D.M. (1983) Biochemical assessment of neurotransmitter and metabolic dysfunction and cerebral atrophy in Alzheimer's disease. In R. Katzman (Ed.), Banbury Report 15: Biological Aspects of Alzheimer's Disease, Cold Spring Harbor Laboratory, New York, pp. 219 – 231.

Coyle, J.T., Price, D.L. and Delong, M.R. (1983) Alzheimer's disease: a disorder of cortical cholinergic innervation. Science, 219: 1184 – 1190.

Danielsson, E., Eckernas, S.-A., Westlind-Danielsson, A., Nordstrom, O., Bartfai, T., Gottfries, C.-G. and Wallin, A. (1988) VIP-sensitive adenylate cyclase, guanylate cyclase, muscarinic receptors, choline acetyltransferase and acetylcholinesterase, in brain tissue afflicted by Alzheimer's disease/senile dementia of the Alzheimer type. Neurobiol. Aging, 9: 153 – 162.

Davies, P. (1979) Neurotransmitter-related enzymes in senile dementia of the Alzheimer type. Brain Res., 171: 319 – 327.

Davies, P. and Feisullin, S. (1981) Postmortem stability of alpha-bungarotoxin binding sites in mouse and human brain. Brain Res, 216: 449 – 454.

DeSarno, P., Giacobini, E. and Clark, B. (1988) Changes in nicotinic receptors in human and rat CNS. Fed. Proc., 2: 364.

Drachman, D.A. (1978) Memory, dementia and the cholinergic system. In R. Kartzman, R. Terry and K. Bick (Eds.), Alzheimer's Disease: Senile Dementia and Related Disorders: Aging, Vol. 7, Raven Press, New York, pp. 141 – 148.

Drachman, D.A. (1983) Aging and dementia: insights from the study of anticholinergic drugs. In: R. Kartzman (Ed.), Banbury Report 15: Biological Aspects of Alzheimer's Disease, Cold Spring Harbor Laboratory, New York, pp. 363 – 370.

Drachman, D.A. and Leavitt, J. (1974) Human memory and the cholinergic system: a relationship to aging? Arch. Neurol., 30: 113 – 121.

Drachman, D.A., Glosser, G., Fleming P. and Longenecker,

342

G. (1982) Memory decline in the aged: treatment with lecithin and physostigmine. *Neurology,* 37: 674–675.

Fahn, S. (1986) Parkinson disease and other basal ganglion disorders. In E. Asbury, G. McKhann and S. McDonald (Eds.), *Diseases of the Nervous System,* Andmore Medical Books, W.B. Saunders, Philadelphia, pp. 1217–1228.

Gauthier, S., Robitaille, Y., Quirion, R. and LeBlanc, R. (1986) Antemortem laboratory diagnosis of Alzheimer's disease. *Prog. Neuro-Psychopharmacol. Biol. Psychiatr.,* 10: 391–403.

Giacobini, E. (1987) Neurochemical analysis of single neurons: a mini-review dedicated to Oliver H. Lowry. *J. Neurosci. Res.,* 18: 632–637.

Giacobini, R., Becker, R., McIlhany, M. and Kumar, V. (1988) Intracerebroventricular administration of cholinergic drugs: preclinical trials and clinical experience in Alzheimer patients. In E. Giacobini and R. Becker (Eds.), *Current Research in Alzheimer Therapy: Cholinesterase Inhibitors,* Taylor and Francis, New York, pp. 113–122.

Hallak, M. and Giacobini, E. (1987) A comparison of the effects of two inhibitors on brain cholinesterase. *Neuropharmacology,* 26: 521–530.

Hardy, J.A., Wester, P., Winblad, B., Gezelius, C., Bring, G. and Eriksson, A. (1985) The patients dying after long terminal phase have acidotic brains: implications for biochemical measurements on autopsy tissue. *J. Neural. Trans.,* 61: 253–264.

Jenni-Eiermann, S., von Hahn, H.P., Honegger, C.G. and Ulrich, J. (1984) Studies on neurotransmitter binding in senile dementia. *Gerontology,* 30: 350–358.

Kellar, K.J., Whitehouse, P.J., Martino-Barrows, A.M., Marcus, K. and Price, D.L. (1987) Muscarinic and nicotinic cholinergic binding sites in Alzheimer's disease cerebral cortex. *Brain Res.,* 436: 62–68.

Lang, H. and Henke, H. (1983) Cholinergic receptor binding and autoradiography in brains of non-neurological and senile dementia of Alzheimer-type patients. *Brain Res.,* 267: 271–280.

London, E.D. and Waller, S.B. (1986) Relationships between choline acetyltransferase and muscarinic binding in aging rodent brain and in Alzheimer's disease. In I. Hanin (Ed.), *Dynamics of Cholinergic Function,* Plenum Press, New York, pp. 215–224.

Martin, E.M., Wilson, R.S., Penn, R.D., Fox, J.H., Clasen, R.A. and Savoy, S.M. (1987) Cortical biopsy results in Alzheimer's disease: correlation with cognitive deficits. *Neurology,* 37: 1201–1204.

Mash, D.C., Flynn, D.D. and Potter, L.T. (1985) Loss of M_2 muscarine receptors in the cerebral cortex in Alzheimer's disease and experimental cholinergic denervation. *Science,* 228: 115–117.

Neary, D. and Snowden, J.S. (1988) Neuropathological and neurochemical changes in ascending projection pathways; relationship to clinical symptoms in Alzheimer's disease,

Proc. Natl. Symp. Alzheimer Disease (Kuopio, Finland), pp. 23–25.

Nilsson, L., Adem, A., Hardy, J., Winblad, B. and Nordberg, A. (1987) Do tetrahydroaminoacridine (THA) and physostigmine restore acetylcholine release in Alzheimer brains via nicotinic receptors? *J. Neural. Trans.,* 70: 357–368.

Nordberg, A. and Winblad, B. (1986) Reduced number of ^{3}H-nicotine and ^{3}H-acetylcholine binding sites in the frontal cortex of Alzheimer brains. *Neurosci. Lett.,* 72: 115–119.

Parent, A., Cxonka, C. and Etienne, P. (1984) The occurrence of large acetylcholinesterase-containing neurons in human neostriatum as disclosed in normal and Alzheimer-diseased brains. *Brain Res.,* 291: 154–158.

Raiteri, M., Leardi, R. and Marchi, M. (1984) Heterogeneity of presynaptic muscarinic receptors regulating neurotransmitter release in the rat brain. *J. Pharmacol. Exp. Ther.,* 228: 209–214.

Reinikainen, K.J., Riekkinen, P.F., Halonen, T. and Laakso, M. (1987) Decreased muscarinic receptor binding in cerebral cortex and hippocampus in Alzheimer's disease. *Life Sci.,* 41: 453–461.

Rinne, J.O., Laakso, K., Lonneberg, P., Molsa, P., Paljarvi, L., Rinne, J.K., Sako, E. and Rinne, U.K. (1985) Brain muscarinic receptors in senile dementia. *Brain Res.,* 336: 19–25.

Rossor, M.N., Garrett, N.J., Johnson, A.L., Mountjoy, C.Q., Roth, M. and Iversen, L.L. (1982) A post-mortem study of the cholinergic and GABA systems in senile dementia. *Brain,* 105: 313–330.

Rowell, P.R. and Winkler, D.L. (1984) Nicotinic stimulation of (^{3}H)acetylcholine release from mouse cerebral cortex synaptosomes. *J. Neurochem.,* 43: 1593–1598.

Shimohama, S., Taniguchi, T., Fujiwara, M. and Kameyama, M. (1986) Changes in nicotinic and muscarinic cholinergic receptors in Alzheimer-type dementia. *J. Neurochem.,* 46: 288–293.

Sims, N.R., Bowen, D.M., Allen, S.J., Smith, C.C.T., Neary, D., Thomas, D.J. and Davison, A.N. (1983) Presynaptic cholinergic dysfunction in patients with dementia. *J. Neurochem.,* 40: 503–509.

Szerb, J.C. (1979) Autoregulation of acetylcholine release. In S.Z. Langer, K. Starke and M.L. Dubocovich (Eds.), *Presynaptic Receptors,* Pergamon Press, New York, pp. 293–298.

Suzuki, T., Fujimoto, K., Oohata, H. and Kawashima, K. (1988) Presynaptic M_1 muscarinic receptor modulates spontaneous release of acetylcholine from rat basal forebrain slices. *Neurosci. Lett.,* 84: 209–212.

Vanderheyden, P., Ebinger, G., Dierckx, R. and Vauquelin, G. (1987) Muscarinic cholinergic receptor subtypes in normal human brain and Alzheimer's presenile dementia. *J. Neurol. Sci.,* 82: 257–269.

Waller, S.B., Ball, M.J., Reynolds, M.A. and London, E.D. (1986) Muscarinic binding and choline acetyltransferase in postmortem brains of demented patients. *Can. J. Neurol. Sci.,* 13: 528 – 532.

Washington, E. and McCarthy, R.E. (1986) Parkinson disease and other basal ganglion disorders. In Asbury, McKhann and McDonald (Eds.), *Diseases of the Nervous System,* Andmore Medical Books, W.B. Saunders, Philadelphia, pp. 828 – 847.

Whitehouse, P.J., Martino, A.M., Antuono, P.G., Lowen-stein, P.R., Coyle, J.T., Price, D.L. and Kellar, K.J. (1986) Nicotinic acetylcholine binding sites in Alzheimer's disease. *Brain Res.,* 371: 146 – 151.

Wood, P.L., Etienne, P., Lal, S., Nair, N.P.V., Finlayson, M.H., Gauthier, S., Palo, J., Haltia, M., Paetau, A. and Bird, E.D. (1983) A post-mortem comparison of the cortical cholinergic system in Alzheimer's disease and Pick's disease. *J. Neurol. Sci.,* 62: 211 – 217.

A. Nordberg, K. Fuxe, B. Holmstedt and A. Sundwall (Eds.)
Progress in Brain Research, Vol. 79
© 1989 Elsevier Science Publishers B.V. (Biomedical Division)

CHAPTER 34

N-[3H]methylcarbamylcholine binding sites in the rat and human brain: relationship to functional nicotinic autoreceptors and alterations in Alzheimer's disease

Dalia M. Araujo[a], Paul A. Lapchak[b], Brian Collier[b] and Remi Quirion[a]

[a] Douglas Hospital Research Center and Departments of [a] Psychiatry, and [b] Pharmacology, McGill University, Montreal, Quebec, Canada

Introduction

N-[3H]methylcarbamylcholine (MCC) has recently been used to characterize nicotinic receptors in the rat CNS (Abood and Grassi, 1986; Boksa and Quirion, 1987; Yamada et al., 1987; Araujo et al., 1988a; Lapchak et al., 1988). The distribution of [3H]MCC binding sites in rat brain is similar to that of other nicotinic ligands such as [3H]nicotine and [3H]acetylcholine (ACh) (see Clarke et al., 1985; Clarke, 1987a; Boksa and Quirion, 1987). However, there are some advantages in using [3H]MCC as a nicotinic receptor probe rather than other ligands such as [3H]nicotine and [3H]ACh. For example, [3H]MCC is a more stable compound and it appears to bind only to one class of nicotinic sites in brain (Boksa and Quirion, 1987). In human brain, [3H]MCC also binds to nicotinic sites with high affinity (Wang et al., 1987; Araujo et al., 1988b). Therefore, it is evident that [3H]MCC is a good probe to study neuronal nicotinic receptors in the mammalian CNS.

Despite the extensive characterization of nicotinic high affinity agonist sites in brain (see Wonnacott, 1987), there is conflicting evidence as

to the precise localization and the physiological role of these sites. Although it has been suggested that these high affinity agonist sites may represent functional nicotinic autoreceptors (see Whitehouse et al., 1986; Araujo et al., 1988a), there is some disagreement on this point. Thus, the significance of [3H]MCC/nicotine sites, especially in relation to their possible role in disease states, will be briefly discussed.

[3H]MCC binding sites in the rat brain

[3H]MCC binds specifically, saturably, and with high affinity to a single class (mean n_H close to 1.0) of nicotinic sites in the rat cerebral cortex (Abood and Grassi, 1986; Boksa and Quirion, 1987; Yamada et al., 1987; Araujo et al., 1988a), hippocampus and striatum (Araujo et al., 1988a). Ligand selectivity studies further showed that unlabelled MCC, nicotine, and other nicotinic drugs have a high affinity for the [3H]MCC sites, whereas muscarinic agonists and antagonists are weak competitors for these sites (Boksa and Quirion, 1987; Araujo et al., 1988a). The results from these membrane binding experiments demonstrate a nicotinic profile for [3H]MCC binding, which is similar to those shown previously for [3H]nicotine and [3H]ACh (Romano and Goldstein, 1980; Marks and Collins, 1982;

Address for correspondence: Remi Quirion, Douglas Hospital Research Center, Heinz Lehmann Pavilion, 6875 Lasalle Blvd., Verdun, Quebec, Canada H4H 1R3

346

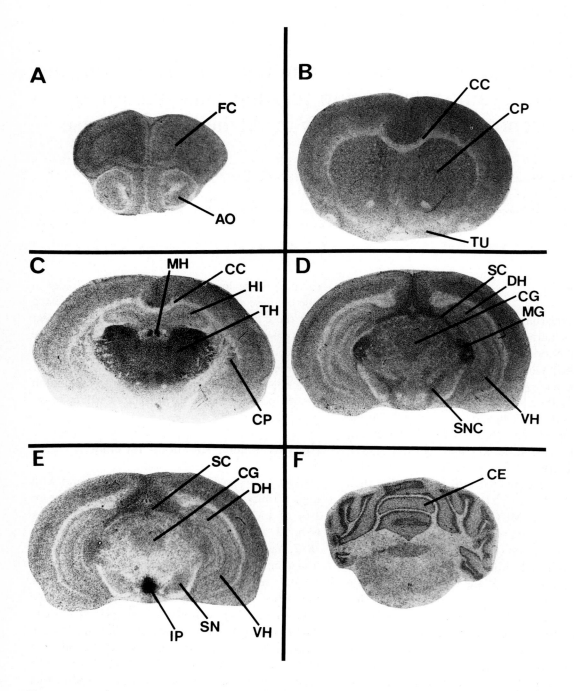

Fig. 1. Photomicrographs of the autoradiographic distribution of [³H]MCC binding sites in the rat brain. Sections were incubated with 10 nM [³H]MCC. Abbreviations: AO = accessory olfactory nucleus; CC = corpus callosum; CE = cerebellum; CG = central gray; CP = caudate-putamen; DH = dorsal hippocampus; FC = frontal cortex; HI = hippocampus; IP = interpeduncular nucleus; MG = medial geniculate nucleus; MH = medial habenula; SC = superior colliculus; SN = substantia nigra; SNC = substantia nigra pars compacta; TH = thalamus; TU = olfactory tubercle; VH = ventral hippocampus.

Schwartz et al., 1982; Schwartz and Kellar, 1983a,b; Sloan et al., 1985; Martino-Barrows and Kellar, 1987).

The autoradiographic distribution of [3H]MCC sites shows a pattern of localization which is similar to that of other nicotinic agonist receptor ligands (see Clarke et al., 1985; Clarke, 1987a; Harfstrand et al., 1987). Very high densities of sites are present in various thalamic nuclei, the interpeduncular nucleus, superior colliculus, medial habenula, and layers III/IV of the cerebral cortex (Fig. 1). Comparison of the pattern of distribution of [3H]MCC sites with those of cholinergic non-nicotinic ligands (e.g. [3H]quinuclidinyl benzilate, [3H]hemicholinium-3, [3H]AH-5183) provides further evidence that [3H]MCC binds selectively to nicotinic sites in the rat brain (see Discussion in Boksa and Quirion, 1987). Thus, both membrane binding and receptor autoradiographic studies demonstrate that [3H]MCC is a selective nicotinic receptor probe.

Possible function of [3H]MCC/nicotinic sites in the rat brain

As mentioned in the Introduction, there is some disagreement as to whether brain nicotinic binding sites, which have been characterized using [3H]nicotine, [3H]ACh, and [3H]MCC, represent functional nicotinic receptors. Clearly, nicotinic agonists exert powerful central effects, at least as can be inferred from electrophysiological (Krnjevic, 1975), biochemical (Giorguieff-Chesselet et al., 1979; Yoshida et al., 1980; Wonnacott, 1987), and behavioral (Clarke and Kumar, 1983; Clarke, 1987b) studies. Furthermore, systemic administration of nicotine has been shown to alter the release of ACh from the mammalian CNS (Armitage et al., 1969; Balfour, 1982). However, it is not clear whether these effects of nicotine on CNS function are mediated by pre- or post-synaptic nicotinic receptors. More recent studies have suggested a direct action of nicotine on cholinergic nerve terminals (Rowell and Winkler, 1984; Araujo et al., 1988a; Lapchak et al., 1988), whereas others have

concluded that nicotine's effect is an indirect one on nicotinic receptors located distal to the terminals (Beani et al., 1985; Meyer et al., 1987). Our work shows that the effect of a nicotinic agonist, MCC, to enhance ACh release from slices of rat hippocampus and frontal cortex was not altered by tetrodotoxin (TTX) (Table I). Therefore, it appears that in rat cerebral cortex and hippocampus, the site of action of nicotinic agonists is at the cholinergic terminal, and does not require the initiation of impulse activity. Results from experiments which showed that nicotine enhanced the release of [3H]ACh from rat hippocampal synaptosomes (Thorne et al., 1988) further support this hypothesis.

The effect of nicotinic agonists to increase ACh release from rat brain appears to be specific to certain regions. For example, although there is a high density of [3H]MCC/nicotinic sites in the rat striatum (Boksa and Quirion, 1987; Araujo et al.,

TABLE I

Effect of TTX on the MCC-induced increase in spontaneous ACh release from rat brain slices

Brain region	Drug	Spontaneous ACh release (% control)
Hippocampus	MCC (1 μM)	211 $\pm$ 33
	MCC (1 μM) + TTX (1 μM)	191 $\pm$ 16
	(Control ACh release = 100% = 3.8 $\pm$ 0.8 pmol/mg tissue/20 min)	
Frontal cortex	MCC (10 μM)	177 $\pm$ 25
	MCC (10 μM) + TTX (1 μM)	164 $\pm$ 12
	(Control ACh release = 100% = 2.1 $\pm$ 0.3 pmol/mg tissue/20 min)	

Slices were incubated in Krebs medium with or without drug (control). Test slices were incubated with the indicated concentrations of MCC alone or MCC with TTX. Release in the presence of MCC was significantly different from control ($p < 0.01$); release in the presence of MCC and TTX was significantly different from control ($p < 0.01$). Release in the presence of MCC and TTX was not significantly different from that in the presence of MCC alone ($p > 0.50$, by Student's t-test).

1988a), MCC did not alter the spontaneous release of ACh from striatal slices (Araujo et al., 1988a; Lapchak et al., 1988). Similar results were obtained by Weiler et al. (1984), who found that nicotine did not affect ACh release from slices of rat striatum. Thus, it is reasonable to conclude that not all nicotinic receptors in the rat brain are located on cholinergic terminals and therefore do not function as autoreceptors. In the striatum, for example, there is evidence for the existence of nicotinic receptor sites on dopaminergic terminals (Giorguieff-Chesselet et al., 1979).

[³H]MCC binding sites in Alzheimer's disease

Alzheimer's disease (AD) is characterized by various neurotransmitter and receptor deficits. In particular, there is a degeneration of certain cholinergic pathways (see Bartus et al., 1982; Candy et al., 1986), which correlates with the severity of the dementia in AD (Perry et al., 1978). In support of this cholinergic hypothesis of AD has been the consistent finding that the activity of choline acetyltransferase (ChAT) is markedly reduced in the neocortex and the hippocampus in AD brains

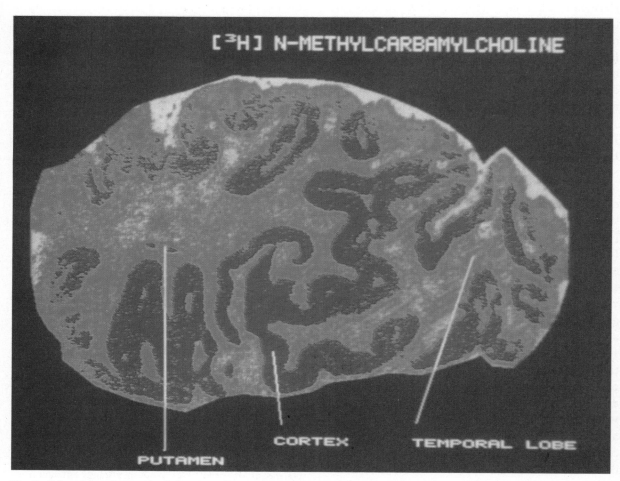

Fig. 2. Color photomicrograph of the autoradiographic distribution of [³H]MCC/nicotinic sites in the human brain. This section (20 μm) was incubated with 10 nM [³H]MCC.

(Bowen et al., 1976; Reisine et al., 1978; Davies, 1979; McGeer, 1984; Etienne et al., 1986; Araujo et al., 1988b). Other cholinergic markers, such as muscarinic and nicotinic receptor ligands, have also been used to determine the status of the cholinergic synapse in AD. However, these studies have obtained discrepant results, with some investigators reporting increases, decreases, or no change in receptor density and/or affinity (see Araujo et al., 1988b, for discussion).

Recent studies have shown that the density of nicotinic receptor sites is decreased in certain brain regions in AD. When $[^{125}I]\alpha$-bungarotoxin was used as the ligand, only small deficits in the number of nicotinic sites were reported (Davies and Feisullin, 1981). With other more selective ligands for CNS nicotinic sites, such as $[^3H]$nicotine (Flynn and Mash, 1986; Nordberg and Winblad, 1986; Quirion et al., 1986; Shimohama et al., 1986; Whitehouse et al., 1986) and $[^3H]$ACh (Nordberg and Winblad, 1986; Whitehouse et al., 1986), marked reductions in the density of nicotinic receptor sites in various cortical regions in AD brains have been demonstrated.

Recently, $[^3H]$MCC has been used to characterize nicotine binding sites in the human brain (Araujo et al., 1988b). Preliminary autoradiographic data of $[^3H]$MCC binding in human brain demonstrates a moderate to high density of sites in the cerebral cortex, striatum, and various thalamic nuclei. A color autoradiogram of a representative section of human brain is shown in Fig. 2. The distribution observed with $[^3H]$MCC in human brain closely resembles that seen in autoradiograms obtained using either $[^3H]$ACh (Kellar et al., 1987) or $[^3H]$nicotine (Adem et al., 1988), as ligand. In the former study, it was also apparent that there was a reduction in the density of nicotinic sites in some cortical regions in AD brains (Kellar et al., 1987).

Results from membrane binding studies in human brain, using $[^3H]$MCC, have provided additional insight into the regional loss of nicotinic receptor sites in AD. As in the rat brain, $[^3H]$MCC binds specifically and saturably to a single class of high affinity sites in both control and AD brains (Araujo et al., 1988b). Moreover, we found a large decrease in the maximal density (B_{max}) of $[^3H]$MCC/nicotinic sites in cerebral cortical areas and in hippocampus in AD brains (Araujo et al., 1988b). Similarly, using a single concentration of $[^3H]$MCC (5 nM), Wang et al. (1987) reported a decrease in the binding of $[^3H]$MCC to nicotinic sites in the cortex of AD brains. We found the reduction in the density of $[^3H]$MCC/nicotinic sites to be most pronounced in the frontal and temporal cortices, and in the hippocampus (Fig. 3). In all three areas, but particularly in the hippocampus, there was a significant correlation ($r > 0.80$) between the magnitude of the reduction in the density of nicotinic receptor sites and that in ChAT activity (Fig. 3). No alterations in either ChAT activity or $[^3H]$MCC/nicotinic sites were seen in sub-cortical structures

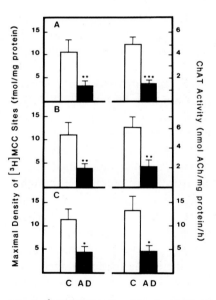

Fig. 3. $[^3H]$MCC binding (B_{max}) and ChAT activity in human brain cortical regions. Binding data are derived from full saturation analysis. The B_{max} for $[^3H]$MCC binding (left-hand axis) and the ChAT activity (right-hand axis) in frontal (A) temporal (B) cortices, and in hippocampus (C), in AD and age-matched control brains was assessed. Results are mean ± S.E.M. of 5–9 brains. Both the B_{max} for $[^3H]$MCC binding to homogenates and the ChAT activity were significantly reduced in AD compared to control brains (*$p < 0.005$, **$p < 0.01$).

such as the thalamus, striatum, and globus pallidus (Araujo et al., 1988b).

The parallel decrease in ChAT activity and in the density of [^{3}H]MCC/nicotinic sites in AD brains further supports the hypothesis that in certain regions of the mammalian brain, nicotinic receptors may be located presynaptically, on cholinergic nerve terminals (see also Whitehouse et al., 1986). Assuming that these receptors represent functional nicotinic autoreceptors, which might normally enhance spontaneous ACh release (see above), then the regulation of ACh release may be severely compromised in AD patients. Thus, stimulation of the remaining functional nicotinic autoreceptors in the cortex of AD patients may increase ACh release, and perhaps ameliorate the memory dysfunctions associated with cholinergic impairment in the disease. Alternatively, it may be of some therapeutic benefit to increase the number of nicotinic sites in the brain. Studies in rodent (Marks et al., 1983; Schwartz and Kellar, 1983a,b,1985; Martino-Barrows and Kellar, 1987; Lapchak et al., 1988) and human (Benwell et al., 1988) brain, which have shown that chronic nicotine upregulates nicotinic receptors, suggest that such manipulation may be feasible. Whether the observed increase in nicotine receptor number with chronic nicotine represents an increase in functional nicotinic autoreceptors remains to be investigated. However, animal studies have shown that chronic administration of nicotine, although it may upregulate nicotinic receptor number, may also "downregulate" receptor function, perhaps by shifting the receptor to a desensitized state (Lapchak et al., 1988). Thus, if any amelioration of cholinergic dysfunction is to be attained with nicotinic agonist therapy, treatment regimens must be approached with caution and closely monitored.

Acknowledgements

This work was supported by the Medical Research Council of Canada. D.M.A. is a Fellow and R.Q. is a Chercheur-Boursier of the Fonds de Recherche en Sante du Quebec. P.A.L. is a recipient of a studentship from F.C.A.R., Quebec.

References

Abood, L.G. and Grassi, S. (1986) [^{3}H]Methylcarbamylcholine, a new radioligand for studying brain nicotinic receptors. *Biochem. Pharmacol.*, 35: 4199–4202.

Adem, A., Jossan, S.S., D'Argy, R., Brandt, R., Winblad, B. and Nordberg, A. (1988) Distribution of nicotinic receptors in the human thalamus as visualized by ^{3}H-nicotine and ^{3}H-acetylcholine receptor autoradiography. *J. Neural Trans.* (in press).

Araujo, D.M., Lapchak, P.A., Collier, B. and Quirion, R. (1988a) Characterization of [^{3}H]N-methylcarbamylcholine binding sites and effect of N-methylcarbamylcholine on acetylcholine release in rat brain. *J. Neurochem.*, 51: 292–299.

Araujo, D.M., Lapchak, P.A., Robitaille, Y., Gauthier, S. and Quirion, R. (1988b) Differential alteration of various cholinergic markers in cortical and subcortical regions of human brain in Alzheimer's disease. *J. Neurochem.*, 50: 1914–1923.

Balfour, D.J.K. (1982) The effects of nicotine on brain neurotransmitter systems. *Pharmacol. Ther.*, 16: 269–282.

Bartus, R.T., Dean, R.L., III, Beer, B. and Lippa, A.S. (1982) The cholinergic hypothesis of geriatric memory dysfunction. *Science,* 217: 408–417.

Beani, L., Bianchi, C., Nilsson, L., Nordberg, A., Romanelli, L. and Sivilotti, L. (1985). The effect of nicotine and cytisine on ^{3}H-acetylcholine release from cortical slices of guinea-pig brain. *Naunyn Schmiedebergs Arch. Pharmacol.*, 331: 293–296.

Benwell, M.E.M., Balfour, D.J.K. and Anderson, J.M. (1988) Evidence that tobacco smoking increases the density of (−)-[^{3}H]nicotine binding sites in human brain. *J. Neurochem.*, 50: 1243–1247.

Boksa, P. and Quirion, R. (1987) [^{3}H]N-methyl-carbamylcholine, a new radioligand specific for nicotinic acetylcholine receptors in brain. *Eur. J. Pharmacol.*, 139: 323–333.

Bowen, D.M., Smith, C.B., White, P. and Davison, A.N. (1976) Neurotransmitter-related enzymes and indices of hypoxia in senile dementia and other abiotrophies. *Brain,* 99: 459–496.

Candy, J.M., Perry, E.K., Perry, R.H., Court, J.A., Oakley, A.E. and Edwardson, J.A. (1986) The current status of the cortical cholinergic system in Alzheimer's disease and Parkinson's disease. In D.F. Swaab, E. Fliers, M. Mirmiram, W.A. Van Gool, and F. van Haaren (Eds), *Progress in Brain Research, Vol. 70,* Elsevier Science Publishers, B.V., Amsterdam, pp. 105–132.

Clarke, P.B.S. (1987a) Recent progress in identifying nicotinic cholinoceptors in mammalian brain. *Trends Pharmacol. Sci.,* 8: 32–35.

Clarke, P.B.S. (1987b) Nicotine and smoking: a perspective from animal studies. *Psychopharmacology*, 92: 135 – 143.

Clarke, P.B.S. and Kumar, R. (1983) Characterization of the locomotor stimulant action of nicotine in tolerant rats. *Br. J. Pharmacol.*, 80: 587 – 594.

Clarke, P.B.S., Schwartz, R.D., Paul, S.M., Pert, C.B. and Pert, A. (1985) Nicotinic binding in rat brain: autoradiographic comparison of [3H]acetylcholine, [3H]nicotine, and [125I]-α-bungarotoxin. *J. Neurosci.*, 5: 1307 – 1315.

Davies, P. (1979) Neurotransmitter-related enzymes in senile dementia of the Alzheimer type. *Brain Res.*, 171: 319 – 327.

Davies, P. and Feisullin, S. (1981) Postmortem stability of alpha-bungarotoxin binding sites in mouse and human brain. *Brain Res.*, 216: 449 – 454.

Etienne, P., Robitaille, Y., Wood, P., Gauthier, S., Nair, N.P.V. and Quirion, R. (1986) Nucleus basalis neuronal loss, neuritic plaques and choline acetyltransferase activity in advanced Alzheimer's disease. *Neuroscience*, 19: 1279 – 1291.

Flynn, D.D. and Mash, D.C. (1986) Characterization of L-[3H]nicotine binding in human cerebral cortex: comparison between Alzheimer's disease and the normal. *J. Neurochem.*, 47: 1948 – 1954.

Giorguieff-Chesselet, M.F., Kemel, M.L., Wandscheer, D. and Glowinski, J. (1979) Regulation of dopamine release by presynaptic nicotinic receptors in rat striatal slices: effects of nicotine in a low concentration. *Life Sci.*, 25: 1257 – 1262.

Harfstrand, A., Adem, A., Fuxe, K., Agnati, L., Andersson, K. and Nordberg, A. (1988) Distribution of nicotinic cholinergic receptors in the rat tel- and diencephalon: a quantitative autoradiographical study using [3H]acetylcholine, alpha[125I]bungarotoxin and [3H]nicotine. *Acta Physiol. Scand.*, 132: 1 – 14.

Kellar, K.J., Whitehouse, P.J., Martino-Barrows, A.M., Marcus, K. and Price, D.L. (1987) Muscarinic and nicotinic cholinergic binding sites in Alzheimer's disease cerebral cortex. *Brain Res.*, 436: 62 – 68.

Krnjevic, K. (1975) Acetylcholine receptors in vertebrate central nervous system. In L.L. Iversen, S.C. Iversen and S.H. Snyder (Eds.), *Handbook of Psychopharmacology, Vol. 6*, Plenum Press, New York, pp. 97 – 126.

Lapchak, P.A., Araujo, D.M., Quirion, R. and Collier, B. (1988) Distribution of [3H]N-methylcarbamylcholine binding sites in rat brain: correlation with nicotinic autoreceptors. In F. Clementi et al. (Eds.), *Nicotinic Acetylcholine Receptors in the Nervous System*, NATO ASI Series, Springer Verlag, New York, pp. 149 – 158.

Marks, M.J. and Collins, A.C. (1982) Characterization of nicotine binding in mouse brain and comparison with the binding of alpha-bungarotoxin and quinuclidinyl benzilate. *Mol. Pharmacol.*, 22: 554 – 564.

Marks, M.J., Burch, J.B. and Collins, A.C. (1983) Effects of chronic nicotine infusion on tolerance development and nicotinic receptors. *J. Pharmacol. Exp. Ther.*, 225: 817 – 825.

Marks, M.J., Stitzel, J.A. and Collins, A.C. (1985) Time course of the effects of chronic nicotine infusion on drug responses and brain receptors. *J. Pharmacol. Exp. Ther.*, 235: 619 – 628.

Martino-Barrows, A.M. and Kellar, K.J. (1987) [3H]Acetylcholine and [3H](–)nicotine label the same recognition site in rat brain. *Mol. Pharmacol.*, 31: 160 – 174.

McGeer, P.L. (1984) Aging, Alzheimer's disease, and the cholinergic system. *Can. J. Physiol. Pharmacol.*, 62: 741 – 754.

Meyer, E.M., Arendash, G.W., Judkins, J.H., Ying, L., Wade, C. and Kem, W.R. (1987) Effects of nucleus basalis lesions on the muscarinic and nicotinic modulation of [3H]acetylcholine release in rat cerebral cortex. *J. Neurochem.*, 49: 1758 – 1762.

Nordberg, A. and Winblad, B. (1986) Reduced number of [3H]nicotine and [3H]acetylcholine binding sites in the frontal cortex of Alzheimer brains. *Neurosci. Lett.*, 72: 115 – 119.

Perry, E.K., Tomlinson, B.E., Blessed, G., Bergman, K., Gibson, P.H. and Perry, R.H. (1978) Correlation of cholinergic abnormalities with senile plaques and mental test scores in senile dementia. *Br. Med. J.*, 2: 1457 – 1459.

Quirion, R., Martel, J.C., Robitaille, Y., Etienne, P., Wood, P., Nair, N.P.V. and Gauthier, S. (1986) Neurotransmitter and receptor deficits in senile dementia of the Alzheimer type. *Can. J. Neurol. Sci.*, 13: 503 – 510.

Reisine, T.D., Yamamura, H.I., Bird, E.D., Spokes, E. and Enna, S.J. (1978) Pre- and postsynaptic neurochemical alterations in Alzheimer's disease. *Brain Res.*, 159: 477 – 481.

Romano, C. and Goldstein, A. (1980) Stereospecific nicotine receptors on rat brain membranes. *Science*, 210: 647 – 650.

Rowell, P.P. and Winkler, D.L. (1984) Nicotinic stimulation of [3H]acetylcholine release from mouse cerebral cortical synaptosomes. *J. Neurochem.*, 43: 1593 – 1598.

Schwartz, R.D. and Kellar, K.J. (1983a) [3H]Acetylcholine binding sites in brain: effect of disulfide bond modification. *Mol. Pharmacol.*, 24: 387 – 391.

Schwartz, R.D. and Kellar, K.J. (1983b) Nicotinic cholinergic receptor binding sites in the brain: regulation in vivo. *Science*, 220: 214 – 216.

Schwartz, R.D. and Kellar, K.J. (1985) In vivo regulation of [3H]-acetylcholine recognition sites in brain by nicotinic cholinergic drugs. *J. Neurochem.*, 46: 239 – 246.

Schwartz, R.D., McGee, R., Jr., and Kellar, K.J. (1982) Nicotinic cholinergic receptors labeled by [3H]-acetylcholine in rat brain. *Mol. Pharmacol.*, 22: 55 – 62.

Shimohama, S., Taniguchi, T., Fujiwara, M. and Kameyama, M. (1986) Changes in nicotinic and muscarinic cholinergic receptors in Alzheimer-type dementia. *J. Neurochem.*, 46: 288 – 293.

Sloan, J.W., Martin, W.R., Hernandez, J. and Hook, R. (1985) Binding characteristics of (–)- and (+)-nicotine to the rat brain in P_2 fraction. *Pharmacol. Biochem. Behav.,* 23: 987 – 992.

Thorne, B., Lunt, G.G., Wonnacott, S. and Gallagher, T.C. (1988) Presynaptic nicotinic autoreceptors in the hippocampus. *Int. Symp. on Nicotinic Receptors in the CNS – Their Role in Synaptic Transmission Abst.,* Uppsala, Sweden.

Wang, J.X., Roeske, W.R., Mei, L., Malatynska, E., Wang, W., Perry, E.K., Perry, R.H. and Yamamura, H.I. (1987) Nicotinic and muscarinic M_2 receptor alterations in the cerebral cortex of patients with senile dementia of the Alzheimer's type (SDAT). In M.J. Rand and C. Raper (eds.), *Excerpta Medica 750,* Elsevier Science Publications, Amsterdam, pp. 83 – 86.

Weiler, M.H., Misgeld, U. and Cheong, D.K. (1984) Presynaptic muscarinic modulation of nicotinic excitation in the rat neostriatum. *Brain Res.,* 296: 11 – 20.

Whitehouse, P.J., Kopajtic, T., Jones, B.E., Kuhar, M.J. and Price, D.L. (1985) An in vitro receptor autoradiographic study of muscarinic cholinergic receptor subtypes in the amygdala and neocortex of patients with Alzheimer's disease. *Neurology,* 35 (Suppl. 1), 217.

Wonnacott, S. (1987) Brain nicotine binding sites. *Human Toxicol.,* 6: 343 – 353.

Yamada, S., Gehlert, D.R., Hawkins, K.N., Nakayama, K., Roeske, W.R. and Yamamura, H.I. (1987) Autoradiographic localization of nicotinic receptor binding in rat brain using [^{3}H]methylcarbamylcholine, a novel radioligand. *Life Sci.,* 41: 2851 – 2861.

Yoshida, K., Kato, Y. and Imura, H. (1980) Nicotine-induced release of noradrenaline from hypothalamic synaptosomes. *Brain Res.,* 182: 361 – 368.

A. Nordberg, K. Fuxe, B. Holmstedt and A. Sundwall (Eds.)
Progress in Brain Research, Vol. 79
© 1989 Elsevier Science Publishers B.V. (Biomedical Division)

CHAPTER 35

The role of nicotinic receptors in the pathophysiology of Alzheimer's disease

A. Nordberg[a, b], L. Nilsson-Håkansson[a], A. Adem[a, b], J. Hardy[e], I. Alafuzoff[c],
Z. Lai[a], M. Herrera-Marschitz[c] and B. Winblad[b]

[a] *Department of Pharmacology, Uppsala University, Uppsala, Departments of* [b] *Geriatric Medicine,* [c] *Pathology and*
[d] *Pharmacology, Karolinska Institutet, Stockholm, Sweden, and* [e] *Department of Biochemistry, St. Mary's Hospital Medical
School, London, U.K.*

Introduction

Alzheimer's disease, senile dementia of Alzheimer type (AD/SDAT) the most common dementia disorder is characterized by its typical histopathology (senile plaques and neurofibrillary tangles) and multiple transmitter changes in the brain (for recent reviews see Gottfries, 1985; Hardy et al., 1985). The cholinergic system is severely affected in AD/SDAT brains. Initially the acetylcholine (ACh) synthesizing enzyme choline acetyltransferase (ChAT) was reported to be markedly decreased in AD/SDAT brains (Davies and Maloney, 1976; Perry et al., 1977; White et al., 1977). Later the high affinity choline (Ch) uptake (Rylett et al., 1983) and the synthesis (Sims et al., 1980) and release of ACh (Nilsson et al., 1986) were found to be decreased while the muscarinic cholinergic receptors remained intact (for review see Nordberg and Winblad, 1986). A decrease in cortical M_2 muscarinic receptors has later been described in AD/SDAT brains (Mash et al., 1985; Araujo et al., 1988) but other reports are conflicting (Caulfield et al., 1982; Nordberg et al., 1986; Kellar et al., 1987). In 1986, three research groups

independently reported a marked decrease in the number of high affinity cholinergic nicotinic receptors in AD/SDAT brains compared to control brains (Flynn and Mash, 1986; Nordberg and Winblad, 1986b; Whitehouse et al., 1986). This drastic reduction of the nicotinic receptors in AD/SDAT brains has later been confirmed by other research groups both in autopsy (Perry et al., 1987; Araujo et al., 1988) and biopsy tissue (De Sarno et al., 1988). There is also growing evidence that AD/SDAT is a more generalized disease which is not only confined to the brain. A decrease in number of nicotinic receptors in peripheral lymphocytes from AD/SDAT patients has been observed (Adem et al., 1986; Nordberg et al., 1987).

Disturbances in nicotinic receptor function of AD/SDAT brains will be further analysed in this chapter in relation to other cholinergic mechanisms and other dementia states.

Nicotinic receptors in human brain

The nicotinic receptor binding sites in human brain have been identified and characterized by various radioligands such as the nicotinic antagonists [^{3}H]α-bungarotoxin ([^{3}H]Btx) (Volpe et al., 1979; Davies and Feisullin, 1981), [^{3}H]tubocurarine (Nordberg and Winblad, 1981; Nordberg et al.,

Address for correspondence: Agneta Nordberg, Department of Pharmacology, Box 591, S-751 24 Uppsala, Sweden.

1982) and the agonists [³H]ACh (Larsson et al., 1987; Adem et al., 1987), [³H]nicotine (Shimohama et al., 1985; Flynn and Mash, 1986; Nordberg et al., 1987; Adem and Nordberg, 1988; Adem et al., 1988) and [³H]methylcarbamylcholine (Araujo et al., 1988). In rodent brain receptor binding experiments in tissue homogenates or thin tissue slices (autoradiography) have demonstrated a different localization of the antagonist [³H]Btx sites compared to the agonist

sites labelled by [³H]ACh and [³H]nicotine (Nordberg and Larsson, 1980; Clarke et al., 1985; Larsson and Nordberg, 1985; Härfstrand et al., 1988). The difference in antagonist/agonist sites is evident at molecular level (different receptors/receptor sites) (Wonnacott, 1986; Wada et al., 1988) as well as for the regional distribution of antagonist/agonist nicotinic binding sites in the brain (Clarke et al., 1985; Härfstrand, 1988). The [³H]ACh and [³H]nicotine high affinity binding

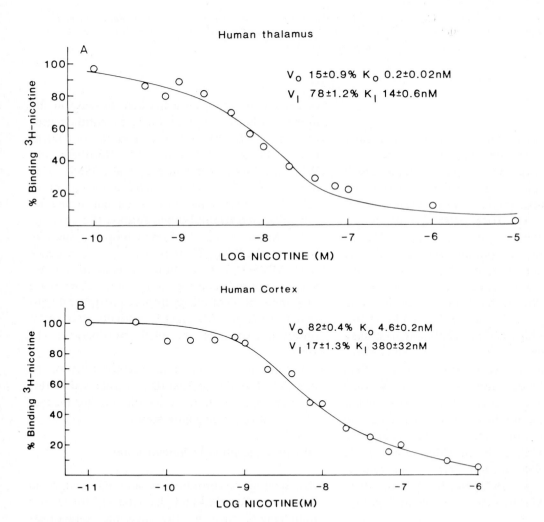

Human thalamus

V_0 15±0.9% K_0 0.2±0.02nM
V_1 78±1.2% K_1 14±0.6nM

Human Cortex

V_0 82±0.4% K_0 4.6±0.2nM
V_1 17±1.3% K_1 380±32nM

Fig. 1(A,B) (−)-[³H]nicotine/nicotine competition experiments in thalamus and cortex of human brain. [³H]nicotine (5 nM) binding to P_2 fractions was performed at 4°C for 40 min in the presence of different concentrations of unlabelled nicotine. Computer assisted least square analysis of the competition data revealed best fit to a two site model. $V_{\%}$ = percent of binding sites; K = affinity constant.

sites seem to colocalize in various regions of the human brain and in addition [³H]nicotine also binds to low affinity sites (Shimohama et al., 1985; Nordberg et al., 1988a). It is possible from experiments obtained in human brain to rationalize the nicotinic agonist binding into terms of three populations of sites termed superhigh, high and low-affinity nicotinic binding sites (Adem et al., 1987; Nordberg et al., 1988a; Adem and Nordberg, 1988; Adem et al., 1988; Nordberg et al., 1988a,b). Figure 1 illustrates different subtypes of nicotinic binding sites in the thalamus (superhigh-high) and cortex (high-low) as revealed by [³H]nicotine/nicotine competition experiments. For [³H]ACh at least two nicotinic binding sites (superhigh-high affinity) are found in the human brain while only a high affinity site can be observed in rodent brain (Larsson et al., 1987; Adem et al., 1987).

The nicotinic agonist ([³H]nicotine) and antagonist ([³H]tubocurarine, [³H]-α-bungarotoxin) binding sites in human brain decrease or increase with age depending on which brain region is studied (Nordberg et al., 1982, 1988a, 1988b; Flynn and Mash, 1986; Nordberg and Winblad, 1986a; Perry et al., 1987). Thus there is a significant decrease in number of high affinity [³H]nicotine binding sites with age in the human cortex (Flynn and Mash, 1986; Nordberg et al., 1987), hippocampus (Perry et al., 1987) while in the thalamus an increase in the [³H]nicotine binding with increasing age is observed (Nordberg et al., 1988a). A significant decrease in the [³H]-α-bungarotoxin binding is found in the thalamus while the [³H]ACh and [³H]tubocurarine binding are unchanged with age (Nordberg et al., 1988c). These different changes during normal aging might indicate interconversion of receptor subtypes and/or different changes in presynaptic and postsynaptic nicotinic receptors.

Studies on nicotinic receptors in human brain have mostly been performed in autopsy brain tissue. As illustrated in Table I, a sightly higher number of high affinity [³H]nicotine binding sites was measured in frontal cortical biopsy material compared to cortical autopsy tissue (postmortem

TABLE I

High affinity nicotinic and muscarinic receptors in human frontal cortex: a comparison between biopsy and autopsy tissue

	Biopsy	Autopsy
Age (yrs)	66.2 ± 2.0	71.5 ± 1.7
Time delay (h) until freezing of the tissue	< 1	11.1 ± 1.7
[³H]nicotine pmol/g protein	28.7 ± 4.1 ($n = 6$)	19.9 ± 3.0 ($n = 12$)
[³H]QNB pmol/g protein	291 ± 27 ($n = 6$)	361 ± 26 ($n = 12$)

[³H]nicotine (5 nM) and [³H]QNB (0.2 nM) were incubated with P_2 fractions prepared from human frontal cortex. Correction for unspecific binding was performed in parallel experiments in the presence of unlabelled 10^{-3} M nicotine or 10^{-6} M atropine. The results are expressed as means ± S.E.M.; n = number of individuals.

delay 11 ± 2 h) (Table I) but the difference was not significant. A similar observation was also made for the muscarinic receptors. In general, autopsy tissue is easier to get access of than biopsy tissue and the absence of drugs can easier be controlled. We have not found any significant change in the number of cortical [³H]nicotine or [³H]ACh binding sites at autopsy delays between 3 and 24 h (Fig. 2a,b).

The nicotinic receptors are widely distributed in the human brain. As illustrated in Table II, the number of high affinity [³H]nicotine binding sites is high in the thalamus, caudate nucleus, putamen, substantia nigra, intermediate in the hypothalamus, different cortical areas and low in the medulla oblongata, pons and globus pallidus. A similar regional distribution has been reported for the high affinity nicotinic binding sites labelled by the radioligands [³H]ACh and [³H]methylcarbamylcholine (Adem et al., 1987; Araujo et al., 1988). Hopefully, molecular biology studies on the characterization of nicotinic receptors in human brain will give further valuable information as it

356

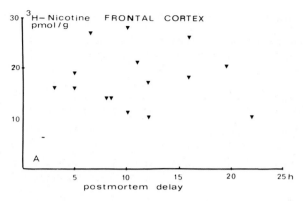

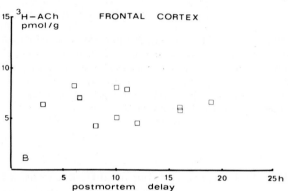

Fig. 2(A,B) [³H]nicotine (20 nM) and [³H]ACh (10 nM) binding in human frontal cortex at autopsy delays between 3 and 24 h. Each point represents data from one individual. The [³H]ACh binding was performed in the presence of atropine 10^{-6} M.

already has in rodent brain (see Patrick et al., this volume).

Changes in nicotinic receptors in Alzheimer brains

A pronounced decrease in the number of nicotinic binding sites has been measured in Alzheimer brains using [³H]nicotine (Flynn and Mash, 1986; Nordberg and Winblad, 1986b; Whitehouse et al., 1986; De Sarno et al., 1987; Perry et al., 1987), [³H]ACh (Whitehouse et al., 1986; Nordberg and Winblad, 1986b) and [³H]methylcarbamylcholine (Araujo et al., 1988). Earlier studies with nicotinic antagonists such as α-bungarotoxin and tubo-curarine did not show any receptor changes in

AD/SDAT brains compared to controls (Davies and Feisullin, 1981; Lang and Henke, 1982; Nordberg et al., 1982) except for a decrease in the Btx binding in the midtemporal gyrus (Davies and Feisullin, 1981) and an increase in [³H]tubo-curarine binding in the thalamus (Nordberg et al., 1982). In different cortical areas we obtained a significant decreased number of high affinity [³H]nicotine binding sites in the frontal and temporal cortex while no change was found in the occipital cortex (Table III).

Further analysis of the nicotinic agonist receptor changes in AD/SDAT by (−)-[³H]nicotine/(−)nicotine competition experiments revealed a shift in the competition curves compared to controls. The change in competition curves in the temporal cortex and hippocampus of AD/SDAT brains compared to controls is shown in Fig. 3. Computer-assisted least square analysis of competition data reveal a change in the proportion of nicotinic

TABLE II

Regional distribution of high affinity [³H]nicotine binding sites in human brain

Brain region	[³H]nicotine pmol/g protein
Thalamus	24.0 ± 1.94
Caudate nucleus	18.9 ± 1.61
Putamen	18.7 ± 1.47
Substantia nigra	17.3 ± 2.23
Cingulate cortex	14.1 ± 2.69
Hypothalamus	13.6 ± 1.83
Olfactory tubercle	13.0 ± 0.70
Temporal cortex	12.0 ± 1.90
Frontal cortex	11.4 ± 1.77
Occipital cortex	10.0 ± 1.40
Hippocampus	8.8 ± 1.03
Medulla oblongata	8.7 ± 1.51
Pons	7.1 ± 1.87
Globus pallidus	3.8 ± 0.72

5 nM [³H]nicotine was incubated with P_2 fractions prepared from different parts of the human brain (14 areas). Correction for unspecific binding was performed in parallel experiments in the presence of 10^{-3} M nicotine. The results are expressed as means ± S.E.M. for 4 – 7 individuals.

TABLE III

High affinity nicotinic receptors in different cortical areas of AD/SDAT brains

	[³H]nicotine pmol/g protein	
	Control	AD/SDAT
Frontal cortex[a]	20.9 ± 2.2 (n = 16)	11.6 ± 1.3** (n = 7)
Temporal cortex	12.0 ± 1.9 (n = 7)	7.0 ± 0.6* (n = 3)
Occipital cortex	9.8 ± 1.1 (n = 5)	9.4 ± 1.0 (n = 4)

[a] 20 nM [³H]nicotine; ** $p < 0.01$; * $p < 0.05$

(−)-[³H]nicotine (5 nM) was incubated with P_2 fractions prepared from cortical tissue of AD/SDAT and control brains. Correction for unspecific binding was performed in parallel experiments in the presence of 10^{-3} M unlabelled nicotine. The results are expressed as means ± S.E.M.; n = number of individuals.

receptor subtypes in AD/SDAT brains compared to controls (Fig. 4).

As illustrated in Fig. 4, the proportion of high and low affinity nicotinic binding sites are changed in the temporal cortex of AD/SDAT brains compared to control brains. There is also a loss in receptor affinity especially for the low affinity sites (Nordberg et al., 1988b). The physiological consequence of a loss in nicotinic receptor affinity in AD/SDAT brains can only be speculated upon. It might, however, provide to keep the functionality of the remaining cholinergic neurons (transmitter release) in a neurodegenerative disease with a general neuronal loss.

In AD/SDAT brain cortical tissue there is in addition to a reduced ChAT activity and a decreased number of high affinity nicotinic receptors also a decreased release of ACh (Nilsson et al., 1986). A significant correlation between number of high affinity nicotinic binding sites and in vitro

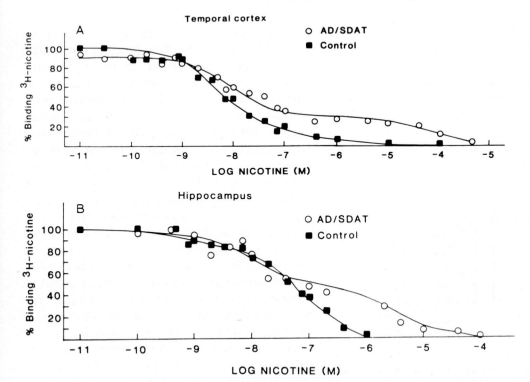

Fig. 3(A,B) (−)[³H]nicotine/nicotine competition experiments in temporal cortex and hippocampus of human control and AD/SDAT brains. Computer assisted least square analysis of the competition data revealed a fit to a two site model (—) for both control and AD/SDAT group.

potassium-induced ACh release has been observed (Nordberg et al., 1987). The reduced in vitro ACh release in autopsy AD/SDAT tissue is restored to control level in the presence of acetylcholinesterase (AChE) inhibitors such as physostigmine and tetra-hydroaminoacridine (THA) (Nilsson et al., 1987). Interestingly, nicotinic antagonists such as tubocurarine (Nilsson et al., 1987), dihydro-β-erythroidine and mecamylamine (Fig. 5) prevent the effect of the cholinesterase inhibitors on ACh release in AD/SDAT tissue. This observation indicates that the effect might be induced via nicotinic receptors (presynaptically located?).

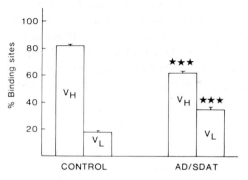

Fig. 4. Proportion of high affinity (V_H) and low affinity (V_L) binding sites in the temporal cortex of AD/SDAT cases and aged-matched controls, ***$p < 0.001$.

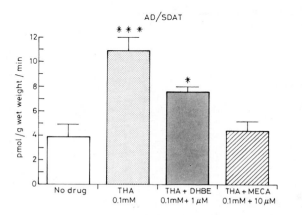

Fig. 5. Effect of THA 10^{-4} M, THA 10^{-4} M plus dihydro-β-erythroidine 10^{-6} M, THA 10^{-4} M plus mecamylamine 10^{-5} M on in vitro release of [^{3}H]ACh in AD/SDAT frontal cortex, Mean $\pm$ S.E., ***$p < 0.001$; *$p < 0.05$.

Nicotinic agonists are known from animal experiments to induce ACh release via stimulation of nicotinic receptors (see Beani et al., this volume). Since the nicotinic receptor binding sites in AD/SDAT brains have a lower binding affinity than in control brains (higher proportion of low affinity sites), the desensitization phenomena might be less pronounced. It is therefore plausible that the cholinesterase inhibitors due to their indirect agonistic effect (increase of ACh in the synaptic cleft) cause an enhanced ACh release via nicotinic receptor stimulation in AD/SDAT brains. A direct receptor effect cannot be excluded since THA competes with [^{3}H]nicotine for the nicotinic receptors in human cortical tissue (Nilsson et al., 1987) while physostigmine has much less affinity. These effects might be of therapeutic value in the treatment of AD/SDAT.

Changes in nicotinic receptors in other dementia disorders

Are the nicotinic receptor changes in AD/SDAT brains selective only for that dementia state or can similar changes be found in other dementia disorders as well? A decreased number of [^{3}H]nicotine binding sites has been observed in the hippocampus (Perry et al., 1987) and caudate nucleus (Nordberg et al., 1985) of Parkinsonian brains. A decreased number of [^{3}H]ACh binding sites was also reported in the frontal cortex of Parkinson brains (Whitehouse et al., 1988). A decrease in brain nicotinic receptors is probably not a general phenomena in neurodegenerative diseases since in Huntington's disease a preserved number of [^{3}H]nicotine binding sites has been observed in the hippocampus (Perry et al., 1988).

Multi-infarct dementia (MiD) is characterized by microscopic infarcts in the brain. In the mixed type of dementia, AD/MiD, a coexistence of degenerative changes (plaques and tangles) and microscopic infarcts are found in the hippocampus and cortex (Alafuzoff et al., 1987). In MiD brains a significant decreased release of ACh and a decreased number of high affinity nicotinic receptors were found in the frontal cortex (Table IV). The change in ACh release was somewhat less pronounced than in the AD/SDAT cases but an increased number of muscarinic ([³H]QNB) cholinergic receptor binding sites was measured as in the

AD/SDAT brains (Table IV). Thus there was no sign for muscarine receptor compensation in the MiD brains. Further studies are needed in order to neurochemically differentiate between different dementia disorders such as AD/SDAT, AD/MiD and MiD.

Experimental approaches to an animal model for AD/SDAT

TABLE IV

In vitro ACh release and number of nicotinic and muscarinic receptors in frontal cortex of AD/SDAT and AD/MiD brains

	[³H]ACh release pmol/ g/min	[³H]nicotine (20 nM) pmol/g	[³H]QNB (0.20 nM) pmol/g
Control	15 ± 2.0 (n = 9)	19.2 ± 3.0 (n = 16)	361 ± 26 (n = 12)
AD/SDAT	5.8 ± 2.8* (n = 4)	11.6 ± 1.3* (n = 7)	435 ± 21* (n = 4)
MiD	7.7 ± 2.7 (n = 3)	9.7 ± 2.3 (n = 3)	322 ± 51 (n = 3)

In vitro potassium evoked [³H]ACh release was measured in human cortical slices preloaded with [³H]Ch (Nilsson et al., 1986). In parallel experiments P_2 fractions prepared from human cortex were incubated with [³H]nicotine (20 nM) and [³H]QNB (0.20 nM). Correction for unspecific binding was performed in the presence of unlabelled 10^{-3} M nicotine or 10^{-6} M atropine. The results are expressed as means ± S.E.M.; n = number of experiments; * $p < 0.05$.

The cholinergic innervation of cerebral cortex in mammalians derives mainly from cell bodies in the basal forebrain known as nucleus basalis magnocellularis, nucleus of Meynert or substantia innominata. The number of cell bodies in that area are reduced in AD/SDAT and Parkinson's disease (Whitehouse et al., 1982) and cortical deficits in the cholinergic system have been presumed to be associated with degeneration of nucleus basalis. However, recently it has been claimed that the cortical lesion is the primary. Experimental lesions of the nucleus basalis of rodents cause a decrease in cortical ChAT activity, high affinity Ch uptake (Pedata et al., 1982). ACh content (Pepeu et al., 1985) and ACh release (Le Conte et al., 1982) which indicate lesions of the cholinergic terminals

TABLE V

Effect of unilateral nucleus basalis lesions on the number of cortical nicotinic and muscarinic receptors in the rat brain

	Nicotinic receptors [³H]ACh	Muscarinic receptors [³H]QNB	[³H]QNB + pirenzepine (M₂)
n = 15	pmol/g protein		
Lesioned side	17.8 ± 0.8*	723 ± 62	307 ± 35
Unlesioned side	21.0 ± 0.9	766 ± 58	295 ± 37

Ibotenic acid (5 μg) was injected into the left nucleus basalis. 15 days after the injection the rats were killed and the brain tissue incubated with [³H]ACh (10 nM plus atropine 10^{-6} M), [³H]QNB (0.20 nM) or [³H]QNB + pirenzepine (300 nM). Correction for unspecific binding was performed in parallel experiments in the presence of unlabelled 10^{-3} M nicotine or 10^{-6} M atropine. The results are expressed as means ± S.E.M.; n = number of experiments; * $p < 0.05$.

360

of the cortex. Two weeks after unilateral lesions of the nucleus basalis with ibotenic acid we obtained a significant reduced number of high affinity nicotinic receptors in the rat parietal cortex (Table V) while the number of muscarinic receptors (both total number and M_2 type) were unchanged.

These findings support the view that at least some of the nicotinic receptors are presynaptically located (presumably the high affinity sites), but does not exclude that the reduced number of nicotinic receptors measured in lesioned animals, may be due to interconversion of nicotinic binding sites with different affinity constants. Interestingly, a similar picture was obtained in the lesioned animals as in AD/SDAT brains. Studies are now in progress to investigate whether the cholinesterase inhibitors can restore the cortical ACh release in lesioned animals as they do in AD/SDAT postmortem brain tissue.

Conclusions

Subtypes of nicotinic receptors are widely distributed in the human brain. Some of the nicotinic receptors are presynaptically located. In AD/SDAT brains there is a loss of high affinity to low affinity nicotinic receptors which partly might be due to an interconversion of high affinity to low affinity nicotinic sites with a simultaneous reduction in binding affinity for the low affinity sites. By lowering the affinity of the nicotinic sites, a quick desensitization phenomena can be devoided and the nicotinic receptors might influence the release of acetylcholine. The finding that cholinesterase inhibitors such as THA restore the ACh release in AD/SDAT brain tissue via interaction with the nicotinic receptors open up new therapeutic strategies in the treatment of AD/SDAT and related disorders.

Acknowledgements

The financial support by the Swedish Medical Research Council, the Swedisch Tobacco Company, Stiftelsen för gamla tjänarinnor, Osterman's fund is highly appreciated.

References

Adem, A. and Nordberg, A. (1988) Nicotinic cholinergic receptor heterogeneity in mammalian brain. In M.J. Rand and K. Thurau (Eds.), *The Pharmacology of Nicotine,* IRL Press, Oxford, pp. 227–247.

Adem, A., Nordberg, A., Bucht, G. and Winblad, B. (1986) Extra-neuronal cholinergic markers in Alzheimer's and Parkinson's disease. *Prog. Neuro-Psycho-Pharmacol. Biol. Psychiatr.,* 10: 247–257.

Adem, A., Synnergren, B., Botros, M., Öhman, B., Winblad, B. and Nordberg, A. (1987) ^{3}H-Acetylcholine nicotinic recognition sites in human brain: characterization of agonist binding. *Neurosci. Lett.,* 83: 298–302.

Adem, A., Sing Jossan, S., Brandt, I., Winblad, B. and Nordberg, A. (1988) Distribution of nicotinic receptors in human thalamus as visualized by ^{3}H-nicotine and ^{3}H-acetylcholine receptor autoradiography. *J. Neural Trans.,* 73: 77–83.

Alafuzoff, I., Iqbal, K., Friden, H., Adolfsson, R. and Winblad, B. (1987) Histopathological criteria for progressive dementia disorder; clinical-pathological correlation and classification by multi-variate data analysis. *Acta Neuropathol. (Berl).,* 74: 209–225.

Araujo, D.M., Lapchak, P.A., Robitaille, Y., Gauthier, S. and Quirion, R. (1988) Differential alteration of various cholinergic markers in cortical and subcortical regions of human brain in Alzheimer's disease. *J. Neurochem,* 50: 1914–1923.

Clarke, P.B.S., Schwartz, R.D., Pert, S.M., Pert, C. and Pert, A. (1985) Nicotine binding in rat brain: autoradiographical comparison of ^{3}H-acetylcholine, ^{3}H-nicotine and ^{125}I-alpha-bungarotoxin. *J. Neurosci.,* 5: 1307–1315.

Caulfield, M.P., Straughan, D.W., Cross, A.J., Crow, T. and Birdsall, N.J.M. (1982) Cortical muscarinic receptor subtypes and Alzheimer's disease. *Lancet,* 11: 1277.

Davies, P. and Feisullin (1981) Postmortem stability of α-bungarotoxin binding sites in mouse and human brain. *Brain Res.,* 216: 449–454.

Davies, P. and Maloney, A.J.F. (1976) Selective loss of central cholinergic neurons in Alzheimer's disease. *Lancet,* II: 1403.

De Sarno, P., Giacobini, E. Clark, B. (1988) Changes in nicotinic receptors in human and rat CNS. *Fed. Proc.,* 2: 364.

Flynn, D.D. and Mash, D.C. (1986) Characterization of L-(^{3}H)-nicotine binding in human cerebral cortex; comparison between Alzheimer's disease and the normal. *J. Neurochem.,* 47: 1948–1954.

Hardy, J.A., Adolfsson, R., Alafuzoff, I., Bucht, G., Marcusson, J., Nyberg, P., Perdahl, E., Wester, P. and Winblad, B (1985) Transmitter deficits in Alzheimer's disease. *Neurochem. Int.,* 7: 545–563.

Härfstrand, A., Adem, A., Fuxe, K., Agnati, L., Andersson, K. and Nordberg, A. (1988) Distribution of nicotinic cholinergic receptors in the rat tel- and diencephalon: a quantitative receptor autoradiographical study using ³H-acetylcholine, ¹²⁵I-bungarotoxin and ³H-nicotine. *Acta Physiol. Scand.,* 132: 1 – 14.

Lang, H. and Henke, H. (1983) Cholinergic receptor binding and autoradiography in brains of non-neurological and senile dementia of Alzheimer-type patients. *Brain Res.,* 267: 271 – 280.

Larsson, C. and Nordberg, A. (1985) Comparative analysis of nicotine-like receptor ligand interaction in rodent brain homogenate. *J. Neurochem* 45: 24 – 31.

Larsson, C., Lundberg, P-Å., Halén, A., Adem, A. and Nordberg, A. (1987) in vitro binding of ³H-acetylcholine to nicotinic receptors in rodent and human brain. *J. Neural. Transm.,* 69: 3 – 18.

Le Conte, G., Bartolini, L., Cesamenti, F., Marconcini-Pepeu, I. and Pepeu, G. (1982) Lesions of cholinergic forebrain nuclei: changes in avoidance behavior and scopolamine actions. *Pharmacol. Biochem. Behav.,* 17: 933 – 937.

Kellar, K.J., Whitehouse, P.J., Martino-Barrows, A.M., Marcus, K. and Price, D.L. (1987) Muscarinic and nicotinic cholinergic binding sites in Alzheimer's cerebral cortex. *Brain Res.,* 436: 62 – 68.

Nilsson, L., Nordberg, A., Hardy, J., Wester, P. and Winblad, B. (1986) Physostigmine restores ³H-acetylcholine efflux from Alzheimer brain slices to normal level. *J. Neural Trans.,* 67: 275 – 285.

Nilsson, L., Adem, A., Hardy, J., Winblad, B. and Nordberg, A. (1987) Do tetrahydroaminoacridine (THA) and physostigmine restore acetylcholine release in Alzheimer brains via nicotinic receptors? *J. Neural Trans.,* 70: 357 – 368.

Nordberg, A. and Larsson, C. (1980) Studies on muscarinic and nicotinic binding sites in brain. *Acta Physiol. Scand. Suppl.,* 497: 19 – 23.

Nordberg, A. and Winblad, B. (1981) Cholinergic receptors in human hippocampus – regional distribution and variance with age. *Life Sci.,* 29: 1937 – 1944.

Nordberg, A. and Winblad, B. (1986a) Brain nicotinic and muscarinic receptors in normal aging and dementia. In A. Fisher and I. Hanin (Eds.), *Alzheimer's and Parkinson's Disease – Strategies for Research and Development: Advances in Behavioral Biology,* Vol. 29, Plenum Press, N.Y., pp. 95 – 108.

Nordberg, A. and Winblad, B. (1986b) Reduced number of ³H-nicotine and ³H-acetylcholine binding sites in the frontal cortex of Alzheimer brains. *Neurosci. Lett.,* 72: 115 – 119.

Nordberg, A., Adolfsson, R., Marcusson, J. and Winblad, B. (1982) Cholinergic receptors in the hippocampus in normal aging and dementia of Alzheimer type. In E. Giacobini, G. Filogama, G. Giacobini and A. Vernadakis (Eds.), *The Aging Brain: Cellular and Molecular Mechanisms of Aging in the Nervous System,* Raven Press, N.Y. pp. 231 – 246.

Nordberg, A., Nyberg, P. and Winblad, B. (1985) Topographic distribution of choline acetyltransferase activity and muscarinic and nicotinic receptors in Parkinson brains.

Neurochem. Pathol., 3: 223 – 236.

Nordberg, A., Alafuzoff, I. and Winblad, B. (1986) Muscarinic receptor subtypes in hippocampus in Alzheimer's disease and mixed dementia type. *Neurosci. Lett.,* 70: 160 – 164.

Nordberg, A., Adem, A., Nilsson, L. and Winblad, B. (1987) Cholinergic deficits in CNS and peripheral nonneuronal tissue in Alzheimer dementia. In M.J. Dowdall and J.N. Hawthorne (Eds.), *Cellular and Molecular Basis of Cholinergic Function,* Ellis Horwood, Ltd, Chichester, U.K. pp. 858 – 868.

Nordberg, A., Adem, A., Nilsson, L. and Winblad, B. (1988a) Nicotinic and muscarinic cholinergic receptor heterogeneity in the human brain at normal aging and dementia of Alzheimer type. In B. Tomlinson, G. Pepeu, C.M. Wischik (Eds.), *New Trends in Aging Research, Fidia Research Series, Vol. 15,* Liviana Press, Italy, pp. 27 – 36.

Nordberg, A., Adem, A., Hardy, J. and Winblad, B. (1988b) Change in nicotinic receptor subtypes in temporal cortex of Alzheimer brains. *Neurosci. Lett.,* 86: 317 – 321.

Nordberg, A., Adem, A., Nilsson, L., Romanelli, L. and Zhang, X. (1988c) Heterogenous cholinergic nicotinic receptors in the CNS. In F. Clementi, C. Gotti and E. Sher (Eds.), *Nicotinic Acetylcholine Receptors in the Nervous System, NATO Asi Series, Vol. H25,* Springer Verlag, Berlin, pp. 331 – 350.

Pedata F., Lo Conte, G., Sorbi, S., Marconcini Pepeu, I. and Pepeu, G. (1982) Changes in high affinity choline uptake in rat cortex following lesions of the magnocellular forebrain nuclei. *Brain Res.,* 233: 359 – 367.

Pepeu, G., Casamenti, F., Bracco, L., Ladinsky, H. and Consolo, S. (1985) Lesions of the nucleus basalis in the rat: functional changes. In J. Traber and W.H. Gispen (Eds.), *Senile Dementia of Alzheimer Type,* Springer-Verlag, Berlin, pp. 305 – 315.

Perry, E.K., Perry, R.H., Gibson, R.H., Blessed, G. and Tomlinson, B.E. (1977) A cholinergic connection between normal aging and senile dementia in the human hippocampus. *Neurosci. Lett.,* 6: 85 – 89.

Perry, E.K., Perry, R., Smith, C.J., Dick, D.J., Candy, J.M., Edwardson, J.A., Fairbairn, A. and Blessed, G. (1987) Nicotinic receptor abnormalities in Alzheimer's and Parkinson's diseases. *J. Neurosurg. Psychiatry,* 50: 806 – 809.

Rylett, R.J., Ball, M.J. and Cohoun, E.H. (1983) Evidence for high affinity choline transport in synaptosomes prepared from hippocampus and neocortex of patients with Alzheimer's disease. *Brain Res.,* 289: 169 – 175.

Shimohama, S., Taniguchi, T., Fujiwara, M. and Kameyama, M. (1985) Biochemical characterization of the nicotinic cholinergic receptors in human brain: binding of (–)-³H-nicotine. *J. Neurochem.,* 45: 604 – 610.

Sims, N.R., Smith, C.C.T., Davison, A.N., Bowen, D.M., Flack, R.H.A. and Snowden, J.S. (1980) Glucose metabolism and acetylcholine synthesis in relation to neuronal activity in Alzheimer's disease. *Lancet,* I:

333 – 336.

Volpe, B.T., Francis, A., Gazzaniga, S. and Schechter, N. (1979) Regional concentration of putative nicotinic cholinergic receptor sites in human brain. *Exp. Neurol.*, 66: 737 – 744.

Wada, K., Ballivet, M., Boulter, J., Connolly, J., Wada, E., Deneris, E.S., Swanson, L.W., Heinemann, S. and Patrick, J. (1988) Functional expression of a new pharmacological subtype of brain nicotinic acetylcholine receptor. *Science*, 240: 330 – 334.

White, P., Goodhardt, M.J., Keet, J., Hiley, C.R., Carrasco, L.H. and Williams, I.E.I. (1977) Neocortical cholinergic neurons in elderly people. *Lancet*, I: 668 – 670.

Whitehouse, P.J., Price, D.L., Struble, R.G., Clark, A.W., Coyle, J.T. and De Long, M.R. (1982) Alzheimer's disease and senile dementia: loss of neurons in the basal forebrain. *Science*, 215: 1237 – 1239.

Whitehouse, P.J., Martino, A.M., Antuono, P.G., Lowenstein, P.R., Coyle, J.T., Price, D.L. and Kellar, K.J. (1986) Nicotinic acetylcholine binding sites in Alzheimer's disease. *Brain Res.*, 371: 146 – 151.

Whitehouse, P.J., Martino, A.M., Wagster, M.V., Price, D.J., Mayeux, R., Atack, J.R. and Kellar, K.J. (1988) Reduction in ^{3}H-nicotinic acetylcholine binding in Alzheimer's disease and Parkinson's disease: an autoradiographic study. *Neurology*, 38: 720 – 723.

Wonnacott, S. (1986) *a*-Bungarotoxin binds to low affinity nicotine binding sites in rat brain. *J. Neurochem.*, 47: 1706 – 1712.

Subject Index